A First Course in General Relativity
Third Edition

Clarity, readability, and rigor combine in the third edition of this widely used textbook to provide the first step into general relativity for advanced undergraduate students with a minimal background in mathematics. Topics within relativity that fascinate astrophysics researchers and students alike are covered with Schutz's characteristic ease and authority, from black holes to relativistic objects, from pulsars to the study of the Universe as a whole. This third edition contains discoveries by astronomers that require general relativity for their explanation; two chapters on gravitational waves, including direct detections of gravitational waves and their observations' impact on cosmological measurements; new information on black holes and neutron stars; and greater insight into the expansion of the Universe. Over 300 exercises, many new to this edition, give students the confidence to work with general relativity and the necessary mathematics, while the informal writing style and worked examples make the subject matter easily accessible.

Bernard Schutz is a Professor in the School of Physics and Astronomy at Cardiff University, Wales, and Director (ret.) at the Max Planck Institute for Gravitational Physics (Albert Einstein Institute), Germany. In 2019 he was elected a member of the US National Academy of Sciences, and in 2021, he was elected a fellow of the Royal Society, in recognition of his seminal contributions to relativistic astrophysics.

A First Course in General Relativity

THIRD EDITION

BERNARD SCHUTZ

Cardiff University

and

Max Planck Institute for Gravitational Physics

(Albert Einstein Institute)

CAMBRIDGE
UNIVERSITY PRESS

CAMBRIDGE
UNIVERSITY PRESS

University Printing House, Cambridge CB2 8BS, United Kingdom

One Liberty Plaza, 20th Floor, New York, NY 10006, USA

477 Williamstown Road, Port Melbourne, VIC 3207, Australia

314–321, 3rd Floor, Plot 3, Splendor Forum, Jasola District Centre, New Delhi – 110025, India

103 Penang Road, #05–06/07, Visioncrest Commercial, Singapore 238467

Cambridge University Press is part of the University of Cambridge.

It furthers the University's mission by disseminating knowledge in the pursuit of
education, learning, and research at the highest international levels of excellence.

www.cambridge.org
Information on this title: www.cambridge.org/highereducation/isbn/9781108492676
DOI: 10.1017/9781108610865

First edition © Cambridge University Press 1985
Second edition © B. Schutz 2009
Third edition © Cambridge University Press 2022

First published 1985
Second edition 2009
14th printing 2019
Third edition 2022

Printed in the United Kingdom by TJ Books Limited, Padstow Cornwall 2022

A catalogue record for this publication is available from the British Library.

ISBN 978-1-108-49267-6 Hardback

Additional resources for this publication at www.cambridge.org/schutz3ed

Contents

Preface to the third edition

This third edition follows the second after 12 years, half of the interval between the first and second editions. The need for a new edition is a happy one: the direct detections of gravitational waves, beginning in 2015, have already made much of the second edition obsolete. These detections have surprised astronomers by revealing how unexpectedly plentiful and massive are the black holes that form when stars die; they have demonstrated that neutron stars are in fact our direct ancestors because their collisions created most of the heaviest elements in our bodies; and they have given us a new and potentially powerful way of measuring the expansion rate and age of the Universe.

This new edition incorporates much of what we have learned since 2015, but more importantly it aims to provide a foundation that should enable the reader to understand future observations and the discoveries they bring. In the preface to the previous edition I noted that general relativity had generated "a host of applications, some of which were not even imagined in 1985 when the first edition appeared." Now there has been another big step: with gravitational wave detections, general relativity has added a radical new way of observing the Universe, a way of learning about things that we could not have investigated before.

Gravitational wave astronomy is now the fastest expanding branch of observational astronomy: the volume of space accessible to the current detectors has doubled on average every 1.5 years since 2015, as their sensitivity has improved. This rate looks set to continue for some years as new detectors in Japan and India join and steadily improve to full sensitivity. The space-based LISA detector is under construction; it will be able to detect events at distances far greater than the range of any optical telescope. And new detectors are being proposed that can register every black-hole and neutron-star merger back to times equal to the age of our Universe. At the same time, gravitational wave detection is being integrated into other observing fields, as large survey telescopes like the Vera Rubin Observatory provide the capability of searching for explosions that may accompany gravitational wave events, as the Athena X-ray satellite observatory mission is being developed to do coincident observing with LISA, as radio astronomers use pulsars as gravitational wave detectors, and as microwave astronomers search for the signature of gravitational waves in the very early Universe.

The changes in this textbook for the third edition have been designed in recognition of this sea-change, anticipating what students want to know and what future researchers need to be prepared for. The final four chapters have been expanded to five, and all have been rewritten. They embrace the new role of general relativity in astronomy in several ways.

- There are now two chapters on gravitational waves. They treat the theory, survey the detectors current and planned, provide details of the most significant detections so far, and explain the basics of how to extract and interpret information from observations now and in the future.
- The chapters on stars and black holes now incorporate what we have learned from the gravitational wave observations to date. In addition, there is much new material on the astronomical context of these observations, such as how stellar evolution produces neutron stars and black holes. The aim is to help readers to fit gravitational wave observations of relativistic objects into the bigger picture.
- The chapter on cosmology – the study of the Universe as a whole – has been completely reorganized and greatly expanded. Cosmology is another rapidly moving and fundamental part of astronomy today and, on top of that, gravitational wave observations have provided it with a new way of measuring distances over vast reaches of space. The chapter now starts with what we observe, because cosmology is now a high-precision observational science. It then goes on to apply Einstein's equations to explain the observations and ends with cosmology's frontiers in fundamental physics.

The mathematical foundations of general relativity, treated in the first eight chapters, remain as solid as before: general relativity is now a rigorously tested theory, with some experiments and observations unable to find deviations from its predictions even at levels of one part in 10^{15} or better. So the revisions for this edition of the first eight chapters have been minor: modernizing some of the treatment, correcting typos, simplifying some explanations.

One aspect preserved from earlier editions is that this textbook remains one of the few at this level that offers readers accessible insights into a selection of key topics that are normally avoided or simply quoted without derivation. These include a physical derivation of the energy flux carried by gravitational waves (an elaboration of Feynman's intuitive 1957 bead-on-a-wire argument for the physical reality of gravitational waves); an order-of-magnitude derivation of the Hawking radiation from the starting point of quantum fluctuations outside the horizon of a black hole; a full discussion of equatorial orbits in the Kerr metric, using a factorizable potential; and a treatment of the elegant Forward–Berman description of inspiralling binaries that shows that they are standard sirens because general relativity is a scale-free theory.

This book remains, of course, a beginner's introduction to the science of general relativity. All the new developments can be studied in much greater depth in many new books, and so the bibliography at the end of each chapter has been updated to point readers to references that can take them further in whatever direction they may find interesting.

I have benefitted from helpful conversations and input from many colleagues, both within the gravitational wave community and in the wider astronomy community. I thank them all, too many to name. But I also wish to thank those readers of previous editions who have very kindly drawn my attention to typos and other issues, particularly Tom Bartholet, Alon Brook-Ray, Sami Gara, Aleksandr Pargamotnikas, and Leo Schirber. Any remaining errors are, of course, my own responsibility. I thank also my editors at Cambridge University Press, Simon Capelin, Vince Higgs, and Ilaria Tassistro, for their patience and encouragement this third time around.

Preface to the second edition

In the twenty-three years between the first edition of this textbook and the present revision, the field of general relativity has blossomed and matured. Upon its solid mathematical foundations have grown a host of applications, some of which were not even imagined in 1985 when the first edition appeared. The study of general relativity has therefore moved from the periphery to the core of the education of a professional theoretical physicist, and more and more undergraduates expect to learn at least the basics of general relativity before they graduate.

My readers have been patient. Students have continued to use the first edition of this book to learn about the mathematical foundations of general relativity, even though it has become seriously out of date on applications like the astrophysics of black holes, the detection of gravitational waves, and the exploration of the Universe. This extensively revised second edition will, I hope, finally bring the book back into balance and give readers a consistent and unified introduction to modern research in classical gravitation.

The first eight chapters have seen little change. Recent references for further reading have been included, and a few sections have been expanded, but in general the geometrical approach to the mathematical foundations of the theory seems to have stood the test of time. By contrast, the final four chapters, which deal with general relativity in the astrophysical arena, have been updated, expanded, and in some cases completely re-written.

In Chapter 9, on gravitational radiation, there is now an extensive discussion of detection with interferometers like LIGO and the planned space-based detector LISA. I have also included a discussion of likely gravitational wave sources, and what we can expect to learn from detections. This is a field that is rapidly changing, and the first-ever direct detection could come at any time. Chapter 9 is intended to provide a durable framework for understanding the implications of these detections.

In Chapter 10, the discussion of the structure of spherical stars remains robust, but I have inserted material on real neutron stars, which we see as pulsars and which are potential sources of detectable gravitational waves.

Chapter 11, on black holes, has also gained extensive material about the astrophysical evidence for black holes, both for stellar-mass black holes and for the supermassive black holes that astronomers have astonishingly discovered in the centers of most galaxies. The discussion of the Hawking radiation has also been slightly amended.

Finally, Chapter 12 on cosmology is completely rewritten. In the first edition I essentially ignored the cosmological constant. In this I followed the prejudice of the time, which assumed that the expansion of the Universe was slowing down, even though it had not yet been accurately enough measured. We now know, from a variety of mutually consistent observations, that the expansion is accelerating. This is probably the biggest challenge to theoretical physics today, having an impact as great on fundamental theories of particle physics as on cosmological questions. I have organized Chapter 12 around this fact, developing mathematical models of an expanding Universe that include the cosmological constant, then discussing in detail how astronomers measure the kinematics of the Universe, and finally exploring the way that the physical constituents of the Universe evolved after the Big Bang. The roles of inflation, of dark matter, and of dark energy all affect the structure of the Universe today, and even our very existence. In this chapter it is possible only to give a brief taste of what astronomers have learned about these issues, but I hope it is enough to encourage readers to go on to learn more.

I have included more exercises in various chapters, where it was appropriate, but I have removed the exercise solutions from the book. They are available now on the website for the book.

The subject of this book remains classical general relativity: apart from a brief discussion of the Hawking radiation, there is no reference to quantization effects. While quantum gravity is one of the most active areas of research in theoretical physics today, there is still no clear direction to point a student who wants to learn how to quantize gravity. Perhaps by the third edition it will be possible to include a chapter on how gravity is quantized!

I want to thank many people who have helped me with this second edition. Several have generously supplied me with lists of misprints and errors in the first edition; I especially want to mention Frode Appel, Robert D'Alessandro, J. A. D. Ewart, Steve Fulling, Toshi Futamase, Gerald Quinlan, and B. Sathyaprakash. Any remaining errors are, of course, my own responsibility. I thank also my editors at Cambridge University Press, Rufus Neal, Simon Capelin, and Lindsay Barnes, for their patience and encouragement. And of course I am deeply indebted to my wife Sîan for her generous patience during all the hours, days, and weeks I spent working on this revision.

Preface to the first edition

This book has evolved from lecture notes for a full-year undergraduate course in general relativity which I taught from 1975 to 1980, an experience which firmly convinced me that general relativity is not significantly more difficult for undergraduates to learn than the standard undergraduate-level treatments of electromagnetism and quantum mechanics. The explosion of research interest in general relativity in the past 20 years, largely stimulated by astronomy, has not only led to a deeper and more complete understanding of the theory; it has also taught us simpler, more physical ways of understanding it. Relativity is now in the mainstream of physics and astronomy, so that no theoretical physicist can be regarded as broadly educated without some training in the subject. The formidable reputation relativity acquired in its early years (Interviewer: 'Professor Eddington, is it true that only three people in the world understand Einstein's theory?' Eddington: 'Who is the third?') is today perhaps the chief obstacle that prevents it being more widely taught to theoretical physicists. The aim of this textbook is to present general relativity at a level appropriate for undergraduates, so that the student will understand the basic physical concepts and their experimental implications, will be able to solve elementary problems, and will be well prepared for the more advanced texts on the subject.

In pursuing this aim, I have tried to satisfy two competing criteria: first, to assume a minimum of prerequisites; and second, to avoid watering down the subject matter. Unlike most introductory texts, this one does not assume that the student has already studied electromagnetism in its manifestly relativistic formulation, the theory of electromagnetic waves, or fluid dynamics. The necessary fluid dynamics is developed in the relevant chapters. The main consequence of not assuming a familiarity with electromagnetic waves is that gravitational waves have to be introduced slowly: the wave equation is studied from scratch. A full list of prerequisites appears below.

The second guiding principle, that of not watering down the treatment, is very subjective and rather more difficult to describe. I have tried to introduce differential geometry fully, not being content to rely only on analogies with curved surfaces, but I have left out subjects that are not essential to general relativity at this level, such as nonmetric manifold theory, Lie derivatives, and fiber bundles.[1] I have introduced the full nonlinear field equations, not just those of linearized theory, but I solve them only in the plane and spherical cases, quoting and examining, in addition, the Kerr solution. I study gravitational waves mainly in the linear approximation, but go slightly beyond it to derive the energy in the waves and the reaction effects in the wave emitter. I have tried in each topic to supply enough

---

[1] The treatment here is therefore different in spirit from that in my book *Geometrical Methods of Mathematical Physics* (Cambridge University Press 1980b), which may be used to supplement this one.

foundation for the student to be able to go to more advanced treatments without having to start over again at the beginning.

The first part of the book, up to Chapter 8, introduces the theory in a sequence which is typical of many treatments: a review of special relativity, development of tensor analysis and continuum physics in special relativity, study of tensor calculus in curvilinear coordinates in Euclidean and Minkowski spaces, geometry of curved manifolds, physics in a curved spacetime, and finally the field equations. The remaining four chapters study a few topics which I have chosen because of their importance in modern astrophysics. The chapter on gravitational radiation is more detailed than usual at this level because the observation of gravitational waves may be one of the most significant developments in astronomy in the next decade. The chapter on spherical stars includes, besides the usual material, a useful family of exact compressible solutions due to Buchdahl. A long chapter on black holes studies in some detail the physical nature of the horizon, going as far as the Kruskal coordinates, then exploring the rotating (Kerr) black hole, and concluding with a simple discussion of the Hawking effect, the quantum mechanical emission of radiation by black holes. The concluding chapter on cosmology derives the homogeneous and isotropic metrics and briefly studies the physics of cosmological observation and evolution. There is an appendix summarizing the linear algebra needed in the text, and another appendix containing hints and solutions for selected exercises. One subject I have decided not to give as much prominence to as other texts traditionally have is experimental tests of general relativity and of alternative theories of gravity. Points of contact with experiment are treated as they arise, but systematic discussions of tests now require whole books (Will 1981, 2nd edn 2018). Physicists today have far more confidence in the validity of general relativity than they had a decade or two ago, and I believe that an extensive discussion of alternative theories is therefore almost as out of place in a modern elementary text on gravity as it would be in one on electromagnetism.

The student is assumed already to have studied: special relativity, including the Lorentz transformation and relativistic mechanics; Euclidean vector calculus; ordinary and simple partial differential equations; thermodynamics and hydrostatics; Newtonian gravity (simple stellar structure would be useful but not essential); and enough elementary quantum mechanics to know what a photon is.

The notation and conventions are essentially the same as in Misner, Thorne, & Wheeler, *Gravitation* (W. H. Freeman 1973), which may be regarded as one possible follow-on text after this one. The physical point of view and development of the subject are also inevitably influenced by that book, partly because Thorne was my teacher and partly because *Gravitation* has become such an influential text. But because I have tried to make the subject accessible to a much wider audience, the style and pedagogical method of the present book are very different.

Regarding the use of the book, it is designed to be studied sequentially as a whole, in a one-year course, but it can be shortened to accommodate a half-year course. Half-year courses probably should aim at restricted goals. For example, it would be reasonable to aim to teach gravitational waves and black holes in half a year to students who have already studied electromagnetic waves, by carefully skipping some of Chapters 1–3 and most of Chapters 4, 7, and 10. Students with preparation in special relativity and fluid dynamics could learn stellar structure and cosmology in half a year, provided they could go quickly through the first four chapters and then skip Chapters 9 and 11. A graduate-level course can, of course, go much more quickly, and it should be possible to cover the whole text in half a year.

Each chapter is followed by a set of exercises, which range from trivial ones (filling in missing steps in the body of the text, manipulating newly introduced mathematics) to advanced problems that considerably extend the discussion in the text. Some problems require programmable calculators or computers. I cannot overstress the importance of doing a selection of problems. The easy and medium-hard ones in the early chapters give essential practice, without which the later chapters will be much less comprehensible. The medium-hard and hard problems of the later chapters are a test of the student's understanding. It is all too common in relativity for students to find the conceptual framework so interesting that they relegate problem solving to second place. Such a separation is false and dangerous: a student who can't solve problems of reasonable difficulty doesn't really understand the concepts of the theory either. There are generally more problems than one would expect a student to solve; several chapters have more than 30. The teacher will have to select them judiciously. Another rich source of problems is the *Problem Book in Relativity and Gravitation*, Lightman *et al.* (Princeton University Press 1975).

I am indebted to many people for their help, direct and indirect, with this book. I would like especially to thank my undergraduates at University College, Cardiff, whose enthusiasm for the subject and whose patience with the inadequacies of the early lecture notes encouraged me to turn them into a book. And I am certainly grateful to Suzanne Ball, Jane Owen, Margaret Vallender, Pranoat Priesmeyer and Shirley Kemp for their patient typing and retyping of the successive drafts.

1 Special relativity

1.1 Fundamental principles of special relativity theory (SR)

The way in which special relativity is taught at an elementary undergraduate level – the level at which the reader is assumed to be competent – is often close in spirit to the way in which it was first understood by physicists. This is an algebraic approach, based on the Lorentz transformation (§ 1.9 below). At this basic level, one learns how to use the Lorentz transformation to convert between one observer's measurements and another's, to verify and understand such remarkable phenomena as time dilation and Lorentz contraction, and to make elementary calculations of the conversion of mass into energy.

This purely algebraic point of view began to change, to widen, less than four years after Einstein proposed the theory.[1] Minkowski pointed out that it is very helpful to regard (t, x, y, z) as simply four coordinates in a four-dimensional space which we now call spacetime. This was the beginning of the geometrical point of view which ultimately led Einstein to general relativity in 1915. We will develop this geometric point of view as our first step along the same path.

As we shall see, special relativity can be deduced from two fundamental postulates:

(1) *Principle of relativity* (Galileo). No experiment can measure the absolute velocity of an observer; the results of any experiment performed by an observer do not depend on the observer's speed relative to other observers who are not involved in the experiment.
(2) *Universality of the speed of light* (Einstein). The speed of light relative to any unaccelerated observer is $c = 3 \times 10^8$ m s^{-1}, regardless of the motion of the light source relative to the observer. Let us be quite clear about this postulate's meaning: two different unaccelerated observers measuring the speed of the *same photon* will each find it to be moving at 3×10^8 m s^{-1} relative to themselves, regardless of their state of motion relative to each other.

As noted above, the principle of relativity is not at all a modern concept: it goes back all the way to Galileo's hypothesis that a body in a state of uniform motion remains in that state unless acted upon by some external agency. It is fully embodied in Newton's second

---

[1] Einstein's original paper was published in 1905, while Minkowski's discussion of the geometry of spacetime was given in 1908. Einstein's and Minkowski's papers are reprinted (in English translation) in *The Principle of Relativity* by A. Einstein, H. A. Lorentz, H. Minkowski & H. Weyl (Dover).

law, which contains only accelerations, not velocities themselves. Newton's laws are, in fact, all invariant under the replacement

$$v(t) \rightarrow v'(t) = v(t) - V,$$

where V is any *constant* velocity. This equation says that the velocity $v(t)$ of an object relative to one observer becomes $v'(t)$ when measured by a second observer whose velocity relative to the first is V. This is called the Galilean law of addition of velocities.

By saying that Newton's laws are *invariant* under the Galilean law of addition of velocities, we are making a statement of a sort which we will often make in our study of relativity, so it is as well to start by making it very precise. Newton's first law, that a body moves at a constant velocity in the absence of external forces, is unaffected by the replacement above, since if $v(t)$ is really a constant, say v_0, then the new velocity $v_0 - V$ is also a constant. Newton's second law,

$$F = ma = m \, dv/dt,$$

is also unaffected, since

$$a' = dv'/dt = d(v - V)/dt = dv/dt = a.$$

Therefore the second law will be valid according to the measurements of both observers, provided that we add to the Galilean transformation law the statement that F and m are themselves invariant, i.e. the same regardless of which of the two observers measures them. Newton's third law, that the force exerted by one body on another is equal and opposite to that exerted by the second on the first, is clearly unaffected by a change of observer, again because we assume the forces to be invariant.

So, there is no absolute velocity. Is there an absolute acceleration? Newton argued that there was. Suppose, for example, that I am in a train on a perfectly smooth track,[2] eating a bowl of soup in the dining car. Then if the train moves at constant velocity the soup remains level, thereby offering me no information about what my velocity is. But if the train changes its velocity then the soup climbs up one side of the bowl, and I can tell by looking at it how large and in what direction is the acceleration.[3]

Therefore, it is reasonable and useful to single out a class of preferred observers: those who are unaccelerated. They are called *inertial observers*, and each has a constant velocity with respect to any other. These inertial observers are fundamental in special relativity, and when we use the term 'observer' from now on we will mean an inertial observer.

The postulate of the universality of the speed of light was Einstein's great and radical contribution to relativity. It smashes the Galilean law of addition of velocities because it says that if v has magnitude c then so does v', regardless of V. The earliest direct evidence for this postulate was the Michelson–Morley experiment, although it is not clear whether Einstein himself was influenced by it. The counterintuitive predictions of special relativity all flow from this postulate, and they are amply confirmed by experiment. In fact it is probably fair to say that special relativity has a firmer experimental basis than any

[2] Physicists frequently find it helpful to make idealizations!
[3] For Newton's discussion of this point, see the excerpt from his *Principia* in Williams (1968).

other of our laws of physics, since it has been tested countless times in our giant particle accelerators, which send particles nearly to the speed of light.

Although the concept of relativity is old, it is customary to refer to Einstein's theory simply as 'relativity'. The adjective 'special' is applied in order to distinguish it from Einstein's theory of gravitation, which acquired the name 'general relativity' because it permits one to describe physics from the point of view of both accelerated and inertial observers and is in that respect a more general form of relativity. But the real physical distinction between these two theories is that special relativity (SR) is capable of describing physics only in the absence of gravitational fields, while general relativity (GR) extends SR to describe gravitation itself.

We can understand why gravitational fields cause problems for SR by considering astronauts in orbit about Earth, where they are weightless. An astronaut holding a bowl of soup does not see the soup climb up the side of the bowl, despite the gravitational 'force' that holds the spacecraft in orbit. Similarly, two astronauts in different orbits accelerate relative to one another, but neither *feels* an acceleration. Problems like this make gravity different from other forces, and we will have to wait until Chapter 5 to resolve them. Until then, the word 'force' will refer only to a nongravitational force. One can wish that an earlier generation of physicists had chosen more appropriate names for these theories !

1.2 Definition of an inertial observer in SR

It is important to realize that an 'observer' is in fact a huge information-gathering system, not simply one person with binoculars. In fact, we shall remove the human element entirely from our definition and say that an inertial observer is simply a spacetime coordinate system, which makes an observation simply by recording the location (x, y, z) and time (t) of any event. This coordinate system must satisfy the following three properties to be called *inertial*:

(1) The distance between point P_1 (coordinates x_1, y_1, z_1) and point P_2 (coordinates x_2, y_2, z_2) is independent of time.
(2) The clocks that sit at every point ticking off the time coordinate t are synchronized and all run at the same rate.
(3) The geometry of space for any constant time t is Euclidean.

Notice that this definition does not mention whether the observer accelerates or not. That will come later. It will turn out that only an unaccelerated observer can keep its clocks synchronized. But we prefer to start out with the above geometrical definition of an inertial observer.

It is a matter for experiment to decide whether an observer can really exist: it is not self-evident that any of the properties listed above *must* be realizable, although we would probably expect a 'nice' Universe to permit them! We will see later, however, that the presence of a gravitational field does generally make it impossible to construct such a coordinate system, and this is why GR is then required. But let us not get ahead of the story. At the moment we are assuming that we *can* construct such a coordinate system (that, if you like, the gravitational fields around us are so weak that they do not really matter). One can envision this coordinate system, rather fancifully, as a lattice of rigid rods filling space, with a clock at every intersection of the rods. Some convenient system, such as a collection of GPS satellites and receivers, is used to ensure that all the clocks are synchronized. The clocks are supposed to be very densely spaced, so that there is a clock next to every event of interest, ready to record its time of occurrence without any delay. We shall now define how we use this coordinate system to make observations.

An *observation* made by the inertial observer is the act of assigning to any event the coordinates x, y, z of the location of its occurrence, and the time read by the clock at (x, y, z) when the event occurred. It is *not* the time t on the wristwatch worn by a scientist located at $(0, 0, 0)$ when news of the event first arrives there. A *visual* observation is of this second type: the eye regards as simultaneous all events it *sees* at the same time; an inertial observer regards as simultaneous all events that *occur* at the same time as that recorded by the clock nearest them when they occurred. This distinction is important and should always be borne in mind; it is at the root of many confusions about SR. Sometimes we will say 'an observer sees...' but this will only be shorthand for 'measures'. We will never mean a *visual* observation unless we say so explicitly.

An inertial observer is also called an *inertial reference frame*, which we will often abbreviate to 'reference frame' or simply 'frame'.

1.3 New units

Since the speed of light c is so fundamental, we shall from now on adopt a new system of units for measurements in which c simply has the value 1! It is perfectly reasonable when driving a car or shopping for food to be content with the SI units: m, s, kg. But it seems silly in SR to use units in which the fundamental constant c has the ridiculous value 3×10^8. The SI units evolved historically. Meters and seconds are not fundamental; they are simply convenient for human use. We shall now take advantage of the universality of the speed of light to adopt a new unit for time, the meter.

One meter of time is the time it takes light to travel one meter.[4] The speed of light in these units is:

$$c = \frac{\text{distance light travels in any given time interval}}{\text{the given time interval}}$$
$$= \frac{1\text{ m}}{\text{the time it takes light to travel one meter}}$$
$$= \frac{1\text{ m}}{1\text{ m}} = 1.$$

---

[4] You are probably more familiar with an alternative approach: a year of distance – called a 'light year' – is the distance light travels in one year.

So if we consistently measure time in meters, then c is not merely 1, it is also dimensionless! In converting from SI units to these 'natural' units, you can use any of the following relations:

$$3 \times 10^8 \,\mathrm{m\,s}^{-1} = 1,$$

$$1\,\mathrm{s} = 3 \times 10^8 \,\mathrm{m},$$

$$1\,\mathrm{m} = \frac{1}{3 \times 10^8}\,\mathrm{s}.$$

The SI units contain many 'derived' units, such as joules and newtons, which are defined in terms of the basic three: m, s, kg. By converting from s to m these units become considerably simpler: energy and momentum are measured in kg, acceleration in m^{-1}, force in $\mathrm{kg\,m}^{-1}$, etc. Do the exercises on this. With practice, these units will seem as natural to you as they do to most modern theoretical physicists.

1.4 Spacetime diagrams

A very important part of learning the geometrical approach to SR is mastering spacetime diagrams. In the rest of this chapter we will derive SR from its postulates by using spacetime diagrams, because they provide a very powerful guide for threading one's way among the many pitfalls SR presents to the beginner. Figure 1.1 shows a two-dimensional slice of spacetime, the t–x plane, in which the basic concepts are illustrated. A single point in this space[5] is a point of fixed x *and* fixed t, and is called an *event*. A line in the space gives a relation $x = x(t)$, and so can represent the position of a particle at different times. This is called the particle's *world line*. Its slope is related to its velocity,

$$\text{slope} = \mathrm{d}t/\mathrm{d}x = 1/v.$$

Notice that a light ray (photon) always travels on a 45° line in this diagram.

We shall adopt the following notational conventions:

(1) *Events* will be denoted by calligraphic capitals, e.g. $\mathcal{A}, \mathcal{B}, \mathcal{P}$. However, the letter $\mathcal{O}$ is reserved to denote observers.
(2) The *coordinates* will be called t, x, y, z. Any quadruple of numbers, for example $(5, -3, 2, 10^{16})$, denotes an event; in this case the coordinates are $t = 5$, $x = -3$, $y = 2$, $z = 10^{16}$. Thus, we always put t first. All coordinates are measured in meters.
(3) It is often convenient to refer to the coordinates (t, x, y, z) as a whole, or to each indifferently. That is why we give them the *alternative names* (x^0, x^1, x^2, x^3). These superscripts are *not* exponents but just labels, called indices. Thus $(x^3)^2$ denotes the

---

[5] We use the word 'space' in a more general way than you may be used to. We do not mean a Euclidean space, in which Euclidean distances are necessarily physically meaningful. Rather, we mean just a set of points which are continuous (rather than discrete, as a lattice is). This is the first example of what we will define in Chapter 5 to be a 'manifold'.

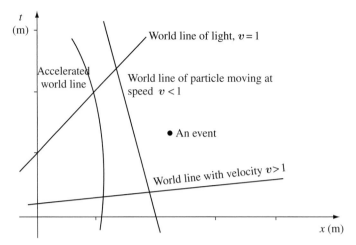

Figure 1.1   A spacetime diagram in natural units.

square of coordinate 3 (which is z), not the square of the cube of x. *Generically*, the coordinates x^0, x^1, x^2, and x^3 are referred to as x^α. A *Greek index* (e.g. α, β, μ, ν) will be assumed to take a value from the set (0, 1, 2, 3). If α is not given a value, then x^α can be *any* of the four coordinates. If we want to refer to all the coordinates we use curly brackets: $\{x^\alpha\}$.

(4) There are occasions when we want to distinguish between t on the one hand and (x, y, z) on the other. We use *Latin indices* to refer to the spatial coordinates alone. Thus a Latin index (e.g. a, b, i, j, k, l) will be assumed to take a value from the set (1, 2, 3). If i is not given a value then x^i is *any* of the three spatial coordinates.[6]

1.5 Construction of the coordinates used by another observer

Since any observer is simply a coordinate system for spacetime, and since all observers are looking at the same events (they are in the same spacetime), it should be possible to draw the coordinate lines of one observer on the spacetime diagram drawn by another observer. To do this we have to make use of the postulates of SR.

Suppose that an observer $\mathcal{O}$ uses the coordinates t, x as above and that another observer $\bar{\mathcal{O}}$, with coordinates $\bar{t}, \bar{x}$, is moving in the x direction with velocity v relative to $\mathcal{O}$. Where do the coordinate axes for $\bar{t}$ and $\bar{x}$ go in the spacetime diagram of $\mathcal{O}$?

$\bar{t}$ *axis* This is the locus of events at constant $\bar{x} = 0$ (and $\bar{y} = \bar{z} = 0$, too, but we shall ignore them here) and is the locus of the origin of $\bar{\mathcal{O}}$'s spatial coordinates. This locus is $\bar{\mathcal{O}}$'s world line, and it looks like that shown in Figure 1.2.

---

[6] Our conventions on the use of Greek and Latin indices are by no means universally used by physicists. Some books reverse them, using Latin for $\{0, 1, 2, 3\}$ and Greek for $\{1, 2, 3\}$; others use $a, b, c, \ldots$ for one set and i, j, k for the other. You should always check the conventions used in whatever work you may be reading.

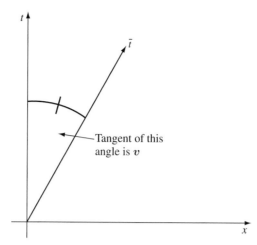

Figure 1.2   The time axis of a frame whose velocity is *v*.

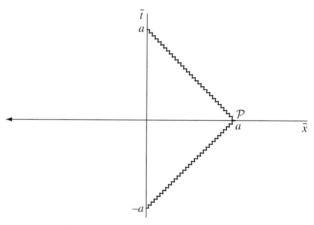

Figure 1.3   Light reflected at *a*, as measured by $\bar{\mathcal{O}}$.

$\bar{x}$ *axis* To locate this we make a construction designed to determine the locus of events at $\bar{t} = 0$, i.e. those that $\bar{\mathcal{O}}$ measures to be simultaneous with the event $\bar{t} = \bar{x} = 0$.

Consider the picture in $\bar{\mathcal{O}}$'s spacetime diagram, shown in Figure 1.3. Events on the $\bar{x}$ axis all have the following property. A light ray emitted at event $\mathcal{E}$ from $\bar{x} = 0$ at, say, time $\bar{t} = -a$ will reach the $\bar{x}$ axis at $\bar{x} = a$ (we call this event $\mathcal{P}$); if reflected, it will return to the point $\bar{x} = 0$ at $\bar{t} = +a$, called event $\mathcal{R}$. The $\bar{x}$ axis can be *defined*, therefore, as the locus of events that reflect light rays in such a manner that they return to the $\bar{t}$ axis at $+a$ if they left it at $-a$, for any a. Now look at this in the spacetime diagram of $\mathcal{O}$, Figure 1.4.

We know where the $\bar{t}$ axis lies, since we constructed it in Figure 1.2. The events of emission and reception, $\bar{t} = -a$ and $\bar{t} = +a$, are shown in Figure 1.4. Since a is arbitrary, it does not matter where along the negative $\bar{t}$ axis we place event $\mathcal{E}$, so no assumption need yet be made about the calibration of the $\bar{t}$ axis relative to the t axis. All that matters for the moment is that the event $\mathcal{R}$ on the $\bar{t}$ axis must be as far from the origin as event $\mathcal{E}$. Having

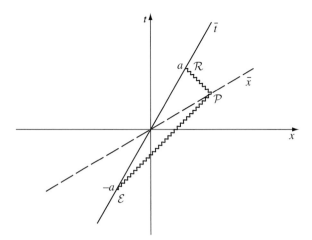

Figure 1.4   The reflection in Figure 1.3, as measured by $\mathcal{O}$.

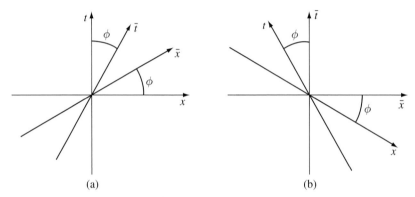

(a)   (b)

Figure 1.5   Spacetime diagrams for $\mathcal{O}$ (left) and $\bar{\mathcal{O}}$ (right).

drawn them in Figure 1.4, we next draw in the same light beam as before, emitted from $\mathcal{E}$, and traveling on a 45° line in this diagram. The reflected light beam must arrive at $\mathcal{R}$, so it is the 45° line with negative slope through $\mathcal{R}$. The intersection of these two light beams must be the event of reflection $\mathcal{P}$. This establishes the location of $\mathcal{P}$ in our diagram. The line joining it with the origin – the dashed line – must be the $\bar{x}$ axis: it does *not* coincide with the x axis. If you compare Figure 1.4 with Figure 1.3 you will see why: in both diagrams light moves on a 45° line, while the t and $\bar{t}$ axes change slope from one diagram to the other. This is the embodiment of the second fundamental postulate of SR: that a light beam has speed $c = 1$ (and hence slope = 1) with respect to *every* observer. When we apply this to these geometrical constructions we immediately find that events simultaneous in $\bar{\mathcal{O}}$ (along a line parallel to $\bar{t} = 0$, i.e. observer $\bar{\mathcal{O}}$'s $\bar{x}$ axis) are not simultaneous in $\mathcal{O}$ (they are not parallel to the line $t = 0$, the x axis). This *failure of simultaneity* is inescapable.

The diagrams in Figure 1.5 represent the same physical situation. The one on the left is the spacetime diagram for $\mathcal{O}$, in which $\bar{\mathcal{O}}$ moves to the right. The one on the right is drawn from the point of view of $\bar{\mathcal{O}}$, in which $\mathcal{O}$ moves to the left. The four angles to the axes are all $\phi = \arctan|v|$, where $|v|$ is the relative speed of $\mathcal{O}$ and $\bar{\mathcal{O}}$.

1.6 Invariance of the interval

We have not quite finished the construction of $\bar{\mathcal{O}}$'s coordinates, of course. We have found the positions of the axes but not the length scale along them. We shall find this scale by proving what is probably the most important theorem of SR, the invariance of intervals.

Consider two events on the world line of the same light beam, such as $\mathcal{E}$ and $\mathcal{P}$ in Figure 1.4. The differences $(\Delta t, \Delta x, \Delta y, \Delta z)$ between the coordinates of $\mathcal{E}$ and $\mathcal{P}$ in some frame $\mathcal{O}$ satisfy the relation $(\Delta x)^2 + (\Delta y)^2 + (\Delta z)^2 - (\Delta t)^2 = 0$, remembering that the speed of light is 1. But, by the universality of the speed of light, the coordinate differences between the same two events in the coordinates of $\bar{\mathcal{O}}, (\Delta \bar{t}, \Delta \bar{x}, \Delta \bar{y}, \Delta \bar{z})$, also satisfy $(\Delta \bar{x})^2 + (\Delta \bar{y})^2 + (\Delta \bar{z})^2 - (\Delta \bar{t})^2 = 0$. We shall define the *interval* between *any* two events (not necessarily on the same light beam's world line) that are separated by coordinate increments $(\Delta t, \Delta x, \Delta y, \Delta z)$ to be

$$\Delta s^2 = -(\Delta t)^2 + (\Delta x)^2 + (\Delta y)^2 + (\Delta z)^2. \tag{1.1}$$

It follows that if $\Delta s^2 = 0$ for two events, using their coordinates in $\mathcal{O}$, then $\Delta \bar{s}^2 = 0$ for the same two events using their coordinates in $\bar{\mathcal{O}}$. What does this imply about the relation between the coordinates of the two frames? To answer this question, we shall assume that the relation between the coordinates of $\mathcal{O}$ and $\bar{\mathcal{O}}$ is *linear* and that we choose their origins to coincide (i.e. the events $\bar{t} = \bar{x} = \bar{y} = \bar{z} = 0$ and $t = x = y = z = 0$ are the same). Then, in the expression for $\Delta \bar{s}^2$,

$$\Delta \bar{s}^2 = -(\Delta \bar{t})^2 + (\Delta \bar{x})^2 + (\Delta \bar{y})^2 + (\Delta \bar{z})^2,$$

the numbers $(\Delta \bar{t}, \Delta \bar{x}, \Delta \bar{y}, \Delta \bar{z})$ are linear combinations of their unbarred counterparts, which means that $\Delta \bar{s}^2$ is a *quadratic* function of the unbarred coordinate increments. We can therefore write

$$\Delta \bar{s}^2 = \sum_{\alpha=0}^{3} \sum_{\beta=0}^{3} M_{\alpha\beta} (\Delta x^\alpha)(\Delta x^\beta) \tag{1.2}$$

for some numbers $\{M_{\alpha\beta}; \alpha, \beta = 0, \ldots, 3\}$, which may be functions of v, the relative velocity of the two frames. Note that we can suppose that $M_{\alpha\beta} = M_{\beta\alpha}$ for all α and β, since only the sum $M_{\alpha\beta} + M_{\beta\alpha}$ ever appears in Eq. 1.2 when $\alpha \neq \beta$. Now we again suppose that $\Delta s^2 = 0$, so that from Eq. 1.1 we have

$$\Delta t = \Delta r, \quad \Delta r = [(\Delta x)^2 + (\Delta y)^2 + (\Delta z)^2]^{1/2}.$$

(We have assumed that $\Delta t > 0$ for convenience.) Putting this into Eq. 1.2 gives

$$\Delta \bar{s}^2 = M_{00}(\Delta r)^2 + 2 \left(\sum_{i=1}^{3} M_{0i} \Delta x^i \right) \Delta r$$

$$+ \sum_{i=1}^{3} \sum_{j=1}^{3} M_{ij} \Delta x^i \Delta x^j. \tag{1.3}$$

However, we have already observed that $\Delta\bar{s}^2$ must vanish if Δs^2 does, and this must be true for *arbitrary* $\{\Delta x^i; i = 1, 2, 3\}$. It is easy to show (see Exercise 1.8) that this implies

$$M_{0i} = 0, \quad i = 1, 2, 3 \tag{1.4a}$$

and

$$M_{ij} = -(M_{00})\delta_{ij}, \quad i, j = 1, 2, 3 \tag{1.4b}$$

where δ_{ij} is the Kronecker delta, defined by

$$\delta_{ij} = \begin{cases} 1 & \text{if } i = j, \\ 0 & \text{if } i \neq j. \end{cases} \tag{1.4c}$$

From this and Eq. 1.2 we conclude that

$$\Delta\bar{s}^2 = M_{00}[(\Delta t)^2 - (\Delta x)^2 - (\Delta y)^2 - (\Delta z)^2].$$

If we define a function

$$\phi(v) = -M_{00},$$

then we have proved the following theorem: *The universality of the speed of light implies that the intervals Δs^2 and $\Delta\bar{s}^2$ between any two events as computed by different observers satisfy the relation*

$$\Delta\bar{s}^2 = \phi(v)\Delta s^2. \tag{1.5}$$

We shall now show that, in fact, $\phi(v) = 1$, which is the statement that the interval is independent of the observer. The proof of this has two parts. The first part shows that $\phi(v)$ depends only on $|v|$. Consider a rod which is oriented perpendicular to the velocity v of $\bar{\mathcal{O}}$ relative to $\mathcal{O}$. Suppose the rod is at rest in $\mathcal{O}$, lying on the y axis. In the spacetime diagram of $\mathcal{O}$ (Figure 1.6), the world lines of its ends are drawn and the region between is hatched. It is easy to see that the square of its length is just the interval between two events $\mathcal{A}$ and $\mathcal{B}$ that are simultaneous in $\mathcal{O}$ (at $t = 0$) and occur at the ends of the rod. This is so because, for these events, $(\Delta x)_{\mathcal{A}\mathcal{B}} = (\Delta z)_{\mathcal{A}\mathcal{B}} = (\Delta t)_{\mathcal{A}\mathcal{B}} = 0$. Now comes the key point of the first part

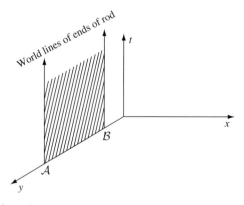

Figure 1.6   A rod at rest in $\mathcal{O}$, lying on the y axis.

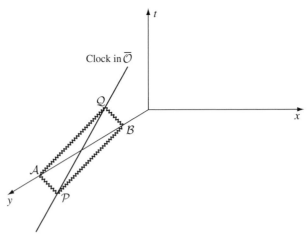

Figure 1.7 A clock in $\bar{\mathcal{O}}$'s frame, moving in the x direction in $\mathcal{O}$'s frame.

of the proof: the events $\mathcal{A}$ and $\mathcal{B}$ are simultaneous as measured by $\bar{\mathcal{O}}$ as well. The reason is most easily seen by the construction shown in Figure 1.7, which is the same spacetime diagram as Figure 1.6, but in which the world line of a clock in $\bar{\mathcal{O}}$ is drawn. This line is perpendicular to the y axis and parallel to the t–x plane, i.e. it is parallel to the $\bar{t}$ axis shown in Figure 1.5(a).

Suppose this clock emits light rays at event $\mathcal{P}$ which reach events $\mathcal{A}$ and $\mathcal{B}$. (Not every clock can do this, so we have chosen the one clock in $\bar{\mathcal{O}}$ which passes through the y axis at $t = 0$ and can send out such light rays.) The light rays reflect from $\mathcal{A}$ and $\mathcal{B}$, and one can see from the geometry (if you allow for the perspective in the diagram) that they arrive back at $\bar{\mathcal{O}}$'s clock at the *same* event $\mathcal{Q}$. Therefore, from $\bar{\mathcal{O}}$'s point of view, the two events occur at the same time. (This is the *same* construction as we used to determine the $\bar{x}$ axis.) But if $\mathcal{A}$ and $\mathcal{B}$ are simultaneous in $\bar{\mathcal{O}}$, then the interval between them in $\bar{\mathcal{O}}$ is also the square of their length in $\bar{\mathcal{O}}$. The result is:

$$(\text{length of rod in } \bar{\mathcal{O}})^2 = \phi(v)(\text{length of rod in } \mathcal{O})^2.$$

On the other hand, the length of the rod cannot depend on the *direction* of the velocity, because the rod is perpendicular to it and there are no preferred directions of motion (the principle of relativity). Hence the first part of the proof concludes that

$$\phi(\boldsymbol{v}) = \phi(|\boldsymbol{v}|). \tag{1.6}$$

The second step of the proof is easier. It uses the principle of relativity to show that $\phi(|\boldsymbol{v}|) = 1$. Consider three frames, $\mathcal{O}$, $\bar{\mathcal{O}}$, and $\bar{\bar{\mathcal{O}}}$. Frame $\bar{\mathcal{O}}$ moves with speed v in, say, the x direction relative to $\mathcal{O}$. Frame $\bar{\bar{\mathcal{O}}}$ moves with speed v in the negative x direction relative to $\bar{\mathcal{O}}$. It is clear that $\bar{\bar{\mathcal{O}}}$ is in fact identical to $\mathcal{O}$, but for the sake of clarity we shall keep separate notation for the moment. We have, from Eqs. 1.5 and 1.6,

$$\left.\begin{aligned}\Delta \bar{\bar{s}}^2 &= \phi(v)\Delta \bar{s}^2 \\ \Delta \bar{s}^2 &= \phi(v)\Delta s^2\end{aligned}\right\} \Rightarrow \Delta \bar{\bar{s}}^2 = [\phi(v)]^2 \Delta s^2.$$

However, since $\mathcal{O}$ and $\bar{\bar{\mathcal{O}}}$ are identical, $\Delta \bar{\bar{s}}^2$ and Δs^2 are equal. It follows that

$$\phi(v) = \pm 1.$$

We must choose the plus sign, since in the first part of this proof the square of the length of a rod must be positive. We have therefore proved the fundamental theorem that *the interval between any two events is the same when calculated by any inertial observer:*

$$\Delta \bar{s}^2 = \Delta s^2. \tag{1.7}$$

Notice that from the first part of this proof we can also conclude now that *the length of a rod oriented perpendicular to the relative velocity of two frames is the same when measured by either frame*. It is also worth reiterating that the construction in Figure 1.7 also proves a related result, that *two events which are simultaneous in one frame are simultaneous in any frame moving in a direction perpendicular to their separation relative to the first frame*.

Because Δs^2 is a property only of the two events and not of the observer, it can be used to classify the relation between the events. If Δs^2 is positive (so that the spatial increments dominate Δt) the events are said to be *spacelike separated*, and we call $\sqrt{(\Delta s^2)}$ the *proper distance* between the events. If Δs^2 is negative then the events are said to be *timelike separated*, and we call $\sqrt{|\Delta s^2|}$ the *proper time* between them. If Δs^2 is zero (so the events are on the same light path) the events are said to be *lightlike* or *null separated*.[7]

Events that are lightlike separated from any particular event $\mathcal{A}$ lie on a cone whose apex is $\mathcal{A}$. This cone is illustrated in Figure 1.8. This is called the *light cone of $\mathcal{A}$*. All events within the light cone are timelike separated from $\mathcal{A}$; all events outside it are spacelike separated. Therefore all events inside the cone can be reached from $\mathcal{A}$ on a world line which everywhere moves in a timelike direction. Since we will see later that nothing can move faster than light, all world lines of physical objects move in a timelike direction. Therefore events inside the light cone are reachable from $\mathcal{A}$ by a physical object, whereas those outside are not.

For this reason the events inside the 'future' or 'forward' light cone are sometimes called the *absolute future* of the apex; those within the 'past' or 'backward' light cone are called the *absolute past*; and those outside are called the *absolute elsewhere*. The cone is the boundary of the absolute past and future. This is a causal boundary: because no influences can move faster than light, anything that can later be affected by event $\mathcal{A}$ must be inside its absolute future, and anything that could earlier have contributed to conditions at $\mathcal{A}$ must lie within its absolute past.

---

[7] Note that some authors call Δs^2 the *squared interval* and use the word *interval* for $\sqrt{|\Delta s^2|}$. The terms proper time and proper distance are universally defined as we have done. Note also that the notation Δs^2 is *never* taken to mean $\Delta(s^2)$ for some quantity s^2; it always means $(\Delta s)^2$.

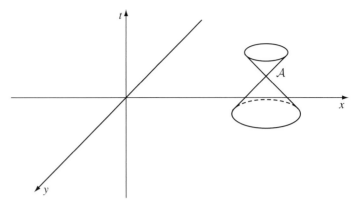

Figure 1.8 The light cone of an event $\mathcal{A}$, lying at the apex of the double cone. The z-dimension is suppressed.

Thus, although different observers will define 'time' and 'space' differently, it is important to understand that we can still talk about 'future' and 'past' in an invariant manner. To Galileo and Newton, the past was everything 'earlier' than 'now'; all of spacetime was the union of the past and the future, whose boundary was 'now'. In SR, the past is only everything inside the past light cone, and spacetime has *three* invariant divisions: SR adds the notion of 'elsewhere'. What is more, although *all* observers agree on what constitutes the past, future and elsewhere of a given event (because intervals are invariant), each different event has a *different* past and future; no two events have identical pasts and futures, even though they can overlap. These ideas are illustrated in Figure 1.9.

1.7 Invariant hyperbolae

We can now calibrate the axes of $\bar{\mathcal{O}}$'s coordinates in the spacetime diagram for $\mathcal{O}$, Figure 1.5. We restrict ourselves to the t–x plane. Consider a curve whose equation is

$$-t^2 + x^2 = a^2,$$

where a is a real constant. This is a hyperbola in the spacetime diagram for $\mathcal{O}$, and it passes through all events whose interval from the origin is a^2. By the invariance of intervals, these same events have interval a^2 from the origin in $\bar{\mathcal{O}}$, so they also lie on the curve $-\bar{t}^2 + \bar{x}^2 = a^2$. This is a hyperbola which is spacelike separated from the origin. Similarly, the events on the curve

$$-t^2 + x^2 = -b^2$$

all have timelike interval $-b^2$ from the origin, and also lie on the curve $-\bar{t}^2 + \bar{x}^2 = -b^2$. These hyperbolae are drawn in Figure 1.10. They are all asymptotic to the lines having slope ±1, which are of course the light paths through the origin. In a three-dimensional diagram (in which we would add the y axis, as in Figure 1.8), hyperbolae of revolution would be asymptotic to the light cone.

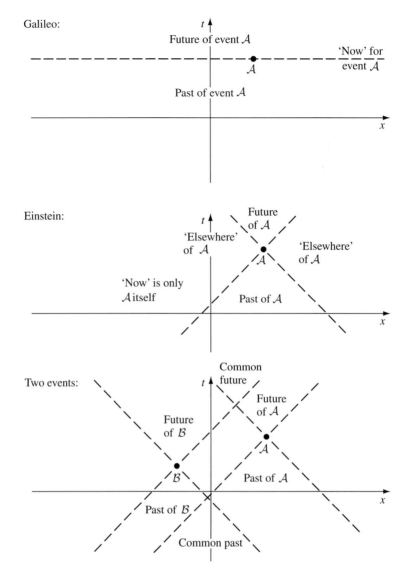

Figure 1.9 Old and new concepts of spacetime.

We can now calibrate the axes of $\bar{\mathcal{O}}$. In Figure 1.11 are drawn the axes of $\mathcal{O}$ and $\bar{\mathcal{O}}$, and an invariant hyperbola of timelike interval -1 from the origin. Event $\mathcal{A}$ is on the t axis, so has $x = 0$. Since the hyperbola has the equation

$$-t^2 + x^2 = -1,$$

it follows that event $\mathcal{A}$ has $t = 1$. Similarly, event $\mathcal{B}$ lies on the $\bar{t}$ axis so has $\bar{x} = 0$. Since the hyperbola also has the equation

$$-\bar{t}^2 + \bar{x}^2 = -1,$$

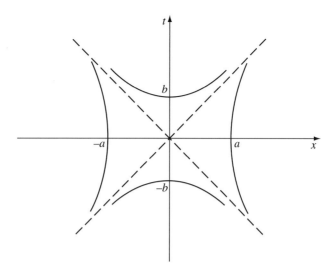

Figure 1.10   Invariant hyperbolae, for $a > b$.

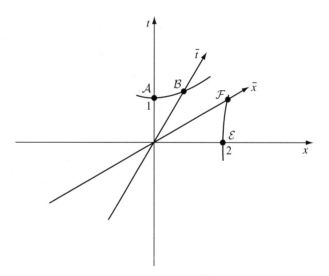

Figure 1.11   Using the hyperbolae through events $\mathcal{A}$ and $\mathcal{E}$ to calibrate the $\bar{x}$ and $\bar{t}$ axes.

it follows that event $\mathcal{B}$ has $\bar{t} = 1$. We have therefore used the hyperbolae to calibrate the $\bar{t}$ axis. In the same way, the invariant hyperbola

$$-t^2 + x^2 = 4$$

shows that event $\mathcal{E}$ has coordinates $t = 0$, $x = 2$ and that event $\mathcal{F}$ has coordinates $\bar{t} = 0$, $\bar{x} = 2$. This kind of hyperbola calibrates the spatial axes of $\bar{\mathcal{O}}$.

Notice that event $\mathcal{B}$ looks as though it is 'further' from the origin than $\mathcal{A}$. This again shows the inappropriateness of using geometrical intuition based upon Euclidean geometry. Here the important physical quantity is the interval $-(\Delta t)^2 + (\Delta x)^2$, not the Euclidean distance $(\Delta t)^2 + (\Delta x)^2$: when we learn relativity we have to learn to use Δs^2 as the physical measure of 'distance' in spacetime, and we have to adapt our intuition accordingly. This is not, of course, in conflict with everyday experience. Everyday experience asserts that 'space' (e.g. the section of spacetime with $t = 0$) is Euclidean. For events that have $\Delta t = 0$ (i.e. that are simultaneous to observer $\mathcal{O}$), the interval is

$$\Delta s^2 = (\Delta x)^2 + (\Delta y)^2 + (\Delta z)^2.$$

This is just their Euclidean distance. The new feature of SR is that time can (and must) be brought into the computation of distance. It is not possible to define 'space' uniquely since different observers identify different sets of events to be simultaneous (Figure 1.5). But there is still a distinction between space and time, since temporal increments enter Δs^2 with the opposite sign from spatial ones.

In order to use the hyperbolae to derive the effects of time dilation and Lorentz contraction, as we do in the next section, we must point out a simple but important property

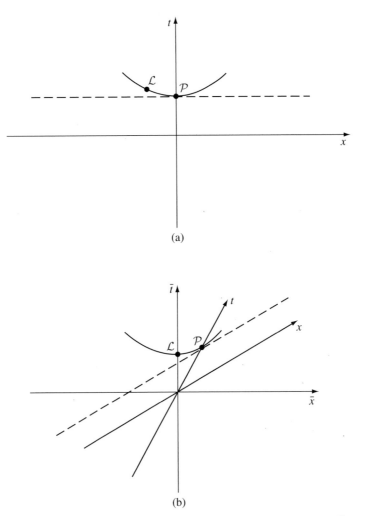

(a)

(b)

Figure 1.12   (a) A line of simultaneity in $\mathcal{O}$ is tangent to the hyperbola at $\mathcal{P}$. (b) The same tangency as seen by $\bar{\mathcal{O}}$.

of the tangents to the hyperbolae. In Figure 1.12(a) we have drawn a hyperbola and its tangent at $x = 0$, which is obviously a line of simultaneity, $t = $ const. In Figure 1.12(b) we have drawn the same curves from the point of view of an observer $\bar{\mathcal{O}}$ who is moving to the left relative to $\mathcal{O}$. The event $\mathcal{P}$ has been shifted to the right: it could be shifted anywhere on the hyperbola by choosing the velocity of $\bar{\mathcal{O}}$ relative to $\mathcal{O}$ appropriately. The lesson of Figure 1.12(b) is that the tangent to a hyperbola at any event $\mathcal{P}$ is a line of simultaneity of the inertial frame whose time axis joins $\mathcal{P}$ to the origin. If this frame has velocity v, the tangent has slope v.

1.8 Particularly important results

Time dilation. From Figure 1.11 and the calculation relating to it, we can deduce that when a clock moving on the $\bar{t}$ axis reaches $\mathcal{B}$ it has a reading of $\bar{t} = 1$, but that event $\mathcal{B}$ has coordinate $t = 1/\sqrt{(1 - v^2)}$ in $\mathcal{O}$. So, to $\mathcal{O}$ it appears to run slowly:

$$(\Delta t)_{\text{measured in } \mathcal{O}} = \frac{(\Delta \bar{t})_{\text{measured in } \mathcal{O}}}{\sqrt{(1 - v^2)}}. \tag{1.8}$$

Notice that $\Delta \bar{t}$ is the time actually measured by a single clock which moves on a world line from the origin to $\mathcal{B}$, while Δt is the difference in the readings of two clocks at rest in $\mathcal{O}$, one on a world line through the origin and one on a world line through $\mathcal{B}$. We shall show below that this is the reason for the disagreement between the observers on the rate of running of clocks.

For now, it is important to understand that the *proper time* between an event $\mathcal{B}$ and the origin is the time ticked by the clock which actually passes through both events. In this case, that clock is at rest in $\bar{\mathcal{O}}$, so in that frame it is easy to compute the interval:

$$\Delta s^2 = -\Delta \bar{t}^2 = -\Delta \tau^2. \tag{1.9}$$

Alternatively, by expressing the interval in terms of $\mathcal{O}$'s coordinates, we get

$$\begin{aligned} \Delta \tau &= [(\Delta t)^2 - (\Delta x)^2 - (\Delta y)^2 - (\Delta z)^2]^{1/2} \\ &= \Delta t \sqrt{(1 - v^2)}. \end{aligned} \tag{1.10}$$

This is time dilation all over again. But we have learned something more: *the interval between two events in SR is a physically measurable quantity.* It is the time elapsed on a clock that moves (at constant velocity) between the two events.

Lorentz contraction. In Figure 1.13 we show the *world path* of a rod at rest in $\bar{\mathcal{O}}$ but moving with velocity **v** with respect to $\mathcal{O}$. Each observer assigns it a length by measuring the distance between the locations of its front and rear ends at the same time, as determined by that observer. The lack of agreement on simultaneity translates into a lack of agreement

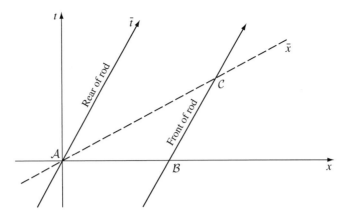

Figure 1.13 The proper length of $\mathcal{AC}$ is the length of the rod in its rest frame, while that of $\mathcal{AB}$ is its Lorentz-contracted length in $\mathcal{O}$.

on the length of the rod. The fact that the measurement involves two locations at the same time means that the length is the same as the proper distance between the two events used in the measurement.

To see what this means, look at the figure. The length of the rod in the rod's rest frame $\bar{\mathcal{O}}$ is the square root of Δs^2_{AC}, because the events A and C are simultaneous in this frame. In contrast, its length in $\mathcal{O}$ is the square root of Δs^2_{AB}. If event C has coordinates $\bar{t} = 0$, $\bar{x} = l$, then, by a calculation identical to that used before, its x coordinate in $\mathcal{O}$ is

$$x_C = l/\sqrt{(1 - v^2)},$$

and since the $\bar{x}$ axis is the line $t = vx$, we have

$$t_C = vl/\sqrt{(1 - v^2)}.$$

The line $\mathcal{BC}$ has slope (relative to the t axis)

$$\Delta x/\Delta t = v,$$

and so we have

$$\frac{x_C - x_B}{t_C - t_B} = v,$$

and we want to know x_B when $t_B = 0$:

$$x_B = x_C - vt_C$$
$$= \frac{l}{\sqrt{(1 - v^2)}} - \frac{v^2 l}{\sqrt{(1 - v^2)}} = l\sqrt{(1 - v^2)}. \qquad (1.11)$$

This is the Lorentz contraction.

Conventions. The interval Δs^2 is one of the most important mathematical concepts of SR but there is no universal agreement on its definition: many authors define $\Delta s^2 = (\Delta t)^2 - (\Delta x)^2 - (\Delta y)^2 - (\Delta z)^2$. This overall sign is a matter of convention (like the use of Latin and Greek indices we referred to earlier), since the invariance of Δs^2 implies the invariance of $-\Delta s^2$. The physical result of importance is just this invariance, which arises from the difference in sign between $(\Delta t)^2$ and $[(\Delta x)^2 + (\Delta y)^2 + (\Delta z)^2]$. As with other conventions, anyone consulting another reference work should ensure that they know which sign is being used: it affects all sorts of formulae, for example Eq. 1.9.

Failure of relativity? Newcomers to SR, and others who don't understand it well enough, often worry at this point that the theory is inconsistent. We began by assuming the principle of relativity, which asserts that all observers are equivalent. Now we have shown that if $\bar{\mathcal{O}}$ moves relative to $\mathcal{O}$, the clocks of $\bar{\mathcal{O}}$ will be measured by $\mathcal{O}$ to be running more slowly than those of $\mathcal{O}$. So isn't it therefore the case that $\bar{\mathcal{O}}$ will measure $\mathcal{O}$'s clocks to be running faster than those of $\bar{\mathcal{O}}$? If so, this violates the principle of relativity, since we could as easily have begun with $\bar{\mathcal{O}}$ and deduced that $\mathcal{O}$'s clocks run more slowly than $\bar{\mathcal{O}}$'s.

This is what is known as a 'paradox', but like all 'paradoxes' in SR, it comes from not having reasoned correctly. We will now demonstrate, using spacetime diagrams, that $\bar{\mathcal{O}}$ measures $\mathcal{O}$'s clocks to be running more slowly. Clearly, one could simply draw the spacetime diagram from $\bar{\mathcal{O}}$'s point of view, and the result would follow. But it is more instructive to stay in $\mathcal{O}$'s spacetime diagram.

Different observers will agree on the outcome of certain kinds of experiments. For example, if A flips a coin, *every* observer will agree on the result. Similarly, if two clocks are right next to each other, all observers will agree which is reading an earlier time than the other. But the question of the *rate* at which clocks run can only be settled by comparing the same two clocks on two different occasions, and if the clocks are moving relative to one another then they can be next to each other on only one of these occasions. On the other occasion they must be compared over some distance, and this is why different observers may draw different conclusions. The reason for this is that they actually perform different and inequivalent experiments. In the following analysis, we will see that each observer uses *two* of its own clocks and one of the other's. They get different results because they do different experiments.

Let us analyze $\mathcal{O}$'s measurement first, using Figure 1.14. This consists of comparing the reading on a single clock in $\bar{\mathcal{O}}$ (which travels from $\mathcal{A}$ to $\mathcal{B}$) with *two* clocks at rest in $\mathcal{O}$: the first is $\mathcal{O}$'s clock at the origin, which reads $\bar{\mathcal{O}}$'s clock at event $\mathcal{A}$; and the second is the clock in $\mathcal{O}$ that is at $\mathcal{F}$ at $t = 0$ and that coincides with $\bar{\mathcal{O}}$'s clock at $\mathcal{B}$. This second clock in $\mathcal{O}$ moves parallel to the first one, on the vertical dashed line. What $\mathcal{O}$ says is that both clocks at $\mathcal{A}$ read $t = 0$, while at $\mathcal{B}$ the clock in $\bar{\mathcal{O}}$ reads $\bar{t} = 1$ and that in $\mathcal{O}$ reads a later time, $t = (1 - v^2)^{-1/2}$. Clearly, $\bar{\mathcal{O}}$ agrees with this, because these are just statements of fact about the readings of the clock dials. Of course, for $\mathcal{O}$'s measurement to be correct, the two clocks used by $\mathcal{O}$ must be synchronized, for otherwise there is no particular significance in observing that at $\mathcal{B}$ the clock at rest in $\bar{\mathcal{O}}$ lags behind that at rest in $\mathcal{O}$. Now, from $\mathcal{O}$'s point of view, these clocks *are* synchronized, and the measurement and its conclusion are valid. Indeed, they are the only conclusions that $\mathcal{O}$ can properly make.

But $\bar{\mathcal{O}}$ need not accept them because, according to $\bar{\mathcal{O}}$, $\mathcal{O}$'s clocks are *not* synchronized. The dashed line joining $\mathcal{B}$ and $\mathcal{E}$ is the locus of events that $\bar{\mathcal{O}}$ regards as simultaneous to $\mathcal{B}$. Event $\mathcal{E}$ is also on the world line of $\mathcal{O}$'s first clock, so for $\bar{\mathcal{O}}$ the event $\mathcal{E}$ is simultaneous with

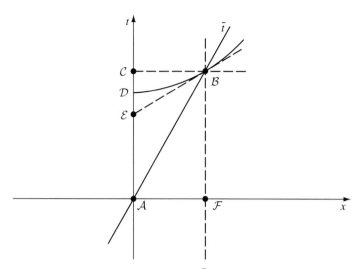

Figure 1.14 The proper length of $\mathcal{AB}$ is the time ticked by a clock at rest in $\bar{\mathcal{O}}$, while that of $\mathcal{AC}$ is the time ticked by a clock at rest in $\mathcal{O}$.

event $\mathcal{B}$. A simple calculation shows that this clock of $\mathcal{O}$'s reads the time $t = (1 - v^2)^{1/2}$ at this event. However, $\mathcal{O}$'s second clock, which passes through $\mathcal{B}$, reads $(1 - v^2)^{-1/2}$ at $\mathcal{B}$. So $\bar{\mathcal{O}}$ can reject $\mathcal{O}$'s measurement since the two clocks of $\mathcal{O}$ that were involved aren't synchronized with each other. Moreover, $\bar{\mathcal{O}}$'s own clock was synchronised with the first clock of $\mathcal{O}$ when they were both at event $\mathcal{A}$, and yet when the clock of $\bar{\mathcal{O}}$ reads $\bar{t} = 1$ at event $\mathcal{B}$, the first clock of $\mathcal{O}$ only reads $t = (1-v^2)^{1/2}$ (because $\bar{\mathcal{O}}$ regards $\mathcal{E}$ as simultaneous with $\mathcal{B}$). Therefore $\bar{\mathcal{O}}$ measures – in the only way that is reasonable – that the clocks in $\mathcal{O}$ run more slowly than all the clocks in $\bar{\mathcal{O}}$.

It follows that the principle of relativity is not contradicted: each observer measures the other's clocks to be running slowly. The reason they seem to disagree is that they perform different experiments. Observer $\mathcal{O}$ compares the interval from $\mathcal{A}$ to $\mathcal{B}$ with that from $\mathcal{A}$ to $\mathcal{C}$. The other observer compares that from $\mathcal{A}$ to $\mathcal{B}$ with that from $\mathcal{A}$ to $\mathcal{E}$. Both observers agree on the values of the intervals involved. What they disagree about is which pair to use in order to decide on the rate at which a clock is running. This disagreement arises directly from the fact that the observers do not agree on which events are simultaneous. Simultaneity (clock synchronization) is at the heart of clock comparisons: $\mathcal{O}$ uses *two* clocks to 'time' the rate of $\bar{\mathcal{O}}$'s one clock, and similarly $\bar{\mathcal{O}}$ uses two $\bar{\mathcal{O}}$ clocks to time one clock of $\mathcal{O}$.

Is this apparent disagreement worrisome? It should not be, but it should make us very cautious about importing our normal intuitions about time and space into SR. The fact that different observers disagree on clock rates or simultaneity just means that such concepts are not invariant: they are coordinate dependent. It does *not* prevent any given observer from using such concepts consistently for all observed phenomena. For example, $\mathcal{O}$ correctly says that $\mathcal{A}$ and $\mathcal{F}$ are simultaneous, in the sense that they have the same value of the coordinate t. This is a useful thing to know, as it helps to locate the events in spacetime in this frame. Any single observer can make consistent observations using concepts which are valid in that frame but which may not transfer to other observers. All the so-called paradoxes of relativity involve, not the inconsistency of a single observer's deductions, but the inconsistency of assuming that certain concepts are independent of the observer when they are in fact very observer dependent.

Two more points should be made before we turn to the calculation of the Lorentz transformation. The first is that we have not had to define a 'clock', so our statements apply to any good timepiece: atomic clocks, wrist watches, circadian rhythm, or the half-life of the decay of an elementary particle. Truly, all time is 'slowed' by the effects that we have been discussing. Put more properly, since time dilation is a consequence of the failure of simultaneity, it has nothing to do with the physical construction of the clock. It applies equally to physical particles and to biological processes or psychological perception: *all* these phenomena, if they are at rest in $\bar{\mathcal{O}}$, run more slowly as measured by $\mathcal{O}$. This leads to the twin 'paradox', which we discuss later.

The second point is that these effects are *not* optical illusions, since our observers exercise as much care as possible in performing their experiments. Beginning students sometimes convince themselves that the problem arises in the finite transmission speed of signals, but this is incorrect. Observers define 'now' as described in § 1.5 for observer $\bar{\mathcal{O}}$, and this is the most reasonable way to do it. The problem is that when two different observers each define 'now' in the most reasonable way, they don't agree. This is an inescapable consequence of the universality of the speed of light.

1.9 The Lorentz transformation

We shall now make our transition from the geometric formulation of SR to its algebraic formulation. The Lorentz transformation expresses the coordinates of $\bar{\mathcal{O}}$ in terms of those of $\mathcal{O}$. Without losing generality, we orient our axes so that $\bar{\mathcal{O}}$ moves with speed v on the positive x axis relative to $\mathcal{O}$. We know that lengths perpendicular to the x axis are the same when measured by $\mathcal{O}$ or $\bar{\mathcal{O}}$. For simplicity we assume that the spatial axes of both frames are aligned with one another (no spatial rotation). The most general linear transformation we need consider is then

$$\bar{t} = \alpha t + \beta x, \qquad \bar{y} = y,$$
$$\bar{x} = \gamma t + \sigma x, \qquad \bar{z} = z,$$

where α, β, γ, and σ are functions that depend only on v.[8]

---

[8] Not every use of Greek letters will be for indices. Make sure you do not confuse variables that happen to have Greek letter names with indices.

From our construction in § 1.5 (Figure 1.4) it is clear that the $\bar{t}$ and $\bar{x}$ axes have the following equations:

$$\bar{t} \text{ axis } (\bar{x} = 0): \quad vt - x = 0,$$
$$\bar{x} \text{ axis } (\bar{t} = 0): \quad vx - t = 0.$$

The equations of the axes imply, respectively,

$$\gamma/\sigma = -v, \quad \beta/\alpha = -v,$$

which gives the transformation

$$\bar{t} = \alpha(t - vx),$$
$$\bar{x} = \sigma(x - vt).$$

Figure 1.4 gives us one other bit of information: the events $(\bar{t} = 0, \bar{x} = a)$ and $(\bar{t} = a, \bar{x} = 0)$ are connected by a light ray. This can easily be shown to imply that $\alpha = \sigma$. Therefore we have, just from the geometry,

$$\bar{t} = \alpha(t - vx),$$
$$\bar{x} = \alpha(x - vt).$$

Now we use the invariance of intervals:

$$-(\Delta \bar{t})^2 + (\Delta \bar{x})^2 = -(\Delta t)^2 + (\Delta x)^2.$$

This gives, after some straightforward algebra,

$$\alpha = \pm 1/\sqrt{(1 - v^2)}.$$

We must select the $+$ sign so that when $v = 0$ we get an identity rather than an inversion of the coordinates. The complete Lorentz transformation is, therefore,

$$\begin{aligned}
\bar{t} &= \frac{t}{\sqrt{(1 - v^2)}} - \frac{vx}{\sqrt{(1 - v^2)}}, \\
\bar{x} &= \frac{-vt}{\sqrt{(1 - v^2)}} + \frac{x}{\sqrt{(1 - v^2)}}, \\
\bar{y} &= y, \\
\bar{z} &= z.
\end{aligned} \tag{1.12}$$

This is called a *boost* of velocity v in the x direction.

This gives the simplest form of the relation between the coordinates of $\bar{\mathcal{O}}$ and $\mathcal{O}$. To get this form, we assumed that the spatial coordinates are oriented in a particular way: $\bar{\mathcal{O}}$ must move with speed v in the positive x direction as seen by $\mathcal{O}$, and the axes of $\bar{\mathcal{O}}$ must be parallel to the corresponding axes in $\mathcal{O}$. Spatial rotations of the axes relative to one another produce more complicated sets of equations than Eq. 1.12, but we will normally be able to get away with Eq. 1.12.

1.10 The velocity-addition law

The Lorentz transformation contains all the information one needs to derive the standard formulae such as those for time dilation and Lorentz contraction. As an example of its use we will generalize the Galilean law of addition of velocities (§ 1.1).

Suppose a particle has speed W in the $\bar{x}$ direction of $\bar{\mathcal{O}}$, i.e. $\Delta\bar{x}/\Delta\bar{t} = W$. In another frame $\mathcal{O}$ its velocity will be $W' = \Delta x/\Delta t$, and we can deduce Δx and Δt from the Lorentz transformation. If $\bar{\mathcal{O}}$ moves with velocity v with respect to $\mathcal{O}$ then Eq. 1.12 implies that

$$\Delta x = (\Delta\bar{x} + v\Delta\bar{t})/(1 - v^2)^{1/2}$$

and

$$\Delta t = (\Delta\bar{t} + v\Delta\bar{x})/(1 - v^2)^{1/2}.$$

Then we have

$$W' = \frac{\Delta x}{\Delta t} = \frac{(\Delta\bar{x} + v\Delta\bar{t})/(1 - v^2)^{1/2}}{(\Delta\bar{t} + v\Delta\bar{x})/(1 - v^2)^{1/2}}$$

$$= \frac{\Delta\bar{x}/\Delta\bar{t} + v}{1 + v\Delta\bar{x}/\Delta\bar{t}} = \frac{W + v}{1 + Wv}. \tag{1.13}$$

This is the Einstein law for the composition of velocities. The important point is that $|W'|$ never exceeds 1 if $|W|$ and $|v|$ are both smaller than 1. To see this, set $W' = 1$. Then Eq. 1.13 implies

$$(1 - v)(1 - W) = 0,$$

i.e. that either v or W must also equal 1. Therefore, two 'subluminal' velocities produce another subluminal velocity. Moreover, if $W = 1$ then $W' = 1$ independently of v: this is due to the universality of the speed of light. What is more, if $|W| \ll 1$ and $|v| \ll 1$ then, to first order, Eq. 1.13 gives

$$W' = W + v.$$

This is the Galilean law of velocity addition, which we know to be valid for small velocities. This was true for our previous formulae in § 1.8: the relativistic 'corrections' to the Galilean expressions are of order v^2 and so are negligible for small v.

1.11 Paradoxes and physical intuition

Elementary introductions to SR often try to illustrate the physical differences between Galilean relativity and SR by posing certain problems called 'paradoxes'. The commonest ones include the 'twin paradox', the 'pole-in-the-barn paradox', and the 'space-war paradox'. The idea is to pose these problems in language that makes the predictions of SR seem inconsistent or paradoxical, and then to resolve them by showing that a careful application of the fundamental principles of SR leads to no inconsistencies at all: the paradoxes are apparent, not real, and result invariably from mixing Galilean concepts with modern ones.

Unfortunately, the careless student (or the attentive student of a careless teacher) sometimes comes away with the idea that SR does in fact lead to paradoxes. This is pure nonsense. It is important to understand that all such 'paradoxes' are really mathematically ill-posed problems, that SR is a perfectly consistent picture of spacetime which has been experimentally verified countless times in situations where gravitational effects can be neglected, and that SR forms the framework in which every modern physicist must construct theories. (If you really wants to study a paradox in depth, see a discussion of the twin paradox in the appendix to this chapter.)

Psychologically, the reason that newcomers to SR have trouble and perhaps give 'paradoxes' more weight than they deserve is that we have so little direct experience of velocities comparable with that of light. The only remedy is to solve problems in SR and to study carefully its 'counterintuitive' predictions. One of the best methods for developing a modern intuition is to be completely familiar with the geometrical picture of SR: Minkowski space, the effect of Lorentz transformations on axes, and 'pictures' of such things as time dilation and Lorentz contraction. This geometrical picture should be in the back of your mind as we go on from here to study vector and tensor calculus; we shall bring it to the front again when we study GR.

1.12 Bibliography

There are many good introductions to SR. A very readable one which has guided our own treatment, and is far more detailed, is Taylor & Wheeler (1992), which is now available free online. Another classic is French (1968). A highly respected, more recent, book is Morin (2017).

For treatments that take a more thoughtful look at the fundamentals of the theory, consult Arzeliès (1966), Bohm (2008), Dixon (1978), or Geroch (1978). Paradoxes are discussed in some detail by Arzeliès (1966), Marder (1971), and Terletskii (1968). For a scientific biography of Einstein, see Pais (1982).

Our interest in SR in this text is primarily because it is a simple special case of GR in which it is possible to develop the mathematics we shall later need. But SR is itself the underpinning of all the other fundamental theories of physics, such as electromagnetism and quantum theory, and as such it deserves much more study than we shall give it in this book. See the classic discussions in Synge (1965), Schrödinger (1950), and Møller (1972), and more modern treatments in Rindler (1991), Schwarz & Schwarz (2004), and Woodhouse (2003).

The original papers on SR may be found in Kilmister (1970).

The twin paradox, the subject of the appendix to this chapter (below), is brought vividly to life in the feature film *Interstellar*, which appeared in 2014. In this film, the time dilation is produced by gravity near a black hole rather than by relativistic space travel. The film was conceived and its production guided by the well-known relativist Kip Thorne, who in 2017 shared the Nobel Prize in Physics for his part in the first detection of gravitational waves.

1.13 Appendix: The twin 'paradox' dissected

The problem. Diana leaves her twin Artemis behind on Earth and travels in her rocket for 2.2×10^8 s (≈ 7 yr) of *her* time at $24/25 = 0.96$ the speed of light. She then instantaneously reverses her direction (fearlessly braving those gs) and returns to Earth in the same manner. Who is older at the reunion of the twins? A spacetime diagram can be very helpful.

Brief solution. Refer to Figure 1.15. Diana travels out on the line $\mathcal{PB}$. In Diana's frame, Artemis' event $\mathcal{A}$ is simultaneous with event $\mathcal{B}$, so Artemis is indeed ageing slowly. But as soon as Diana turns around she changes inertial reference frames: now she regards $\mathcal{B}$ as simultaneous with Artemis' event $\mathcal{C}$! Effectively, Diana sees Artemis age incredibly quickly for a moment. This one spurt more than makes up for the slowness Diana observed all along. Numerically, Artemis ages 50 years for Diana's 14.

Fuller discussion. For readers who are unsatisfied with the statement 'Diana sees Artemis age incredibly quickly for a moment', or who wonder what physics lies beneath such a statement, we will discuss this in more detail, bearing in mind that the statement 'Diana sees' really means 'Diana observes', using the rods, clocks, and data bank that every good relativistic observer has.

Diana might make her measurements in the following way. Blasting off from Earth, she leaps on to an inertial frame called $\bar{\mathcal{O}}$ rushing away from the Earth at $v = 0.96$. As soon

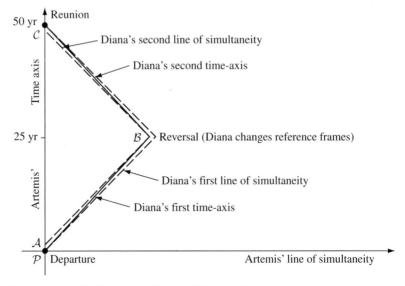

Figure 1.15 The idealized twin 'paradox' in the spacetime diagram of the stay-at-home twin.

as she gets settled in this new frame she orders all clocks synchronized with hers, which read $\bar{t} = 0$ upon leaving Earth. She further places a graduate student on every one of her clocks and orders those riding a clock that passes Earth to note the time on Earth's clock at the event of passage. After traveling seven years by her own watch, she leaps off inertial frame $\bar{\mathcal{O}}$ and grabs hold of another one $\mathcal{O}$ that is flying *toward* Earth at $v = 0.96$ (measured in Earth's frame, of course). When she settles into this frame she again distributes her graduate students on the clocks and orders all clocks to be synchronized with hers, which read $\bar{\bar{t}} = 7$ yr at the changeover. (All clocks were already synchronized with each other – she simply adjusts their zero of time.) She further orders that every graduate student who passes Earth from $\bar{\bar{t}} = 7$ yr until she gets there herself should record the time of passage and the reading of Earth's clocks at that event.

Diana finally arrives home after ageing 14 years. Knowing a little about time dilation, she expects Artemis to have aged much less, but to her surprise Artemis is a full 50 years older than her! Diana keeps her surprise to herself and runs over to the computer room to check out the data. She reads the dispatches from the graduate students riding the clocks of the outgoing frame. Sure enough, Artemis seems to have aged very slowly by their reports. At Diana's time $\bar{t} = 7$ yr, the graduate student passing Earth recorded that Earth's clocks read only slightly less than two years of elapsed time. But then Diana checks the information from her graduate students riding the clocks of the ingoing frame. She finds that at her time $\bar{\bar{t}} = 7$ yr, the graduate student reported a reading of Earth's clocks at more than *48 years* of elapsed time! How could one student see Earth to be at $t = 2$ yr, and another student, *at the same time*, see it at $t = 48$ yr? Diana leaves the computer room muttering about the declining standards of undergraduate education today.

We know the mistake Diana made, however. Her two messengers did *not* pass Earth at the same time. Their clocks read the same amount, because they had been synchronized with hers at the moment she switched frames. But they encountered Earth at the very different events $\mathcal{A}$ and $\mathcal{C}$. Diana should have asked the first frame's students to continue recording information until they saw the second frame's $\bar{\bar{t}} = 7$ yr student pass Earth. What does it matter, after all, that they would have sent her dispatches dated $\bar{t} = 171$ yr? Time is only a coordinate. One must be sure to catch *all* the events.

What Diana really did was to use a bad coordinate system. By demanding information only before $\bar{t} = 7$ yr in the outgoing frame and only after $\bar{\bar{t}} = 7$ yr in the ingoing frame, she left the whole interior of the triangle $\mathcal{ABC}$ out of her coordinate patches (Figure 1.16(a)). Small wonder that a lot happened that she did not discover! Had she allowed the first frame's students to gather data until $\bar{t} = 171$ yr, she could have covered the interior of that triangle.

One can devise an analogy with rotations in the plane (Figure 1.16(b)). Consider trying to measure the length of the curve $ABCD$, but being forced to rotate coordinates in the middle of the measurement, say after you have measured from A to B in the $x-y$ system. If you then rotate to the $\bar{x}-\bar{y}$ system, you must resume the measuring at B, which might be at a coordinate $\bar{y} = -5$, whereas originally B had coordinate $y = 2$. If you were to measure the curve's length starting at whatever point had $\bar{y} = 2$ (the same $\bar{y}$ as the y value you ended at in the other frame), you would begin at C and get much too short a length for the curve.

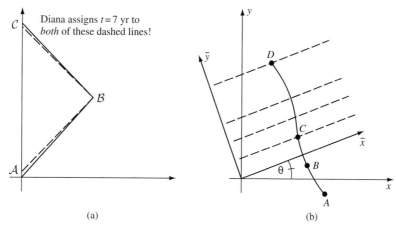

Figure 1.16   Diana's change of frame is analogous to a rotation of coordinates in Euclidean geometry.

Now, nobody would make that error in measurements in a plane. But lots of people would if they were confronted by the twin paradox. This comes from our refusal to see time as simply a coordinate. We are used to thinking of a universal time, the same everywhere to everyone regardless of their motion. But it is not the same to everyone, and one must treat it as a coordinate, and make sure that one's coordinates cover all of spacetime.

Coordinates that do not cover all of spacetime have caused a lot of problems in GR. When we study gravitational collapse and black holes we will see that the usual coordinates for the spacetime outside the black hole do not reach inside the black hole. For this reason, a particle falling into a black hole takes infinite coordinate time to go a finite distance. This is purely the fault of the coordinates: the particle falls, in a finite proper time, into a region not covered by the 'outside' coordinates. A coordinate system that covers both inside and outside completely was not discovered until the mid-1950s.

Exercises

1.1   Convert the following to units in which $c = 1$, expressing everything in terms of m and kg:

(a) Worked example: 10 J. In SI units, $10\,\mathrm{J} = 10\,\mathrm{kg\,m^2\,s^{-2}}$. Since $c = 1$, we have $1\,\mathrm{s} = 3 \times 10^8\,\mathrm{m}$, and so $1\,\mathrm{s^{-2}} = (9 \times 10^{16})^{-1}\,\mathrm{m^{-2}}$. Therefore we get $10\,\mathrm{J} = 10\,\mathrm{kg\,m^2}(9 \times 10^{16})^{-1}\,\mathrm{m^{-2}} = 1.1 \times 10^{-16}\,\mathrm{kg}$. Alternatively, treat c as a conversion factor:

$$1 = 3 \times 10^8\,\mathrm{m\,s^{-1}},$$
$$1 = (3 \times 10^8)^{-1}\,\mathrm{m^{-1}\,s},$$
$$10\,\mathrm{J} = 10\,\mathrm{kg\,m^2\,s^{-2}} = 10\,\mathrm{kg\,m^2\,s^{-2}} \times (1)^2$$

$$= 10 \, \text{kg} \, \text{m}^2 \, \text{s}^{-2} \times (3 \times 10^8)^{-2} \, \text{s}^2 \, \text{m}^{-2}$$
$$= 1.1 \times 10^{-16} \, \text{kg}.$$

One is allowed to multiply or divide by as many factors of c as are necessary to cancel out the seconds.

(b) The power output of 100 W.

(c) Planck's reduced constant, $\hbar = 1.05 \times 10^{-34}$ J s. (Note the definition of $\hbar$ in terms of Planck's constant h: $\hbar = h/2\pi$.)

(d) Velocity of a car, $v = 30 \, \text{m} \, \text{s}^{-1}$.

(e) Momentum of a car, $3 \times 10^4 \, \text{kg} \, \text{m} \, \text{s}^{-1}$.

(f) Pressure of one atmosphere $= 10^5 \, \text{N} \, \text{m}^{-2}$.

(g) Density of water, $10^3 \, \text{kg} \, \text{m}^{-3}$.

(h) Luminosity flux $10^6 \, \text{J} \, \text{s}^{-1} \, \text{cm}^{-2}$.

1.2   Convert the following from natural units ($c = 1$) to SI units:

(a) Velocity $v = 10^{-2}$.

(b) Pressure $10^{19} \, \text{kg} \, \text{m}^{-3}$.

(c) Time $t = 10^{18}$ m.

(d) Energy density $u = 1 \, \text{kg} \, \text{m}^{-3}$.

(e) Acceleration $10 \, \text{m}^{-1}$.

1.3   Draw the t and x axes of the spacetime coordinates of an observer $\mathcal{O}$ and then draw:

(a) The world line of $\mathcal{O}$'s clock at $x = 1$ m.

(b) The world line of a particle, moving with velocity $dx/dt = 0.1$, which is at $x = 0.5$ m when $t = 0$.

(c) The $\bar{t}$ and $\bar{x}$ axes of an observer $\bar{\mathcal{O}}$ who moves with velocity $v = 0.5$ in the positive x direction relative to $\mathcal{O}$ and whose origin ($\bar{x} = \bar{t} = 0$) coincides with that of $\mathcal{O}$.

(d) The locus of events whose interval Δs^2 from the origin is $-1 \, \text{m}^2$.

(e) The locus of events whose interval Δs^2 from the origin is $+1 \, \text{m}^2$.

(f) The calibration ticks at one meter intervals along the $\bar{x}$ and $\bar{t}$ axes.

(g) The locus of events whose interval Δs^2 from the origin is 0.

(h) The locus of events, all of which occur at the time $t = 2$ m (they are simultaneous as seen by $\mathcal{O}$).

(i) The locus of events, all of which occur at the time $\bar{t} = 2$ m (they are simultaneous as seen by $\bar{\mathcal{O}}$).

(j) The event which occurs at $\bar{t} = 0$ and $\bar{x} = 0.5$ m.

(k) The locus of events $\bar{x} = 1$ m.

(l) The world line of a photon which is emitted from the event $t = -1$ m, $x = 0$, travels in the negative x direction, is reflected when it encounters a mirror located at $\bar{x} = -1$ m, and is absorbed when it encounters a detector located at $x = 0.75$ m.

1.4   Write out all the terms of the following sums, substituting the coordinate names (t, x, y, z) for (x^0, x^1, x^2, x^3):

(a) $\sum_{\alpha=0}^{3} V_\alpha \Delta x^\alpha$, where $\{V_\alpha, \alpha = 0, \ldots, 3\}$ is a collection of four arbitrary numbers.

(b) $\sum_{i=1}^{3} (\Delta x^i)^2$.

1.5 (a) Use the spacetime diagram of an observer $\mathcal{O}$ to describe the following experiment performed by $\mathcal{O}$. Two bursts of particles of speed $v = 0.5$ are emitted from $x = 0$ at $t = -2\,$m, one traveling in the positive x direction and the other in the negative x direction. These encounter detectors located at $x = \pm 2\,$m. After a delay of $0.5\,$m of time, the detectors send signals back to $x = 0$ at speed $v = 0.75$.

(b) The signals arrive back at $x = 0$ at the same event. (Make sure your spacetime diagram shows this!) From this the experimenter concludes that the particle detectors did indeed send out their signals simultaneously, since it is known that they are equal distances from $x = 0$. Explain why this conclusion is valid.

(c) A second observer $\bar{\mathcal{O}}$ moves with speed $v = 0.75$ in the *negative* x direction relative to $\mathcal{O}$. Draw the spacetime diagram of $\bar{\mathcal{O}}$ and in it depict the experiment performed by $\mathcal{O}$. Does $\bar{\mathcal{O}}$ conclude that particle detectors sent out their signals simultaneously? If not, which signal was sent first?

(d) Compute the interval Δs^2 between the events at which the detectors emitted their signals, using both the coordinates of $\mathcal{O}$ and those of $\bar{\mathcal{O}}$.

1.6 Show that Eq. 1.2 contains only the sum $M_{\alpha\beta} + M_{\beta\alpha}$ when $\alpha \neq \beta$, not $M_{\alpha\beta}$ and $M_{\beta\alpha}$ independently. Argue that this enables us to set $M_{\alpha\beta} = M_{\beta\alpha}$ without loss of generality.

1.7 In the discussion leading up to Eq. 1.2, assume that the coordinates of $\bar{\mathcal{O}}$ are given as the following linear combinations of those of $\mathcal{O}$:

$$\bar{t} = \alpha t + \beta x,$$

$$\bar{x} = \mu t + vx,$$

$$\bar{y} = ay,$$

$$\bar{z} = bz,$$

where α, β, μ, v, a, and b may be functions of the velocity v of $\bar{\mathcal{O}}$ relative to $\mathcal{O}$ but do not depend on the coordinates. Find the numbers $\{M_{\alpha\beta}; \alpha, b = 0, \ldots, 3\}$ of Eq. 1.2 in terms of α, β, μ, v, a, and b.

1.8 (a) Derive Eq. 1.3 from Eq. 1.2, for general $\{M_{\alpha\beta}; \alpha, \beta = 0, \ldots, 3\}$.

(b) Since $\Delta\bar{s}^2 = 0$ in Eq. 1.3 for any $\{\Delta x^i\}$, replace Δx^i by $-\Delta x^i$ in Eq. 1.3 and subtract the resulting equation from Eq. 1.3 to establish that $M_{0i} = 0$ for $i = 1, 2, 3$.

(c) Use Eq. 1.3 with $\Delta\bar{s}^2 = 0$ to establish Eq. 1.4b. (Hint: $\Delta x, \Delta y$, and Δz are arbitrary.)

1.9 Explain why the line $\mathcal{PQ}$ in Figure 1.7 is drawn as shown.

1.10 For the pairs of events whose coordinates (t, x, y, z) in some frame are given below, classify their separations as timelike, spacelike, or null.

(a) $(0, 0, 0, 0)$ and $(-1, 1, 0, 0)$,

(b) $(1, 1, -1, 0)$ and $(-1, 1, 0, 2)$,

(c) $(6, 0, 1, 0)$ and $(5, 0, 1, 0)$,

(d) $(-1, 1, -1, 1)$ and $(4, 1, -1, 6)$.

1.11 Show that the hyperbolae $-t^2 + x^2 = a^2$ and $-t^2 + x^2 = -b^2$ are asymptotic to the lines $t = \pm x$, regardless of the values of a and b.

1.12 (a) Use the fact that the tangent to the hyperbola $\mathcal{DB}$ in Figure 1.14 is the line of simultaneity for $\bar{\mathcal{O}}$ to show that the time interval $\mathcal{AE}$ is shorter than the time recorded on $\bar{\mathcal{O}}$'s clock as it moved from $\mathcal{A}$ to $\mathcal{B}$.

(b) Calculate that

$$(\Delta s^2)_{AC} = (1 - v^2)(\Delta s^2)_{AB}.$$

(c) Use (b) to show that $\bar{\mathcal{O}}$ regards $\mathcal{O}$'s clocks to be running slowly, at just the 'right' rate.

1.13 The half-life of the elementary particle called the pion is 2.5×10^{-8} s when it is at rest relative to the observer measuring its decay time. Show, by the principle of relativity, that pions moving at speed $v = 0.999$ must have a half-life of 5.6×10^{-7} s, as measured by an observer at rest.

1.14 Suppose that the velocity v of $\bar{\mathcal{O}}$ relative to $\mathcal{O}$ is small, $|v| \ll 1$. Show that the time dilation, Lorentz contraction, and velocity-addition formulae can be approximated by, respectively:

(a) $\Delta t \approx (1 + \frac{1}{2}v^2)\Delta \bar{t}$.

(b) $\Delta x \approx (1 - \frac{1}{2}v^2)\Delta \bar{x}$.

(c) $w' \approx w + v - wv(w + v)$ (with $|w| \ll 1$ as well).

(d) What are the relative errors in these approximations when $|v| = w = 0.1$?

1.15 Suppose that the velocity v of a frame $\bar{\mathcal{O}}$ relative to a frame $\mathcal{O}$ is nearly that of light, $|v| = 1 - \varepsilon, 0 < \varepsilon \ll 1$.

Show that the formulae of Exercise 1.14 become

(a) $\Delta t \approx \Delta \bar{t}/\sqrt{(2\varepsilon)}$,

(b) $\Delta x \approx \Delta \bar{x}/\sqrt{(2\varepsilon)}$,

(c) $w' \approx 1 - \varepsilon(1 - w)/(1 + w)$.

(d) What are the relative errors on these approximations when $\varepsilon = 0.1$ and $w = 0.9$?

1.16 Use the Lorentz transformation, Eq. 1.12, to derive (a) the time dilation and (b) the Lorentz contraction formulae. Do this by identifying pairs of events whose separations (in time or space) are to be compared and then using the Lorentz transformation to accomplish the algebra corresponding to the invariant hyperbolae in the text.

1.17 A lightweight pole 20 m long lies on the ground next to a barn 15 m long. An Olympic athlete picks up the pole, carries it far away, and runs with it toward the end of the barn at a speed $0.8c$. A physicist friend remains at rest, standing by the door of the barn. Attempt all parts of this question, even if you can't answer some.

(a) How long does the friend measure the pole to be, as it approaches the barn?

(b) The barn door is initially open, and immediately after the runner and pole are entirely inside the barn, the friend shuts the door. How long after the door is shut does the front of the pole hit the other end of the barn, as measured by the friend? Compute the interval between the events of shutting the door and hitting the wall. Is it spacelike, timelike, or null?

(c) In the reference frame of the runner, what is the length of the barn and the pole?

(d) Does the runner believe that the pole is entirely inside the barn when its front hits the end of the barn? Can you explain why?

(e) After the collision, the pole and runner come to rest relative to the barn. From the friend's point of view, the 20 m pole is now inside a 15 m barn, since the barn door was shut before the pole stopped. How is this possible? Alternatively, from the runner's point of view, the collision should have stopped the pole *before* the door closed, so the door could not be closed at all. Was or was not the door closed with the pole inside?

(f) Draw a spacetime diagram from the friend's point of view and use it to illustrate and justify all your conclusions.

1.18 (a) The Einstein velocity-addition law, Eq. 1.13, has a simpler form if we introduce the concept of the *velocity parameter u*, defined by the equation

$$v = \tanh u.$$

Notice that for $-\infty < u < \infty$, the velocity is confined to the acceptable limits $-1 < v < 1$. Show that if

$$v = \tanh u$$

and

$$w = \tanh U,$$

then Eq. 1.13 implies

$$w' = \tanh(u + U).$$

This means that velocity parameters add linearly.

(b) Use this to solve the following problem. A star measures a second star to be moving away at speed $v = 0.9c$. The second star measures a third to be receding in the same direction at $0.9c$. Similarly, the third measures a fourth, and so on, up to some large number N of stars. What is the velocity of the Nth star relative to the first? Give an exact answer and an approximation that is useful for large N.

1.19 (a) Using the velocity parameter introduced in Exercise 1.18, show that the Lorentz transformation equations, Eq. 1.12, can be put into the form

$$\bar{t} = t \cosh u - x \sinh u, \quad \bar{y} = y,$$
$$\bar{x} = -t \sinh u + x \cosh u, \quad \bar{z} = z.$$

(b) Use the identity $\cosh^2 u - \sinh^2 u = 1$ to demonstrate the invariance of intervals from these equations.

(c) Draw as many parallels as you can between the geometry of spacetime and ordinary two-dimensional Euclidean geometry, where the coordinate transformation analogous to the Lorentz transformation is

$$\bar{x} = x \cos \theta + y \sin \theta,$$
$$\bar{y} = -x \sin \theta + y \cos \theta.$$

What is the analog of an interval? Of the invariant hyperbolae?

1.20 Write the Lorentz transformation equations in matrix form.

1.21 (a) Show that if two events are timelike separated then there is a Lorentz frame in which they occur at the same point, i.e. at the same spatial coordinate values.

(b) Similarly, show that if two events are spacelike separated, there is a Lorentz frame in which they are simultaneous.

2 Vector analysis in special relativity

2.1 Definition of a vector

For the moment we will use the notion of a vector that carries over from Euclidean geometry: that a vector is something whose components transform like coordinates do under a coordinate transformation. Later on we shall define vectors in a more satisfactory manner.

A typical vector is the displacement vector, which points from one event to another and has components equal to the coordinate differences:

$$\Delta \vec{x} \underset{\mathcal{O}}{\rightarrow} (\Delta t, \Delta x, \Delta y, \Delta z). \tag{2.1}$$

Here we have introduced some new notation: an *arrow* over a symbol denotes a vector (so that $\vec{x}$ is a vector having nothing in particular to do with the coordinate x); the arrow after $\Delta \vec{x}$ means 'has components' and the $\mathcal{O}$ underneath it means 'in the frame $\mathcal{O}$'; the components will always be in the order t, x, y, z (equivalently, indices in the order 0, 1, 2, 3). The notation $\underset{\mathcal{O}}{\rightarrow}$ is used in order to emphasize the distinction between the vector and its components. The vector $\Delta \vec{x}$ can be regarded as an arrow between two events, while the collection of components is a set of four coordinate-dependent numbers. We shall always emphasize the notion of a vector (and, later, any tensor) as a *geometrical object*: something which can be defined and (sometimes) visualized without referring to a specific coordinate system. Another important notation is

$$\Delta \vec{x} \underset{\mathcal{O}}{\rightarrow} \{\Delta x^{\alpha}\}, \tag{2.2}$$

where by $\{\Delta x^{\alpha}\}$ we mean all of Δx^0, Δx^1, Δx^2, Δx^3. If we ask for this vector's components in another coordinate system, say the frame $\bar{\mathcal{O}}$, we write

$$\Delta \vec{x} \underset{\bar{\mathcal{O}}}{\rightarrow} \{\Delta x^{\bar{\alpha}}\}.$$

That is, we put a bar over the *index* to denote the new coordinates. The vector $\Delta \vec{x}$ is the *same*, and no new notation is needed for it when the frame is changed. Only its components change.[1] What *are* the new components $\Delta x^{\bar{\alpha}}$? We get them from the Lorentz transformation:

$$\Delta x^{\bar{0}} = \frac{\Delta x^0}{\sqrt{(1-v^2)}} - \frac{v\Delta x^1}{\sqrt{(1-v^2)}}, \quad \text{etc.}$$

Since this is a linear transformation, it can be written

$$\Delta x^{\bar{0}} = \sum_{\beta=0}^{3} \Lambda^{\bar{0}}{}_{\beta} \Delta x^{\beta},$$

where $\{\Lambda^{\bar{0}}{}_{\beta}\}$ are four numbers, one for each value of β. In this case

$$\Lambda^{\bar{0}}{}_0 = 1/\sqrt{(1-v^2)}, \quad \Lambda^{\bar{0}}{}_1 = -v/\sqrt{(1-v^2)},$$
$$\Lambda^{\bar{0}}{}_2 = \Lambda^{\bar{0}}{}_3 = 0.$$

A similar equation holds for $\Delta x^{\bar{1}}$, and so in general we write

$$\Delta x^{\bar{\alpha}} = \sum_{\beta=0}^{3} \Lambda^{\bar{\alpha}}{}_{\beta} \Delta x^{\beta}, \quad \text{for arbitrary } \bar{\alpha}. \tag{2.3}$$

Now $\{\Lambda^{\bar{\alpha}}{}_{\beta}\}$ is a collection of 16 numbers, which constitute the Lorentz transformation matrix. The reason why we have written one index up and the other down will become clear when we study differential geometry. For now, it enables us to introduce the final bit of notation, the *Einstein summation convention*: whenever an expression contains one index as a superscript and the *same* index as a subscript, a summation is implied over all the values that index can take. That is,

$$A_{\alpha} B^{\alpha} \text{ and } T^{\gamma} E_{\gamma\alpha}$$

are shorthand for the summations

$$\sum_{\alpha=0}^{3} A_{\alpha} B^{\alpha} \text{ and } \sum_{\gamma=0}^{3} T^{\gamma} E_{\gamma\alpha},$$

while

$$A_{\alpha} B^{\beta}, \quad T^{\gamma} E_{\beta\alpha}, \quad \text{and } A_{\beta} A_{\beta}$$

do *not* represent sums on any index. The Lorentz transformation Eq. 2.3 can now be abbreviated to

$$\Delta x^{\bar{\alpha}} = \Lambda^{\bar{\alpha}}{}_{\beta} \Delta x^{\beta}, \tag{2.4}$$

saving some messy writing.

---

[1] This is what some books on linear algebra call a 'passive' transformation: the coordinates change, but the vector does not.

Notice that Eq. 2.4 is identically equal to

$$\Delta x^{\tilde{\alpha}} = \Lambda^{\tilde{\alpha}}{}_{\gamma} \Delta x^{\gamma}.$$

Since the repeated index (β in one case, γ in the other) merely denotes a summation from 0 to 3, it doesn't matter what letter is used. Such a summed index is called a *dummy index*, and relabeling a dummy index (as we have done, replacing β by γ) is often a useful tool in tensor algebra. There is only one thing one should *not* replace the dummy index β with: a Latin index. The reason is that Latin indices can (by our convention) only take the values 1, 2, 3, whereas β must be able to equal zero as well. Thus, the expressions

$$\Lambda^{\tilde{\alpha}}{}_{\beta} \Delta x^{\beta} \text{ and } \Lambda^{\tilde{\alpha}}{}_{i} \Delta x^{i}$$

are not the same; in fact we have

$$\Lambda^{\tilde{\alpha}}{}_{\beta} \Delta x^{\beta} = \Lambda^{\tilde{\alpha}}{}_{0} \Delta x^{0} + \Lambda^{\tilde{\alpha}}{}_{i} \Delta x^{i}. \tag{2.5}$$

Equation 2.4 is really four different equations, one for each value that $\tilde{\alpha}$ can assume. An index like $\tilde{\alpha}$, on which no sum is performed, is called a *free index*. Whenever an equation is written down with one or more free indices, it is valid if and only if it is true for *all* possible values that the free indices can assume. As with a dummy index, the name given to a free index is largely arbitrary. Thus, Eq. 2.4 can be rewritten as

$$\Delta x^{\tilde{\gamma}} = \Lambda^{\tilde{\gamma}}{}_{\beta} \Delta x^{\beta}.$$

This is equivalent to Eq. 2.4 because $\tilde{\gamma}$ can assume the same four values that $\tilde{\alpha}$ could assume. If a free index is renamed, it must be renamed everywhere. For example, the following modification of Eq. 2.4,

$$\Delta x^{\tilde{\gamma}} = \Lambda^{\tilde{\alpha}}{}_{\beta} \Delta x^{\beta},$$

makes no sense and should never be written. The difference between these last two expressions is that the first guarantees that, whatever value $\tilde{\gamma}$ assumes, both $\Delta x^{\tilde{\gamma}}$ on the left and $\Lambda^{\tilde{\gamma}}{}_{\beta}$ on the right will have the *same* free index. The second expression does not link the indices in this way, so it is not equivalent to Eq. 2.4.

A *general vector*[2] is defined by a collection of numbers (its components in some frame, say $\mathcal{O}$),

$$\vec{A} \underset{\mathcal{O}}{\rightarrow} (A^0, A^1, A^2, A^3) = \{A^{\alpha}\}, \tag{2.6}$$

and by the rule that its components in a frame $\bar{\mathcal{O}}$ are

$$A^{\tilde{\alpha}} = \Lambda^{\tilde{\alpha}}{}_{\beta} A^{\beta}. \tag{2.7}$$

---

[2] Such a vector, with four components, is sometimes called a *four-vector* to distinguish it from the three-component vectors one is used to in elementary physics, which we shall call *three-vectors*. Unless we say otherwise, by a 'vector' we always mean a four-vector. We denote four-vectors by arrows, e.g. $\vec{A}$, and three-vectors by boldface, e.g. *A*.

That is, its components transform in the same way as the coordinates do. Remember that a vector can be defined by giving four numbers, e.g. $(10^8, -10^{-16}, 5.8368, \pi)$, in some frame; then its components in all other frames are uniquely determined. Vectors in spacetime obey the usual rules: if $\vec{A}$ and $\vec{B}$ are vectors and μ is a number, then $\vec{A} + \vec{B}$ and $\mu \vec{A}$ are also vectors, with components

$$\left. \begin{aligned} \vec{A} + \vec{B} &\underset{\mathcal{O}}{\to} (A^0 + B^0, A^1 + B^1, A^2 + B^2, A^3 + B^3), \\ \mu \vec{A} &\underset{\mathcal{O}}{\to} (\mu A^0, \mu A^1, \mu A^2, \mu A^3). \end{aligned} \right\} \tag{2.8}$$

Thus, vectors add by the usual parallelogram rule. Notice that any four numbers make a vector, except that if the numbers are not dimensionless they must all have the same dimensions since under a transformation they will be added together.

2.2 Vector algebra

Basis vectors. In any frame $\mathcal{O}$ there are four special vectors, defined by their components as follows:

$$\left. \begin{aligned} \vec{e}_0 &\underset{\mathcal{O}}{\to} (1,0,0,0), \\ \vec{e}_1 &\underset{\mathcal{O}}{\to} (0,1,0,0), \\ \vec{e}_2 &\underset{\mathcal{O}}{\to} (0,0,1,0), \\ \vec{e}_3 &\underset{\mathcal{O}}{\to} (0,0,0,1). \end{aligned} \right\} \tag{2.9}$$

These are the basis vectors of frame $\mathcal{O}$. Similarly, $\bar{\mathcal{O}}$ has basis vectors

$$\vec{e}_{\bar{0}} \underset{\bar{\mathcal{O}}}{\to} (1,0,0,0), \text{ etc.}$$

Generally, $\vec{e}_{\bar{0}} \neq \vec{e}_0$ since they are defined in different frames. The reader should verify that the definition of the basis vectors is equivalent to

$$(\vec{e}_\alpha)^\beta = \delta_\alpha{}^\beta, \tag{2.10}$$

that is, the β component of $\vec{e}_\alpha$ is the Kronecker delta, which is 1 if $\beta = \alpha$ and 0 if $\beta \neq \alpha$.

Any vector can be expressed in terms of the basis vectors. If

$$\vec{A} \underset{\mathcal{O}}{\to} (A^0, A^1, A^2, A^3)$$

then

$$\vec{A} = A^0 \vec{e}_0 + A^1 \vec{e}_1 + A^2 \vec{e}_2 + A^3 \vec{e}_3,$$
$$\vec{A} = A^\alpha \vec{e}_\alpha. \tag{2.11}$$

In the last line we have used the summation convention (remember always to write the index on $\vec{e}$ as a *subscript* in order to employ the convention in this manner). The meaning of Eq. 2.11 is that $\vec{A}$ is the linear sum of the four vectors $A^0 \vec{e}_0$, $A^1 \vec{e}_1$, etc.

Transformation of basis vectors. The discussion leading up to Eq. 2.11 could have been applied to any frame, so it is equally true in $\bar{\mathcal{O}}$:

$$\vec{A} = A^{\bar{\alpha}}\vec{e}_{\bar{\alpha}}.$$

This says that $\vec{A}$ is also the sum of the four vectors $A^{\bar{0}}\vec{e}_{\bar{0}}$, $A^{\bar{1}}\vec{e}_{\bar{1}}$, etc. These are not the same four vectors as in Eq. 2.11, since they are parallel to the basis vectors of $\bar{\mathcal{O}}$ and not of $\mathcal{O}$, but they add up to the same vector $\vec{A}$. It is important to understand that the expressions $A^{\alpha}\vec{e}_{\alpha}$ and $A^{\bar{\alpha}}\vec{e}_{\bar{\alpha}}$ are not obtained from one another merely by relabeling dummy indices. Barred and unbarred indices cannot be interchanged, since they have different meanings. Thus, $\{A^{\bar{\alpha}}\}$ is a different set of numbers from $\{A^{\alpha}\}$, just as the set of vectors $\{\vec{e}_{\bar{\alpha}}\}$ is different from $\{\vec{e}_{\alpha}\}$. But, by definition, the two sums are the same:

$$A^{\alpha}\vec{e}_{\alpha} = A^{\bar{\alpha}}\vec{e}_{\bar{\alpha}}, \tag{2.12}$$

and this has an important consequence: from it we can deduce the transformation law for basis vectors, i.e. the relation between $\{\vec{e}_{\alpha}\}$ and $\{\vec{e}_{\bar{\alpha}}\}$. Using Eq. 2.7 for $A^{\bar{\alpha}}$, we write Eq. 2.12 as

$$\Lambda^{\bar{\alpha}}{}_{\beta}A^{\beta}\vec{e}_{\bar{\alpha}} = A^{\alpha}\vec{e}_{\alpha}.$$

On the left we have *two* sums. Since they are finite sums their order doesn't matter. Since the numbers $\Lambda^{\bar{\alpha}}{}_{\beta}$ and A^{β} *are* just numbers, their order doesn't matter, and we can write

$$A^{\beta}\Lambda^{\bar{\alpha}}{}_{\beta}\vec{e}_{\bar{\alpha}} = A^{\alpha}\vec{e}_{\alpha}.$$

Now we use the fact that β and $\bar{\alpha}$ are dummy indices: we change β to α and $\bar{\alpha}$ to $\bar{\beta}$,

$$A^{\alpha}\Lambda^{\bar{\beta}}{}_{\alpha}\vec{e}_{\bar{\beta}} = A^{\alpha}\vec{e}_{\alpha}.$$

This equation must be true for *all* sets $\{A^{\alpha}\}$, since $\vec{A}$ is an arbitrary vector. Writing it as

$$A^{\alpha}(\Lambda^{\bar{\beta}}{}_{\alpha}\vec{e}_{\bar{\beta}} - \vec{e}_{\alpha}) = 0$$

we deduce that

$$\Lambda^{\bar{\beta}}{}_{\alpha}\vec{e}_{\bar{\beta}} - \vec{e}_{\alpha} = 0 \quad \text{for every value of } \alpha,$$

or

$$\vec{e}_{\alpha} = \Lambda^{\bar{\beta}}{}_{\alpha}\vec{e}_{\bar{\beta}}. \tag{2.13}$$

This gives the law by which basis vectors change. It is *not* a component transformation: it gives the basis $\{\vec{e}_{\alpha}\}$ of $\mathcal{O}$ as a linear sum over the basis $\{\vec{e}_{\bar{\alpha}}\}$ of $\bar{\mathcal{O}}$. Comparing this with the law for components, Eq. 2.7,

$$A^{\bar{\beta}} = \Lambda^{\bar{\beta}}{}_{\alpha}A^{\alpha},$$

we see that it is indeed different.

The above discussion introduced many new techniques, so study it carefully. Notice that the omission of the summation signs keeps things neat. Notice also that a step of key importance was relabeling the dummy indices: this allowed us to isolate the arbitrary A^α from the rest of the quantities in the equation.

An example. Let $\bar{\mathcal{O}}$ move with velocity v in the x direction relative to $\mathcal{O}$. Then the matrix $[\Lambda^{\bar{\beta}}{}_\alpha]$ is given by

$$[\Lambda^{\bar{\beta}}{}_\alpha] = \begin{bmatrix} \gamma & -v\gamma & 0 & 0 \\ -v\gamma & \gamma & 0 & 0 \\ 0 & 0 & 1 & 0 \\ 0 & 0 & 0 & 1 \end{bmatrix},$$

where we use the standard notation

$$\gamma := 1/\sqrt{(1 - v^2)}.$$

Then, if $\vec{A} \underset{\mathcal{O}}{\to} (5, 0, 0, 2)$, we find its components in $\bar{\mathcal{O}}$ from

$$A^{\bar{0}} = \Lambda^{\bar{0}}{}_0 A^0 + \Lambda^{\bar{0}}{}_1 A^1 + \cdots$$
$$= \gamma \cdot 5 + (-v\gamma) \cdot 0 + 0 \cdot 0 + 0 \cdot 2$$
$$= 5\gamma.$$

Similarly,

$$A^{\bar{1}} = -5v\gamma,$$
$$A^{\bar{2}} = 0,$$
$$A^{\bar{3}} = 2.$$

Therefore $\vec{A} \underset{\bar{\mathcal{O}}}{\to} (5\gamma, -5v\gamma, 0, 2)$.

The basis vectors are expressible as

$$\vec{e}_\alpha = \Lambda^{\bar{\beta}}{}_\alpha \vec{e}_{\bar{\beta}}$$

or

$$\vec{e}_0 = \Lambda^{\bar{0}}{}_0 \vec{e}_{\bar{0}} + \Lambda^{\bar{1}}{}_0 \vec{e}_{\bar{1}} + \cdots$$
$$= \gamma \vec{e}_{\bar{0}} - v\gamma \vec{e}_{\bar{1}}.$$

Similarly,

$$\vec{e}_1 = -v\gamma \vec{e}_{\bar{0}} + \gamma \vec{e}_{\bar{1}},$$
$$\vec{e}_2 = \vec{e}_{\bar{2}},$$
$$\vec{e}_3 = \vec{e}_{\bar{3}}.$$

This gives $\mathcal{O}$'s basis in terms of $\bar{\mathcal{O}}$'s, so let us draw the situation (Figure 2.1) in $\bar{\mathcal{O}}$'s frame. The above transformation is of course exactly what is needed to keep the basis vectors pointing along the axes of their respective frames. Compare this with Figure 1.5(b).

Notice that the matrix $[\Lambda^{\bar{\beta}}{}_\alpha]$ is *symmetric*. This is also true if the boost velocity v is in a more general direction. This symmetry turns out to be a general property of pure boosts, i.e. of Lorentz transformations that do not involve rotated spatial axes.

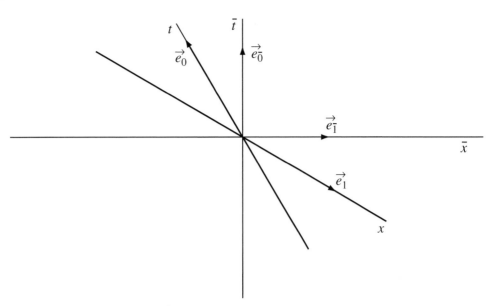

Figure 2.1 Basis vectors of $\mathcal{O}$ and $\bar{\mathcal{O}}$ as drawn by $\bar{\mathcal{O}}$.

Inverse transformations. The Lorentz transformation $\Lambda^{\bar{\beta}}{}_{\alpha}$ depends only on the relative velocity of the two frames. Let us for the moment show this explicitly by writing

$$\Lambda^{\bar{\beta}}{}_{\alpha} = \Lambda^{\bar{\beta}}{}_{\alpha}(v).$$

Then

$$\vec{e}_{\alpha} = \Lambda^{\bar{\beta}}{}_{\alpha}(v)\vec{e}_{\bar{\beta}}. \tag{2.14}$$

If the basis of $\mathcal{O}$ is obtained from that of $\bar{\mathcal{O}}$ by the transformation corresponding to velocity v, then the reverse must be true if we use $-v$. Thus we must have

$$\vec{e}_{\bar{\mu}} = \Lambda^{\nu}{}_{\bar{\mu}}(-v)\vec{e}_{\nu}. \tag{2.15}$$

In this equation I have used $\bar{\mu}$ and ν as indices to avoid confusion with the previous formula. The bars still refer, of course, to the frame $\bar{\mathcal{O}}$. The matrix $[\Lambda^{\nu}{}_{\bar{\mu}}]$ is exactly the matrix $[\Lambda^{\bar{\beta}}{}_{\alpha}]$ except with v changed to $-v$. The bars on the indices serve only to indicate the *names* of the observers involved: they affect the entries in the matrix $[\Lambda]$ only in that the matrix is *always* constructed using the velocity of the upper-index frame relative to the lower-index frame. This is made explicit in Eqs. 2.14 and 2.15. Since v is the velocity of $\bar{\mathcal{O}}$ (the upper-index frame in Eq. 2.14) relative to $\mathcal{O}$, then $-v$ is the velocity of $\mathcal{O}$ (the upper-index frame in Eq. 2.15) relative to $\bar{\mathcal{O}}$. Exercise 2.11 will help you understand this point.

We can rewrite the last expression as

$$\vec{e}_{\bar{\beta}} = \Lambda^{\nu}{}_{\bar{\beta}}(-v)\vec{e}_{\nu}.$$

Here we have just changed $\bar{\mu}$ to $\bar{\beta}$. This doesn't change anything: it is still the same four equations, one for each value of $\bar{\beta}$. In this form we can put it into the expression for $\vec{e}_\alpha$, Eq. 2.14:

$$\vec{e}_\alpha = \Lambda^{\bar{\beta}}{}_\alpha(\boldsymbol{v})\vec{e}_{\bar{\beta}} = \Lambda^{\bar{\beta}}{}_\alpha(\boldsymbol{v})\Lambda^\nu{}_{\bar{\beta}}(-\boldsymbol{v})\vec{e}_\nu. \tag{2.16}$$

In this equation only the basis of $\mathcal{O}$ appears. It must therefore be an identity *for all* $\boldsymbol{v}$. On the right-hand side there are two sums, one over $\bar{\beta}$ and one over ν. If we imagine performing the $\bar{\beta}$ sum first, then the right-hand side is a sum over the basis $\{\vec{e}_\nu\}$ in which each basis vector $\vec{e}_\nu$ has coefficient

$$\sum_{\bar{\beta}} \Lambda^{\bar{\beta}}{}_\alpha(\boldsymbol{v})\Lambda^\nu{}_{\bar{\beta}}(-\boldsymbol{v}). \tag{2.17}$$

Imagine evaluating Eq. 2.16 for some fixed value of the index α. If the right-hand side of Eq. 2.16 is equal to the left, the coefficient of $\vec{e}_\alpha$ on the *right* must be 1 and all other coefficients must vanish. The mathematical way of saying this is

$$\Lambda^{\bar{\beta}}{}_\alpha(\boldsymbol{v})\Lambda^\nu{}_{\bar{\beta}}(-\boldsymbol{v}) = \delta^\nu{}_\alpha,$$

where again $\delta^\nu{}_\alpha$ is the Kronecker delta. This would imply

$$\vec{e}_\alpha = \delta^\nu{}_\alpha\vec{e}_\nu,$$

which is an identity.

Let us change the order of multiplication above and write down the key formula

$$\Lambda^\nu{}_{\bar{\beta}}(-\boldsymbol{v})\Lambda^{\bar{\beta}}{}_\alpha(\boldsymbol{v}) = \delta^\nu{}_\alpha. \tag{2.18}$$

This expresses the fact that the matrix $[\Lambda^\nu{}_{\bar{\beta}}(-\boldsymbol{v})]$ is the *inverse* of $[\Lambda^{\bar{\beta}}{}_\alpha(\boldsymbol{v})]$, because the sum on $\bar{\beta}$ is exactly the operation that is performed when we multiply two matrices. The matrix $(\delta^\nu{}_\alpha)$ is, of course, the identity matrix.

The expression for the change of a vector's components,

$$A^{\bar{\beta}} = \Lambda^{\bar{\beta}}{}_\alpha(\boldsymbol{v})A^\alpha,$$

also has its inverse. Let us multiply both sides by $\Lambda^\nu{}_{\bar{\beta}}(-\boldsymbol{v})$ and sum over $\bar{\beta}$. We get

$$\Lambda^\nu{}_{\bar{\beta}}(-\boldsymbol{v})A^{\bar{\beta}} = \Lambda^\nu{}_{\bar{\beta}}(-\boldsymbol{v})\Lambda^{\bar{\beta}}{}_\alpha(\boldsymbol{v})A^\alpha$$
$$= \delta^\nu{}_\alpha A^\alpha$$
$$= A^\nu.$$

This says that the components of $\vec{A}$ in $\mathcal{O}$ are obtained from those in $\bar{\mathcal{O}}$ by a transformation with $-\boldsymbol{v}$, which is, of course, correct.

The operations we have performed should be familiar to you, in concept, from vector algebra in Euclidean space. The new element we have introduced here is the index notation, which will be a permanent and powerful tool in the rest of the book. Make sure that you understand the geometrical meaning of all our results as well as their algebraic justification.

2.3 The four-velocity

A particularly important vector is the four-velocity of a world line. In the three-geometry of Galileo, the velocity was a vector tangent to a particle's path. In our four-geometry we define the four-velocity $\vec{U}$ to be a vector tangent to the world line of the particle, and of such a length that it stretches for one unit of time in that particle's frame. For a uniformly moving particle, let us look at this definition in the inertial frame in which it is at rest. Then the four-velocity points parallel to the time axis and is one unit of time long. That is, it is identical with the basis vector $\vec{e}_0$ of that frame. Thus we could also use as our definition of the four-velocity of a uniformly moving particle that it is the vector $\vec{e}_0$ in its inertial rest frame. The word 'velocity' is justified by the fact that the spatial components of $\vec{U}$ are closely related to the particle's ordinary velocity v, which is called the three-velocity. This will be demonstrated in the example below, Eq. 2.21.

An *accelerated particle* has no inertial frame in which it is always at rest. However, there *is* an inertial frame which momentarily has the same velocity as the particle, but which a moment later is of course no longer comoving with it. This frame is the *momentarily comoving reference frame* (MCRF) and is an important concept. (Actually, there is an infinity of MCRFs for a given accelerated particle at a given event; they all have the same velocity, but their spatial axes are obtained from one another by rotations. This ambiguity will usually not be important.) The four-velocity of an accelerated particle is *defined* as the $\vec{e}_0$ basis vector of its MCRF at that event. This vector is tangent to the (curved) world line of the particle. In Figure 2.2 the particle at event $\mathcal{A}$ has MCRF $\bar{\mathcal{O}}$, whose basis vectors are shown. The vector $\vec{e}_{\bar{0}}$ is identical to the particle's four-velocity $\vec{U}$ there.

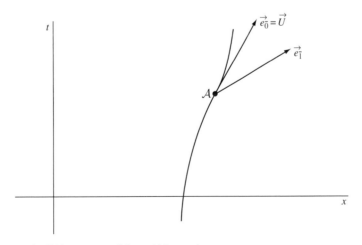

The four-velocity and MCRF basis vectors of the world line at $\mathcal{A}$.

2.4 The four-momentum and its conservation

The four-momentum $\vec{p}$ is defined as

$$\vec{p} = m\vec{U},\tag{2.19}$$

where m is the *rest mass* of the particle, which is its mass as measured in its rest frame. In some frame $\mathcal{O}$ it has components conventionally denoted by

$$\vec{p} \underset{\mathcal{O}}{\rightarrow} (E, p^1, p^2, p^3).\tag{2.20}$$

We call p^0 the *energy* E of the particle in the frame $\mathcal{O}$. The other components are its spatial momentum p^i.

An example. A particle of rest mass m moves with velocity v in the x direction of frame $\mathcal{O}$. What are the components of the four-velocity and four-momentum? Its rest frame $\bar{\mathcal{O}}$ has time basis vector $\vec{e}_{\bar{0}}$, so, by definition of $\vec{p}$ and $\vec{U}$, we have

$$\vec{U} = \vec{e}_{\bar{0}}, \qquad\qquad \vec{p} = m\vec{U},$$
$$U^\alpha = \Lambda^\alpha{}_{\bar{\beta}}(\vec{e}_{\bar{0}})^{\bar{\beta}} = \Lambda^\alpha{}_{\bar{0}}, \qquad p^\alpha = m\Lambda^\alpha{}_{\bar{0}}.\tag{2.21}$$

Therefore we have

$$\begin{aligned}
U^0 &= (1 - v^2)^{-1/2}, & p^0 &= m(1 - v^2)^{-1/2}, \\
U^1 &= v(1 - v^2)^{-1/2}, & p^1 &= mv(1 - v^2)^{-1/2}, \\
U^2 &= 0, & p^2 &= 0, \\
U^3 &= 0, & p^3 &= 0.
\end{aligned}$$

For small v the spatial components of $\vec{U}$ are $(v, 0, 0)$, which justifies calling it the four-velocity, while the spatial components of $\vec{p}$ are $(mv, 0, 0)$, justifying its name. For small v the energy is

$$E := p^0 = m(1 - v^2)^{-1/2} \simeq m + \tfrac{1}{2}mv^2.$$

This is the rest-mass energy plus the Galilean kinetic energy.

Conservation of four-momentum. The interactions of particles in Galilean physics are governed by the laws of conservation of energy and of momentum. Since the components of $\vec{p}$ reduce in the nonrelativistic limit to the familiar Galilean energy and momentum, it is natural to postulate that the correct relativistic law is that the four-vector $\vec{p}$ is conserved. That is, if several particles interact, then the four-momentum

$$\vec{p} := \sum_{\substack{\text{all} \\ \text{particles} \\ (i)}} \vec{p}_{(i)},\tag{2.22}$$

where $\vec{p}_{(i)}$ is the ith particle's momentum, is the same before and after each interaction.

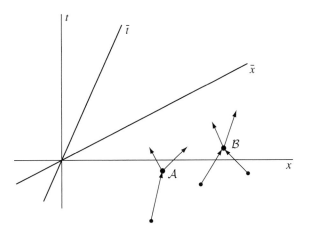

Figure 2.3 When several collisions are involved, the individual four-momentum vectors contributing to the total four-momentum at any particular time may depend upon the frame, but the total four-momentum is the same four-vector in all frames; its components transform from frame to frame by the Lorentz transformation.

This law has the status of an extra *postulate*, since it is only one of many whose nonrelativistic limit is correct. However, like the two fundamental postulates of SR, this one is amply verified by experiment. Not the least of its new predictions is that the energy conservation law must include rest mass: rest mass can be decreased and the difference turned into kinetic energy and hence into heat. This happens every day in nuclear power stations.

There is an important point glossed over in the above statement of the conservation of four-momentum. What is meant by 'before' and 'after' a collision? Suppose there are two collisions, involving different particles, which occur at spacelike separated events, as below. When adding up the total four-momentum, should one take them as they are on the line of constant time t or on the line of constant $\bar{t}$? As measured by $\mathcal{O}$, event $\mathcal{A}$ in Figure 2.3 occurs before $t = 0$ and $\mathcal{B}$ after, so the total momentum at $t = 0$ is the sum of the momenta after $\mathcal{A}$ and before $\mathcal{B}$. On the other hand, to $\bar{\mathcal{O}}$ they both occur before $\bar{t} = 0$ and so the total momentum at $\bar{t} = 0$ is the sum of the momenta after $\mathcal{A}$ and after $\mathcal{B}$. There is even a frame in which $\mathcal{B}$ is *earlier* than $\mathcal{A}$ and the adding-up may be the reverse of $\mathcal{O}$'s. There is really no problem here, though. Since each collision conserves momentum, the sum of the momenta before $\mathcal{A}$ is the same as that after $\mathcal{A}$, and likewise for $\mathcal{B}$. So *every* inertial observer will get the same total four-momentum vector $\vec{p}$. (Its components will be different in different frames, but it will be the same vector.) This is an important point: *any* observer can define a line of constant time (this is actually a three-space of constant time, which is called a *hypersurface* of constant time in four-dimensional spacetime), at that time add up all the momenta, and get the same vector as any other observer does. It is important to understand this, because such conservation laws will appear again.

Center of momentum (CM) frame. This is defined as the inertial frame where

$$\sum_i \vec{p}_{(i)} \xrightarrow[CM]{} (E_{\text{total}}, 0, 0, 0). \tag{2.23}$$

As with MCRFs, any other frame at rest relative to a CM frame is also a CM frame.

2.5 Scalar product

Magnitude of a vector. By analogy with the definition of an interval we define

$$\vec{A}^2 = -(A^0)^2 + (A^1)^2 + (A^2)^2 + (A^3)^2 \tag{2.24}$$

to be the *magnitude* of the vector $\vec{A}$. Because we *defined* the components to transform under a Lorentz transformation in the same manner as $(\Delta t, \Delta x, \Delta y, \Delta z)$, we are *guaranteed* that

$$-(A^0)^2 + (A^1)^2 + (A^2)^2 + (A^3)^2 = -(A^{\bar{0}})^2 + (A^{\bar{1}})^2 + (A^{\bar{2}})^2 + (A^{\bar{3}})^2. \tag{2.25}$$

The magnitude so defined is a frame-independent number, i.e. a scalar under Lorentz transformations.

   This magnitude doesn't have to be positive, of course. As with intervals we adopt the following names: if $\vec{A}^2$ is positive, $\vec{A}$ is a *spacelike* vector; if zero, a *null* vector; and if negative, a *timelike* vector. Thus, spatially pointing vectors have positive magnitude, as is usual in Euclidean space. It is particularly important to understand that a null vector is *not* a zero vector. That is, a null vector has $\vec{A}^2 = 0$, but not all its components A^α vanish; a zero vector is defined as a vector, all of whose components vanish. Only in a space where $\vec{A}^2$ is positive-definite does $\vec{A}^2 = 0$ require $A^\alpha = 0$ for all α.

Scalar product of two vectors. We define

$$\vec{A} \cdot \vec{B} = -A^0 B^0 + A^1 B^1 + A^2 B^2 + A^3 B^3 \tag{2.26}$$

in some frame $\mathcal{O}$. We now prove that this is the same number in all other frames. We note first that $\vec{A} \cdot \vec{A}$ is just $\vec{A}^2$, which we know is invariant. Therefore $(\vec{A} + \vec{B}) \cdot (\vec{A} + \vec{B})$, which is the magnitude of $\vec{A} + \vec{B}$, is also invariant. But from Eqs. 2.24 and 2.26 it follows that

$$(\vec{A} + \vec{B}) \cdot (\vec{A} + \vec{B}) = \vec{A}^2 + \vec{B}^2 + 2\vec{A} \cdot \vec{B}.$$

Since the left-hand side and the first two terms on the right-hand side are the same in all frames, the last term on the right must be as well. This proves the frame invariance of the scalar product.

   Two vectors $\vec{A}$ and $\vec{B}$ are said to be *orthogonal* if $\vec{A} \cdot \vec{B} = 0$. The minus sign in the definition of the scalar product means that two vectors orthogonal to one another are not necessarily at right angles in the spacetime diagram (see examples below). An extreme example is the null vector, which is orthogonal to *itself*! Such a phenomenon is not encountered in spaces where the scalar product is positive-definite.

Example. The basis vectors of a frame $\mathcal{O}$ satisfy

$$\vec{e}_0 \cdot \vec{e}_0 = -1,$$
$$\vec{e}_1 \cdot \vec{e}_1 = \vec{e}_2 \cdot \vec{e}_2 = \vec{e}_3 \cdot \vec{e}_3 = +1,$$
$$\vec{e}_\alpha \cdot \vec{e}_\beta = 0 \quad \text{if } \alpha \neq \beta.$$

They thus make up a tetrad of mutually orthogonal vectors: an *orthonormal* tetrad, which means *ortho*gonal and *normal*ized to unit magnitude. (A timelike vector has 'unit magnitude' if its magnitude is −1.) The relations above can be summarized as

$$\vec{e}_\alpha \cdot \vec{e}_\beta = \eta_{\alpha\beta}, \tag{2.27}$$

where $\eta_{\alpha\beta}$ is similar to a Kronecker delta in that it is zero when $\alpha \neq \beta$, but it differs in that $\eta_{00} = -1$, while $\eta_{11} = \eta_{22} = \eta_{33} = +1$. We will see later that in fact $\eta_{\alpha\beta}$ is of central importance: it is the metric tensor. But for now we treat it as a generalized Kronecker delta.

Example. The basis vectors of $\bar{\mathcal{O}}$ also satisfy

$$\vec{e}_{\bar\alpha} \cdot \vec{e}_{\bar\beta} = \eta_{\bar\alpha\bar\beta},$$

so that, in particular, $\vec{e}_{\bar 0} \cdot \vec{e}_{\bar 1} = 0$. Look at this in the spacetime diagram of $\mathcal{O}$, Figure 2.4: the two vectors certainly are not perpendicular in the picture. Nevertheless, their scalar product is zero. The rule is that two vectors are orthogonal if they make equal angles with the 45° line representing the path of a light ray. Thus, a vector tangent to the light ray is orthogonal to itself. This is just another way in which SR cannot be 'visualized' in terms of notions we have developed in Euclidean space.

Example. The four-velocity $\vec{U}$ of a particle is just the time basis vector of its MCRF, so from Eq. 2.27 we have

$$\vec{U} \cdot \vec{U} = -1. \tag{2.28}$$

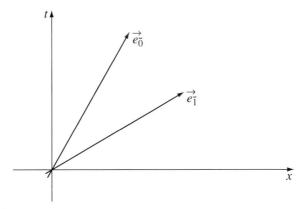

Figure 2.4 The basis vectors of $\bar{\mathcal{O}}$ are not 'perpendicular' (in the Euclidean sense) when drawn in $\mathcal{O}$, but they *are* orthogonal with respect to the dot product of Minkowski spacetime.

2.6 Applications

Four-velocity and acceleration as derivatives. Suppose a particle makes an infinitesimal displacement $d\vec{x}$, whose components in $\mathcal{O}$ are (dt, dx, dy, dz). The magnitude of this displacement is, by Eq. 2.24, just $-dt^2 + dx^2 + dy^2 + dz^2$. Comparing this with Eq. 1.1, we see that this is just the interval ds^2:

$$ds^2 = d\vec{x} \cdot d\vec{x}. \tag{2.29}$$

Since the world line is timelike, this is negative. This led us (Eq. 1.9) to define the proper time $d\tau$ by

$$(d\tau)^2 = -d\vec{x} \cdot d\vec{x}. \tag{2.30}$$

Now consider (Figure 2.5) the vector $d\vec{x}/d\tau$, where $d\tau$ is the square root of Eq. 2.30. This vector is tangent to the world line since it is a multiple of $d\vec{x}$. Its magnitude is

$$\frac{d\vec{x}}{d\tau} \cdot \frac{d\vec{x}}{d\tau} = \frac{d\vec{x} \cdot d\vec{x}}{(d\tau)^2} = -1.$$

It is therefore a timelike vector of unit magnitude tangent to the world line. In an MCRF,

$$d\vec{x} \underset{\substack{\text{MCRF} \\ d\tau = dt}}{\longrightarrow} (dt, 0, 0, 0).$$

so that

$$\frac{d\vec{x}}{d\tau} \underset{\text{MCRF}}{\longrightarrow} (1, 0, 0, 0)$$

or

$$\frac{d\vec{x}}{d\tau} = (\vec{e}_0)_{\text{MCRF}}.$$

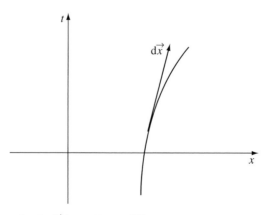

Figure 2.5   The infinitesimal displacement vector $d\vec{x}$ tangent to a world line.

This was the *definition* of the four-velocity. So we have the useful expression

$$\vec{U} = d\vec{x}/d\tau. \tag{2.31}$$

Moreover, let us examine

$$\frac{d\vec{U}}{d\tau} = \frac{d^2\vec{x}}{d\tau^2},$$

which is some sort of four-acceleration. First we differentiate Eq. 2.28 and use Eq. 2.26:

$$\frac{d}{d\tau}(\vec{U} \cdot \vec{U}) = 2\vec{U} \cdot \frac{d\vec{U}}{d\tau}.$$

But since $\vec{U} \cdot \vec{U} = -1$ is a constant we have

$$\vec{U} \cdot \frac{d\vec{U}}{d\tau} = 0.$$

Since, in the MCRF, $\vec{U}$ has only a 0-component, this orthogonality means that

$$\frac{d\vec{U}}{d\tau} \xrightarrow[\text{MCRF}]{} (0, a^1, a^2, a^3).$$

This vector is defined as the *acceleration* four-vector $\vec{a}$:

$$\vec{a} = \frac{d\vec{U}}{d\tau}, \qquad \vec{U} \cdot \vec{a} = 0. \tag{2.32}$$

Exercise 2.19 justifies the name 'acceleration'.

Energy and momentum. Consider a particle whose momentum is $\vec{p}$. Then

$$\vec{p} \cdot \vec{p} = m^2 \vec{U} \cdot \vec{U} = -m^2. \tag{2.33}$$

But

$$\vec{p} \cdot \vec{p} = -E^2 + (p^1)^2 + (p^2)^2 + (p^3)^2.$$

Therefore

$$E^2 = m^2 + \sum_{i=1}^{3}(p^i)^2. \tag{2.34}$$

This is the familiar expression for the total energy of a particle.

Suppose an observer $\bar{\mathcal{O}}$ moves with four-velocity $\vec{U}_{\text{obs}}$ that is not necessarily equal to the particle's four-velocity. Then

$$\vec{p} \cdot \vec{U}_{\text{obs}} = \vec{p} \cdot \vec{e}_{\bar{0}},$$

where $\vec{e}_{\bar{0}}$ is the basis vector of the frame of the observer. In that frame the four-momentum has components

$$\vec{p} \underset{\bar{\mathcal{O}}}{\rightarrow} (\bar{E}, p^{\bar{1}}, p^{\bar{2}}, p^{\bar{3}}).$$

Therefore we obtain, from Eq. 2.26,

$$-\vec{p} \cdot \vec{U}_{\text{obs}} = \bar{E}. \tag{2.35}$$

This is an important equation. It says that the energy of the particle relative to the observer, $\bar{E}$, can be computed by any one in any frame by taking the scalar product $\vec{p} \cdot \vec{U}_{\text{obs}}$. This expression for the energy relative to the observer is frame invariant. It is almost always helpful in calculations to use such expressions.

2.7 Photons

No four-velocity. Photons move on null lines, so, for a photon path,

$$d\vec{x} \cdot d\vec{x} = 0. \tag{2.36}$$

Therefore $d\tau$ is *zero* and Eq. 2.31 shows that the *four-velocity cannot be defined.* Another way of saying the same thing is to note that there is no frame in which light is at rest (the second postulate of SR), so there is no MCRF for a photon. Thus, no $\vec{e}_0$ in any frame will be tangent to a photon's world line.

Note carefully that it is still possible to find vectors tangent to a photon's path (which, being a straight line, has the same tangent everywhere): $d\vec{x}$ is one such vector. The problem is finding a tangent of *unit magnitude*, since they all have vanishing magnitude.

Four-momentum. The four-momentum of a particle is *not* a unit vector. Instead, it is a vector whose components in some frame give the particle's energy and momentum relative to that frame. If a photon carries energy E in some frame, then in that frame $p^0 = E$. If it moves in the x direction, then $p^y = p^z = 0$, and in order for the four-momentum to be parallel to its world line (and hence be null) we must have $p^x = E$. This ensures that

$$\vec{p} \cdot \vec{p} = -E^2 + E^2 = 0. \tag{2.37}$$

So we conclude that photons have *spatial momentum* equal to their energy.

We know from quantum mechanics that a photon has energy

$$E = h\nu, \tag{2.38}$$

where ν is its frequency and h is Planck's constant, $h = 6.6256 \times 10^{-34} \, \text{J s}^{-1}$.

This relation and the Lorentz transformation of the four-momentum immediately give us the Doppler-shift formula for photons. Suppose, for instance, that in frame $\mathcal{O}$ a photon has frequency ν and moves in the x direction. Then, in $\bar{\mathcal{O}}$, which has velocity v in the x direction relative to $\mathcal{O}$, the photon's energy is

$$\bar{E} = E/\sqrt{(1-v^2)} - p^x v/\sqrt{(1-v^2)}$$
$$= h\nu/\sqrt{(1-v^2)} - h\nu v/\sqrt{(1-v^2)}.$$

Setting this equal to $h\bar{\nu}$ gives $\bar{\nu}$, the frequency in $\bar{\mathcal{O}}$:

$$\bar{\nu}/\nu = (1-v)/\sqrt{(1-v^2)} = \sqrt{[(1-v)/(1+v)]}. \tag{2.39}$$

This is generalized in Exercise 2.25.

Zero rest-mass particles. The rest mass of a photon must be zero, since

$$m^2 = -\vec{p} \cdot \vec{p} = 0. \tag{2.40}$$

Any particle whose four-momentum is null must have rest mass zero, and conversely. The only known zero-rest-mass particle is the photon. Neutrinos are very light, but not massless. (Sometimes the 'graviton' is added to this list since gravitational waves also travel at the speed of light, as we shall see later. But 'photon' and 'graviton' are concepts that come from quantum mechanics, and there is as yet no satisfactory quantized theory of gravity, so that 'graviton' is not really a well-defined notion yet.) The idea that only particles with zero rest mass can travel at the speed of light is reinforced by the fact that no particle of finite rest mass can be accelerated to the speed of light, since then its energy would be infinite. Put another way, a particle traveling at the speed of light (in, say, the x direction) has $p^1/p^0 = 1$, while, from the equation $\vec{p} \cdot \vec{p} = -m^2$, a particle of rest mass m moving in the x direction has $p^1/p^0 = [1 - m^2/(p^0)^2]^{1/2}$, which is always less than one, no matter how much energy the particle is given. Although it may seem to get close to the speed of light, there is an important distinction: the particle with $m \neq 0$ always has an MCRF, a Lorentz frame in which it is at rest, namely that whose velocity v is p^1/p^0 relative to the old frame. A photon has *no* rest frame.

2.8 Bibliography

We have only scratched the surface of relativistic kinematics and particle dynamics. These are particularly important in particle physics, which in turn provides the most stringent tests of SR. See Hagedorn (1963) or Wiedemann (2007).

Exercises

2.1 Given the numbers $\{A^0 = 5, A^1 = 0, A^2 = -1, A^3 = -6\}, \{B_0 = 0, B_1 = -2, B_2 = 4, B_3 = 0\}, \{C_{00} = 1, C_{01} = 0, C_{02} = 2, C_{03} = 3, C_{30} = -1, C_{10} = 5, C_{11} = -2, C_{12} = -2, C_{13} = 0, C_{21} = 5, C_{22} = 2, C_{23} = -2, C_{20} = 4, C_{31} = -1, C_{32} = -3, C_{33} = -3, C_{33} = 0\}$, find:

(a) $A^\alpha B_\alpha$; (b) $A^\alpha C_{\alpha\beta}$ for all β; (c) $A^\gamma C_{\gamma\sigma}$ for all σ; (d) $A^\nu C_{\mu\nu}$ for all μ; (e) $A^\alpha B_\beta$ for all α, β; (f) $A^i B_i$; (g) $A^j B_k$ for all j, k.

2.2 Identify the free and dummy indices in the following equations and change them into equivalent expressions with different indices. How many different equations does each expression represent?

(a) $A^\alpha B_\alpha = 5$; (b) $A^{\bar{\mu}} = \Lambda^{\bar{\mu}}{}_\nu A^\nu$; (c) $T^{\alpha\mu\lambda} A_\mu C_\lambda{}^\gamma = D^{\gamma\alpha}$; (d) $R_{\mu\nu} - \frac{1}{2}g_{\mu\nu}R = G_{\mu\nu}$.

2.3 Prove Eq. 2.5.

2.4 Given the vectors $\vec{A} \rightarrow_{\mathcal{O}} (5, -1, 0, 1)$ and $\vec{B} \rightarrow_{\mathcal{O}} (-2, 1, 1, -6)$, find the components in $\mathcal{O}$ of (a) $-6\vec{A}$; (b) $3\vec{A} + \vec{B}$; (c) $-6\vec{A} + 3\vec{B}$.

2.5 A collection of vectors $\{\vec{a}, \vec{b}, \vec{c}, \vec{d}\}$ is said to be linearly independent if no linear combination of them is zero except the trivial combination, $0\vec{a} + 0\vec{b} + 0\vec{c} + 0\vec{d} = 0$.

(a) Show that the basis vectors in Eq. 2.9 are linearly independent.

(b) Is the following set linearly independent?: $\{\vec{a}, \vec{b}, \vec{c}, 5\vec{a} + 3\vec{b} - 2\vec{c}\}$.

2.6 In the $t-x$ spacetime diagram of $\mathcal{O}$, draw the basis vectors $\vec{e}_0$ and $\vec{e}_1$. Draw the corresponding basis vectors of $\bar{\mathcal{O}}$, which moves with speed 0.6 in the positive x direction relative to $\mathcal{O}$. Draw the corresponding basis vectors of $\bar{\bar{\mathcal{O}}}$, which moves with speed 0.6 in the positive x direction relative to $\bar{\mathcal{O}}$.

2.7 (a) Verify Eq. 2.10 for all α, β.

(b) Prove Eq. 2.11 from Eq. 2.9.

2.8 (a) Prove that the zero vector $(0, 0, 0, 0)$ has these same components in *all* reference frames.

(b) Use (a) to prove that if two vectors have equal components in one frame, they have equal components in all frames.

2.9 Prove, by writing out all the terms, that

$$\sum_{\bar{\alpha}=0}^{3}\left(\sum_{\beta=0}^{3} \Lambda^{\bar{\alpha}}{}_\beta A^\beta \vec{e}_{\bar{\alpha}}\right) = \sum_{\beta=0}^{3}\left(\sum_{\bar{\alpha}=0}^{3} \Lambda^{\bar{\alpha}}{}_\beta A^\beta \vec{e}_{\bar{\alpha}}\right).$$

Since the order of summation doesn't matter, we are justified in using the Einstein summation convention to write simply $\Lambda^{\bar{\alpha}}{}_\beta A^\beta \vec{e}_{\bar{\alpha}}$, which doesn't specify the order of summation.

2.10 Prove Eq. 2.13 from the equation $A^\alpha (\Lambda^{\bar{\beta}}{}_\alpha \vec{e}_{\bar{\beta}} - \vec{e}_\alpha) = 0$ by making specific choices for the components of the arbitrary vector $\vec{A}$.

2.11 Let $\Lambda^{\bar{\alpha}}{}_\beta$ be the matrix of the Lorentz transformation from $\mathcal{O}$ to $\bar{\mathcal{O}}$ given in Eq. 1.12. Let $\vec{A}$ be an arbitrary vector with components (A^0, A^1, A^2, A^3) in frame $\mathcal{O}$.

(a) Write down the matrix of $\Lambda^\nu{}_{\bar\mu}(-v)$.

(b) Find $A^{\bar\alpha}$ for all $\bar\alpha$.

(c) Verify Eq. 2.18 by performing the indicated sum for all values of ν and α.

(d) Write down the Lorentz transformation matrix from $\bar{\mathcal{O}}$ to $\mathcal{O}$, justifying each entry.

(e) Use (d) to find A^β from $A^{\bar\alpha}$. How is this related to Eq. 2.18?

(f) Verify, in the same manner as (c), that

$$\Lambda^\nu{}_{\bar\beta}(v)\Lambda^{\bar\alpha}{}_\nu(-v) = \delta^{\bar\alpha}{}_{\bar\beta}.$$

(g) Establish that

$$\vec{e}_\alpha = \delta^\nu{}_\alpha \vec{e}_\nu$$

and

$$A^{\bar\beta} = \delta^{\bar\beta}{}_{\bar\mu} A^{\bar\mu}.$$

2.12   Given $\vec{A} \to_{\mathcal{O}} (0, -2, 3, 5)$, find

(a) the components of $\vec{A}$ in $\bar{\mathcal{O}}$, which moves at speed 0.8 relative to $\mathcal{O}$ in the positive x direction;

(b) the components of $\vec{A}$ in $\bar{\bar{\mathcal{O}}}$, which moves at speed 0.6 relative to $\bar{\mathcal{O}}$ in the positive x direction;

(c) the magnitude of $\vec{A}$ from its components in $\mathcal{O}$;

(d) the magnitude of $\vec{A}$ from its components in $\bar{\mathcal{O}}$.

2.13   Consider three observers called $\mathcal{O}$, $\bar{\mathcal{O}}$, and $\bar{\bar{\mathcal{O}}}$, defined so that their origins coincide at the same event. Suppose that $\bar{\mathcal{O}}$ moves with velocity v relative to $\mathcal{O}$, and that $\bar{\bar{\mathcal{O}}}$ moves with velocity v' relative to $\bar{\mathcal{O}}$.

(a) Show that the Lorentz transformation from $\mathcal{O}$ to $\bar{\bar{\mathcal{O}}}$ is

$$\Lambda^{\bar{\bar\alpha}}{}_\mu = \Lambda^{\bar{\bar\alpha}}{}_{\bar\gamma}(v')\Lambda^{\bar\gamma}{}_\mu(v) \qquad\qquad (2.41)$$

(b) Show that Eq. 2.41 is just the matrix product of the matrices of the individual Lorentz transformations.

(c) Let $v = 0.6\vec{e}_x$, $v' = 0.8\vec{e}_{\bar{y}}$. Now assume that the spatial (x, y, z) axes of $\mathcal{O}$ and of $\bar{\mathcal{O}}$ are aligned with one another, as measured by both observers, and that the spatial axes of $\bar{\mathcal{O}}$ and of $\bar{\bar{\mathcal{O}}}$ are likewise aligned, as measured in their frames. (These alignments ensure that the Lorentz transformations between each pair of frames are pure boosts.) Write down the individual Lorentz transformation matrices and then find $\Lambda^{\bar{\bar\alpha}}{}_\mu$ for all μ and $\bar{\bar\alpha}$.

(d) Verify that the transformation found in (c) is indeed a Lorentz transformation by showing explicitly that $\Delta\bar{\bar{s}}^2 = \Delta s^2$ for any $(\Delta t, \Delta x, \Delta y, \Delta z)$.

(e) The transformation matrix that you should have found in (c) is not symmetric. We have seen earlier that pure boosts have symmetric transformation matrices. Interpret this result.

(f) Compute

$$\Lambda^{\bar{\bar{\alpha}}}{}_{\bar{\gamma}}(v)\Lambda^{\bar{\gamma}}{}_{\mu}(v')$$

for v and v', as given in (c), and show that the result does not equal that of (c). Interpret this physically.

2.14 The following matrix gives a Lorentz transformation from $\mathcal{O}$ to $\bar{\mathcal{O}}$:

$$\begin{pmatrix} 1.25 & 0 & 0 & 0.75 \\ 0 & 1 & 0 & 0 \\ 0 & 0 & 1 & 0 \\ 0.75 & 0 & 0 & 1.25 \end{pmatrix}$$

(a) What is the velocity (speed and direction) of $\bar{\mathcal{O}}$ relative to $\mathcal{O}$?
(b) What is the inverse of the given matrix?
(c) Find the components in $\mathcal{O}$ of a vector $\vec{A} \underset{\bar{\mathcal{O}}}{\rightarrow} (1, 2, 0, 0)$.

2.15 (a) Compute the four-velocity components in $\mathcal{O}$ of a particle whose speed in $\mathcal{O}$ is v in the positive x direction, by using the Lorentz transformation from the rest frame of the particle.
(b) Generalize this result to find the four-velocity components when the particle has arbitrary velocity v, with $|v| < 1$.
(c) Use your result in (b) to express v in terms of the components $\{U^{\alpha}\}$.
(d) Find the three-velocity v of a particle whose four-velocity components are (2, 1, 1, 1).

2.16 Derive the Einstein velocity-addition formula by performing a Lorentz transformation with velocity v on the four-velocity of a particle whose speed in the original frame was W.

2.17 (a) Prove that any timelike vector $\vec{U}$ for which $U^0 > 0$ and $\vec{U} \cdot \vec{U} = -1$ is the four-velocity of *some* world line.
(b) Use this to prove that for any timelike vector $\vec{V}$ there is a Lorentz frame in which $\vec{V}$ has zero spatial components.

2.18 (a) Show that the sum of any two orthogonal spacelike vectors is spacelike.
(b) Show that a timelike vector and a null vector cannot be orthogonal.

2.19 A body is said to be *uniformly accelerated* if its acceleration four-vector $\vec{a}$ has constant spatial direction and magnitude, say $\vec{a} \cdot \vec{a} = \alpha^2 \geqslant 0$.

(a) Show that this implies that $\vec{a}$ always has the same components in the body's MCRF, and that these components constitute what one would call the 'acceleration' in Galilean terms. (This would be the physical situation for a rocket whose engine always gave the same acceleration.)
(b) Suppose a body is uniformly accelerated with $\alpha = 10\,\mathrm{m\,s^{-2}}$ (about the acceleration of gravity on Earth). If the body starts from rest, find its speed after time t. (Be sure to use the correct units.) How far has it traveled in this time? How long does it take to reach $v = 0.999$?

(c) Find the elapsed *proper* time for the body in (b), as a function of t. (Integrate d_τ along its world line.) How much proper time has elapsed by the time its speed is $v = 0.999$? How much would a person accelerated as in (b) age on a trip from Earth to the center of our Galaxy, a distance of about 2×10^{20} m?

2.20 The world line of a particle is described by the equations

$$x(t) = at + b \sin \omega t, \qquad y(t) = b \cos \omega t,$$
$$z(t) = 0, \qquad |a| + |b\omega| < 1,$$

in some inertial frame. Describe the motion and compute the components of the particle's four-velocity and four-acceleration.

2.21 The world line of a particle is described by the following parametric equations in some Lorentz frame:

$$t(\lambda) = a \sinh\left(\frac{\lambda}{a}\right), \qquad x(\lambda) = a \cosh\left(\frac{\lambda}{a}\right),$$

where λ is the parameter and a is a constant. Describe the motion and compute the particle's four-velocity and four-acceleration components. Show that λ is the proper time along the world line and that the four-acceleration is uniform. Interpret a.

2.22 (a) Find the energy, rest mass, and three-velocity v of a particle whose four-momentum has the components $(4, 1, 1, 0)$ kg.

(b) The collision of two particles of four-momenta

$$\vec{p}_1 \underset{\mathcal{O}}{\rightarrow} (3, -1, 0, 0) \, \text{kg}, \qquad \vec{p}_2 \underset{\mathcal{O}}{\rightarrow} (2, 1, 1, 0) \, \text{kg}$$

results in the destruction of the two particles and the production of three new particles, two of which have four-momenta

$$\vec{p}_3 \underset{\mathcal{O}}{\rightarrow} (1, 1, 0, 0) \, \text{kg}, \qquad \vec{p}_4 \underset{\mathcal{O}}{\rightarrow} (1, -\tfrac{1}{2}, 0, 0) \, \text{kg}.$$

Find the four-momentum, energy, rest mass, and three-velocity of the third particle produced. Find the CM frame's three-velocity.

2.23 A particle of rest mass m has three-velocity v. Find its energy correct to terms of order $|v|^4$. At what speed $|v|$ does the absolute value of the $0(|v|^4)$ term equal half the kinetic energy term $\frac{1}{2}m|v|^2$?

2.24 Prove that the conservation of four-momentum forbids a reaction in which an electron and positron annihilate and produce a single photon (γ-ray). Prove that the production of two photons is not forbidden.

2.25 (a) Let frame $\bar{\mathcal{O}}$ move with speed v in the x direction relative to $\mathcal{O}$. Let a photon have frequency ν in $\mathcal{O}$ and move at an angle θ with respect to $\mathcal{O}$'s x axis. Show that its frequency $\bar{\nu}$ in $\bar{\mathcal{O}}$ is given by

$$\bar{\nu}/\nu = (1 - v \cos \theta)/\sqrt{(1 - v^2)}. \tag{2.42}$$

(b) Even when the motion of the photon is perpendicular to the x axis ($\theta = \pi/2$) there is a frequency shift. This is called the *transverse Doppler shift*, and arises because of time dilation. At what angle θ does the photon have to move so that there is *no* Doppler shift between $\mathcal{O}$ and $\bar{\mathcal{O}}$?

(c) Use Eqs. 2.35 and 2.38 to calculate Eq. 2.42.

2.26 Calculate the energy that is required to accelerate a particle of rest mass $m \neq 0$ from speed v to speed $v + \delta v$ ($\delta v \ll v$), to first order in δv. Show that it would take an infinite amount of energy to accelerate the particle to the speed of light.

2.27 Two identical bodies of mass 10 kg are at rest at the same temperature. One of them is heated by the addition of 100 J of heat. Both are then subjected to the same force. Which accelerates faster, and by how much?

2.28 Let $\vec{A} \to_{\mathcal{O}} (5, 1, -1, 0), \vec{B} \to_{\mathcal{O}} (-2, 3, 1, 6), \vec{C} \to_{\mathcal{O}} (2, -2, 0, 0)$. Let $\bar{\mathcal{O}}$ be a frame moving at speed $v = 0.6$ in the positive x direction relative to $\mathcal{O}$, with its spatial axes oriented parallel to $\mathcal{O}$'s.

(a) Find the components of $\vec{A}, \vec{B}$, and $\vec{C}$ in $\bar{\mathcal{O}}$.

(b) Form the dot products $\vec{A} \cdot \vec{B}, \vec{B} \cdot \vec{C}, \vec{A} \cdot \vec{C}$, and $\vec{C} \cdot \vec{C}$ using the components in $\bar{\mathcal{O}}$. Verify the frame-independence of these numbers.

(c) Classify $\vec{A}, \vec{B}$, and $\vec{C}$ as timelike, spacelike, or null.

2.29 Prove, using the component expressions Eqs. 2.24 and 2.26, that

$$\frac{\mathrm{d}}{\mathrm{d}\tau}(\vec{U} \cdot \vec{U}) = 2\vec{U} \cdot \frac{\mathrm{d}\vec{U}}{\mathrm{d}\tau}.$$

2.30 The four-velocity of a rocket ship is $\vec{U} \to_{\mathcal{O}} (2, 1, 1, 1)$. It encounters a high-velocity cosmic ray whose momentum is $\vec{P} \to_{\mathcal{O}} (300, 299, 0, 0) \times 10^{-27}$ kg. Compute the energy of the cosmic ray as measured by the rocket ship's passengers, using each of the two following methods.

(a) Find the Lorentz transformation from $\mathcal{O}$ to the MCRF of the rocket ship, and use it to transform the components of $\vec{P}$.

(b) Use Eq. 2.35.

(c) Which method is quicker? Why?

2.31 A photon of frequency ν is reflected without change of frequency from a mirror, with an angle of incidence θ. Calculate the momentum transferred to the mirror. What momentum would be transferred if the photon were absorbed rather than reflected?

2.32 Let a particle of charge e and rest mass m, initially at rest in a laboratory, scatter a photon of initial frequency ν_i. This is called *Compton scattering*. Suppose the scattered photon comes off at an angle θ from the incident direction. Use the conservation of four-momentum to deduce that the photon's final frequency ν_f is given by

$$\frac{1}{\nu_f} = \frac{1}{\nu_i} + h\left(\frac{1 - \cos\theta}{m}\right). \tag{2.43}$$

2.33 Space is filled with cosmic rays (high-energy protons) and the cosmic microwave background radiation. These can Compton-scatter off one another. Suppose a photon of energy $h\nu = 2 \times 10^{-4}$ eV scatters off a proton of energy $10^9 m_P = 10^{18}$ eV, where energies are measured in the Sun's rest frame. Use Eq. 2.43 in the proton's initial rest frame to calculate the maximum final energy that the photon can have in the solar rest frame after the scattering. What energy range is this (X-ray, visible, etc.)?

2.34 Show that, if $\vec{A}$, $\vec{B}$, and $\vec{C}$ are any vectors and α and β are any real numbers,

$$(\alpha\vec{A}) \cdot \vec{B} = \alpha(\vec{A} \cdot \vec{B}),$$
$$\vec{A} \cdot (\beta\vec{B}) = \beta(\vec{A} \cdot \vec{B}),$$
$$\vec{A} \cdot (\vec{B} + \vec{C}) = \vec{A} \cdot \vec{B} + \vec{A} \cdot \vec{C},$$
$$(\vec{A} + \vec{B}) \cdot \vec{C} = \vec{A} \cdot \vec{C} + \vec{B} \cdot \vec{C}.$$

2.35 Show that the vectors $\{\vec{e}_{\bar{\beta}}\}$ obtained from $\{\vec{e}_\alpha\}$ by Eq. 2.15 satisfy $\vec{e}_{\bar{\alpha}} \cdot \vec{e}_{\bar{\beta}} = \eta_{\bar{\alpha}\bar{\beta}}$ for all $\bar{\alpha}$, $\bar{\beta}$.

3 Tensor analysis in special relativity

3.1 The metric tensor

Consider the representation of two vectors $\vec{A}$ and $\vec{B}$ in the basis $\{\vec{e}_\alpha\}$ of some frame $\mathcal{O}$:

$$\vec{A} = A^\alpha \vec{e}_\alpha, \qquad \vec{B} = B^\beta \vec{e}_\beta.$$

Their scalar product is

$$\vec{A} \cdot \vec{B} = (A^\alpha \vec{e}_\alpha) \cdot (B^\beta \vec{e}_\beta).$$

(Note the importance of using *different* indices α and β to distinguish the first summation from the second.) Following Exercise 2.34, we can rewrite this as

$$\vec{A} \cdot \vec{B} = A^\alpha B^\beta (\vec{e}_\alpha \cdot \vec{e}_\beta),$$

which, by Eq. 2.27, is

$$\vec{A} \cdot \vec{B} = A^\alpha B^\beta \eta_{\alpha\beta}. \tag{3.1}$$

This is a *frame-invariant* way of writing

$$-A^0 B^0 + A^1 B^1 + A^2 B^2 + A^3 B^3.$$

The numbers $\eta_{\alpha\beta}$ are called 'components of the metric tensor'. We will justify this name later. Right now we observe that they essentially give a rule for associating with two vectors $\vec{A}$ and $\vec{B}$ a single *number*, which we call their scalar product. The rule is that the number is the double sum $A^\alpha B^\beta \eta_{\alpha\beta}$. Such a rule is at the heart of the meaning of the word 'tensor', as we now discuss.

3.2 Definition of tensors

We give the following definition of a tensor:

A tensor of type $\binom{0}{N}$ is a function of N vectors into the real numbers which is linear in each of its N arguments.

Let us see what this definition means. For the moment, we will just accept the notation $\binom{0}{N}$; its justification will come later in this chapter. The rule for the scalar product, Eq. 3.1, satisfies our definition of a $\binom{0}{2}$ tensor. It is a rule which takes two vectors, $\vec{A}$ and $\vec{B}$, and produces a single real number $\vec{A} \cdot \vec{B}$. To say that it is linear in its arguments means what is proved in Exercise 2.34. Linearity in the first argument means that

$$\left.\begin{array}{l} (\alpha\vec{A}) \cdot \vec{B} = \alpha(\vec{A} \cdot \vec{B}), \\ (\vec{A} + \vec{B}) \cdot \vec{C} = \vec{A} \cdot \vec{C} + \vec{B} \cdot \vec{C}, \end{array}\right\} \tag{3.2}$$

while linearity in the second argument means that

$$\vec{A} \cdot (\beta\vec{B}) = \beta(\vec{A} \cdot \vec{B}),$$
$$\vec{A} \cdot (\vec{B} + \vec{C}) = \vec{A} \cdot \vec{B} + \vec{A} \cdot \vec{C}.$$

This definition of linearity is of central importance for tensor algebra, and you should study it carefully.

To give concreteness to this notion of the dot product being a tensor, we introduce a name and notation for it. We let **g** be the *metric tensor* and define

$$\mathbf{g}(\vec{A}, \vec{B}) := \vec{A} \cdot \vec{B}. \tag{3.3}$$

Then we regard **g**(,) as a function which can take two arguments and which is linear in that

$$\mathbf{g}(\alpha\vec{A} + \beta\vec{B}, \vec{C}) = \alpha\mathbf{g}(\vec{A}, \vec{C}) + \beta\mathbf{g}(\vec{B}, \vec{C}), \tag{3.4}$$

and similarly for the second argument. The value of **g** for two arguments, denoted by $\mathbf{g}(\vec{A}, \vec{B})$, is their dot product, a real number.

Notice that the definition of a tensor does not mention components of the vector arguments. A tensor must amount to a rule which gives the same real number independently of the reference frame in which the vectors' components are calculated. We showed in the previous chapter that Eq. 3.1 satisfies this requirement. This enables us to regard a tensor as a function of the vectors themselves rather than of their components, and this can sometimes be helpful conceptually.

Notice that an ordinary function of position, $f(t, x, y, z)$, is a real-valued function of no vectors at all. It is therefore classified as a $\binom{0}{0}$ tensor.

Aside on the usage of the term 'function'. The most familiar notion of a function is expressed in the equation

$$y = f(x),$$

where y and x are real numbers. But this can be written more precisely as follows: f is a 'rule' (called a mapping) which associates a real number (symbolically called y, above) with another real number, which is the argument of f (symbolically called x, above). The function itself is *not* $f(x)$, since $f(x)$ is y, which is a real number called the 'value' of the function. The function itself is f, which we can write as $f(\ )$ in order to show that it has one argument. In algebra this seems like hair-splitting since we unconsciously think of x

and y as two things at once: they are, on the one hand, specific real numbers and, on the other hand, *names* for general and arbitrary real numbers. In tensor calculus we will make this distinction explicit: $\vec{A}$ and $\vec{B}$ are *specific* vectors, $\vec{A} \cdot \vec{B}$ is a specific real number, and **g** is the name of the function that associates $\vec{A} \cdot \vec{B}$ with $\vec{A}$ and $\vec{B}$.

Components of a tensor. Just like a vector, a tensor has components. They are defined as follows:

> The components in a frame $\mathcal{O}$ of a tensor of type $\binom{0}{N}$ are the values of the function when its arguments are the basis vectors $\{\vec{e}_\alpha\}$ of the frame $\mathcal{O}$.

Thus we have the notion of components as frame-dependent numbers (frame-dependent because the basis refers to a specific frame). For the metric tensor this gives the components as

$$\mathbf{g}(\vec{e}_\alpha, \vec{e}_\beta) = \vec{e}_\alpha \cdot \vec{e}_\beta = \eta_{\alpha\beta}. \tag{3.5}$$

So, the matrix $\eta_{\alpha\beta}$ that we introduced before is to be thought of as an array of the components of **g** in this basis. In another basis the components could be different. We will have many more examples of this later. First we will study a particularly important class of tensors.

3.3 The $\binom{0}{1}$ tensors: one-forms

A tensor of the type $\binom{0}{1}$ is called a covector, a covariant vector, or a one-form. Often these names are used interchangeably, even in a single textbook or reference.

General properties. Let an arbitrary one-form be called $\tilde{p}$. (We adopt the notation that a tilde above a symbol denotes a one-form, just as an arrow above a symbol denotes a vector.) Then $\tilde{p}$, supplied with a single vector argument, gives a real number: $\tilde{p}(\vec{A})$ is a real number. Suppose $\tilde{q}$ is another one-form. Then we can define

$$\left.\begin{aligned} \tilde{s} &= \tilde{p} + \tilde{q}, \\ \tilde{r} &= \alpha\tilde{p}, \end{aligned}\right\} \tag{3.6a}$$

to be the one-forms whose values for an argument $\vec{A}$ are

$$\left.\begin{aligned} \tilde{s}(\vec{A}) &= \tilde{p}(\vec{A}) + \tilde{q}(\vec{A}), \\ \tilde{r}(\vec{A}) &= \alpha\tilde{p}(\vec{A}). \end{aligned}\right\} \tag{3.6b}$$

With these rules, the set of all one-forms satisfies the axioms for a vector space, which accounts for their other names. This space is called the 'dual vector space' to distinguish it from the space of all vectors like $\vec{A}$.

When discussing vectors we relied heavily on components and their transformations. Let us look at those of $\tilde{p}$. The components of $\tilde{p}$ are written as p_α:

$$p_\alpha := \tilde{p}(\vec{e}_\alpha). \tag{3.7}$$

Any component with a single lower index is, by convention, the component of a one-form; an upper index denotes the component of a vector. In terms of components, $\tilde{p}(\vec{A})$ is

$$\tilde{p}(\vec{A}) = \tilde{p}(A^\alpha \vec{e}_\alpha)$$
$$= A^\alpha \tilde{p}(\vec{e}_\alpha),$$
$$\tilde{p}(\vec{A}) = A^\alpha p_\alpha. \tag{3.8}$$

The second step follows from the linearity which is at the heart of the definition we gave of a tensor. So, the real number $\tilde{p}(\vec{A})$ is easily found to be the sum $A^0 p_0 + A^1 p_1 + A^2 p_2 + A^3 p_3$. Notice that *all* terms have plus signs: this operation is called the *contraction* of $\vec{A}$ and $\tilde{p}$ and is *more* fundamental in tensor analysis than the scalar product because it can be performed between any one-form and vector without reference to other tensors. We have seen that two vectors cannot produce a scalar (their dot product) without the help of a third tensor, the metric.

The components of $\tilde{p}$ in a basis $\{\vec{e}_{\bar{\beta}}\}$ are

$$p_{\bar{\beta}} := \tilde{p}(\vec{e}_{\bar{\beta}}) = \tilde{p}(\Lambda^\alpha{}_{\bar{\beta}} \vec{e}_\alpha)$$
$$= \Lambda^\alpha{}_{\bar{\beta}} \tilde{p}(\vec{e}_\alpha) = \Lambda^\alpha{}_{\bar{\beta}} p_\alpha. \tag{3.9}$$

Comparing this with

$$\vec{e}_{\bar{\beta}} = \Lambda^\alpha{}_{\bar{\beta}} \, \vec{e}_\alpha,$$

we see that components of one-forms transform in exactly the same manner as basis vectors and in the opposite manner to components of vectors. By 'opposite', we mean using the inverse transformation. This use of the inverse guarantees that $A^\alpha p_\alpha$ is frame independent for any vector $\vec{A}$ and one-form $\tilde{p}$. This is such an important observation that we shall prove it explicitly:

$$A^{\bar{\alpha}} p_{\bar{\alpha}} = (\Lambda^{\bar{\alpha}}{}_\beta A^\beta)(\Lambda^\mu{}_{\bar{\alpha}} p_\mu), \tag{3.10a}$$
$$= \Lambda^\mu{}_{\bar{\alpha}} \Lambda^{\bar{\alpha}}{}_\beta A^\beta p_\mu, \tag{3.10b}$$
$$= \delta^\mu{}_\beta A^\beta p_\mu, \tag{3.10c}$$
$$= A^\beta p_\beta. \tag{3.10d}$$

(This is the same way in which the vector $A^\alpha \vec{e}_\alpha$ is kept frame independent.) This inverse transformation gives rise to the word 'dual' in 'dual vector space'. The property of transforming *with* basis vectors gives rise to the 'co' in 'covariant vector' and its shorter form 'covector'. Since components of ordinary vectors transform in an opposite way to basis vectors (in order to keep $A^\beta \vec{e}_\beta$ frame independent), they are often called 'contravariant' vectors. Most of these names are old-fashioned; 'vectors' and 'dual vectors' or 'one-forms' are the modern names. The reason that 'co' and 'contra' have been abandoned is that they mix up two very different things: the transformation of a basis is the expression of *new* vectors in terms of *old* ones; the transformation of components is the expression of the *same* object in terms of the new basis. It is important for you to be sure of these distinctions before proceeding further.

Basis one-forms. Since the set of all one-forms is a vector space, one can use any set of four linearly independent one-forms as a basis. (As with any vector space, one-forms are said to be linearly independent if no nontrivial linear combination equals the zero one-form. The zero one-form is the one-form whose value for any vector is zero.) However, in the previous section we have already used the basis vectors $\{\vec{e}_\alpha\}$ to define the components of a one-form. This suggests that we should be able to use the basis vectors to define an associated one-form basis $\{\tilde{\omega}^\alpha, \alpha = 0, \ldots, 3\}$, which we shall call the basis *dual* to $\{\vec{e}_\alpha\}$, in which a one-form has the components defined above. That is, we require a set $\{\tilde{\omega}^\alpha\}$ such that

$$\tilde{p} = p_\alpha \tilde{\omega}^\alpha. \tag{3.11}$$

(Notice that using a raised index on $\tilde{\omega}^\alpha$ permits the summation convention to operate.) The $\{\tilde{\omega}^\alpha\}$ are *four distinct* one-forms, just as the $\{\vec{e}_\alpha\}$ are four distinct vectors. This equation must imply Eq. 3.8 for any vector $\vec{A}$ and one-form $\tilde{p}$:

$$\tilde{p}(\vec{A}) = p_\alpha A^\alpha.$$

However, from Eq. 3.11 we get

$$\begin{aligned}
\tilde{p}(\vec{A}) &= p_\alpha \tilde{\omega}^\alpha(\vec{A}) \\
&= p_\alpha \tilde{\omega}^\alpha(A^\beta \vec{e}_\beta) \\
&= p_\alpha A^\beta \tilde{\omega}^\alpha(\vec{e}_\beta).
\end{aligned}$$

(Notice the use of β as an index in the second line, in order to distinguish its summation from the summation over α.) Now, this final line can only equal $p_\alpha A^\alpha$ for all A^β and p_α if

$$\tilde{\omega}^\alpha(\vec{e}_\beta) = \delta^\alpha{}_\beta. \tag{3.12}$$

Comparing with Eq. 3.7, we see that this equation gives the βth component of the αth basis one-form. It therefore *defines* the αth basis one-form. We can write out these components as

$$\tilde{\omega}^0 \underset{\mathcal{O}}{\rightarrow} (1, 0, 0, 0),$$

$$\tilde{\omega}^1 \underset{\mathcal{O}}{\rightarrow} (0, 1, 0, 0),$$

$$\tilde{\omega}^2 \underset{\mathcal{O}}{\rightarrow} (0, 0, 1, 0),$$

$$\tilde{\omega}^3 \underset{\mathcal{O}}{\rightarrow} (0, 0, 0, 1).$$

It is important to understand two points here. One is that Eq. 3.12 defines the basis $\{\tilde{\omega}^\alpha\}$ in terms of the basis $\{\vec{e}_\beta\}$. The vector basis induces a unique and convenient one-form basis. This is not the only possible one-form basis, but it is so useful to have the relationship Eq. 3.12 between the bases that we will always use it. The relationship Eq. 3.12 is between the two bases, not between individual pairs such as $\tilde{\omega}^0$ and $\vec{e}_0$. That is, if we change $\vec{e}_0$, while leaving $\vec{e}_1$, $\vec{e}_2$, and $\vec{e}_3$ unchanged, then in general this induces changes not only in $\tilde{\omega}^0$ but also in $\tilde{\omega}^1$, $\tilde{\omega}^2$, and $\tilde{\omega}^3$. The second point to understand is that, although we can describe both vectors and one-forms by giving a set of four components, their geometrical significance is very different. Do not lose sight of the fact that the components tell only part of the story. The basis contains the rest of the information. That is, a set of numbers $(0, 2, -1, 5)$ alone does not define anything; to make it into something, one must say whether these are components in a vector basis or a one-form basis and, indeed, which of the infinite number of possible bases is being used.

It remains to determine how $\{\tilde{\omega}^\alpha\}$ transforms under a change of basis. That is, each frame has its own unique set $\{\tilde{\omega}^\alpha\}$; how are those of two frames related? The derivation here is analogous to that for the basis vectors. It leads to the only equation one can write down with the indices in their correct positions:

$$\tilde{\omega}^{\bar{\alpha}} = \Lambda^{\bar{\alpha}}{}_\beta \tilde{\omega}^\beta. \tag{3.13}$$

This is the same as for the components of a vector, and opposite to that for the components of a one-form.

Picture of a one-form. For vectors we usually imagine an arrow, if we need a picture. It is helpful to have an image of a one-form as well. First of all, it is not an arrow. Its picture must reflect the fact that it maps vectors into real numbers. A vector itself does not automatically map another vector into a real number. To do this it needs a metric tensor to define the scalar product. With a different metric, the *same* two vectors will produce a *different* scalar product. So two vectors by themselves don't give a number. We need a picture of a one-form which doesn't depend on any other tensors having been defined. The picture generally used by mathematicians is shown in Figure 3.1. The one-form consists of a series of surfaces. The 'magnitude' of the one-form is given by the spacing between the surfaces: the larger the spacing the *smaller* the magnitude. In this picture, the number produced when a one-form acts on a vector is the number of surfaces that the arrow of the vector pierces. So the closer their spacing, the larger the number (compare (b) and (c) in Figure 3.1). In a four-dimensional space, the surfaces are three-dimensional.

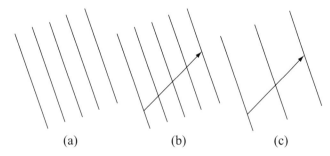

(a)   (b)   (c)

Figure 3.1   (a) The picture of a one-form complementary to that of a vector as an arrow. (b) The value of a one-form on a given vector is the number of surfaces that the vector arrow pierces. (c) The value of a smaller one-form on the same vector is a smaller number of surfaces. The larger the one-form, the more 'intense' the slicing of space in its picture.

The one-form doesn't define a unique direction, since it is not a vector. Rather, it defines a way of 'slicing' the space. In order to justify this picture we shall look at a particular one-form, the gradient.

3.4 Gradient of a function is a one-form

In elementary vector calculus, students are introduced to the gradient of a function; the gradient is described as a vector field. That description works well in Cartesian coordinates in Euclidean space, but it fails in more general situations, including in SR. In fact, the gradient is a one-form, and understanding why helps us to develop an intuitive understanding of one-forms.

Consider a scalar field $\phi(\vec{x})$ defined at every event $\vec{x}$. The world line of some particle (or person) encounters a value of ϕ at each event on it (see Figure 3.2), and this value changes from event to event. If we label (parametrize) each point on the curve by the value of proper time τ along it (i.e. the reading of a clock moving on the line), then we can express the coordinates of events on the curve as functions of τ:

$$t = t(\tau), \ x = x(\tau), \ y = y(\tau), \ z = z(\tau).$$

The four-velocity has components

$$\vec{U} \rightarrow \left(\frac{\mathrm{d}t}{\mathrm{d}\tau}, \frac{\mathrm{d}x}{\mathrm{d}\tau}, \ldots \right).$$

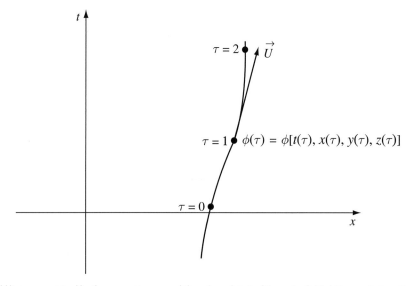

Figure 3.2 A world line parametrized by the proper time τ, and the values $\phi(\tau)$ of the scalar field $\phi(t, x, y, z)$ along it.

Since ϕ is a function of t, x, y, and z, it is implicitly a function of τ on the curve:

$$\phi(\tau) = \phi[t(\tau),\ x(\tau),\ y(\tau),\ z(\tau)],$$

and its rate of change on the curve is

$$\frac{d\phi}{d\tau} = \frac{\partial\phi}{\partial t}\frac{dt}{d\tau} + \frac{\partial\phi}{\partial x}\frac{dx}{d\tau} + \frac{\partial\phi}{\partial y}\frac{dy}{d\tau} + \frac{\partial\phi}{\partial z}\frac{dz}{d\tau}$$

$$= \frac{\partial\phi}{\partial t}U^t + \frac{\partial\phi}{\partial x}U^x + \frac{\partial\phi}{\partial y}U^y + \frac{\partial\phi}{\partial z}U^z. \tag{3.14}$$

It is clear from this that in the second line we have devised a means of producing from the vector $\vec{U}$ the number $d\phi/d\tau$ that represents the rate of change of ϕ on a curve on which $\vec{U}$ is the tangent. This number $d\phi/d\tau$ is clearly a linear function of $\vec{U}$, so we have defined a one-form.

By comparison with Eq. 3.8 we see that the components of this one-form are just the partial derivatives of ϕ with respect to the four coordinates. This is a familiar combination. The name of this one-form is the *gradient* of ϕ, denoted by $\tilde{d}\phi$:

$$\tilde{d}\phi \underset{\mathcal{O}}{\rightarrow} \left(\frac{\partial\phi}{\partial t},\ \frac{\partial\phi}{\partial x},\ \frac{\partial\phi}{\partial y},\ \frac{\partial\phi}{\partial z} \right). \tag{3.15}$$

It is clear that the gradient fits our definition of a one-form.

The gradient enables us to justify our picture of a one-form. In Figure 3.3 we have drawn part of a topographical map, showing contours of equal elevation. If h is the elevation, then the gradient $\tilde{d}h$ is clearly largest in an area like A, where the lines are closest together, and smallest near B, where the lines are spaced far apart. Moreover, suppose one wanted to know how much elevation a walk between two points would involve. One would lay out on the map a line (vector $\Delta\vec{x}$) between the points. Then the number of contours the line crossed would give the change in elevation. For example, line 1 crosses $1\frac{1}{2}$ contours, while line 2 crosses two contours. Line 3 starts near line 2 but goes in a different direction, winding up only half a contour higher. But these numbers are just Δh, which is the contraction of $\tilde{d}h$ with $\Delta\vec{x}$: Δh equals $\sum_i (\partial h/\partial x^i)\Delta x^i$, the *value* of $\tilde{d}h$ on $\Delta\vec{x}$ (see Eq. 3.8).

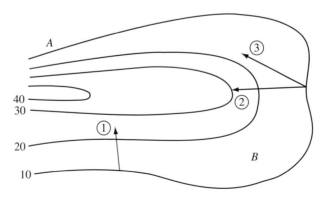

Figure 3.3 A topographical map giving local contours of constant elevation illustrates the gradient one-form. The change of height along any path, shown by an arrow, is the number of contours crossed by the arrow.

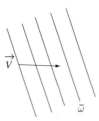

The value $\tilde{\omega}(\vec{V})$ is 2.5.

Therefore, a one-form is represented by a series of surfaces (Figure 3.4), and its contraction with a vector $\vec{V}$ is the number of surfaces that $\vec{V}$ crosses. The closer the surfaces, the larger is $\tilde{\omega}$. Properly, just as a vector is straight, the one-form's surfaces are straight and parallel. This is so because we are dealing with one-forms at a point, not over an extended region: 'tangent' one-forms, in the same sense as tangent vectors.

These pictures show why one in general *cannot* call a gradient a vector. One would like to identify the *vector* gradient as that vector pointing 'up' the slope, i.e. in such a way that it crosses the greatest number of contours per unit length. The key phrase is 'per unit length'. If there is a metric, a measure of distance in the space, then a vector *can* be associated with a gradient. But the metric must intervene here in order to produce a vector. Geometrically, on its own, the gradient is a one-form.

Let us be sure that Eq. 3.15 is a consistent definition. How do the components transform? For a one-form we must have

$$(\tilde{d}\phi)_{\bar{\alpha}} = \Lambda^\beta{}_{\bar{\alpha}} (\tilde{d}\phi)_\beta. \tag{3.16}$$

But we know how to transform partial derivatives:

$$\frac{\partial \phi}{\partial x^{\bar{\alpha}}} = \frac{\partial \phi}{\partial x^\beta} \frac{\partial x^\beta}{\partial x^{\bar{\alpha}}},$$

which means that

$$(\tilde{d}\phi)_{\bar{\alpha}} = \frac{\partial x^\beta}{\partial x^{\bar{\alpha}}} (\tilde{d}\phi)_\beta. \tag{3.17}$$

Are Eqs. 3.16 and 3.17 consistent? The answer, of course, is yes. The reason: since

$$x^\beta = \Lambda^\beta{}_{\bar{\alpha}} x^{\bar{\alpha}},$$

and since the $\Lambda^\beta{}_{\bar{\alpha}}$ are just constants, then

$$\partial x^\beta / \partial x^{\bar{\alpha}} = \Lambda^\beta{}_{\bar{\alpha}}. \tag{3.18}$$

This identity is fundamental. Components of the gradient transform according to the *inverse* of the components of vectors. So the gradient is the 'archetypal' one-form.

Notation for derivatives. From now on we shall employ the usual subscripted notation to indicate derivatives:

$$\frac{\partial \phi}{\partial x} := \phi_{,x}$$

and, more generally,

$$\frac{\partial \phi}{\partial x^\alpha} := \phi_{,\alpha}. \qquad (3.19)$$

Note that the index α appears as a superscript in the denominator of the left-hand side of Eq. 3.19 and as a subscript on the right-hand side. As we have seen, this placement of indices is consistent with the transformation properties of the expression.[1]

In particular, we have

$$x^\alpha{}_{,\beta} \equiv \delta^\alpha{}_\beta,$$

which we can compare with Eq. 3.12 to conclude that

$$\tilde{\mathrm{d}}x^\alpha := \tilde{\omega}^\alpha. \qquad (3.20)$$

This is a useful result, that the basis one-form is just $\tilde{\mathrm{d}}x^\alpha$. We can use it to write, for any function f,

$$\tilde{\mathrm{d}}f = \frac{\partial f}{\partial x^\alpha}\tilde{\mathrm{d}}x^\alpha.$$

This looks very much like the physicist's 'sloppy-calculus' way of writing differentials or infinitesimals. The notation $\tilde{\mathrm{d}}$ has been chosen partly to suggest this comparison, but this choice makes it doubly important to avoid confusion on this point. The object $\tilde{\mathrm{d}}f$ is a tensor, not a small increment in f; it *can* have a small ('infinitesimal') value if it is contracted with a small vector.

Normal one-forms. Like the gradient, the concept of a normal vector – a vector orthogonal to a surface – is one which is more naturally replaced by that of a normal one-form. For a normal vector to be defined we need to have a scalar product: the normal vector must be orthogonal to all vectors tangent to the surface. This can be defined only by using the metric tensor. But a normal one-form can be defined without reference to the metric. A one-form is said to be normal to a surface if its value is zero on every vector tangent to the surface. If the surface is closed and divides spacetime into an 'inside' and 'outside', a normal is said to be an *outward* normal one-form if it is a normal one-form and if its value on vectors which point outwards from the surface is positive. In Exercise 3.13 you are asked to prove that $\tilde{\mathrm{d}}f$ is normal to surfaces of constant f.

---

[1] Another commonly used notation that allows derivatives to be written more compactly is $\partial \phi / \partial x^\alpha := \partial_\alpha \phi$. We shall not use this notation in this book.

3.5 The $\binom{0}{2}$ tensors

These are tensors that have two vector arguments. We have encountered the metric tensor already, but the simplest tensor of this type is the product of two one-forms, formed according to the following rule: if $\tilde{p}$ and $\tilde{q}$ are one-forms then $\tilde{p} \otimes \tilde{q}$ is the $\binom{0}{2}$ tensor which, when supplied with vectors $\vec{A}$ and $\vec{B}$ as arguments, produces the number $\tilde{p}(\vec{A})\,\tilde{q}(\vec{B})$, i.e. the product of the numbers produced by the $\binom{0}{1}$ tensors. The symbol $\otimes$ is called an 'outer product sign' and is a formal notation to show how the $\binom{0}{2}$ tensor is formed from the one-forms. Notice that $\otimes$ is *not* commutative: $\tilde{p} \otimes \tilde{q}$ and $\tilde{q} \otimes \tilde{p}$ are *different* tensors. The first gives the value $\tilde{p}(\vec{A})\,\tilde{q}(\vec{B})$, the second the value $\tilde{q}(\vec{A})\,\tilde{p}(\vec{B})$.

Components. The most general $\binom{0}{2}$ tensor is not a simple outer product, but it can always be represented as a sum of outer products. To see this we must first consider the components of an arbitrary $\binom{0}{2}$ tensor $\mathbf{f}$:

$$f_{\alpha\beta} := \mathbf{f}(\vec{e}_\alpha, \vec{e}_\beta). \tag{3.21}$$

Since each index can have four values, there are 16 components, and they can be thought of as being arrayed in a matrix. The value of $\mathbf{f}$ on arbitrary vectors is

$$\mathbf{f}(\vec{A}, \vec{B}) = \mathbf{f}(A^\alpha \vec{e}_\alpha, B^\beta \vec{e}_\beta)$$

$$= A^\alpha B^\beta \mathbf{f}(\vec{e}_\alpha, \vec{e}_\beta)$$

$$= A^\alpha B^\beta f_{\alpha\beta}. \tag{3.22}$$

(Again notice that two different dummy indices are used to keep the different summations distinct.) Can we form a basis for these tensors? That is, can we define a set of 16 $\binom{0}{2}$ tensors $\tilde{\omega}^{\alpha\beta}$ such that, analogously to Eq. 3.11,

$$\mathbf{f} = f_{\alpha\beta} \tilde{\omega}^{\alpha\beta}? \tag{3.23}$$

For this to be the case we would have to have

$$f_{\mu\nu} = \mathbf{f}(\vec{e}_\mu, \vec{e}_\nu) = f_{\alpha\beta} \tilde{\omega}^{\alpha\beta}(\vec{e}_\mu, \vec{e}_\nu)$$

and this would imply, as before, that

$$\tilde{\omega}^{\alpha\beta}(\vec{e}_\mu, \vec{e}_\nu) = \delta^\alpha{}_\mu \delta^\beta{}_\nu. \tag{3.24}$$

But $\delta^\alpha{}_\mu$ is (by Eq. 3.12), the value of $\tilde{\omega}^\alpha$ on $\vec{e}_\mu$, and analogously for $\delta^\beta{}_\nu$. Therefore, $\tilde{\omega}^{\alpha\beta}$ is a tensor whose value is just the product of the values of two basis one-forms, and we therefore conclude that

$$\tilde{\omega}^{\alpha\beta} = \tilde{\omega}^\alpha \otimes \tilde{\omega}^\beta. \tag{3.25}$$

So, the tensors $\tilde{\omega}^\alpha \otimes \tilde{\omega}^\beta$ form a basis for all $\binom{0}{2}$ tensors, and we can write

$$\mathbf{f} = f_{\alpha\beta}\tilde{\omega}^\alpha \otimes \tilde{\omega}^\beta. \tag{3.26}$$

This is one way in which it can be seen that a general $\binom{0}{2}$ tensor is a sum over simple outer-product tensors.

Symmetries. A $\binom{0}{2}$ tensor takes two arguments, and their order is important, as we have seen. The behavior of the value of a tensor under an interchange of its arguments is an important property of the value. A tensor $\mathbf{f}$ is called *symmetric* if

$$\mathbf{f}(\vec{A}, \vec{B}) = \mathbf{f}(\vec{B}, \vec{A}), \quad \forall \vec{A}, \vec{B}. \tag{3.27}$$

Setting $\vec{A} = \vec{e}_\alpha$ and $\vec{B} = \vec{e}_\beta$, this implies for its components that

$$f_{\alpha\beta} = f_{\beta\alpha}. \tag{3.28}$$

This is the same as the condition that the matrix array of the elements is symmetric. From an arbitrary $\binom{0}{2}$ tensor $\mathbf{h}$ we can define a new symmetric $\mathbf{h}_{(s)}$ by the rule

$$\mathbf{h}_{(s)}(\vec{A}, \vec{B}) = \tfrac{1}{2}\mathbf{h}(\vec{A}, \vec{B}) + \tfrac{1}{2}\mathbf{h}(\vec{B}, \vec{A}). \tag{3.29}$$

Make sure you understand that $\mathbf{h}_{(s)}$ satisfies Eq. 3.27 above. For the components this implies

$$h_{(s)\alpha\beta} = \tfrac{1}{2}(h_{\alpha\beta} + h_{\beta\alpha}). \tag{3.30}$$

This is such an important mathematical property that a special notation is used for it:

$$h_{(\alpha\beta)} := \tfrac{1}{2}(h_{\alpha\beta} + h_{\beta\alpha}). \tag{3.31}$$

Therefore the numbers $h_{(\alpha\beta)}$ are the components of the symmetric tensor formed from $\mathbf{h}$.

Similarly, a tensor $\mathbf{f}$ is called *antisymmetric* if

$$\mathbf{f}(\vec{A}, \vec{B}) = -\mathbf{f}(\vec{B}, \vec{A}), \quad \forall \vec{A}, \vec{B}, \tag{3.32}$$

$$f_{\alpha\beta} = -f_{\beta\alpha}. \tag{3.33}$$

An antisymmetric $\binom{0}{2}$ tensor can always be formed:

$$\mathbf{h}_{(A)}(\vec{A}, \vec{B}) = \tfrac{1}{2}\mathbf{h}(\vec{A}, \vec{B}) - \tfrac{1}{2}\mathbf{h}(\vec{B}, \vec{A}),$$

$$h_{(A)\alpha\beta} = \tfrac{1}{2}(h_{\alpha\beta} - h_{\beta\alpha}).$$

The notation for this uses square brackets on the indices:

$$h_{[\alpha\beta]} = \tfrac{1}{2}(h_{\alpha\beta} - h_{\beta\alpha}). \tag{3.34}$$

Notice that

$$h_{\alpha\beta} = \tfrac{1}{2}(h_{\alpha\beta} + h_{\beta\alpha}) + \tfrac{1}{2}(h_{\alpha\beta} - h_{\beta\alpha})$$

$$= h_{(\alpha\beta)} + h_{[\alpha\beta]}. \tag{3.35}$$

Thus any $\binom{0}{2}$ tensor can be split *uniquely* into its symmetric and antisymmetric parts.

The metric tensor **g** is symmetric, as can be deduced from Eq. 3.26:

$$\mathbf{g}(\vec{A}, \vec{B}) = \mathbf{g}(\vec{B}, \vec{A}). \tag{3.36}$$

3.6 Metric as a mapping of vectors into one-forms

We now introduce what we shall later see is the fundamental role of the metric in differential geometry, to act as a mapping between vectors and one-forms. To see how this works, consider **g** and a single vector $\vec{V}$. Since **g** requires two vectorial arguments, the expression $\mathbf{g}(\vec{V}, \ )$ still lacks one: when another one is supplied, **g** becomes a number. Therefore $\mathbf{g}(\vec{V}, \ )$ considered as a function of vectors (which fill in the empty 'slot' in it), is linear and produces real numbers: a one-form. We call it $\tilde{V}$:

$$\mathbf{g}(\vec{V}, \ ) := \tilde{V}(\ ), \tag{3.37}$$

where the blanks inside the parentheses indicate that a vector argument is to be supplied. Then $\tilde{V}$ is the one-form whose value on a vector $\vec{A}$ is $\vec{V} \cdot \vec{A}$:

$$\tilde{V}(\vec{A}) := \mathbf{g}(\vec{V}, \vec{A}) = \vec{V} \cdot \vec{A}. \tag{3.38}$$

Note that, since **g** is symmetric, we also can write

$$\mathbf{g}(\ , \vec{V}) := \tilde{V}(\ ).$$

What are the components of $\tilde{V}$? They are

$$V_\alpha := \tilde{V}(\vec{e}_\alpha) = \vec{V} \cdot \vec{e}_\alpha = \vec{e}_\alpha \cdot \vec{V}$$

$$= \vec{e}_\alpha \cdot (V^\beta \vec{e}_\beta)$$

$$= (\vec{e}_\alpha \cdot \vec{e}_\beta) V^\beta,$$

$$V_\alpha = \eta_{\alpha\beta} V^\beta. \tag{3.39}$$

It is important to notice here that we distinguish the components V^α of $\vec{V}$ from the components V_β of $\tilde{V}$ *only* by the position of the index: on a vector it is up; on a one-form, down. Then, from Eq. 3.39, we have as a special case

$$V_0 = V^\beta \eta_{\beta 0} = V^0 \eta_{00} + V^1 \eta_{10} + \cdots$$
$$= V^0(-1) + 0 + 0 + 0$$
$$= -V^0, \tag{3.40}$$
$$V_1 = V^\beta \eta_{\beta 1} = V^0 \eta_{01} + V^1 \eta_{11} + \cdots$$
$$= +V^1, \tag{3.41}$$

etc. This may be summarized as:

$$\text{if } \vec{V} \to (a, b, c, d),$$
$$\text{then } \tilde{V} \to (-a, b, c, d). \tag{3.42}$$

The components of $\tilde{V}$ are obtained from those of $\vec{V}$ by changing the sign of the time component. (Since this feature depended upon the components $\eta_{\alpha\beta}$, in situations that we encounter later where the metric has more complicated components, the rule of correspondence between $\tilde{V}$ and $\vec{V}$ will also be more complicated.)

The inverse: going from $\tilde{A}$ to $\vec{A}$. Does the metric also provide a way of finding a vector $\vec{A}$ that is related to a given one-form $\tilde{A}$? The answer is yes. Consider Eq. 3.39. It says that $\{V_\alpha\}$ is obtained by multiplying $\{V^\beta\}$ by a matrix $(\eta_{\alpha\beta})$. If this matrix has an inverse, then one could use it to obtain $\{V^\beta\}$ from $\{V_\alpha\}$. This inverse exists if and only if $(\eta_{\alpha\beta})$ has nonvanishing determinant. But, since $(\eta_{\alpha\beta})$ is a diagonal matrix with entries $(-1, 1, 1, 1)$, its determinant is simply -1. An inverse does exist, and we call its components $\eta^{\alpha\beta}$. Then, given $\{A_\beta\}$ we can find $\{A^\alpha\}$:

$$A^\alpha := \eta^{\alpha\beta} A_\beta. \tag{3.43}$$

The use of the inverse guarantees that the two sets of components satisfy Eq. 3.39:

$$A_\beta = \eta_{\beta\alpha} A^\alpha.$$

So, the mapping provided by **g** between vectors and one-forms is one-to-one and invertible.

In particular, with $\tilde{\mathrm{d}}\phi$ we can associate a vector $\vec{\mathrm{d}}\phi$, which is the vector usually associated with the gradient. One can see that this vector is orthogonal to surfaces of constant ϕ as follows. Its inner product with a vector in a surface of constant ϕ is, by this mapping, identical with the value of the one-form $\tilde{\mathrm{d}}\phi$ on that vector. This, in turn, must be zero since $\tilde{\mathrm{d}}\phi(\vec{V})$ is the rate of change of ϕ along $\vec{V}$, which in this case is zero since $\vec{V}$ is taken to be in a surface of constant ϕ.

It is important to know $\{\eta^{\alpha\beta}\}$. You can easily verify that

$$\eta^{00} = -1, \quad \eta^{0i} = 0, \quad \eta^{ij} = \delta^{ij}, \tag{3.44}$$

so that $(\eta^{\alpha\beta})$ is *identical* to $(\eta_{\alpha\beta})$. Thus, to go from a one-form to a vector, simply change the sign of the time component.

Why distinguish one-forms from vectors? Considering Euclidean space, in Cartesian coordinates the metric is just $\{\delta_{ij}\}$, so the components of one-forms and vectors are the same. Therefore no distinction is ever made in elementary vector algebra. But in SR the components differ (by that one change in sign). Therefore, whereas the gradient has components

$$\tilde{d}\phi \rightarrow \left(\frac{\partial \phi}{\partial t}, \frac{\partial \phi}{\partial x}, \cdots \right),$$

the associated vector normal to surfaces of constant ϕ has components

$$\vec{d}\phi \rightarrow \left(-\frac{\partial \phi}{\partial t}, \frac{\partial \phi}{\partial x}, \cdots \right). \tag{3.45}$$

Had we simply tried to *define* the 'vector gradient' of a function as the vector with these components, without first discussing one-forms, you would have been justified in being more than a little skeptical. The nonEuclidean metric of SR forces us to be aware of the basic distinction between one-forms and vectors: it can't be swept under the rug.

As we remarked earlier, vectors and one-forms are dual to one another. Such dual spaces are important and are found elsewhere in mathematical physics. The simplest example is the space of column vectors in matrix algebra,

$$\begin{pmatrix} a \\ b \\ \vdots \end{pmatrix},$$

whose dual space is the space of row vectors $(a \ b \ \cdots)$. Notice that the product

$$(a \ b \ \cdots) \begin{pmatrix} p \\ q \\ \vdots \end{pmatrix} = ap + bq + \cdots \tag{3.46}$$

is a real number, so that a row vector can be considered to be a one-form on column vectors. The operation of finding an element of one space from an element of the other is called taking the 'adjoint' and is one-to-one and invertible. A less trivial example arises in quantum mechanics. A wave function (i.e. a probability amplitude that is a solution to Schrödinger's equation) is a complex scalar field $\psi(\vec{x})$, and is drawn from the *Hilbert space* of all such functions. This Hilbert space is a vector space, since its elements (functions) satisfy the axioms of a vector space. What is the corresponding dual space of one-forms? The crucial hint is that the inner product of any two functions $\phi(\vec{x})$ and $\psi(\vec{x})$ is *not* $\int \phi(\vec{x})\psi(\vec{x}) \, d^3 x$ but, rather, is $\int \phi^*(\vec{x})\psi(\vec{x}) \, d^3 x$, the asterisk denoting complex conjugation. The function $\phi^*(\vec{x})$ acts like a one-form whose value on $\psi(\vec{x})$ is its integral with it (analogous to the sum in Eq. 3.8). The operation of complex conjugation acts like our metric tensor, transforming a vector $\phi(\vec{x})$ (in the Hilbert space) into a one-form $\phi^*(\vec{x})$. The fact that $\phi^*(\vec{x})$ is also a function in the Hilbert space is, at this level, a distraction. (It is equivalent to saying that members of the set $(1, -1, 0, 0)$ can be components of either a vector or a one-form.) The important point is that in the integral $\int \phi^*(\vec{x})\psi(\vec{x}) \, d^3 x$, the function $\phi^*(\vec{x})$ is acting as a one-form, producing a (complex) number from the vector $\psi(\vec{x})$. This dualism is most clearly brought out in the Dirac 'bra' and 'ket' notation. Elements of the space of all states of the system are called $| \ \rangle$ (with identifying labels written inside), while the elements of the dual (adjoint with complex conjugate) space are called $\langle$ $|$. Two 'vectors' $|1\rangle$ and $|2\rangle$ do not form a number, but a vector and a dual vector $|1\rangle$ and $\langle 2|$ do: $\langle 2|1\rangle$ is how this number is written.

In such ways the concept of a dual vector space arises very frequently in advanced mathematical physics.

Magnitudes and scalar products of one-forms. A one-form $\tilde{p}$ is defined to have the same magnitude as its associated vector $\vec{p}$. Thus we write

$$\tilde{p}^2 = \vec{p}^2 = \eta_{\alpha\beta} p^\alpha p^\beta. \tag{3.47}$$

This would seem to involve finding $\{p^\alpha\}$ from $\{p_\alpha\}$ before we can use Eq. 3.47, but we can easily get around this. We use Eq. 3.43 for both p^α and p^β in Eq. 3.47:

$$\tilde{p}^2 = \eta_{\alpha\beta} (\eta^{\alpha\mu} p_\mu)(\eta^{\beta\nu} p_\nu). \tag{3.48}$$

(Notice that each independent summation uses a different dummy index.) But since the matrices $\eta_{\alpha\beta}$ and $\eta^{\beta\nu}$ are inverse to each other, the sum on β collapses:

$$\eta_{\alpha\beta} \eta^{\beta\nu} = \delta^\nu{}_\alpha. \tag{3.49}$$

Using this in Eq. 3.48 gives

$$\tilde{p}^2 = \eta^{\alpha\mu} p_\mu p_\alpha. \tag{3.50}$$

Thus, the inverse metric tensor can be used directly to find the magnitude of $\tilde{p}$ from its components. We can use Eq. 3.44 to write this explicitly as

$$\tilde{p}^2 = -(p_0)^2 + (p_1)^2 + (p_2)^2 + (p_3)^2. \tag{3.51}$$

This is the same rule, in fact, as Eq. 2.24 for vectors. By its definition, it is frame invariant. One-forms are timelike, spacelike, or null, as their associated vectors are.

As with vectors, we can now define an inner product of one-forms. This is

$$\tilde{p} \cdot \tilde{q} := \tfrac{1}{2} \left[(\tilde{p} + \tilde{q})^2 - \tilde{p}^2 - \tilde{q}^2 \right]. \tag{3.52}$$

Its expression in terms of components is, not surprisingly,

$$\tilde{p} \cdot \tilde{q} = -p_0 q_0 + p_1 q_1 + p_2 q_2 + p_3 q_3. \tag{3.53}$$

Normal vectors and unit normal one-forms. A vector is said to be normal to a surface if its associated one-form is a normal one-form. Equation 3.38 shows that this definition is equivalent to the usual one: that the vector must be orthogonal to all tangent vectors. A normal vector or one-form is said to be a *unit normal* if its magnitude is ± 1. (We cannot demand that it be $+1$, since timelike vectors will have negative magnitudes. All we can do is to multiply the vector or form by an overall factor to scale its magnitude to ± 1.) Note that null normals cannot be unit normals.

A three-dimensional surface is said to be timelike, spacelike, or null according to which of these classes its normal falls into. Exercise 3.12 proves that this definition is self-consistent. In Exercise 3.21, we explore the following curious properties that normal vectors have on account of our metric. An outward normal vector is the vector associated with an outward normal one-form, as defined earlier. This ensures that its scalar product with any vector which points outwards is positive. If the surface is spacelike, the outward normal vector points outwards. If the surface is timelike, however, the outward normal vector points *inwards*. And if the surface is null, the outward vector is *tangent* to the surface! These peculiarities simply reinforce the view that it is more natural to regard the normal as a one-form, where the metric doesn't enter the definition.

3.7 Finally: $\binom{M}{N}$ tensors

Vector as a function of one-forms. The dualism discussed above is in fact complete. Although we defined one-forms as functions of vectors, we can now see that vectors can perfectly well be regarded as linear functions that map one-forms into real numbers. Given a vector $\vec{V}$, once we supply a one-form we get a real number:

$$\vec{V}(\tilde{p}) := \tilde{p}(\vec{V}) = p_\alpha V^\alpha := \langle \tilde{p}, \vec{V} \rangle. \tag{3.54}$$

In this way we dethrone vectors from their special position as things 'acted on' by tensors, and regard them as tensors themselves, specifically as linear functions of single one-forms into real numbers. The last notation on Eq. 3.54 is new, and emphasizes the equal status of the two objects.

$\binom{M}{0}$ *Tensors.* Generalizing this, we make the following definition:

An $\binom{M}{0}$ tensor is a linear function of M one-forms into the real numbers.

All our previous discussions of $\binom{0}{N}$ tensors apply here. A simple $\binom{2}{0}$ tensor is $\vec{V} \otimes \vec{W}$, which, when supplied with two arguments $\tilde{p}$ and $\tilde{q}$, gives the number $\vec{V}(\tilde{p})\vec{W}(\tilde{q}) := \tilde{p}(\vec{V})\tilde{q}(\vec{W}) = V^\alpha p_\alpha W^\beta q_\beta$. So $\vec{V} \otimes \vec{W}$ has components $V^\alpha W^\beta$. A basis for $\binom{2}{0}$ tensors is $\vec{e}_\alpha \otimes \vec{e}_\beta$. The components of an $\binom{M}{0}$ tensor are its values when the basis one-forms $\tilde{\omega}^\alpha$ are its arguments. Notice that $\binom{M}{0}$ tensors have components all of whose indices are superscripts.

$\binom{M}{N}$ *tensors.* The final generalization is as follows:

An $\binom{M}{N}$ tensor is a linear function of M one-forms *and* N vectors into the real numbers.

For instance, if **R** is a $\binom{1}{1}$ tensor then it requires a one-form $\tilde{p}$ and a vector $\vec{A}$ to give a number $\mathbf{R}(\tilde{p}; \vec{A})$. It has components $\mathbf{R}(\tilde{\omega}^\alpha; \vec{e}_\beta) := R^\alpha{}_\beta$. In general, the components of a $\binom{M}{N}$ tensor will have M indices up and N down. In a new frame,

$$R^{\bar{\alpha}}{}_{\bar{\beta}} = \mathbf{R}(\tilde{\omega}^{\bar{\alpha}}; \vec{e}_{\bar{\beta}})$$
$$= \mathbf{R}(\Lambda^{\bar{\alpha}}{}_{\mu}\tilde{\omega}^{\mu}; \Lambda^{\nu}{}_{\bar{\beta}}\vec{e}_{\nu})$$
$$= \Lambda^{\bar{\alpha}}{}_{\mu}\Lambda^{\nu}{}_{\bar{\beta}}\mathbf{R}^{\mu}{}_{\nu}. \tag{3.55}$$

So, the transformation of components is simple: each index transforms by bringing in a Λ whose indices are arranged in the only way permitted by the summation convention. As mentioned earlier, some old names that are still in current use are as follows: upper indices are called 'contravariant' (because they transform in a *contrary* way to basis vectors) and lower ones 'covariant'. An $\binom{M}{N}$ tensor is said to be 'M-times contravariant and N-times covariant'.

Circular reasoning? At this point you might begin to worry that all of tensor algebra has become circular: one-forms were defined in terms of vectors, but now we have defined vectors in terms of one-forms. This 'duality' is at the heart of the theory, but is not circularity. It means we can do as physicists do, which is to identify the vectors with displacements $\Delta\vec{x}$ and things like it (such as $\vec{p}$ and $\vec{v}$) and then generate all $\binom{M}{N}$ tensors by the rules of tensor algebra; these tensors inherit a physical meaning from the original meaning we gave vectors. But we could equally well have associated one-forms with some physical objects (gradients, for example) and recovered the whole algebra from that starting point. The power of the mathematics is that it doesn't need (or want) to say *what* the original vectors or one-forms are. It simply gives rules for manipulating them. The association of, say, $\vec{p}$ with a vector is at the interface between physics and mathematics: it is how we make a mathematical model of the physical world. A geometer does the same, adding to the notion of these abstract tensor spaces the idea of what a vector in a curved space is. The modern geometer's idea of a vector is something we shall learn about when we come to curved spaces. For now we will get some practice with tensors in physical situations, where we stick with our (admittedly imprecise) notion of vectors 'like' $\Delta\vec{x}$.

3.8 Index 'raising' and 'lowering'

In the same way that the metric maps a vector $\vec{V}$ into a one-form $\tilde{V}$, it maps an $\binom{N}{M}$ tensor into an $\binom{N-1}{M+1}$ tensor. Similarly, the inverse maps an $\binom{N}{M}$ tensor into an $\binom{N+1}{M-1}$ tensor. Normally these are given the same name, and are distinguished only by the positions of their indices. Suppose that $T^{\alpha\beta}{}_{\gamma}$ are the components of a $\binom{2}{1}$ tensor. Then

$$T^{\alpha}{}_{\beta\gamma} := \eta_{\beta\mu}T^{\alpha\mu}{}_{\gamma} \tag{3.56}$$

are the components of a $\binom{1}{2}$ tensor (obtained by mapping the *second* one-form argument of $T^{\alpha\beta}{}_{\gamma}$ into a vector), and

$$T_{\alpha}{}^{\beta}{}_{\gamma} := \eta_{\alpha\mu}T^{\mu\beta}{}_{\gamma} \tag{3.57}$$

are the components of another (inequivalent) $\binom{1}{2}$ tensor (mapping on the *first* index), while

$$T^{\alpha\beta\gamma} := \eta^{\gamma\mu}T^{\alpha\beta}{}_{\mu} \tag{3.58}$$

are the components of a $\binom{3}{0}$ tensor. These operations are, naturally enough, called index raising and lowering. Whenever we speak of raising or lowering an index we mean this map generated by the metric. The rule in SR is simple: when raising or lowering a '0' index, the sign of the component changes; when raising or lowering a '1' or '2' or '3' index (in general, an 'i' index) the sign of the component is unchanged.

Mixed components of metric. The numbers $\{\eta_{\alpha\beta}\}$ are the components of the metric, and $\{\eta^{\alpha\beta}\}$ those of its inverse. Suppose we raise an index of $\eta_{\alpha\beta}$ using the inverse. Then we get 'mixed' components of the metric,

$$\eta^{\alpha}{}_{\beta} \equiv \eta^{\alpha\mu}\eta_{\mu\beta}. \tag{3.59}$$

But on the right we have just the matrix product of two matrices that are the inverse of each other (readers who aren't sure of this should verify the following equation by direct calculation), so it is the unit identity matrix. Since one index is up and one down, it is the Kronecker delta, written as

$$\eta^{\alpha}{}_{\beta} \equiv \delta^{\alpha}{}_{\beta}. \tag{3.60}$$

By raising the other index we merely obtain an identity, $\eta^{\alpha\beta} = \eta^{\alpha\beta}$. So, we can regard $\eta^{\alpha\beta}$ as the components of the $\binom{2}{0}$ tensor which is mapped from the $\binom{0}{2}$ tensor $\mathbf{g}$ by $\mathbf{g}^{-1}$. Thus, for $\mathbf{g}$, its 'contravariant' components equal the elements of the matrix inverse of the matrix of its 'covariant' components. It is the only tensor for which this is true.

Metric and nonmetric vector algebras. It is of some interest to ask why it is the metric tensor that generates the correspondence between one-forms and vectors. Why not some other $\binom{0}{2}$ tensor that has an inverse? We will explore that idea in stages.

First, why is there a correspondence at all? Suppose we had a 'nonmetric' vector algebra, complete with all the dual spaces and $\binom{M}{N}$ tensors. Why make a correspondence between one-forms and vectors? The answer is that sometimes one needs to do this and sometimes one does not. Without such a correspondence, the inner product of two vectors is undefined, since numbers are produced only when one-forms act on vectors and vice versa. In physics scalar products are useful, so one needs a metric. But there are *some* vector spaces in mathematical physics where metrics are not important. An example is the phase space of classical and quantum mechanics.

Second, why the metric and not another tensor? If a metric were not defined but another symmetric tensor did the mapping, a mathematician would just call the other tensor the metric. That is, the mathematician would define it as the tensor generating a mapping. To a mathematician, the metric is an added bit of *structure* in the vector algebra. Different spaces in mathematics can have different metric structures. A *Riemannian* space is characterized by a metric that gives positive-definite magnitudes of vectors. One like ours, with indefinite sign, is called *pseudo-Riemannian*. One can even define a 'metric' that is *antisymmetric*: a two-dimensional space called *spinor space* has such a metric, and it turns out to be of fundamental importance in physics. But its structure is outside the scope of our lectures. The point here is that we do not end up with SR if we just discuss vectors and tensors. We get SR when we say that we have a metric with the particular components $\eta_{\alpha\beta}$. If we assigned other components we might get other spaces, in particular the curved spacetime of GR.

3.9 Differentiation of tensors

A function f is a $\binom{0}{0}$ tensor, and its gradient $\tilde{d}f$ is a $\binom{0}{1}$ tensor. Differentiation of a function produces a tensor of one higher (covariant) rank. We shall now see that this applies as well to the differentiation of tensors of *any* rank.

Consider a $\binom{1}{1}$ tensor **T** whose components $\{T^\alpha{}_\beta\}$ are functions of position. We can write **T** as

$$\mathbf{T} = T^\alpha{}_\beta \tilde{\omega}^\beta \otimes \vec{e}_\alpha. \tag{3.61}$$

Suppose, as we did for functions, that we move along a world line with parameter τ, the proper time. The rate of change of **T**,

$$\frac{d\mathbf{T}}{d\tau} = \lim_{\Delta\tau \to 0} \frac{\mathbf{T}(\tau + \Delta\tau) - \mathbf{T}(\tau)}{\Delta\tau}, \tag{3.62}$$

is not hard to calculate. Since the basis one-forms and vectors are the same everywhere (i.e. $\tilde{\omega}^\alpha(\tau + \Delta\tau) = \tilde{\omega}^\alpha(\tau)$), it follows that

$$\frac{d\mathbf{T}}{d\tau} = \left(\frac{dT^\alpha{}_\beta}{d\tau}\right)\tilde{\omega}^\beta \otimes \vec{e}_\alpha, \tag{3.63}$$

where $dT^\alpha{}_\beta/d\tau$ is the ordinary derivative of the function $T^\alpha{}_\beta$ along the world line. Using the chain rule, we can express this derivative in terms of derivatives with respect to the coordinates:

$$dT^\alpha{}_\beta/d\tau = (\partial T^\alpha{}_\beta/\partial x^\gamma)\,(dx^\gamma/d\tau) = T^\alpha{}_{\beta,\gamma}\,U^\gamma. \tag{3.64}$$

Now, the object $d\mathbf{T}/d\tau$ is a $\binom{1}{1}$ tensor, since in Eq. 3.62 it is defined to be just the difference between two such tensors. From Eqs. 3.63 and 3.64 we have, for any vector $\vec{U}$,

$$d\mathbf{T}/d\tau = (T^\alpha{}_{\beta,\gamma}\tilde{\omega}^\beta \otimes \vec{e}_\alpha)\,U^\gamma, \tag{3.65}$$

from which we can deduce that

$$\nabla\mathbf{T} := (T^\alpha{}_{\beta,\gamma}\tilde{\omega}^\beta \otimes \tilde{\omega}^\gamma \otimes \vec{e}_\alpha) \tag{3.66}$$

is a $\binom{1}{2}$ tensor, called the *gradient of* **T**. To see why this is so, it helps to use our earlier device of creating a list of arguments for a tensor. If $\nabla\mathbf{T}$ is a $\binom{1}{2}$ tensor, then we can write it as

$$\nabla \mathbf{T}(\quad , \quad , \quad),$$

with spaces for the three arguments. From the way in which its basis one-forms and vectors are associated with indices in Eq. 3.66, the first argument is a one-form and the second and third are vectors. If we let it operate on the vector $\vec{U}$ in the second argument position, then we can write

$$\nabla \mathbf{T}(\quad , \vec{U}, \quad).$$

For a given fixed vector $\vec{U}$, the gradient vector is now a linear function of a one-form and a vector, so it is a $\binom{1}{1}$ tensor. If we put the basis one-form $\tilde{\omega}^\beta$ and the basis vector $\vec{e}_\alpha$ into the two empty arguments, we get the components of the gradient, $T^\alpha{}_{\beta,\gamma}U^\gamma$. So the $\binom{1}{1}$ tensor we created by fixing the argument $\vec{U}$ is just $d\mathbf{T}/d\tau$, as defined in Eq. 3.65. This proves our initial assumption, that $\nabla \mathbf{T}$ is indeed a $\binom{1}{2}$ tensor.

We use the notation $\nabla \mathbf{T}$ rather than $\tilde{d}\mathbf{T}$ because the latter notation is usually reserved by mathematicians for something else. We then have a convenient notation for Eq. 3.65 and its components in Eq. 3.64:

$$d\mathbf{T}/d\tau = \nabla_{\vec{U}} \mathbf{T}, \tag{3.67}$$

$$\nabla_{\vec{U}} \mathbf{T} \rightarrow \left\{ T^\alpha{}_{\beta,\gamma} U^\gamma \right\}. \tag{3.68}$$

This derivation makes use of the fact that the basis vectors (and therefore the basis one-forms) are constant everywhere. We will find that we can't assume this in the curved spacetime of GR, and taking this into account will be our entry point into GR theory !

3.10 Bibliography

Our approach to tensor analysis stresses the geometrical nature of tensors rather than the transformation properties of their components. Students who wish amplification of some of the points here can consult the early chapters of Misner *et al.* (1973) or Schutz (1980b). See also Bishop & Goldberg (1981).

Most introductions to tensors for physicists outside relativity confine themselves to 'Cartesian' tensors, i.e. to tensor components in three-dimensional Cartesian coordinates. See, for example, Bourne & Kendall (1992).

Exercises

3.1   (a) Given an arbitrary set of numbers $\{M_{\alpha\beta}; \alpha = 0, \ldots, 3; \beta = 0, \ldots, 3\}$ and two arbitrary sets of vector components $\{A^\mu; \mu = 0, \ldots, 3\}$ and $\{B^\nu; \nu = 0, \ldots, 3\}$, show that the two expressions

$$M_{\alpha\beta}A^{\alpha}B^{\beta} := \sum_{\alpha=0}^{3}\sum_{\beta=0}^{3} M_{\alpha\beta}A^{\alpha}B^{\beta}$$

and

$$\sum_{\alpha=0}^{3} M_{\alpha\alpha}A^{\alpha}B^{\alpha}$$

are not equivalent.

(b) Show that

$$A^{\alpha}B^{\beta}\eta_{\alpha\beta} = -A^{0}B^{0} + A^{1}B^{1} + A^{2}B^{2} + A^{3}B^{3}.$$

3.2   Prove that the set of all one-forms is a vector space.

3.3   (a) Prove, by writing out all the terms, the validity of the following:

$$\tilde{p}(A^{\alpha}\vec{e}_{\alpha}) = A^{\alpha}\tilde{p}(\vec{e}_{\alpha}).$$

(b) Let the components of $\tilde{p}$ be $(-1, 1, 2, 0)$, those of $\vec{A}$ be $(2, 1, 0, -1)$ and those of $\vec{B}$ be $(0, 2, 0, 0)$. Find (i) $\tilde{p}(\vec{A})$; (ii) $\tilde{p}(\vec{B})$; (iii) $\tilde{p}(\vec{A} - 3\vec{B})$; (iv) $\tilde{p}(\vec{A}) - 3\tilde{p}(\vec{B})$.

3.4   Given the following vectors in $\mathcal{O}$:

$$\vec{A} \underset{\mathcal{O}}{\to} (2, 1, 1, 0), \quad \vec{B} \underset{\mathcal{O}}{\to} (1, 2, 0, 0), \quad \vec{C} \underset{\mathcal{O}}{\to} (0, 0, 1, 1), \quad \vec{D} \underset{\mathcal{O}}{\to} (-3, 2, 0, 0),$$

(a) show that they are linearly independent;

(b) find the components of $\tilde{p}$ if

$$\tilde{p}(\vec{A}) = 1, \quad \tilde{p}(\vec{B}) = -1, \quad \tilde{p}(\vec{C}) = -1, \quad \tilde{p}(\vec{D}) = 0;$$

(c) find the value of $\tilde{p}(\vec{E})$ for

$$\vec{E} \underset{\mathcal{O}}{\to} (1, 1, 0, 0);$$

(d) determine whether the one-forms $\tilde{p}, \tilde{q}, \tilde{r}$, and $\tilde{s}$ are linearly independent if $\tilde{q}(\vec{A}) = \tilde{q}(\vec{B}) = 0$, $\tilde{q}(\vec{C}) = 1$, $\tilde{q}(\vec{D}) = -1$, $\tilde{r}(\vec{A}) = 2$, $\tilde{r}(\vec{B}) = \tilde{r}(\vec{C}) = r(\vec{D}) = 0$, $\tilde{s}(\vec{A}) = -1$, $\tilde{s}(\vec{B}) = -1$, $\tilde{s}(\vec{C}) = \tilde{s}(\vec{D}) = 0$.

3.5   Justify each step leading from Eq. 3.10a to Eq. 3.10d.

3.6   Consider the basis $\{\vec{e}_{\alpha}\}$ of a frame $\mathcal{O}$ and the basis $(\tilde{\lambda}^{0}, \tilde{\lambda}^{1}, \tilde{\lambda}^{2}, \tilde{\lambda}^{3})$ for the space of one-forms, where we have

$$\tilde{\lambda}^{0} \underset{\mathcal{O}}{\to} (1, 1, 0, 0),$$

$$\tilde{\lambda}^{1} \underset{\mathcal{O}}{\to} (1, -1, 0, 0),$$

$$\tilde{\lambda}^{2} \underset{\mathcal{O}}{\to} (0, 0, 1, -1),$$

$$\tilde{\lambda}^{3} \underset{\mathcal{O}}{\to} (0, 0, 1, 1).$$

Note that $\{\tilde{\lambda}^{\beta}\}$ is *not* the basis dual to $\{\vec{e}_{\alpha}\}$.

(a) Show that $\tilde{p} \neq \tilde{p}(\vec{e}_{\alpha})\tilde{\lambda}^{\alpha}$ for arbitrary $\tilde{p}$.

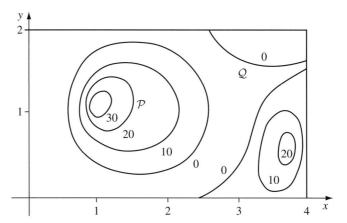

Figure 3.5   Isotherms of an irregularly heated plate.

(b) Let $\tilde{p} \rightarrow_{\mathcal{O}} (1, 1, 1, 1)$. Find numbers l_α such that

$$\tilde{p} = l_\alpha \tilde{\lambda}^\alpha.$$

These are the components of $\tilde{p}$ on $\{\tilde{\lambda}^\alpha\}$, which is to say that they are the values of $\tilde{p}$ on the elements of the vector basis dual to $\{\tilde{\lambda}^\alpha\}$.

3.7 Prove Eq. 3.13.

3.8 Draw the basis one-forms $\tilde{d}t$ and $\tilde{d}x$ of a frame $\mathcal{O}$.

3.9 Figure 3.5 shows curves of equal temperature T (isotherms) on a metal plate. At the points $\mathcal{P}$ and $\mathcal{Q}$ as shown, estimate the components of the gradient $\tilde{d}T$. (Hint: the components are the contractions with the basis vectors, which can be estimated by counting the number of isotherms crossed by the vectors.)

3.10 (a) Given a frame $\mathcal{O}$ whose coordinates are $\{x^\alpha\}$, show that

$$\partial x^\alpha / \partial x^\beta = \delta^\alpha{}_\beta.$$

(b) For any two frames, we have, using Eq. 3.18:

$$\partial x^\beta / \partial x^{\tilde{\alpha}} = \Lambda^\beta{}_{\tilde{\alpha}}.$$

Show that (a) and the chain rule imply

$$\Lambda^\beta{}_{\tilde{\alpha}} \Lambda^{\tilde{\alpha}}{}_\mu = \delta^\beta{}_\mu.$$

This is the inverse property again.

3.11 Use the notation $\partial \phi / \partial x^\alpha = \phi_{,\alpha}$ to rewrite Eqs. 3.14, 3.15, and 3.18.

3.12 Let S be the two-dimensional plane $x = 0$ in three-dimensional Euclidean space. Let $\tilde{n} \neq 0$ be a normal one-form to S.

(a) Show that if $\vec{V}$ is a vector which is not tangent to S, then $\tilde{n}(\vec{V}) \neq 0$.

(b) Show that if $\tilde{n}(\vec{V}) > 0$, then $\tilde{n}(\vec{W}) > 0$ for any $\vec{W}$ which points toward the same side of S as $\vec{V}$ does (i.e. any $\vec{W}$ whose x component has the same sign as V^x).

(c) Show that any normal to S is a multiple of $\tilde{n}$.

(d) Generalize these statements to an arbitrary three-dimensional surface in four-dimensional spacetime.

3.13 Prove, by geometric or algebraic arguments, that $\tilde{d}f$ is normal to surfaces of constant f.

3.14 Let $\tilde{p} \to_{\mathcal{O}} (1,1,0,0)$ and $\tilde{q} \to_{\mathcal{O}} (-1,0,1,0)$ be two one-forms. Prove, by taking two vectors $\vec{A}$ and $\vec{B}$ as arguments, that $\tilde{p} \otimes \tilde{q} \neq \tilde{q} \otimes \tilde{p}$. Then find the components of $\tilde{p} \otimes \tilde{q}$.

3.15 Supply the reasoning leading from Eq. 3.23 to Eq. 3.24.

3.16 (a) Prove that $\mathbf{h}_{(s)}$ defined by

$$\mathbf{h}_{(s)}(\vec{A}, \vec{B}) = \tfrac{1}{2}\mathbf{h}(\vec{A}, \vec{B}) + \tfrac{1}{2}\mathbf{h}(\vec{B}, \vec{A}) \tag{3.69}$$

is a symmetric tensor.

(b) Prove that $\mathbf{h}_{(A)}$ defined by

$$\mathbf{h}_{(A)}(\vec{A}, \vec{B}) = \tfrac{1}{2}\mathbf{h}(\vec{A}, \vec{B}) - \tfrac{1}{2}\mathbf{h}(\vec{B}, \vec{A}) \tag{3.70}$$

is an antisymmetric tensor.

(c) Find the components of the symmetric and antisymmetric parts of $\tilde{p} \otimes \tilde{q}$ defined in Exercise 3.14.

(d) Prove that if $\mathbf{h}$ is an antisymmetric $\binom{0}{2}$ tensor then

$$\mathbf{h}(\vec{A}, \vec{A}) = 0$$

for any vector $\vec{A}$.

(e) Find the number of independent components of $\mathbf{h}_{(s)}$ and $\mathbf{h}_{(A)}$.

3.17 (a) Suppose that $\mathbf{h}$ is a $\binom{0}{2}$ tensor with the property that, for *any* two vectors $\vec{A}$ and $\vec{B}$ (where $\vec{B} \neq 0$)

$$\mathbf{h}(\ ,\vec{A}) = \alpha\mathbf{h}(\ ,\vec{B}),$$

where α is a number which may depend on $\vec{A}$ and $\vec{B}$. Show that there exist one-forms $\tilde{p}$ and $\tilde{q}$ such that

$$\mathbf{h} = \tilde{p} \otimes \tilde{q}.$$

(b) Suppose that $\mathbf{T}$ is a $\binom{1}{1}$ tensor, $\tilde{\omega}$ a one-form, $\vec{v}$ a vector, and $\mathbf{T}(\tilde{\omega}; \vec{v})$ the value of $\mathbf{T}$ on $\tilde{\omega}$ and $\vec{v}$. Prove that $\mathbf{T}(\ ; \vec{v})$ is a vector and $\mathbf{T}(\tilde{\omega}; \ )$ is a one-form, i.e. that a $\binom{1}{1}$ tensor provides a map of vectors to vectors and one-forms to one-forms.

3.18 (a) Find the one-forms mapped by the metric tensor from the vectors

$$\vec{A} \to_{\mathcal{O}} (1,0,-1,0), \qquad \vec{B} \to_{\mathcal{O}} (0,1,1,0),$$
$$\vec{C} \to_{\mathcal{O}} (-1,0,-1,0), \qquad \vec{D} \to_{\mathcal{O}} (0,0,1,1).$$

(b) Find the vectors mapped by the inverse of the metric tensor from the one-forms $\tilde{p} \to_{\mathcal{O}} (3,0,-1,-1)$, $\tilde{q} \to_{\mathcal{O}} (1,-1,1,1)$, $\tilde{r} \to_{\mathcal{O}} (0,-5,-1,0)$, $\tilde{s} \to_{\mathcal{O}} (-2,1,0,0)$.

3.19 (a) Prove that the matrix $\{\eta^{\alpha\beta}\}$ is inverse to $\{\eta_{\alpha\beta}\}$ by performing matrix multiplication.

(b) Derive Eq. 3.53.

3.20 In Euclidean three-space in Cartesian coordinates, one does not normally distinguish between vectors and one-forms, because their components transform identically. Prove this in two steps.

(a) Show that

$$A^{\bar{\alpha}} = \Lambda^{\bar{\alpha}}{}_{\beta} A^{\beta}$$

and

$$P_{\bar{\beta}} = \Lambda^{\alpha}{}_{\bar{\beta}} P_{\alpha}$$

are the same transformation if the matrix $\{\Lambda^{\bar{\alpha}}{}_{\beta}\}$ equals the transpose of its inverse. Such a matrix is said to be *orthogonal*.

(b) The metric of such a space has components $\{\delta_{ij}; i, j = 1, \ldots, 3\}$. Prove that a transformation from one Cartesian coordinate system to another must obey

$$\delta_{\bar{i}\bar{j}} = \Lambda^{k}{}_{\bar{i}} \Lambda^{l}{}_{\bar{j}} \delta_{kl}$$

and that this implies that $\{\Lambda^{k}{}_{\bar{i}}\}$ is an orthogonal matrix. See Exercise 3.32 for the analog of this in SR.

3.21 (a) Let a region of the t–x plane be bounded by the lines $t = 0$, $t = 1$, $x = 0$, $x = 1$. Within the t–x plane, find the unit outward normal one-forms and their associated vectors for each of the boundary lines.

(b) Let another region be bounded by the straight lines joining the events whose coordinates are $(1, 0)$, $(1, 1)$, and $(2, 1)$. Find an outward normal for the null boundary and find its associated vector.

3.22 Suppose that, instead of defining vectors first, we had begun by defining one-forms, aided by pictures like Figure 3.4. Then we could have introduced vectors as linear real-valued functions of one-forms, and defined vector algebra by the analogs of Eqs. 3.6a and 3.6b (i.e. by exchanging arrows for tildes). Prove that, so defined, vectors form a vector space. This is another example of the duality between vectors and one-forms.

3.23 (a) Prove that the set of all $\binom{M}{N}$ tensors for fixed M, N forms a vector space. (You need to define the addition of such tensors and their multiplication by numbers.)

(b) Prove that a basis for this space is the set

$$\underbrace{\{\vec{e}_{\alpha} \otimes \vec{e}_{\beta} \otimes \cdots \otimes \vec{e}_{\gamma}}_{M \text{ vectors}} \otimes \underbrace{\tilde{\omega}^{\mu} \otimes \tilde{\omega}^{\nu} \otimes \cdots \otimes \tilde{\omega}^{\lambda}\}}_{N \text{ one-forms}}$$

(You will need to define the outer product of more than two one-forms.)

3.24 (a) Given the components of a $\binom{2}{0}$ tensor $M^{\alpha\beta}$ as the matrix

$$\begin{pmatrix} 0 & 1 & 0 & 0 \\ 1 & -1 & 0 & 2 \\ 2 & 0 & 0 & 1 \\ 1 & 0 & -2 & 0 \end{pmatrix},$$

find

(i) the components of the symmetric tensor $M^{(\alpha\beta)}$ and the antisymmetric tensor $M^{[\alpha\beta]}$;

(ii) the components of $M^{\alpha}{}_{\beta}$;

(iii) the components of $M_{\alpha}{}^{\beta}$;

(iv) the components of $M_{\alpha\beta}$.

(b) For the $\binom{1}{1}$ tensor whose components are $M^{\alpha}{}_{\beta}$, does it make sense to speak of its symmetric and antisymmetric parts? If so, define them. If not, say why.

(c) Raise an index of the metric tensor to prove that

$$\eta^{\alpha}{}_{\beta} = \delta^{\alpha}{}_{\beta}.$$

3.25 Show that if **A** is a $\binom{2}{0}$ tensor and **B** a $\binom{0}{2}$ tensor then

$$A^{\alpha\beta} B_{\alpha\beta}$$

is frame invariant, i.e. a scalar.

3.26 Suppose **A** is an antisymmetric $\binom{2}{0}$ tensor, **B** a symmetric $\binom{0}{2}$ tensor, **C** an arbitrary $\binom{0}{2}$ tensor, and **D** an arbitrary $\binom{2}{0}$ tensor. Prove that:

(a) $A^{\alpha\beta} B_{\alpha\beta} = 0$;

(b) $A^{\alpha\beta} C_{\alpha\beta} = A^{\alpha\beta} C_{[\alpha\beta]}$;

(c) $B_{\alpha\beta} D^{\alpha\beta} = B_{\alpha\beta} D^{(\alpha\beta)}$.

3.27 (a) Suppose **A** is an antisymmetric $\binom{2}{0}$ tensor. Show that the $\{A_{\alpha\beta}\}$, obtained by lowering indices by using the metric tensor, are components of an antisymmetric $\binom{0}{2}$ tensor.

(b) Suppose $V^{\alpha} = W^{\alpha}$. Prove that $V_{\alpha} = W_{\alpha}$.

3.28 Deduce Eq. 3.66 from Eq. 3.65.

3.29 Prove that tensor differentiation obeys the Leibniz (product) rule:

$$\nabla(\mathbf{A} \otimes \mathbf{B}) = (\nabla\mathbf{A}) \otimes \mathbf{B} + \mathbf{A} \otimes \nabla\mathbf{B}.$$

3.30 In some frame $\mathcal{O}$, the vector fields $\vec{U}$ and $\vec{D}$ have components

$$\vec{U} \rightarrow (1 + t^2, t^2, \sqrt{2}\,t, 0),$$
$$\vec{D} \rightarrow (x, 5tx, \sqrt{2}\,t, 0),$$

and the scalar ρ has the value

$$\rho = x^2 + t^2 - y^2.$$

(a) Find $\vec{U} \cdot \vec{U}, \vec{U} \cdot \vec{D}, \vec{D} \cdot \vec{D}$. Is $\vec{U}$ suitable as a four-velocity field? Is $\vec{D}$?

(b) Find the spatial velocity v of a particle whose four-velocity is $\vec{U}$, for arbitrary t. What happens to it in the limits $t \to 0, t \to \infty$?

(c) Find U_α for all α.

(d) Find $U^\alpha{}_{,\beta}$ for all α, β.

(e) Show that $U_\alpha U^\alpha{}_{,\beta} = 0$ for all β. Show that $U^\alpha U_{\alpha,\beta} = 0$ for all β.

(f) Find $D^\beta{}_{,\beta}$.

(g) Find $(U^\alpha D^\beta)_{,\beta}$ for all α.

(h) Find $U_\alpha (U^\alpha D^\beta)_{,\beta}$ and compare with (f) above. Why are the two answers similar?

(i) Find $\rho_{,\alpha}$ for all α. Find $\rho^{,\alpha}$ for all α. (Recall that $\rho^{,\alpha} := \eta^{\alpha\beta} \rho_{,\beta}$.) Of what tensor are the numbers $\{\rho^{,\alpha}\}$ the components?

(j) Find $\nabla_{\vec{U}} \rho, \nabla_{\vec{U}} \vec{D}, \nabla_{\vec{D}} \rho, \nabla_{\vec{D}} \vec{U}$.

3.31 Consider a timelike unit four-vector $\vec{U}$ and the tensor **P** whose components are given by

$$P_{\mu\nu} = \eta_{\mu\nu} + U_\mu U_\nu.$$

Show that **P** is a projection operator that projects an arbitrary vector $\vec{V}$ into a vector orthogonal to $\vec{U}$. That is, show that the vector $\vec{V}_\perp$ whose components are

$$V^\alpha_\perp = P^\alpha{}_\beta V^\beta = (\eta^\alpha{}_\beta + U^\alpha U_\beta) V^\beta$$

is

(a) orthogonal to $\vec{U}$, and

(b) unaffected by **P**:

$$V^\alpha_{\perp\perp} := P^\alpha{}_\beta V^\beta_\perp = V^\alpha_\perp.$$

(c) Show that for an arbitrary non-null vector $\vec{q}$, the tensor that projects orthogonally to it has components

$$\eta_{\mu\nu} - q_\mu q_\nu / (q^\alpha q_\alpha).$$

How does this fail for null vectors? How does this relate to the definition of **P**?

(d) Show that **P** as defined above is the metric tensor for vectors perpendicular to $\vec{U}$:

$$\mathbf{P}(\vec{V}_\perp, \vec{W}_\perp) = \mathbf{g}(\vec{V}_\perp, \vec{W}_\perp)$$
$$= \vec{V}_\perp \cdot \vec{W}_\perp.$$

3.32 (a) From the definition $f_{\alpha\beta} = \mathbf{f}(\vec{e}_\alpha, \vec{e}_\beta)$ for the components of a $\binom{0}{2}$ tensor, prove that the transformation law is

$$f_{\bar{\alpha}\bar{\beta}} = \Lambda^\mu{}_{\bar{\alpha}} \Lambda^\nu{}_{\bar{\beta}} f_{\mu\nu}$$

and that the matrix version of this is

$$(\bar{f}) = (\Lambda)^T (f)(\Lambda),$$

where (Λ) is the matrix with components $\Lambda^\mu{}_{\bar{\alpha}}$.

(b) Since our definition of a Lorentz frame led us to deduce that the metric tensor has components $\eta_{\alpha\beta}$, this must be true in all Lorentz frames. We are thus led to a more general *definition* of a Lorentz transformation as one whose matrix $\Lambda^{\mu}{}_{\bar{\alpha}}$ satisfies

$$\eta_{\bar{\alpha}\bar{\beta}} = \Lambda^{\mu}{}_{\bar{\alpha}}\Lambda^{\nu}{}_{\bar{\beta}}\eta_{\mu\nu}. \tag{3.71}$$

Prove that the matrix for a boost of velocity $v\vec{e}_x$ satisfies this, so that this new definition includes our older one.

(c) Suppose (Λ) and (L) are two matrices which satisfy Eq. 3.71, i.e. $(\eta) = (\Lambda)^T (\eta)(\Lambda)$ and similarly for (L). Prove that $(\Lambda)(L)$ is also the matrix of a Lorentz transformation.

3.33　The result of Exercise 3.32(c) establishes that Lorentz transformations form a group, represented by the multiplication of their matrices. This is called the *Lorentz group*, denoted by $L(4)$ or $0(1,3)$.

(a) Find the matrices of the identity element of the Lorentz group and of the element inverse to that whose matrix is implicit in Eq. 1.12.

(b) Prove that the determinant of any matrix representing a Lorentz transformation is ± 1.

(c) Prove that those elements whose matrices have determinant $+1$ form a subgroup, while those with -1 do not.

(d) The three-dimensional orthogonal group $O(3)$ is the analogous group for the metric of three-dimensional Euclidean space. In Exercise 3.20(b), we saw that it was represented by the orthogonal matrices. Show that the orthogonal matrices do form a group, and then show that $0(3)$ is (isomorphic to) a subgroup of $L(4)$.

3.34　Consider the coordinates $u = t - x, v = t + x$ in Minkowski space.

(a) Define $\vec{e}_u$ to be the vector connecting the events with coordinates $\{u = 1, v = 0, y = 0, z = 0\}$ and $\{u = 0, v = 0, y = 0, z = 0\}$, and analogously for $\vec{e}_v$. Show that $\vec{e}_u = (\vec{e}_t - \vec{e}_x)/2, \vec{e}_v = (\vec{e}_t + \vec{e}_x)/2$, and draw $\vec{e}_u$ and $\vec{e}_v$ in a spacetime diagram of the t–x plane.

(b) Show that $\{\vec{e}_u, \vec{e}_v, \vec{e}_y, \vec{e}_z\}$ forms a basis for vectors in Minkowski space.

(c) Find the components of the metric tensor in this basis.

(d) Show that $\vec{e}_u$ and $\vec{e}_v$ are null and not orthogonal. (They are called a *null basis* for the t–x plane.)

(e) Compute the four one-forms $\tilde{d}u, \tilde{d}v, \mathbf{g}(\vec{e}_u,), \mathbf{g}(\vec{e}_v,)$ in terms of $\tilde{d}t$ and $\tilde{d}x$.

Perfect fluids in special relativity

4.1 Fluids

In many interesting situations in astrophysical GR, the source of the gravitational field can be taken to be a perfect fluid as a good first approximation. In general, a 'fluid' is a special kind of *continuum*. A continuum is a collection of particles so numerous that the dynamics of individual particles cannot be followed, leaving only a description of the collection in terms of 'average' or 'bulk' quantities: the number of particles per unit volume, the energy density, the momentum density, the pressure, the temperature, etc. The behavior of a lake of water, and the gravitational field it generates, does not depend upon where any one particular water molecule happens to be: it depends only on the average properties of huge collections of molecules.

Nevertheless, these properties can vary from point to point in the lake: the pressure is larger at the bottom than at the top, and the temperature may vary as well. The atmosphere, another fluid, has a density that varies with position. This raises the question of how large a collection of particles should one average over: it must clearly be large enough that the individual particles don't matter but small enough that it is relatively homogeneous: the average velocity, kinetic energy, and interparticle spacing must be the same everywhere in the collection. Such a collection is called an *element*. This is a somewhat imprecise but useful term for a large collection of particles that may be regarded as having a single value of such quantities as density, average velocity, and temperature. If such a collection doesn't exist (e.g. in a *very* rarified gas), then the continuum approximation breaks down.

The continuum approximation assigns to each element a value of density, temperature, etc. Since the elements are regarded as 'small', this approximation is expressed mathematically by assigning to each *point* a value of density, temperature, etc. So, a continuum is defined by various fields, having values at each point and at each time. Each element has an MCRF, an inertial frame that is at least momentarily comoving with it, in the same sense as we defined for particles in § 2.3.

So far, this notion of a continuum embraces rocks as well as gases. A *fluid* is a continuum that 'flows': this definition is not very precise, and so the division between solids and fluids is not very well defined. Most solids will flow under high enough pressure. What makes a substance rigid? After some thought one should be able to see that rigidity comes from forces *parallel* to the interface between two elements. Two adjacent elements can push and pull on each other, but the continuum won't be rigid unless they can also prevent each other from sliding along their common boundary. A *fluid* is characterized by the weakness of such anti-slipping forces compared to the direct push–pull force, which is called pressure. A *perfect* fluid is defined as one in which *all* anti-slipping forces are zero, and the only force between neighboring fluid elements is pressure. We will soon see how to make this mathematically precise.

4.2 Dust: the number-flux vector $\vec{N}$

We will introduce the relativistic description of fluids using the simplest type of fluid: 'dust' is defined to be a collection of particles, all of which are at rest in some Lorentz frame. It isn't very clear how this usage of the term 'dust' evolved from the other meaning as that substance which is at rest on the windowsill, but it has become a standard usage in relativity.

The number density n. The simplest question we can ask about these particles is, how many are there per unit volume? In their rest frame, this is merely an exercise in counting the particles and dividing by the volume they occupy. By doing this in many small regions we could come up with different numbers at different points, since the particles may be distributed more densely in one area than in another. We define this *number density* to be n:

$$n := \text{number density in the MCRF of the element.} \tag{4.1}$$

What is the number density in a frame $\bar{\mathcal{O}}$ in which the particles are not at rest? They will all have the same velocity v in $\bar{\mathcal{O}}$. If we look at the same particles as we counted up in the rest frame, then there are clearly the same *number* of particles, but they do not occupy the same volume. Suppose they were originally in a rectangular solid of dimension $\Delta x \Delta y \Delta z$. The Lorentz contraction will reduce this to $\Delta x \Delta y \Delta z \sqrt{(1 - v^2)}$, since lengths in the direction of motion contract but lengths perpendicular do not (Figure 4.1). Because of this, the number of particles per unit volume is $[\sqrt{(1 - v^2)}]^{-1}$ times its value in the rest frame:

$$\frac{n}{\sqrt{(1 - v^2)}} = \left\{ \begin{array}{l} \text{number density in the frame in} \\ \text{which particles have velocity } v \end{array} \right\}. \tag{4.2}$$

The flux across a surface. When particles move, another question of interest is, how many of them are moving in a certain direction? This is made precise by the definition of flux: *the flux of particles across a surface is the number crossing a unit area of that surface in a unit time.* This clearly depends on the inertial reference frame ('area' and 'time' are frame-dependent concepts) and on the orientation of the surface (a surface parallel to the velocity of the particles will not be crossed by any of them). In the rest frame of the dust the flux is zero, since all particles are at rest. In the frame $\bar{\mathcal{O}}$, suppose the particles all move with velocity v in the $\bar{x}$ direction, and let us for simplicity consider a surface $\mathcal{S}$ perpendicular to $\bar{x}$ (Figure 4.2). The rectangular volume enclosed by the dashed line clearly contains all and only those particles that will cross the area ΔA of $\mathcal{S}$ in the time $\Delta \bar{t}$. It has volume $v \Delta \bar{t} \, \Delta A$,

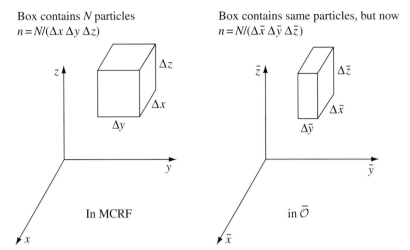

Box contains N particles
$n = N/(\Delta x\, \Delta y\, \Delta z)$

Box contains same particles, but now
$n = N/(\Delta \bar{x}\, \Delta \bar{y}\, \Delta \bar{z})$

In MCRF

in $\bar{\mathcal{O}}$

Figure 4.1 Lorentz contraction causes the density of particles to depend upon the frame in which it is measured.

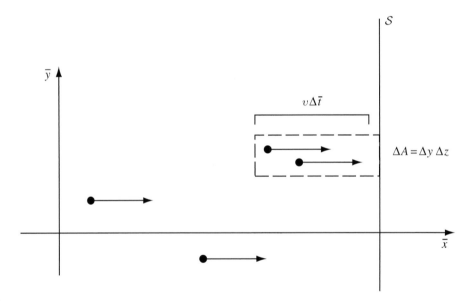

Figure 4.2 Simple illustration of the transformation of flux: if particles are moving only in the x direction, then all those within a distance $v\Delta\bar{t}$ of the surface $\mathcal{S}$ will cross $\mathcal{S}$ in the time $\Delta\bar{t}$.

and contains $[n/\sqrt{(1 - v^2)}]v\Delta\bar{t}\,\Delta A$ particles, since in this frame the number density is $n/\sqrt{(1 - v^2)}$. The number crossing *per unit* time and per unit area is the flux across surfaces of constant $\bar{x}$:

$$(\text{flux})^{\bar{x}} = \frac{nv}{\sqrt{(1 - v^2)}}.$$

Suppose, more generally, that the particles have a y component of velocity in $\bar{\mathcal{O}}$ as well. Then the dashed line in Figure 4.3 encloses all and only those particles that cross ΔA in

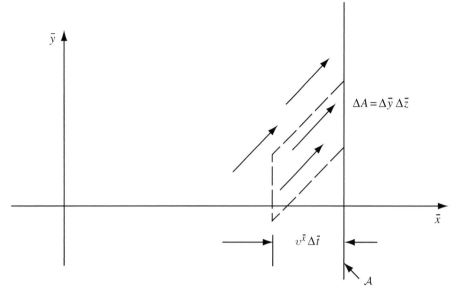

$$\Delta A = \Delta \bar{y}\,\Delta \bar{z}$$

$$v^{\bar{x}}\Delta \bar{t}$$

$$\mathcal{A}$$

Figure 4.3 The general situation for flux: only the x-component of the velocity carries particles across a surface of constant x.

$\mathcal{S}$ in the time $\Delta \bar{t}$. This is a parallelepiped, whose volume is the area of its base times its height. But its height – its extent in the x direction – is just $v^{\bar{x}}\Delta \bar{t}$. Therefore we get

$$(\text{flux})^{\bar{x}} = \frac{nv^{\bar{x}}}{\sqrt{(1-v^2)}}. \tag{4.3}$$

The number-flux four-vector $\vec{N}$. Consider the vector $\vec{N}$ defined by

$$\vec{N} = n\vec{U}, \tag{4.4}$$

where $\vec{U}$ is the four-velocity of the particles. In a frame $\bar{\mathcal{O}}$ in which the particles have velocity (v^x, v^y, v^z), we have

$$\vec{U} \underset{\bar{\mathcal{O}}}{\rightarrow} \left(\frac{1}{\sqrt{(1-v^2)}}, \frac{v^x}{\sqrt{(1-v^2)}}, \frac{v^y}{\sqrt{(1-v^2)}}, \frac{v^z}{\sqrt{(1-v^2)}} \right).$$

It follows that

$$\vec{N} \underset{\bar{\mathcal{O}}}{\rightarrow} \left(\frac{n}{\sqrt{(1-v^2)}}, \frac{nv^x}{\sqrt{(1-v^2)}}, \frac{nv^y}{\sqrt{(1-v^2)}}, \frac{nv^z}{\sqrt{(1-v^2)}} \right). \tag{4.5}$$

Thus, in any frame, the time component of $\vec{N}$ is the number density and the spatial components are the fluxes across surfaces of the various coordinates. This is a very important conceptual result. In Galilean physics, number density is a scalar, the same in all frames (no Lorentz contraction), while flux was quite another thing: a three-vector that was *frame dependent*, since the velocities of particles are a frame-dependent notion.

Our relativistic approach has unified these two notions into a single, frame-independent four-vector. This is progress in our thinking of the most fundamental sort: the union of apparently disparate notions into a single coherent one.

It is worth re-emphasizing the sense in which we use the term 'frame-independent'. The vector $\vec{N}$ is a geometrical object whose existence is independent of any frame; as a tensor, its action on a one-form to give a number is independent of any frame. Its components *do* of course depend on the frame. Since prerelativity physicists regarded flux as a three-vector, they had to settle for it as a frame-dependent vector, in the following sense. As a three-vector it was independent of the orientation of the spatial axes in the same sense that four-vectors are independent of all frames; but the flux three-vector is different in frames that move relative to one another, since the velocity of the particles is different in different frames. Before special relativity, a flux vector had to be defined relative to some inertial frame. To a relativist there is only *one* four-vector, and the frame dependence of the older way of looking at things came from concentrating only on a set of three of the four components of $\vec{N}$. This unification of the Galilean frame-independent number density and the frame-dependent flux into a single frame-independent four-vector $\vec{N}$ is similar to the unification of energy and momentum into the four-momentum.

One final note: it is clear that

$$\vec{N} \cdot \vec{N} = -n^2, \quad n = (-\vec{N} \cdot \vec{N})^{1/2}. \tag{4.6}$$

Thus, n is a scalar. In the same way that rest mass is a scalar, even though energy and inertial mass are frame dependent, here we have that n is a scalar, the 'rest density', even though number density is frame dependent. We will *always* define n to be a scalar number equal to the number density in the MCRF. We will make similar definitions for pressure, temperature, and other quantities characteristic of the fluid. These will be discussed later.

4.3 One-forms and surfaces

Number density as a timelike flux. We can complete the above discussion of the unity of number density and flux by realizing that number density can be regarded as a timelike flux. To see this, let us look at the flux across surfaces of constant x again, this time in a *spacetime* diagram, in which we plot only $\bar{t}$ and $\bar{x}$ (Figure 4.4). The surface $\mathcal{S}$ perpendicular to $\bar{x}$ has the world line shown. At any time $\bar{t}$ it is represented by just one point, since we are suppressing both $\bar{y}$ and $\bar{z}$. The world lines of those particles that go through $\mathcal{S}$ in the time $\Delta\bar{t}$ are also shown. The flux is the number of world lines that cross $\mathcal{S}$ in the interval $\Delta\bar{t} = 1$. Really, since it is a two-dimensional surface, its 'world path' is three-dimensional, of which we have drawn only a section. The *flux* is the number of world lines that cross a unit volume of this three-surface: by volume we of course mean a cube of unit side, $\Delta\bar{t} = 1, \Delta\bar{y} = 1, \Delta\bar{z} = 1$. So we can define flux as the number of world lines crossing a

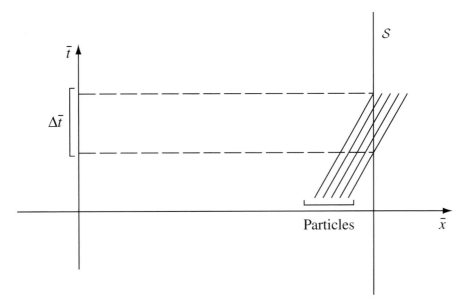

Figure 4.4   Figure 4.2 in a spacetime diagram, with the $\bar{y}$ direction suppressed.

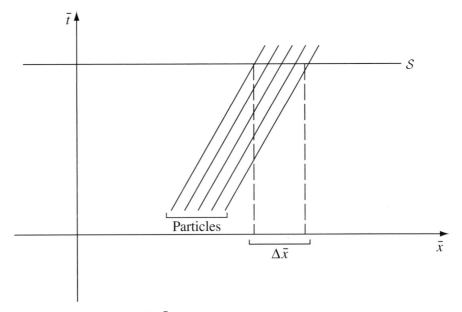

Figure 4.5   Number density as a flux across a surface $\bar{t} = $ const.

unit three-volume. There is no reason why we cannot now define this three-volume to be an ordinary spatial volume $\Delta\bar{x} = 1, \Delta\bar{y} = 1, \Delta\bar{z} = 1$, taken at some particular time $\bar{t}$. This is shown in Figure 4.5. Now the flux is the number crossing in the interval $\Delta\bar{x} = 1$ (since $\bar{y}$ and $\bar{z}$ are suppressed). But this is just the number contained in the unit volume at the given time: the number density. So the timelike flux is the number density.

A one-form defines a surface. The way in which we described surfaces above was somewhat clumsy. To push our invariant picture further we need a somewhat more satisfactory mathematical representation of the surface that these world lines are crossing. This representation is given by one-forms. In general, a surface is described by an equation of the form

$$\phi(t, x, y, z) = \text{const.}$$

The gradient of the function ϕ, $\tilde{d}\phi$, is a normal one-form. In some sense, $\tilde{d}\phi$ *defines* the surface ϕ = const., since it uniquely determines the directions normal to that surface. However, any multiple of $\tilde{d}\phi$ also defines the same surface, so it is customary to use the unit-normal one-form when the surface is not null:

$$\tilde{n} := \tilde{d}\phi/|\tilde{d}\phi|, \tag{4.7}$$

where $|\tilde{d}\phi|$ is the magnitude of $\tilde{d}\phi$:

$$|\tilde{d}\phi| = |\, \eta^{\alpha\beta}\, \phi_{,\alpha}\, \phi_{,\beta}\, |^{1/2}. \tag{4.8}$$

(Do not confuse $\tilde{n}$ with n, the number density in the MCRF: they are completely different, but they have been given, by historical accident, the same letter.)

As in three-dimensional vector calculus (e.g. in Gauss' law), one defines the surface element vector as the unit normal times an area element in the surface. In this case, a volume element in a three-space whose coordinates are x^α, x^β, and x^γ (for some *particular* values of $\alpha, \beta,$ and γ, all distinct) can be represented by

$$\tilde{n}\, dx^\alpha dx^\beta dx^\gamma, \tag{4.9}$$

and a *unit* volume ($dx^\alpha = dx^\beta = dx^\gamma = 1$) is just $\tilde{n}$. (These dx's *are* infinitesimals over which we integrate, not gradient one-forms.)

The flux across the surface. Recall from Gauss' law in three dimensions that the flux across a surface of, say, the electric field is just $\mathbf{E} \cdot \mathbf{n}$, the dot product of $\mathbf{E}$ with the unit normal. The situation here is exactly comparable: the flux (of particles) across a surface of constant ϕ is $\langle \tilde{n}, \vec{N} \rangle$. To see this, let ϕ be a coordinate, say $\bar{x}$. Then the normal to a surface of constant $\bar{x}$ is $\tilde{d}\bar{x}$, which is a unit normal already since $\tilde{d}\bar{x} \to_{\bar{O}} (0, 1, 0, 0)$. Then $\langle \tilde{d}\bar{x}, \vec{N} \rangle = N^\alpha(\tilde{d}\bar{x})_\alpha = N^{\bar{x}}$, which, as we have already seen, is the flux across the $\bar{x}$ surfaces. Clearly, had we chosen $\phi = \bar{t}$, then we would have wound up with $N^{\bar{0}}$, the number density, or flux across surfaces of constant $\bar{t}$.

This is a physical example of our definition of a vector as a function of one-forms into real numbers. Given the vector $\vec{N}$, we can calculate the flux across a surface by finding the unit-normal one-form for that surface, and contracting it with $\vec{N}$. We have, moreover, expressed everything frame invariantly and in a manner that separates the properties of the system of particles $\vec{N}$ from the properties of the surface $\tilde{n}$. All of this will have many parallels in § 4.4 below.

Representation of a frame by a one-form. Before going on to discuss other properties of fluids, we should mention a useful fact. An inertial frame, which up to now has been defined by its four-velocity $\vec{U}$, can be defined also by a one-form, namely that associated with its four-velocity, $\mathbf{g}(\vec{U}, \ )$. This has components

$$U_\alpha = \eta_{\alpha\beta} U^\beta$$

or, in this frame,

$$U_0 = -1, U_i = 0.$$

The one-form in this case is clearly also equal to $-\tilde{d}t$ (since their components are equal). So we could equally well define a frame whose time coordinate is t by giving $\tilde{d}t$. This has a nice picture: $\tilde{d}t$ can be seen as a set of surfaces of constant t, the surfaces of simultaneity. These clearly *do* define the frame up to spatial rotations, which we usually ignore. In fact, in some sense $\tilde{d}t$ is a more natural way to define the frame than $\vec{U}$. For instance, the energy of a particle whose four-momentum is $\vec{p}$ is

$$E = \langle \tilde{d}t, \vec{p} \rangle = p^0. \tag{4.10}$$

Here we don't have the awkward minus sign that one gets in Eq. 2.35,

$$E = -\vec{p} \cdot \vec{U}.$$

4.4 Dust again: the stress–energy tensor

So far we have only discussed how many dust particles there are. But they have energy and momentum, and it will turn out that their energy and momentum are the source of the gravitational field in GR. So we must now ask how to represent them in a frame-invariant manner. We will assume for simplicity that all the dust particles have the same rest mass m.

Energy density. In the MCRF, the energy of each particle is just m, and the number per unit volume is n. Therefore the energy per unit volume is mn. We denote this in general by ρ:

$$\rho := \text{energy density in the MCRF.} \tag{4.11}$$

Thus ρ is a scalar just as n and m are scalars. In our case of dust,

$$\rho = nm \text{ (dust).} \tag{4.12}$$

In more general fluids, where there is random motion of particles and hence kinetic energy of motion, even in an average rest frame Eq. 4.12 will not be valid.

In the frame $\bar{O}$ we again have that the number density is $n/\sqrt{(1-v^2)}$, but now the energy of each particle is $m/\sqrt{(1-v^2)}$, since it is moving. Therefore the energy density is $mn/(1-v^2)$:

$$\frac{\rho}{1-v^2} = \left\{ \begin{array}{l} \text{energy density in a frame in} \\ \text{which particles have velocity } v \end{array} \right\}. \tag{4.13}$$

This transformation involves *two* factors of $(1-v^2)^{-1/2} = \Lambda^{\bar{0}}{}_0$, because *both* volume *and* energy transform. It is impossible, therefore, to represent the energy density as some component of a vector. It is, in fact, a component of a $\binom{2}{0}$ tensor. This is most easily seen from the point of view of our definition of a tensor. To define energy requires a one-form, in order to select the 0 component of the four-vector of energy and momentum; to define a density also requires a one-form, since density is a flux across a constant-time surface. Similarly, an energy flux also requires two one-forms: one to define 'energy' and the other to define the surface. One can also speak of momentum density: again a one-form defines the component of momentum in question, and another one-form defines density. By analogy there is also momentum flux: the rate at which momentum crosses some surface. All these things require two one-forms. So, there is a tensor $\mathbf{T}$, called the stress–energy tensor, which has all these numbers as values when supplied with the appropriate one-forms as arguments.

Stress–energy tensor. The most convenient definition of the stress–energy tensor is in terms of its components in some (arbitrary) frame:

$$\mathbf{T}(\tilde{\mathrm{d}}x^{\alpha}, \tilde{\mathrm{d}}x^{\beta}) = T^{\alpha\beta} := \left\{ \begin{array}{c} \text{flux of } \alpha\text{-momentum across} \\ \text{a surface of constant } x^{\beta} \end{array} \right\}. \tag{4.14}$$

(By 'α-momentum' we mean, of course, the α-component of the four-momentum $p^{\alpha} := \langle \tilde{\mathrm{d}}x^{\alpha}, \vec{p} \rangle$.) That this is truly a tensor is proved in Exercise 4.5.

Let us see how this definition fits in with our discussion above. Consider T^{00}. This is defined as the flux of 0-momentum (i.e. energy) across a surface $t = $ const. This is just the energy density:

$$T^{00} = \text{energy density.} \tag{4.15}$$

Similarly, T^{0i} is the flux of energy across a surface $x^{i} = $ const.:

$$T^{0i} = \text{energy flux across the constant-}x^{i}\text{ surface.} \tag{4.16}$$

Then T^{i0} is the flux of i-momentum across a surface $t = $ const., i.e. the density of i-momentum,

$$T^{i0} = i\text{-momentum density.} \tag{4.17}$$

Finally, T^{ij} is the j-flux of i-momentum:

$$T^{ij} = \text{flux of } i\text{-momentum across the constant-}j\text{ surface.} \tag{4.18}$$

For any particular system, giving the components of $\mathbf{T}$ in some frame defines it completely. For dust, the components of $\mathbf{T}$ in the MCRF are particularly easy. There is no motion of the particles, so all i-momenta are zero and all spatial fluxes are zero. Therefore

$$(T^{00})_{\text{MCRF}} = \rho = mn,$$
$$(T^{0i})_{\text{MCRF}} = (T^{i0})_{\text{MCRF}} = (T^{ij})_{\text{MCRF}} = 0.$$

It is easy to see that the tensor $\vec{p} \otimes \vec{N}$ has exactly these components in the MCRF, where $\vec{p} = m\vec{U}$ is the four-momentum of a particle. Therefore we have

For dust: $\mathbf{T} = \vec{p} \otimes \vec{N} = mn\,\vec{U} \otimes \vec{U} = \rho\,\vec{U} \otimes \vec{U}.$ (4.19)

From this we can conclude that

$$
\begin{aligned}
T^{\alpha\beta} &= \mathbf{T}(\tilde{\omega}^{\alpha}, \tilde{\omega}^{\beta}) \\
&= \rho\vec{U}(\tilde{\omega}^{\alpha})\vec{U}(\tilde{\omega}^{\beta}) \\
&= \rho U^{\alpha}U^{\beta}.
\end{aligned}
$$
(4.20)

In the frame $\bar{\mathcal{O}}$, where

$$
\vec{U} \rightarrow \left(\frac{1}{\sqrt{(1 - v^2)}}, \frac{v^x}{\sqrt{(1 - v^2)}}, \dots \right),
$$

we therefore have

$$
\left.
\begin{aligned}
T^{\bar{0}\bar{0}} &= \rho U^{\bar{0}} U^{\bar{0}} = \rho/(1 - v^2), \\
T^{\bar{0}\bar{i}} &= \rho U^{\bar{0}} U^{\bar{i}} = \rho v^i/(1 - v^2), \\
T^{\bar{i}\bar{0}} &= \rho U^{\bar{i}} U^{\bar{0}} = \rho v^i(1 - v^2), \\
T^{\bar{i}\bar{j}} &= \rho U^{\bar{i}} U^{\bar{j}} = \rho v^i v^j/(1 - v^2).
\end{aligned}
\right\}
$$
(4.21)

These are exactly what one would calculate, from first principles, for energy density, energy flux, momentum density, and momentum flux respectively. (We did the calculation for energy density above.) Notice one important point: $T^{\alpha\beta} = T^{\beta\alpha}$; that is, $\mathbf{T}$ is symmetric. This will turn out to be true in general, not just for dust.

4.5 General fluids

Until now we have dealt with the simplest possible collection of particles. To generalize this to real fluids, we have to take account of the facts that (i) besides the bulk motions of the fluid, each particle has some random velocity; and (ii) there may be various forces between particles that contribute potential energies to the total.

Definition of macroscopic quantities. The concept of a fluid element was discussed in § 4.1. For each fluid element, we will go to the frame in which it is at rest (its total spatial momentum is zero). This is its momentarily comoving reference frame (MCRF). This frame is truly *momentarily* comoving: since a fluid element can be accelerated, a moment later a different inertial frame will be its MCRF. Moreover, two different fluid elements may be moving relative to one another, so that they would not have the same MCRFs. Thus, the MCRF is specific to a single fluid element, and which frame is its MCRF is a function of its position and time. *All scalar quantities associated with a fluid element in relativity* (such as number density, energy density, and temperature) *are defined to be their values in the MCRF.* Thus we make the definitions displayed in Table 4.1. We confine our attention to fluids that consist of only one component, one kind of particle, so that (for example) interpenetrating flows are not considered.

Symbol	Name	Definition
$\vec{U}$	Four-velocity of fluid element	Four-velocity of MCRF
n	Number density	Number of particles per unit volume in MCRF
$\vec{N}$	Flux vector	$\vec{N} := n\vec{U}$
ρ	Energy density	Density of *total* mass energy (rest mass, random kinetic, chemical, . . .)
Π	Internal energy per particle	$\Pi := (\rho/n) - m \Rightarrow \rho = n(m + \Pi)$ Thus Π is a general name for all energies other than the rest mass.
ρ_0	Rest-mass density	$\rho_0 := mn.$ Since m is a constant, this is the 'energy' associated with the rest mass only. Thus, $\rho = \rho_0 + n\Pi$.
T	Temperature	Usual thermodynamic definition in MCRF (see below).
p	Pressure	Usual fluid-dynamical notion in MCRF. More about this later.
S	Specific entropy	Entropy per particle (see below).

Table 4.1 Macroscopic quantities for single-component fluids

First law of thermodynamics. This law is simply a statement of conservation of energy. In the MCRF, we imagine that the fluid element is able to exchange energy with its surroundings in only two ways: by heat conduction (absorbing an amount of heat ΔQ) and by work (doing an amount of work $p\Delta V$, where V is the three-volume of the element). If we let E be the total energy of the element then, since ΔQ is energy gained and $p\Delta V$ is energy lost, we can write (assuming small changes)

$$\left. \begin{aligned} \Delta E &= \Delta Q - p\Delta V, \\ \text{or} \\ \Delta Q &= \Delta E + p\Delta V. \end{aligned} \right\} \tag{4.22}$$

Now, if the element contains a total of N particles, and if this number doesn't change (i.e. there is no creation or destruction of particles), we can write

$$V = \frac{N}{n}, \quad \Delta V = -\frac{N}{n^2}\Delta n. \tag{4.23}$$

Moreover, we also have (from the definition of ρ)

$$E = \rho V = \rho N/n,$$
$$\Delta E = \rho\Delta V + V\Delta\rho.$$

These two results imply that

$$\Delta Q = \frac{N}{n}\Delta\rho - N(\rho + p)\frac{\Delta n}{n^2}.$$

If we write $q := Q/N$, which is the heat absorbed per particle, we obtain

$$n\,\Delta q = \Delta\rho - \frac{\rho + p}{n}\Delta n. \tag{4.24}$$

Now suppose that the changes are infinitesimal. It can be shown in general that a fluid's state can be given by two parameters: for instance, ρ and T or ρ and n. Everything else is a function of, say, ρ and n. That means that the right-hand side of Eq. 4.24 becomes

$$\mathrm{d}\rho - (\rho + p)\mathrm{d}n/n,$$

which depends only on ρ and n. The general theory of first-order differential equations shows that this *always* possesses an *integrating factor*: that is, there exist A and B, functions of only ρ and n, such that

$$\mathrm{d}\rho - (\rho + p)\mathrm{d}n/n \equiv A\,\mathrm{d}B$$

is an identity for all ρ and n. It is customary in thermodynamics to *define* temperature as A/n and specific entropy as B:

$$\mathrm{d}\rho - (\rho + p)\,\mathrm{d}n/n = nT\,\mathrm{d}S, \tag{4.25}$$

or, for finite changes,

$$\Delta q = T\Delta S. \tag{4.26}$$

The heat absorbed by a fluid element is proportional to its increase in entropy.

We have thus introduced T and S as convenient mathematical definitions. A full treatment would show that T is the quantity normally defined as temperature, and that S is the quantity appearing in the second law of thermodynamics, which says that the *total* entropy in any system must increase. We will have nothing more to say about the second law. Entropy appears here only because it is an integral of the first law, which is merely the conservation of energy. In particular, we shall use both Eqs. 4.25 and 4.26 later.

The general stress–energy tensor. The definition of $T^{\alpha\beta}$ in Eq. 4.14 is perfectly general. Let us in particular look at it in the MCRF, where there is no bulk flow of the fluid element and no spatial momentum in the particles. Then *in the MCRF* we have

(1) T^{00} = energy density = ρ.
(2) T^{0i} = energy flux. Although there is no motion in the MCRF, energy may be transmitted to the particles by heat conduction. So T^{0i} is basically a heat-conduction term in the MCRF.
(3) T^{i0} = momentum density. Again the particles themselves have no net momentum in the MCRF, but if heat is being conducted to them then the moving energy will have an associated momentum. We will argue below that $T^{i0} \equiv T^{0i}$.
(4) T^{ij} = momentum flux. This is an interesting and important term. The next section gives a thorough discussion of it. It is called the *stress*.

The spatial components T^{ij} of **T**. By definition, T^{ij} is the flux of i-momentum across the contant-j surface. Consider (Figure 4.6) two adjacent fluid elements, represented as cubes, having common interface S. In general, they exert forces on each other. Shown in the diagram is the force **F** exerted by A on B (B of course exerts an equal and opposite force

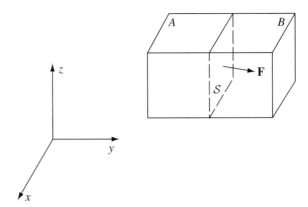

Figure 4.6 The force **F** exerted by element A on its neighbor B may be in any direction depending on the properties of the medium and any external forces.

on A). Since force equals the rate of change of momentum (by Newton's law, which is valid here, since we are in the MCRF where velocities are zero), A is pouring momentum into B at the rate **F** per unit time. Of course, B may or may not acquire a new velocity as a result of this new momentum it acquires; this depends upon how much momentum is being put into B by its other neighbors. Obviously B's motion is the resultant of all the forces. Nevertheless, each force adds momentum to B. There is therefore a flow of momentum across S from A to B at the rate **F**. If S has area $\mathcal{A}$, then the flux of momentum across S is **F**$/\mathcal{A}$. If S is a surface of constant x^j, then T^{ij} for the fluid element A is $F^i/\mathcal{A}$.

This is a brief illustration of the meaning of T^{ij}: it represents forces between adjacent fluid elements. As mentioned before, these forces need not be perpendicular to the surfaces between the elements (i.e. viscosity or other kinds of rigidity give forces parallel to the interface). But if the forces *are* perpendicular to the interfaces, then T^{ij} will be zero unless $i = j$. (Think this through – we'll use it shortly.)

Symmetry of $T^{\alpha\beta}$ in MCRF. We now prove that **T** is a symmetric tensor. We need only prove that its components are symmetric in one frame; this will imply that, for any $\tilde{r}, \tilde{q}$, we have $\mathbf{T}(\tilde{r}, \tilde{q}) = \mathbf{T}(\tilde{q}, \tilde{r})$, which implies the symmetry of its components in any other frame. The easiest frame to consider is the MCRF.

(a) Symmetry of T^{ij}. Consider Figure 4.7, in which we have drawn a fluid element as a cube of side l. The force it exerts on a neighbor across surface (1) (a surface $x = $ const.) is $F_1^i = T^{ix}l^2$, where the factor l^2 gives the area of the face. Here, i runs over 1, 2, and 3, since **F** is not necessarily perpendicular to the surface. Similarly, the force it exerts on a neighbor across (2) is $F_2^i = T^{iy}l^2$. (We shall take the limit $l \to 0$, so bear in mind that the element is small.) The element also exerts a force on its neighbor toward the $-x$ direction, which we call F_3^i. Similarly, there is a force component F_4^i on the face looking in the $-y$ direction. The forces *on* the fluid element are, respectively, $-F_1^i, -F_2^i$, etc. The first point to note is that $F_3^i \approx -F_1^i$ in order that the sum of the forces on the element should vanish when $l \to 0$ (otherwise the tiny mass obtained as $l \to 0$ would have an infinite acceleration). The next point is to compute torques about the z axis through the center of the fluid element. (Since

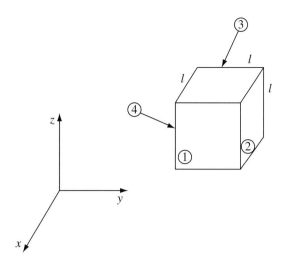

Figure 4.7   A fluid element.

forces on the top and bottom of the cube don't contribute to this, we need not consider them.) For the torque calculation it is convenient to place the origin of coordinates at the center of the cube. The torque due to $-\mathbf{F}_1$ is $-(\mathbf{r} \times \mathbf{F}_1)^z = -xF_1^y = -\frac{1}{2}lT^{yx}l^2$, where we have approximated the force as acting at the center of the face, where $\mathbf{r} \to (l/2, 0, 0)$ (note particularly that $y = 0$ there). The torque due to $-\mathbf{F}_3$ is the *same*, $-\frac{1}{2}l^3 T^{yx}$. The torque due to $-\mathbf{F}_2$ is $-(\mathbf{r} \times \mathbf{F}_2)^z = +yF_2^x = \frac{1}{2}lT^{xy}l^2$. Similarly, the torque due to $-\mathbf{F}_4$ is the same, $\frac{1}{2}l^3 T^{xy}$. Therefore, the total torque around the z axis is

$$\tau_z = l^3(T^{xy} - T^{yx}). \tag{4.27}$$

The moment of inertia of the element about the z axis is proportional to its mass times l^2, or

$$I = \alpha \rho l^2,$$

where α is some numerical constant and ρ is the density (whether of total energy or rest mass doesn't matter in this argument). Therefore the angular acceleration is

$$\ddot{\theta} = \frac{\tau}{I} = \frac{T^{xy} - T^{yx}}{\alpha \rho l^2}. \tag{4.28}$$

Since α is a number and ρ is independent of the size of the element, as are T^{xy} and T^{yx}, this will go to infinity as $l \to 0$ unless

$$T^{xy} = T^{yx}.$$

Thus, since it is obviously not true that fluid elements are whirling around inside fluids, smaller ones whirling ever faster, we have that the stresses are always *symmetric*:

$$T^{ij} = T^{ji}. \tag{4.29}$$

Since we have made no use of any property of the substance, this is true of solids as well as fluids. It is true in Newtonian theory as well as in relativity; in Newtonian theory T^{ij} are the components of a three-dimensional $\binom{2}{0}$ tensor called the stress tensor. It is familiar to any materials engineer; and it contributes its name to its relativistic generalization **T**.

(b) Equality of momentum density and energy flux. This is much easier to demonstrate. The energy flux is the density of energy times the speed it flows at. But since energy and mass are the same, this is the density of mass times the speed it is moving at; in other words, the density of momentum. Therefore $T^{0i} = T^{i0}$.

4.6 Conservation of energy–momentum

Since **T** represents the energy and momentum content of the fluid, there must be some way of using it to express the law of conservation of energy and momentum. In fact it is reasonably easy to do so. In Figure 4.8 we see a cubical fluid element, seen only in cross-section (the z direction has been suppressed). Energy can flow in across all sides. The rate of flow across face 4 is $l^2 T^{0x}(x = 0)$, and across face 2 it is $-l^2 T^{0x}(x = l)$; the second term has a minus sign since T^{0x} represents the energy flowing in the positive x direction, which is out of the volume across face 2. Similarly, the energy flowing in the y direction is $l^2 T^{0y}(y = 0) - l^2 T^{0y}(y = l)$. The sum of these rates must be the rate of increase in the energy inside, $\partial(T^{00}l^3)/\partial t$ (by the conservation of energy). Therefore we have

$$\frac{\partial}{\partial t} l^3 T^{00} = l^2 \left[T^{0x}(x = 0) - T^{0x}(x = l) + T^{0y}(y = 0) \right.$$
$$\left. - T^{0y}(y = l) + T^{0z}(z = 0) - T^{0z}(z = l) \right]. \tag{4.30}$$

Dividing by l^3 and taking the limit $l \to 0$ gives

$$\frac{\partial}{\partial t} T^{00} = -\frac{\partial}{\partial x} T^{0x} - \frac{\partial}{\partial y} T^{0y} - \frac{\partial}{\partial z} T^{0z}. \tag{4.31}$$

Note that in deriving this we have used the definition of the derivative,

$$\lim_{l \to 0} \frac{T^{0x}(x = 0) - T^{0x}(x = l)}{l} \equiv -\frac{\partial}{\partial x} T^{0x}. \tag{4.32}$$

Equation 4.31 can be written as

$$T^{00}{}_{,0} + T^{0x}{}_{,x} + T^{0y}{}_{,y} + T^{0z}{}_{,z} = 0$$

or

$$T^{0\alpha}{}_{,\alpha} = 0. \tag{4.33}$$

This is a statement of the law of conservation of energy.

Similarly, momentum is conserved. The same mathematics applies, with the index '0' changed to whatever spatial index corresponds to the component of momentum whose conservation is being considered. The general conservation law is, then,

$$T^{\alpha\beta}{}_{,\beta} = 0. \tag{4.34}$$

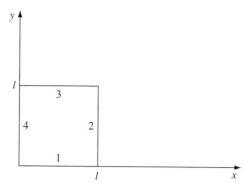

A section $z = $ const. of a cubical fluid element.

This applies to any material in SR. Notice that it is just a four-dimensional divergence. Its relation to Gauss' theorem, which gives an integral form of the conservation law, will be discussed later.

Conservation of particles. It may also happen that, during any flow of the fluid, the number of particles in a fluid element will change, but of course the total number of particles in the fluid will not change. In particular, in Figure 4.8 the rate of change of the number of particles in a fluid element will be due only to loss or gain across the boundaries, i.e. to net fluxes out or in. This conservation law is derivable in the same way as Eq. 4.34. We can then write

$$\frac{\partial}{\partial t} N^0 = -\frac{\partial}{\partial x} N^x - \frac{\partial}{\partial y} N^y - \frac{\partial}{\partial z} N^z$$

or

$$N^\alpha{}_{,\alpha} = (nU^\alpha)_{,\alpha} = 0. \tag{4.35}$$

We will confine ourselves to discussing only fluids that obey this conservation law. This is hardly any restriction, since n can, if necessary, always be taken to be the density of baryons.

The term 'baryon', for those not familiar with high-energy physics, is a general name applied to the more massive particles in physics. The two commonest are the neutron and proton. All other baryons are too unstable to be important in everyday physics – but when they decay they form protons and neutrons, thus conserving the total number of baryons without conserving rest mass or particle identity. Although theoretical physics suggests that baryons may not always be conserved – for instance, so-called 'grand unified theories' of the strong, weak, and electromagnetic interactions may predict a finite lifetime for the proton, and the collapse of a star to, and subsequent evaporation of, a black hole (see Chapter 11) will not conserve baryon number – no such phenomena have yet been observed and, in any case, are unlikely to be important in most situations.

4.7 Perfect fluids

Finally, we come to the type of fluid which is our principal subject of interest. *A perfect fluid in relativity is defined as a fluid that has no viscosity and no heat conduction in the MCRF.* It is a generalization of the ideal gas of ordinary thermodynamics. It is, next to dust, the simplest kind of fluid to deal with. The two restrictions in its definition enormously simplify the stress–energy tensor, as we now see.

No heat conduction. From the definition of **T**, this immediately implies that, in the MCRF, $T^{0i} = T^{i0} = 0$. Energy can flow only if particles flow. Recall that in our discussion of the first law of thermodynamics we showed that if the number of particles was conserved then the specific entropy was related to heat flow by Eq. 4.26. This means that, in a perfect fluid, if Eq. 4.35 for the conservation of particles is obeyed, then we should also have that S is constant in time during the flow of the fluid. We shall soon see how this comes out of the conservation laws.

No viscosity. Viscosity is a force parallel to the interface between particles. Its absence means that the forces should always be perpendicular to the interface, i.e. that T^{ij} should be zero unless $i = j$. This means that T^{ij} should be a diagonal matrix. Moreover, it must be diagonal in *all* MCRF frames, since 'no viscosity' is a statement independent of the spatial axes. The only matrix that is diagonal in all frames is a multiple of the identity matrix: all its diagonal terms are equal. Thus, an x surface will have across it only a force in the x direction, and similarly for y and z surfaces; these forces per unit area are all equal, and are called the *pressure, p.* So we have $T^{ij} = p\delta^{ij}$. From six possible quantities (the number of independent elements in the 3×3 symmetric matrix T^{ij}) the zero-viscosity assumption has reduced the number of functions to one, the pressure.

Form of **T**. In the MCRF, **T** has the components we have just deduced:

$$(T^{\alpha\beta}) = \begin{pmatrix} \rho & 0 & 0 & 0 \\ 0 & p & 0 & 0 \\ 0 & 0 & p & 0 \\ 0 & 0 & 0 & p \end{pmatrix}, \tag{4.36}$$

It is not hard to show that in the MCRF

$$T^{\alpha\beta} = (\rho + p)U^\alpha U^\beta + p\eta^{\alpha\beta}. \tag{4.37}$$

For instance, if $\alpha = \beta = 0$ then $U^0 = 1, \eta^{00} = -1$, and $T^{\alpha\beta} = (\rho + p) - p = \rho$, as in Eq. 4.36. By trying all possible α and β you can verify that Eq. 4.37 gives Eq. 4.36. But Eq. 4.37 is a frame-invariant formula in the sense that it uniquely implies

$$\mathbf{T} = (\rho + p)\vec{U} \otimes \vec{U} + p\mathbf{g}^{-1}. \tag{4.38}$$

This is the stress–energy tensor of a perfect fluid.

Aside on the meaning of pressure. A comparison of Eq. 4.38 with Eq. 4.19 shows that 'dust' is the special case of a pressure-free perfect fluid. This means that a perfect fluid can be pressure free only if its particles have *no* random motion at all. Pressure arises in the random velocities of the particles. Even a gas so dilute as to be virtually collisionless has pressure. This is so because pressure is the flux of momentum; whether this comes from forces or from particles crossing a boundary is immaterial.

The conservation laws. Equation 4.34 gives us

$$T^{\alpha\beta}{}_{,\beta} = \left[(\rho + p) U^\alpha U^\beta + p\eta^{\alpha\beta} \right]_{,\beta} = 0. \tag{4.39}$$

This gives us our first real practice with tensor calculus. There are four equations in Eq. 4.39, one for each α. First, let us also assume

$$(nU^\beta)_{,\beta} = 0 \tag{4.40}$$

and write the first term in Eq. 4.39 as

$$\left[(\rho + p) U^\alpha U^\beta \right]_{,\beta} = \left[\frac{\rho + p}{n} U^\alpha n U^\beta \right]_{,\beta}$$

$$= nU^\beta \left(\frac{\rho + p}{n} U^\alpha \right)_{,\beta}. \tag{4.41}$$

Moreover, $\eta^{\alpha\beta}$ is a constant matrix, so $\eta^{\alpha\beta}{}_{,\gamma} = 0$. This also implies, by the way, that

$$U^\alpha{}_{,\beta} U_\alpha = 0. \tag{4.42}$$

The proof of Eq. 4.42 is as follows:

$$U^\alpha U_\alpha = -1 \implies (U^\alpha U_\alpha)_{,\beta} = 0 \tag{4.43}$$

or

$$(U^\alpha U^\gamma \eta_{\alpha\gamma})_{,\beta} = (U^\alpha U^\gamma)_{,\beta} \eta_{\alpha\gamma} = 2U^\alpha{}_{,\beta} U^\gamma \eta_{\alpha\gamma}. \tag{4.44}$$

The last step follows from the symmetry of $\eta_{\alpha\beta}$, which means that $U^\alpha{}_{,\beta} U^\gamma \eta_{\alpha\gamma} = U^\alpha U^\gamma{}_{,\beta} \eta_{\alpha\gamma}$. Finally, the last expression in Eq. 4.44 converts to

$$2U^\alpha{}_{,\beta} U_\alpha,$$

which is zero by Eq. 4.43. This proves Eq. 4.42. We can make use of Eq. 4.42 in the following way. The original equation now reads, after use of Eq. 4.41,

$$nU^\beta \left(\frac{\rho + p}{n} U^\alpha \right)_{,\beta} + p_{,\beta} \eta^{\alpha\beta} = 0. \tag{4.45}$$

From the four equations here, we can obtain a particularly useful one. Multiply by U_α and sum on α. This gives the time component of Eq. 4.45 in the MCRF:

$$nU^\beta U_\alpha \left(\frac{\rho + p}{n} U^\alpha \right)_{,\beta} + p_{,\beta} \eta^{\alpha\beta} U_\alpha = 0. \tag{4.46}$$

The last term is just

$$p_{,\beta} U^\beta,$$

which we know to be the derivative of p along the world line of the fluid element, $dp/d\tau$. The first term gives zero when the β-derivative operates on U^α (by Eq. 4.42), so we obtain (using $U^\alpha U_\alpha = -1$)

$$U^\beta \left[-n \left(\frac{\rho + p}{n} \right)_{,\beta} + p_{,\beta} \right] = 0. \tag{4.47}$$

A little algebra converts this to

$$-U^\beta \left[\rho_{,\beta} - \frac{\rho + p}{n} n_{,\beta} \right] = 0. \tag{4.48}$$

Written another way, we arrive at

$$\frac{d\rho}{d\tau} - \frac{\rho + p}{n} \frac{dn}{d\tau} = 0. \tag{4.49}$$

This is to be compared with Eq. 4.25. It means that

$$U^\alpha S_{,\alpha} = \frac{dS}{d\tau} = 0. \tag{4.50}$$

Thus, the flow of a particle-conserving perfect fluid conserves specific entropy. This is called an *adiabatic* process. Because entropy is constant in a fluid element as it flows, we shall not normally need to consider it. Nevertheless, it is important to remember that the law of conservation of energy in thermodynamics is embodied in the component of the conservation equations, Eq. 4.39, that is parallel to U^α.

The remaining three components of Eq. 4.39 are derivable in the following way. We write out Eq. 4.45 again:

$$nU^\beta \left(\frac{\rho + p}{n} U^\alpha \right)_{,\beta} + p_{,\beta} \eta^{\alpha\beta} = 0$$

and go to the MCRF, where $U^i = 0$ but $U^i_{,\beta} \neq 0$. In the MCRF, the 0-component of this equation is the same as its contraction with U_α, which we have just examined. So we only need the i-components:

$$nU^\beta \left(\frac{\rho + p}{n} U^i \right)_{,\beta} + p_{,\beta} \eta^{i\beta} = 0. \tag{4.51}$$

Since $U^i = 0$, the β-derivative of $(\rho + p)/n$ contributes nothing, and we get

$$(\rho + p)U^i_{,\beta} U^\beta + p_{,\beta} \eta^{i\beta} = 0. \tag{4.52}$$

Lowering the index i makes this easier to read (and changes nothing). Since $\eta_i{}^\beta = \delta_i{}^\beta$ we get

$$(\rho + p)U_{i,\beta} U^\beta + p_{,i} = 0. \tag{4.53}$$

Finally, we recall that $U_{i,\beta}U^{\beta}$ is the definition of the four-acceleration a_i:

$$(\rho + p)a_i + p_{,i} = 0. \tag{4.54}$$

Those familiar with nonrelativistic fluid dynamics will recognize this as the generalization of

$$\rho\boldsymbol{a} + \boldsymbol{\nabla}p = 0, \tag{4.55}$$

where

$$\boldsymbol{a} = \dot{\boldsymbol{v}} + (\boldsymbol{a} \cdot \boldsymbol{\nabla})\boldsymbol{a}. \tag{4.56}$$

The only difference is the use of $\rho + p$ instead of ρ. In relativity, $\rho + p$ plays the role of the 'inertial mass density' in that, from Eq. 4.54, the larger $\rho + p$, the harder it is to accelerate the object. Equation 4.54 is essentially $\boldsymbol{F} = m\boldsymbol{a}$, with $-p_{,i}$ the force on a fluid element. That is, p is the force a fluid element exerts on its neighbor, so $-p$ is the force on the element. But the neighbor on the opposite side of the element is pushing the other way, so only if there is a change in p across the fluid element will there be a net force causing it to accelerate. That is why $-\boldsymbol{\nabla}p$ is the force.

4.8 Importance for general relativity

General relativity is a relativistic theory of gravity. We were not able to plunge into it immediately because we lacked a good enough understanding of tensors, of fluids in SR, and of curved spaces. We have yet to study curvature (that comes next), but at this point we can look ahead and discern the vague outlines of the theory we shall study.

Our first comment is on the supreme importance of **T** in GR. Newton's theory has as a source of the gravitational field the density ρ. This was understood to be the mass density, and so is closest to our ρ_0. However, a theory that uses only rest mass as its source would be peculiar from a relativistic viewpoint, since rest mass and energy are interconvertible. In fact, one can show that such a theory would violate some very-high-precision experiments (to be discussed later). So the source of the field should be *all* energies, the density of the total mass energy T^{00}. But to have as the source of the field only one component of a tensor would give a noninvariant theory of gravity: one would need to choose a preferred frame in order to calculate T^{00}. Therefore Einstein guessed that the source of the gravitational field ought to be **T**: all stresses and pressures and momenta must also act as sources. Combining this with his insight into curved spaces led him to GR.

The second comment is about pressure, which plays a more fundamental role in GR than in Newtonian theory: first, because it is a source of the field; and second, because of its appearance in the $\rho + p$ term in Eq. 4.54. Consider a dense star, whose strong gravitational field requires a large pressure gradient for stability. The size of this pressure gradient is measured by the acceleration that the fluid element would have, a_i, in the absence of pressure. Given the field, and hence a_i, the required pressure gradient is just that which would cause the opposite acceleration in the absence of gravity:

$$-a_i = \frac{p_{,i}}{\rho + p}.$$

This gives the pressure gradient $p_{,i}$. Since $\rho + p$ is greater than ρ, the pressure gradient must be larger in relativity than in Newtonian theory. Moreover, since all components of **T** are sources of the gravitational field, this larger pressure gradient adds to the gravitational field, causing even larger pressures (compared with Newtonian stars) to be required to hold the star up. For stars where $p \ll \rho$ (see below), this does not make much difference. But when p becomes comparable with ρ, one finds that increasing the pressure is self-defeating: *no* pressure gradient will hold the star up, and gravitational collapse must occur. This description, of course, glosses over much detailed calculation, but it shows that even by just studying fluids in SR we can begin to appreciate some of the fundamental changes GR brings to gravitation.

Let us just remind ourselves of the relative sizes of p and ρ. We saw earlier that $p \ll \rho$ in ordinary situations. In fact, $p \approx \rho$ holds only for very dense material (as in a neutron star) or material so hot that the particles are moving at nearly the speed of light (a 'relativistic' gas).

4.9 Gauss' law

Our final topic on fluids is the integral form of the conservation laws, which were expressed in differential form in Eqs. 4.34 and 4.35. As in three-dimensional vector calculus, the relation giving the conversion of a volume integral of a divergence into a surface integral is called Gauss' law. The proof of the theorem is exactly the same as in three dimensions, so we shall not derive it in detail:

$$\int V^\alpha{}_{,\alpha} \, d^4x = \oint V^\alpha \tilde{n}_\alpha \, d^3S, \tag{4.57}$$

where $\tilde{n}$ is the unit-normal one-form discussed in § 4.3, and d^3S denotes the three-volume of the three-dimensional hypersurface bounding the four-dimensional volume of integration. The sense of the normal is that it is *outward* pointing, of course, just as in three dimensions. In Figure 4.9 a simple volume is drawn, in order to illustrate the meaning of Eq. 4.57. The volume is bounded by four pairs of hypersurfaces, for constant t, x, y and z; only two pairs are shown, since we can only draw two dimensions easily. The normal on the t_2 surface is $\tilde{d}t$. The normal on the t_1 surface is $-\tilde{d}t$, since 'outward' is clearly

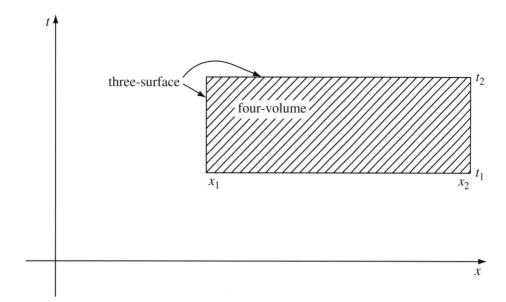

Figure 4.9 The boundary of a region of spacetime.

backwards in time. The normal on the x_2 surface is $\tilde{\mathrm{d}}x$, and that on the x_1 surface is $-\tilde{\mathrm{d}}x$. So the surface integral in Eq. 4.57 is

$$\int_{t_2} V^0 \,\mathrm{d}x\,\mathrm{d}y\,\mathrm{d}z + \int_{t_1} (-V^0)\,\mathrm{d}x\,\mathrm{d}y\,\mathrm{d}z$$

$$+ \int_{x_2} V^x \mathrm{d}t\,\mathrm{d}y\,\mathrm{d}z + \int_{x_1} (-V^x)\,\mathrm{d}t\,\mathrm{d}y\,\mathrm{d}z$$

$$+ \text{ similar terms for the other surfaces in the boundary.}$$

We can rewrite this as

$$\int \left[V^0(t_2) - V^0(t_1) \right] \mathrm{d}x\,\mathrm{d}y\,\mathrm{d}z$$

$$+ \int \left[V^x(x_2) - V^x(x_1) \right] \mathrm{d}t\,\mathrm{d}y\,\mathrm{d}z + \cdots . \tag{4.58}$$

If we replace $\vec{V}$ by the number of particles $\vec{N}$, then $N^\alpha{}_{,\alpha} = 0$ means that the above expression vanishes, which has the interpretation that the change in the number of particles in the three-volume (the first integral) is due to the flux across its boundaries (the second and subsequent terms). If instead we are talking about energy conservation then we replace N^α with $T^{0\alpha}$, and use $T^{0\alpha}{}_{,\alpha} = 0$. Then, obviously, a similar interpretation of Eq. 4.58 applies: Gauss' law gives an integral version of energy conservation.

4.10 Bibliography

Continuum mechanics and conservation laws are treated in most texts on GR. Students whose background in thermodynamics or fluid mechanics is weak are referred to the classic works of Fermi (1956) and Landau & Lifshitz (1987) respectively. Apart from in Exercise 4.25 below, we do not study much about electromagnetism, but it has a stress–energy tensor and illustrates conservation laws particularly clearly. See Landau & Lifshitz (1980) or Jackson (1998). Relativistic fluids with dissipation present their own difficulties, which reward close study. See Israel & Stewart (1980). Another model for continuum systems is the collisionless gas; see Andréasson (2011) for a description of how to treat such systems in GR.

Exercises

4.1 Comment on whether the continuum approximation is likely to apply to the following physical systems: (a) planetary motions in the solar system; (b) lava flow from a volcano; (c) traffic on a major road at rush hour; (d) traffic at an intersection controlled by stop signs for each incoming road; (e) plasma dynamics.

4.2 Flux across a surface of constant x is often loosely called 'flux in the x direction'. Use your understanding of vectors and one-forms to argue that this is an inappropriate way of referring to a flux.

4.3 (a) Describe how the Galilean concept of momentum is frame dependent in a manner in which the relativistic concept is not.

(b) How is this possible, since the relativistic definition is nearly the same as the Galilean one for small velocities? (Define a *Galilean* four-momentum vector.)

4.4 Show that the number density of dust measured by an arbitrary observer whose four-velocity is $\vec{U}_{\text{obs}}$ is $-\vec{N} \cdot \vec{U}_{\text{obs}}$.

4.5 Complete the proof that Eq. 4.14 defines a tensor by arguing that it must be linear in both its arguments.

4.6 Establish Eq. 4.19 from the preceding equations.

4.7 Derive Eq. 4.21.

4.8 (a) Argue that Eqs. 4.25 and 4.26 can be written as relations among one-forms, i.e.

$$\tilde{d}\rho - (\rho + p)\tilde{d}n/n = nT\,\tilde{d}S = n\tilde{\Delta}q.$$

(b) Show that the one-form $\tilde{\Delta}q$ is not a gradient, i.e. it is not $\tilde{d}q$ for any function q.

4.9 Show that Eq. 4.34, when α is any spatial index, is just Newton's second law.

4.10 Take the limit of Eq. 4.35 for $|\mathbf{v}| \ll 1$ to get

$$\partial n/\partial t + \partial(nv^i)/\partial x^i = 0.$$

4.11 (a) Show that the matrix δ^{ij} is unchanged when transformed by a rotation of the spatial axes.

(b) Show that any matrix which has this property is a multiple of δ^{ij}.

4.12 Derive Eq. 4.37 from Eq. 4.36.

4.13 Supply the reasoning in Eq. 4.44.

4.14 Argue that Eq. 4.46 is the time component of Eq. 4.45 in the MCRF.

4.15 Derive Eq. 4.48 from Eq. 4.47.

4.16 In the MCRF, $U^i = 0$. Why can we not assume that $U^i{}_{,\beta} = 0$?

4.17 We have defined $a^\mu = U^\mu{}_{,\beta} U^\beta$. Go to the nonrelativistic limit (small velocity) and show that

$$a^i = \dot{v}^i + (\mathbf{v} \cdot \boldsymbol{\nabla}) v^i = Dv^i/Dt,$$

where the operator D/Dt is the usual 'total' or 'advective' time derivative of fluid dynamics.

4.18 Sharpen the discussion at the end of § 4.7 by showing that $-\boldsymbol{\nabla}p$ is actually the net force per unit volume on the fluid element in the MCRF.

4.19 Show that Eq. 4.58 can be used to prove Gauss' law, Eq. 4.57.

4.20 (a) Show that, if particles are not conserved but are generated locally at a rate ε particles per unit volume per unit time in the MCRF, then the conservation law, Eq. 4.35, becomes

$$N^\alpha{}_{,\alpha} = \varepsilon.$$

(b) Generalize (a) to show that if the energy and momentum of a body are not conserved (e.g. because it interacts with other systems), then we can define a nonzero relativistic force four-vector F^α:

$$T^{\alpha\beta}{}_{,\beta} = F^\alpha.$$

Interpret the components of F^α in the MCRF.

4.21 In an inertial frame $\mathcal{O}$ calculate the components of the stress–energy tensors of the following systems:

(a) A group of particles all moving with the same velocity $\mathbf{v} = \beta \mathbf{e}_x$, as seen in $\mathcal{O}$. Let the rest-mass density of these particles be ρ_0, as measured in their comoving frame. Assume a sufficiently high density of particles to enable treating them as a continuum.

(b) A ring of N similar particles of mass m rotating counterclockwise in the x–y plane about the origin of $\mathcal{O}$, at a radius a from this point, with angular velocity ω. The ring is a torus with circular cross-section of radius $\delta a \ll a$, within which the particles are uniformly distributed with a high enough density for the continuum approximation to apply. Do not include the stress–energy of whatever forces keep them in orbit. (An aspect of the calculation will relate ρ_0 of part (a) to N, a, ω, and δa.)

(c) Two such rings of particles, one rotating clockwise and the other counterclockwise, at the same radius a. The particles do not collide or interact in any way.

4.22 Many physical systems may be idealized as collections of noncolliding particles (for example, black-body radiation, rarified plasmas, galaxies, and globular clusters). By assuming that such a system has a random distribution of velocities at every point, with no bias in any direction in the MCRF, prove that the stress–energy tensor is that of a perfect fluid. If all particles have the same speed v and mass m, express p and ρ as functions of m, v, and n. Show that a photon gas has $p = \frac{1}{3}\rho$.

4.23 Use the identity $T^{\mu\nu}{}_{,\nu} = 0$ to prove the following results for a bounded system (i.e. a system for which $T^{\mu\nu} = 0$ outside a bounded region of space).

(a) $\dfrac{\partial}{\partial t} \int T^{0\alpha}\, \mathrm{d}^3 x = 0$ (conservation of energy and momentum).

(b) $\dfrac{\partial^2}{\partial t^2} \int T^{00} x^i x^j\, \mathrm{d}^3 x = 2 \int T^{ij}\, \mathrm{d}^3 x$ (tensor virial theorem).

(c) $\dfrac{\partial^2}{\partial t^2} \int T^{00} (x^i x_i)^2\, \mathrm{d}^3 x = 4 \int T^i{}_i x^j x_j\, \mathrm{d}^3 x + 8 \int T^{ij} x_i x_j\, \mathrm{d}^3 x$.

4.24 Astronomical observations of the brightness of objects are measurements of the flux of radiation T^{0i} from the object at Earth. This problem calculates how that flux depends on the relative velocity of the object and Earth.

(a) Show that, in the rest frame $\mathcal{O}$ of a star of constant luminosity L (total energy radiated per second), the stress–energy tensor of the radiation from the star at the event $(t, x, 0, 0)$ has components $T^{00} = T^{0x} = T^{x0} = T^{xx} = L/(4\pi x^2)$. The star sits at the origin.

(b) Let $\vec{X}$ be the null vector which separates the events of emission and reception of the radiation. Show that $\vec{X} \rightarrow_{\mathcal{O}} (x, x, 0, 0)$ for radiation observed at the event $(x, x, 0, 0)$. Show that the stress–energy tensor of (a) has the frame-invariant form

$$ \mathbf{T} = \frac{L}{4\pi} \frac{\vec{X} \otimes \vec{X}}{(\vec{U}_s \cdot \vec{X})^4}, $$

where $\vec{U}_s$ is the star's four-velocity, $\vec{U}_s \rightarrow_{\mathcal{O}} (1, 0, 0, 0)$.

(c) Let the Earth-bound observer $\bar{\mathcal{O}}$, traveling with speed v away from the star in the x direction, measure the same radiation, again with the star on the $\bar{x}$ axis. Let $\vec{X} \rightarrow (R, R, 0, 0)$ and find R as a function of x. Express $T^{\bar{0}\bar{x}}$ in terms of R. Explain why R and $T^{\bar{0}\bar{x}}$ depend on v in the way that they do.

4.25 *Electromagnetism in SR.* (This exercise is suitable only for students who have already encountered Maxwell's equations in some form.) Maxwell's equations for the electric and magnetic fields in vacuum, **E** and **B**, in three-vector notation are

$$ \nabla \times \mathbf{B} - \frac{\partial}{\partial t}\mathbf{E} = 4\pi \mathbf{J}, $$
$$ \nabla \times \mathbf{E} + \frac{\partial}{\partial t}\mathbf{B} = 0, \tag{4.59} $$
$$ \nabla \cdot \mathbf{E} = 4\pi \rho, $$
$$ \nabla \cdot \mathbf{B} = 0, $$

in units where $\mu_0 = \varepsilon_0 = c = 1$. (Here ρ is the density of electric charge and **J** the current density.)

(a) An *antisymmetric* $\binom{2}{0}$ tensor **F** can be defined on spacetime by the equations $F^{0i} = E^i \, (i = 1, 2, 3), F^{xy} = B^z, F^{yz} = B^x, F^{zx} = B^y$. From this definition find all other components $F^{\mu\nu}$ in this frame and write them down in a matrix.

(b) A rotation by an angle θ about the z axis is one kind of Lorentz transformation, with matrix

$$\Lambda^{\beta'}{}_{\alpha} = \begin{pmatrix} 1 & 0 & 0 & 0 \\ 0 & \cos\theta & -\sin\theta & 0 \\ 0 & \sin\theta & \cos\theta & 0 \\ 0 & 0 & 0 & 1 \end{pmatrix}.$$

Show that the new components of **F**,

$$F^{\alpha'\beta'} = \Lambda^{\alpha'}{}_{\mu}\Lambda^{\beta'}{}_{\nu}F^{\mu\nu},$$

define new electric and magnetic three-vector components (by the rule given in (a)) that are just the same as the components of the old **E** and **B** in the rotated three-space. (This shows that a spatial rotation of **F** causes a spatial rotation of **E** and **B**.)

(c) Define the current four-vector $\vec{J}$ by $J^0 = \rho, J^i = (J)^i$, and show that two of Maxwell's equations are just given by

$$F^{\mu\nu}{}_{,\nu} = 4\pi J^{\mu}. \tag{4.60}$$

(d) Show that the other two of Maxwell's equations are

$$F_{\mu\nu,\lambda} + F_{\nu\lambda,\mu} + F_{\lambda\mu,\nu} = 0. \tag{4.61}$$

Note that there are only *four* independent equations here. That is, choose one index value, say 0. Then the three other values (1, 2, 3) can be assigned to μ, ν, λ in *any* order, producing the same equation (up to an overall sign) each time. Try it and see: it follows from antisymmetry of $F_{\mu\nu}$.

(e) We have now expressed Maxwell's equations in tensor form. Show that conservation of charge, $J^{\mu}{}_{,\mu} = 0$ (recall Eq. 4.35 for the number–flux vector $\vec{N}$, which is similar), is implied by Eq. 4.60 above. (Hint: use the antisymmetry of $F_{\mu\nu}$.)

(f) The charge density in any frame is J^0. Therefore the total charge in spacetime is $Q = \int J^0 \mathrm{d}x\mathrm{d}y\mathrm{d}z$, where the integral extends over an entire hypersurface $t = $ const. Defining $\tilde{\mathrm{d}t} = \tilde{n}$, a unit normal for this hypersurface, show that

$$Q = \int J^{\alpha} n_{\alpha} \, \mathrm{d}x \, \mathrm{d}y \, \mathrm{d}z. \tag{4.62}$$

(g) Use Gauss' law and Eq. 4.60 to show that the total charge enclosed within any closed two-surface $\mathcal{S}$ in the hypersurface $t = $ const. can be determined by integrating over $\mathcal{S}$ itself:

$$Q = \oint_{\mathcal{S}} F^{0i} n_i \, \mathrm{d}\mathcal{S} = \oint_{\mathcal{S}} \mathbf{E} \cdot \mathbf{n} \, \mathrm{d}\mathcal{S},$$

where **n** is the unit normal to $\mathcal{S}$ in the hypersurface (*not* the same as $\tilde{n}$ in part (f) above).

(h) Perform a Lorentz transformation on $F^{\mu\nu}$ to a frame $\bar{\mathcal{O}}$ moving with velocity v in the x direction relative to the frame used in (a) above. In this frame define a three-vector $\bar{\mathbf{E}}$ with components $\bar{\mathbf{E}}^i = F^{\bar{0}\bar{i}}$, and similarly for $\bar{\mathbf{B}}$ in analogy with (a). In this way discover how $\mathbf{E}$ and $\mathbf{B}$ behave under a Lorentz transformation: they get mixed together! Thus, $\mathbf{E}$ and $\mathbf{B}$ themselves are not Lorentz invariant, but are merely components of $\mathbf{F}$, called the Faraday tensor, which is *the* invariant description of electromagnetic fields in relativity. If you think carefully, you will see that on physical grounds they *cannot* be invariant. In particular, the magnetic field is created by moving charges; but a charge moving in one frame may be at rest in another, so a magnetic field which exists in one frame may not exist in another. What is the same in *all* frames is the Faraday tensor: only its components get transformed.

Preface to curvature

5.1 On the relation of gravitation to curvature

Until now we have discussed only SR. In SR, forces have played a background role and we have never introduced gravitation explicitly as a possible force. One ingredient of SR is the existence of inertial frames that fill all of spacetime: the whole of spacetime can be described by a single frame, all of whose coordinate points are always at rest relative to the origin, and all of whose clocks run at the same rate relative to the origin's clock. From the fundamental postulates we were led to the idea of the interval Δs^2, which gives an invariant geometrical meaning to certain physical statements. For example, a timelike interval between two events is the time elapsed on a clock which passes through the two events; a spacelike interval is the length of a rod that joins two events in a frame in which they are simultaneous. The mathematical function that calculates intervals is the metric, and so the metric of SR is defined physically by the lengths of rods and readings of clocks. This is the power of SR and is one reason for the elegance and compactness of tensor notation in it (for instance the replacement of 'number density' and 'flux' by $\vec{N}$). On a piece of paper on which one had plotted all the events and world lines of interest in some coordinate system, it would always be possible to define *any* metric by just giving its components $g_{\alpha\beta}$ as some arbitrarily chosen set of functions of the coordinates. But this arbitrary metric would be useless in doing physical calculations. The usefulness of $\eta_{\alpha\beta}$ is its close relation to experiment, and our derivation of it drew heavily on the experiments.

This closeness to experiment is, of course, a test. Since $\eta_{\alpha\beta}$ makes certain predictions about rods and clocks, one can ask for their verification. In particular, is it *possible* to construct a frame in which the clocks all run at the same rate? This is a crucial question, and we shall show that in a nonuniform gravitational field the answer, experimentally, is no. In this sense, gravitational fields are incompatible with *global* SR: the ability to construct a global inertial frame. We shall see that in small regions of spacetime – regions small enough that nonuniformities of the gravitational forces are too small to measure – one can always construct a 'local' SR frame. In this sense, we shall have to build local SR into a more general theory. The first step is the proof that clocks don't all run at the same rate in a gravitational field.

The gravitational redshift experiment

Let us imagine performing an idealized experiment, first suggested by Einstein. (i) Let a tower of height h be constructed on the surface of Earth, as in Figure 5.1. Begin with a

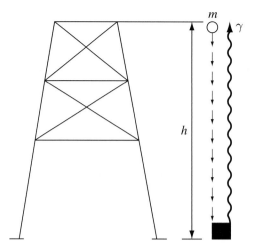

Figure 5.1 A mass m is dropped from a tower of height h. The total mass at the bottom is converted into energy and returned to the top as a photon. Perpetual motion will be performed unless the photon loses as much energy in climbing as the mass gained in falling. Light is therefore redshifted as it climbs in a gravitational field.

particle of rest mass m at the top of the tower. (ii) The particle is dropped and falls freely with acceleration g. It reaches the ground with velocity $v = (2gh)^{1/2}$, so its total energy E, as measured by an experimenter on the ground, is $m + \frac{1}{2}mv^2 + \mathrm{0}(v^4) = m + mgh + \mathrm{0}(v^4)$. (iii) The experimenter on the ground has some magical method of changing all this energy into a single photon of the same energy, which is then directed upwards. (Such a process does not violate conservation laws, since Earth absorbs the photon's momentum but not its energy, just as it does for a bouncing rubber ball. If you are skeptical of 'magic', you may try to show how the argument proceeds if only a fraction ε of the energy is converted into a photon.) (iv) Upon its arrival at the top of the tower with energy E', the photon is again magically changed into a particle of rest mass $m' = E'$. It must be that $m' = m$; otherwise, perpetual motion could result by the gain in energy obtained by operating such an experiment. So we are led by our refusal to allow perpetual motion to *predict* that $E' = m$ or, for the photon,

$$\frac{E'}{E} = \frac{h\nu'}{h\nu} = \frac{m}{m + mgh + \mathrm{O}(v^4)} = 1 - gh + \mathrm{O}(v^4). \tag{5.1}$$

We thus predict that a photon climbing in Earth's gravitational field will lose energy (not surprisingly) and will consequently be redshifted.

Although our thought experiment is too idealized to be practical, it is possible to measure the redshift predicted by Eq. 5.1 directly. This was first done by Pound & Rebka (1960) and improved by Pound & Snider (1965). The experiment used the Mössbauer effect to obtain great precision in the measurement of the difference $\nu' - \nu$ produced in a photon climbing a distance $h = 22.5$ m. Equation 5.1 was verified to approximately 1% precision.

With improvements in technology between 1960 and 1990, the gravitational redshift moved from being a small exotic correction to becoming an effect that is central to society: the GPS navigation system incorporates vital corrections for the redshift, in the absence of which it would not remain accurate for more than a few minutes. The system uses a network of high-precision atomic clocks in orbiting satellites, and navigation by an apparatus on Earth is accomplished by reading the time-stamps on signals received from five or more satellites. But, as we shall see below, the gravitational redshift implies that time itself runs slightly faster at the higher altitude than it does on the Earth. If this were not compensated for, the ground receiver would soon get wrong time-stamps. The successful operation of GPS can be taken to be a very accurate verification of the redshift. See Ashby (2003) for a full discussion of relativity and the GPS system.

This experimental verification of the redshift is comforting from the point of view of energy conservation. But it is the death blow to our chances of finding a simple, special-relativistic theory of gravity, as we shall now show.

Nonexistence of a Lorentz frame at rest on Earth

If SR is to be valid in a gravitational field, it is a natural first guess to assume that the 'laboratory' frame at rest on Earth is a Lorentz frame. The following argument, due originally to Schild (1967), easily shows this assumption to be false. In Figure 5.2 we draw a spacetime diagram in this hypothetical frame, in which the one spatial dimension plotted is the vertical dimension. Consider light as a wave, and look at two successive 'crests' of a light wave as they move upward in the Pound–Rebka–Snider experiment. The top and bottom of the tower have vertical world lines in this diagram, since they are at rest. The light is shown moving on a wiggly line, and it is purposely drawn curved in some arbitrary way. This is to allow for the possibility that gravity may act on light in an unknown way,

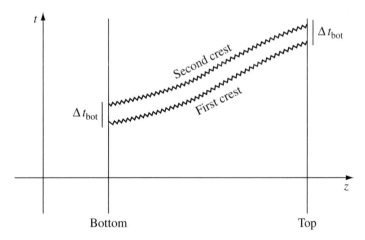

Figure 5.2 In a time-independent gravitational field, two successive 'crests' of an electromagnetic wave must travel identical paths. Because of the redshift (Eq. 5.1) the time between them at the top is larger than at the bottom. An observer at the top therefore 'sees' a clock at the bottom running slowly.

deflecting it from a null path. But no matter how light is affected by gravity the effect must be the same on both wave crests, since the gravitational field does not change from one time to another. Therefore the two crests' paths are *congruent*, and we would conclude from the hypothetical Minkowski geometry that $\Delta t_{\text{top}} = \Delta t_{\text{bottom}}$. On the other hand, the time between two crests is simply the reciprocal of the measured frequency $\Delta t = 1/\nu$. Since the Pound–Rebka–Snider experiment establishes that $\nu_{\text{bottom}} > \nu_{\text{top}}$, we know that $\Delta t_{\text{top}} > \Delta t_{\text{bottom}}$. The conclusion from Minkowski geometry is wrong, and the reference frame at rest on Earth is not a Lorentz frame.

Is this the end, then, of SR? Not quite. We have shown that the Lorentz frame at rest on Earth is not inertial. We have not shown that there are *no* inertial frames. In fact there are certain frames which are inertial in a restricted sense, and in the next paragraph we shall use another physical argument to find them.

The principle of equivalence

One important property of an inertial frame is that a particle at rest in it stays at rest if no forces act on it. In order to use this, we have to have an idea of what a force is. Ordinarily, gravity is regarded as a force. But, as Galileo demonstrated in his famous experiment at the Leaning Tower of Pisa, gravity is distinguished from all other forces in a remarkable way: all bodies given the same initial velocity follow the same trajectory in a gravitational field, regardless of their internal composition. With all other forces, some bodies are affected and others are not: electromagnetism affects charged particles but not neutral ones, and the trajectory of a charged particle depends on the ratio of its charge to its mass, which is not the same for all particles. Similarly, the other two basic forces in physics – the so-called 'strong' and 'weak' interactions – affect different particles differently. With all these forces, it would always be possible to define experimentally the trajectory of a particle unaffected by the force, i.e. a particle that remained at rest in an inertial frame. But, with gravity, this does not work. Attempting to define an inertial frame at rest on Earth, then, is vacuous, since *no* free particle (not even a photon) could possibly be a physical marker for it.

But there is a frame in which particles do keep a uniform velocity. This is a frame which falls freely in the gravitational field. Since this frame accelerates at the same rate as free particles do (at least the low-velocity particles to which Newtonian gravitational physics applies), it follows that all such particles will maintain a uniform velocity relative to this frame. This frame is at least a candidate for an inertial frame. In the next section we will show that photons are not redshifted in this frame, which makes it an even better candidate. Einstein built GR by taking the hypothesis that these frames are inertial.

The argument we have just made, that freely falling frames are inertial, will perhaps be more familiar to you if it is turned around. Consider, in empty space free of gravity, a uniformly accelerating rocket ship. From the point of view of an observer inside, it appears that there is a gravitational field in the rocket: objects dropped accelerate toward the rear of the ship, all with the same acceleration, independent of their internal composition.[1]

---

[1] This has been tested experimentally to extremely high precision in the so-called Eötvös experiment. See Dicke (1964).

Moreover, an object held stationary relative to the ship has 'weight' equal to the force required to keep it accelerating with the ship. Just as in 'real' gravity, this force is proportional to the mass of the object. A true inertial frame is one which falls freely toward the rear of the ship, at the same acceleration as particles. From this it can be seen that uniform gravitational fields are equivalent to frames that accelerate uniformly relative to inertial frames. This is the *principle of equivalence* between gravity and acceleration, and is a cornerstone of Einstein's theory. Although Galileo and Newton would have used different words to describe it, the equivalence principle is one of the foundations of Newtonian gravity.

In more modern terminology, what we have described is called the *weak equivalence principle*, 'weak' because it refers to the way bodies behave only when influenced by gravity. Einstein realized that, in order to create a full theory of gravity, he had to extend this to include the other laws of physics. What we now call the *Einstein equivalence principle* says that one can discover how all the other forces of nature behave in a gravitational field by postulating that the differential equations that describe the laws of physics have the same local form in a freely falling inertial frame as they do in SR, i.e. when there are no gravitational fields. We shall use this stronger form of the principle of equivalence in Chapter 7.

Before we return to the proof that freely falling frames are inertial, even for photons, we must make two important observations. The first is that our arguments are valid only locally – since the gravitational field of Earth is not uniform, particles some distance away do not remain at uniform velocity in a particular freely falling frame. We shall discuss this in some detail below. The second point is that there is of course an infinity of freely falling frames at any point. They differ in their velocities and in the orientation of their spatial axes, but they all accelerate relative to Earth at the same rate.

The redshift experiment again. Let us now take a different point of view on the Pound–Rebka–Snider experiment. Let us view it in a freely falling frame, which we have seen has at least some of the characteristics of an inertial frame. Let us take the particular frame which is at rest when the photon begins its upward journey and which falls freely after that. Since the photon rises a distance h, it takes a time $\Delta t = h$ to arrive at the top. In this time, the frame has acquired velocity gh downward relative to the experimental apparatus. So the photon's frequency relative to the freely falling frame can be obtained by the redshift formula

$$\frac{\nu(\text{freely falling})}{\nu'(\text{apparatus at top})} = \frac{1 + gh}{\sqrt{(1 - g^2 h^2)}} = 1 + gh + \mathrm{O}(v^4). \qquad (5.2)$$

From Eq. 5.2 we see that if we neglect terms of higher order (as we did to derive Eq. 5.1), then we get ν(photon emitted at bottom) = ν(in freely falling frame when photon arrives at top). So there is *no* redshift in a freely falling frame. This gives us a sound basis for postulating that the freely falling frame is an inertial frame.

Local inertial frames. The above discussion makes one suggest that the gravitational redshift experiment really does not render SR and gravity incompatible. Perhaps one simply has to realize that the frame at rest on Earth is not inertial and that a freely falling

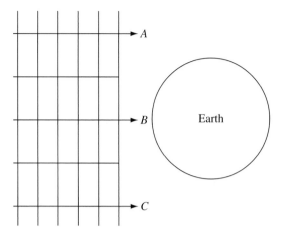

Figure 5.3   A rigid frame cannot fall freely in the Earth's field and still remain rigid.

frame – in which there is no redshift and so Figure 5.2 leads to no contradiction – is a true inertial frame. Unfortunately this does not completely save SR, for the simple reason that freely falling frames on different sides of Earth fall in different directions: there is *no* single global frame which is everywhere freely falling in Earth's gravitational field and which is still rigid, in that the distances between its coordinate points are constant in time. It is still impossible to construct a *global* inertial frame, and so the most we can salvage is a *local* inertial frame, which we now describe.

Consider a freely falling frame in Earth's gravitational field. An inertial frame in SR fills all of spacetime, but the freely falling frame would not be inertial if it extended too far horizontally, because then it would not be falling vertically. In Figure 5.3 the frame is freely falling at *B*, but at *A* and *C* the motion is not along the trajectory of a test particle. Moreover, since the acceleration of gravity changes with height, the frame cannot remain inertial if extended over too large a vertical distance; if it were falling with particles at one height, it would not be at another. Finally, the frame can have only a limited extent in time as well, since, as it falls, both the above limitations become more severe due to the frame's approaching closer to Earth. All these limitations are due to nonuniformities in the gravitational field. Insofar as nonuniformities can be neglected, the freely falling frame can be regarded as inertial. *Any* gravitational field can be regarded as uniform over a small enough region of space and time, and so one can always set up *local* inertial frames. They are analogous to the MCRFs of fluids: in this case the frame is inertial in only a small region for a small time. How small depends on (a) the strength of the nonuniformities of the gravitational field, and (b) the sensitivity of whatever experiment is being used to detect noninertial properties of the frame. Since *any* nonuniformity is, in principle, detectable, a frame can only be regarded mathematically as inertial in a vanishingly small region. But, for current technology, freely falling frames near the surface of Earth can be regarded as inertial to high accuracy. We will be more quantitative in a later chapter. For now, we just emphasize the mathematical notion that any theory of gravity must admit *local inertial frames*: frames that, at a point, are inertial frames of SR.

Tidal forces

Nonuniformities in gravitational fields are called tidal forces, since they are the forces that raise tides. (If Earth were in a uniform gravitational field, it would fall freely and have no tides. Tides bulge due to the *differences* in the Moon's and Sun's gravitational fields across the diameter of Earth.) We have seen that these tidal forces prevent the construction of global inertial frames. It is therefore these forces that are regarded as the fundamental manifestation of gravity in GR.

The role of curvature. The world lines of free particles have been our probe of the possibility of constructing inertial frames. In SR, two such world lines which begin parallel to each other remain parallel, no matter how far they are extended. This is exactly the property that straight lines have in Euclidean geometry. It is natural, therefore, to discuss the *geometry* of spacetime as defined by the world lines of free particles. In these terms, Minkowski space is a *flat* space, because it obeys Euclid's parallelism axiom. It is not a Euclidean space, however, since its metric is different: photons travel on straight world lines of zero proper length. So SR has a flat, nonEuclidean geometry.

Now, in a nonuniform gravitational field the world lines of two nearby particles which begin parallel do not generally remain parallel. Gravitational spacetime is therefore not flat. In Euclidean geometry, when one drops the parallelism axiom, one gets a curved space. For example, the surface of a sphere is curved. Locally straight lines on a sphere extend to great circles, and two great circles always intersect. Nevertheless, sufficiently near to any point, one can pretend that the geometry is flat: the map of a town can be represented on a flat sheet of paper without significant distortion, while a similar attempt for the whole globe fails completely. The sphere is thus locally flat. This is true for all so-called Riemannian[2] spaces: they all are locally flat, but locally straight lines (called *geodesics*) do not usually remain parallel.

Einstein's important advance was to see the similarity between Riemannian spaces and gravitational physics. He identified the trajectories of freely falling particles with the geodesics of a curved geometry: they are locally straight since spacetime admits local inertial frames, in which those trajectories are straight lines, but globally they do not remain parallel.

We shall follow Einstein and look for a theory of gravity which uses a curved spacetime to represent the effects of gravity on particles' trajectories. To do this we shall clearly have to study the mathematics of curvature. The simplest introduction is actually to study curvilinear coordinate systems in a flat space, where our intuition is soundest. We shall see that this will develop nearly all the mathematical concepts we need, and the step to a curved space will be simple. So for the rest of this chapter we will study the Euclidean plane: no more SR (for the time being!) and no more indefinite inner products. What we are looking for in this chapter is parallelism, not metrics. This approach has the added bonus of giving a more sensible derivation to such often mysterious formulae as the expression for ∇^2 in polar coordinates!

---

[2] B. Riemann (1826–1866) was the first to publish a detailed study of the consequences of dropping Euclid's parallelism axiom.

5.2 Tensor algebra in polar coordinates

Consider the Euclidean plane. The usual coordinates are x and y. Sometimes polar coordinates $\{r, \theta\}$ are convenient:

$$r = (x^2 + y^2)^{1/2}, \qquad x = r \cos\theta, \left.\begin{array}{l} \\ \end{array}\right\}$$
$$\theta = \arctan(y/x), \qquad y = r \sin\theta. \qquad (5.3)$$

Small increments Δr and $\Delta\theta$ are produced by small increments Δx and Δy according to

$$\Delta r = \frac{x}{r}\Delta x + \frac{y}{r}\Delta y = \cos\theta\,\Delta x + \sin\theta\,\Delta y,$$

$$\Delta\theta = -\frac{y}{r^2}\Delta x + \frac{x}{r^2}\Delta y = -\frac{1}{r}\sin\theta\,\Delta x + \frac{1}{r}\cos\theta\,\Delta y, \qquad (5.4)$$

which are valid to first order.

It is also possible to use other coordinate systems. Let us denote a general coordinate system by ξ and η:

$$\xi = \xi(x, y), \quad \Delta\xi = \frac{\partial\xi}{\partial x}\Delta x + \frac{\partial\xi}{\partial y}\Delta y,$$

$$\eta = \eta(x, y), \quad \Delta\eta = \frac{\partial\eta}{\partial x}\Delta x + \frac{\partial\eta}{\partial y}\Delta y. \qquad (5.5)$$

In order for (ξ, η) to be useful coordinates, it is necessary that any two distinct points (x_1, y_1) and (x_2, y_2) be assigned different pairs (ξ_1, η_1) and (ξ_2, η_2) by Eq. 5.5. For instance, the definitions $\xi = x$, $\eta = 1$ would not give good coordinates, since the distinct points $(x = 1, y = 2)$ and $(x = 1, y = 3)$ both have $(\xi = 1, \eta = 1)$. Mathematically, this requires that if $\Delta\xi = \Delta\eta = 0$ in Eq. 5.5, then the points must be the same, or $\Delta x = \Delta y = 0$. This will be true if the determinant of Eq. 5.5 is nonzero,

$$\det\begin{pmatrix} \partial\xi/\partial x & \partial\xi/\partial y \\ \partial\eta/\partial x & \partial\eta/\partial y \end{pmatrix} \neq 0. \qquad (5.6)$$

This determinant is called the *Jacobian* of the coordinate transformation, Eq. 5.5. If the Jacobian vanishes at a point, the transformation is said to be *singular* there.

Vectors and one-forms. The old way of defining a vector is to say that it transforms under an *arbitrary* coordinate transformation in the way that the displacement transforms. That is, a vector $\vec{\Delta r}$ can be represented[3] as a displacement $(\Delta x, \Delta y)$, or in polar coordinates $(\Delta r, \Delta\theta)$, or in general $(\Delta\xi, \Delta\eta)$. Then it is clear that for *small* $(\Delta x, \Delta y)$ we have (from Eq. 5.5)

$$\begin{pmatrix} \Delta\xi \\ \Delta\eta \end{pmatrix} = \begin{pmatrix} \partial\xi/\partial x & \partial\xi/\partial y \\ \partial\eta/\partial x & \partial\eta/\partial y \end{pmatrix}\begin{pmatrix} \Delta x \\ \Delta y \end{pmatrix}. \qquad (5.7)$$

---

[3] We shall denote Euclidean vectors by arrows, and we shall use Greek letters for indices (numbered 1 and 2) to denote the fact that the sum is over all possible (i.e. both) values.

By defining the matrix of transformation,

$$(\Lambda^{\alpha'}{}_{\beta}) = \begin{pmatrix} \partial\xi/\partial x & \partial\xi/\partial y \\ \partial\eta/\partial x & \partial\eta/\partial y \end{pmatrix}, \tag{5.8}$$

we can write the transformation for an arbitrary vector $\vec{V}$ in the same manner as in SR:

$$V^{\alpha'} = \Lambda^{\alpha'}{}_{\beta}V^{\beta}, \tag{5.9}$$

where unprimed indices refer to (x, y) and primed indices to (ξ, η), and where indices can only take the values 1 and 2. A vector can be defined as an object whose components transform according to Eq. 5.9. There is a more sophisticated and natural way, however. This is the modern way, which we now introduce.

Consider a scalar field ϕ. Given coordinates (ξ, η) it is always possible to form the derivatives $\partial\phi/\partial\xi$ and $\partial\phi/\partial\eta$. We *define* the one-form $\tilde{\mathrm{d}}\phi$ to be the geometrical object whose components are given by

$$\tilde{\mathrm{d}}\phi \rightarrow (\partial\phi/\partial\xi, \partial\phi/\partial\eta) \tag{5.10}$$

in the (ξ, η) coordinate system. This is a general definition of an infinity of one-forms, each formed from a different scalar field. The transformation of components is automatic from the chain rule for partial derivatives:

$$\frac{\partial\phi}{\partial\xi} = \frac{\partial x}{\partial\xi}\frac{\partial\phi}{\partial x} + \frac{\partial y}{\partial\xi}\frac{\partial\phi}{\partial y}, \tag{5.11}$$

and similarly for $\partial\phi/\partial\eta$. The most convenient way to write this in matrix notation is as a transformation on *row-vectors*,

$$(\partial\phi/\partial\xi \ \ \partial\phi/\partial\eta) = (\partial\phi/\partial x \ \ \partial\phi/\partial y) \begin{pmatrix} \partial x/\partial\xi & \partial x/\partial\eta \\ \partial y/\partial\xi & \partial y/\partial\eta \end{pmatrix}, \tag{5.12}$$

because then the transformation matrix for one-forms is defined by analogy with Eq. 5.8 as a set of derivatives of the (x, y) coordinates by the (ξ, η) coordinates:

$$(\Lambda^{\alpha}{}_{\beta'}) = \begin{pmatrix} \partial x/\partial\xi & \partial x/\partial\eta \\ \partial y/\partial\xi & \partial y/\partial\eta \end{pmatrix}. \tag{5.13}$$

Using this matrix the component-sum version of the transformation in Eq. 5.12 is

$$(\tilde{\mathrm{d}}\phi)_{\beta'} = \Lambda^{\alpha}{}_{\beta'}(\tilde{\mathrm{d}}\phi)_{\alpha}. \tag{5.14}$$

Note that the summation in this equation is on the *first* index of the transformation matrix, as one expects when a row vector premultiplies a matrix.

It is interesting that in SR we did not have to worry about row vectors, because the simple Lorentz transformation matrices that we used were symmetric. But if we want to go beyond even the simplest situations we need to see that one-form components are elements of row vectors. However, matrix notation becomes awkward when one goes beyond tensors with two indices. In GR we need to deal with tensors with four indices, and sometimes even five. As a result, we will normally express transformation equations in their algebraic form, as in Eq. 5.14; you will not see much matrix notation later in this book.

What we have seen in this section is that, in the modern view, the foundation of tensor algebra is the definition of a one-form. This is more natural than the old way, in which a *single* vector $(\Delta x, \Delta y)$ was defined and others were obtained by analogy. Here a whole *class* of one-forms is defined in terms of derivatives, and the transformation properties of one-forms follow automatically.

Now a vector is defined as a linear function of one-forms into real numbers. The implications of this will be explored in the next paragraph. First we just note that all this is the same as in SR, so that vectors do in fact obey the transformation law Eq. 5.9. It is of interest to see explicitly that $(\Lambda^{\alpha'}{}_\beta)$ and $(\Lambda^\alpha{}_{\beta'})$ are inverses of each other. The product of the matrices is

$$\begin{pmatrix} \partial\xi/\partial x & \partial\xi/\partial y \\ \partial\eta/\partial x & \partial\eta/\partial y \end{pmatrix}\begin{pmatrix} \partial x/\partial\xi & \partial x/\partial\eta \\ \partial y/\partial\xi & \partial y/\partial\eta \end{pmatrix}$$
$$= \begin{pmatrix} \dfrac{\partial\xi}{\partial x}\dfrac{\partial x}{\partial\xi} + \dfrac{\partial\xi}{\partial y}\dfrac{\partial y}{\partial\xi} & \dfrac{\partial\xi}{\partial x}\dfrac{\partial x}{\partial\eta} + \dfrac{\partial\xi}{\partial y}\dfrac{\partial y}{\partial\eta} \\ \dfrac{\partial\eta}{\partial x}\dfrac{\partial x}{\partial\xi} + \dfrac{\partial\eta}{\partial y}\dfrac{\partial y}{\partial\xi} & \dfrac{\partial\eta}{\partial x}\dfrac{\partial x}{\partial\eta} + \dfrac{\partial\eta}{\partial y}\dfrac{\partial y}{\partial\eta} \end{pmatrix}. \tag{5.15}$$

By the chain rule this matrix is

$$\begin{pmatrix} \partial\xi/\partial\xi & \partial\xi/\partial\eta \\ \partial\eta/\partial\xi & \partial\eta/\partial\eta \end{pmatrix} = \begin{pmatrix} 1 & 0 \\ 0 & 1 \end{pmatrix}, \tag{5.16}$$

where the equality follows from the definition of a partial derivative.

Curves and vectors. The usual notion of a curve is of a connected series of points in the plane. This we shall call a *path*, and reserve the word curve for a parametrized path. That is, we shall follow modern mathematical terminology and define a *curve* as a mapping of an interval of the real line into a path in the plane. What this means is that a curve is a path with a real number associated with each point on the path. This number is called the parameter s. Each point has coordinates that may then be expressed as a function of s:

$$\text{Curve: } \{\xi = f(s),\ \eta = g(s), a \leqslant s \leqslant b\} \tag{5.17}$$

defines a curve in the plane. If we were to change the parameter (but not the points) to $s' = s'(s)$, a function of the old s, then we would have

$$\{\xi = f'(s'),\ \eta = g'(s'),\ a' \leqslant s' \leqslant b'\}, \tag{5.18}$$

where f' and g' are *new* functions, and where $a' = s'(a)$, $b' = s'(b)$. This is, mathematically, a *new* curve, even though its *image* (the points of the plane that it passes through) is the same. So there is an infinite number of curves having the same path.

The derivative of a scalar field ϕ along a curve is $d\phi/ds$. This depends on s, so by changing the parameter, one changes the derivative. One can write this as

$$d\phi/ds = \langle \tilde{d}\phi, \vec{V}\rangle, \tag{5.19}$$

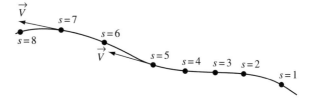

Figure 5.4   A curve, its parametrization, and its tangent vector.

where $\vec{V}$ is the vector whose components are $(d\xi/ds, d\eta/ds)$. This vector depends only on the curve, while $\tilde{d}\phi$ depends only on ϕ. Therefore $\vec{V}$ is a vector characteristic of the curve, called the *tangent* vector. (It clearly lies tangent to the curve: see Figure 5.4.) So a vector may be regarded as a thing which produces $d\phi/ds$, given ϕ. This leads to the most modern view, that the tangent vector to the curve should be *called* d/ds. Some relativity texts occasionally use this notation. For our purposes, however, we shall just let $\vec{V}$ be the tangent vector whose components are $(d\xi/ds, d\eta/ds)$. Notice that a *path* in the plane has, at any point, an infinity of tangents, all of them parallel but differing in length. These are to be regarded as vectors tangent to *different* curves, curves that have different parametrizations in a neighborhood of that point. A *curve* has a *unique* tangent, since the path *and* parameter are given. Moreover, even curves that have identical tangents at a point may not be identical elsewhere. From the Taylor expansion $\xi(s+1) \approx \xi(s) + d\xi/ds$, we see that $\vec{V}(s)$ stretches approximately from s to $s+1$ along the curve.

Now, it is clear that under a coordinate transformation s does not change (its definition has nothing to do with coordinates) but the components of $\vec{V}$ will change, since by the chain rule

$$\begin{pmatrix} d\xi/ds \\ d\eta/ds \end{pmatrix} = \begin{pmatrix} \partial\xi/\partial x & \partial\xi/\partial y \\ \partial\eta/\partial x & \partial\eta/\partial y \end{pmatrix} \begin{pmatrix} dx/ds \\ dy/ds \end{pmatrix}. \tag{5.20}$$

This is the same transformation law as we had for vectors earlier, Eq. 5.7.

To sum up, the modern view is that a vector is a *tangent* to some curve, and is the function that gives $d\phi/ds$ when it takes the one-form $\tilde{d}\phi$ as an argument. Having said this, we are now in a position to treat polar coordinates more thoroughly.

Polar coordinate basis one-forms and vectors. The bases of the coordinates are clearly

$$\vec{e}_{\alpha'} = \Lambda^{\beta}{}_{\alpha'}\vec{e}_{\beta},$$

or

$$\vec{e}_r = \Lambda^x{}_r\vec{e}_x + \Lambda^y{}_r\vec{e}_y \tag{5.21}$$

$$= \frac{\partial x}{\partial r}\vec{e}_x + \frac{\partial y}{\partial r}\vec{e}_y$$

$$= \cos\theta\,\vec{e}_x + \sin\theta\,\vec{e}_y, \tag{5.22}$$

and, similarly,

$$\vec{e}_\theta = \frac{\partial x}{\partial\theta}\vec{e}_x + \frac{\partial y}{\partial\theta}\vec{e}_y$$

$$= -r\sin\theta\,\vec{e}_x + r\cos\theta\vec{e}_y \tag{5.23}$$

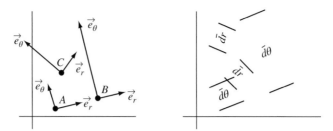

Figure 5.5   Basis vectors and one-forms for polar coordinates.

Notice in this that we have used, among others things

$$\Lambda^x{}_r = \frac{\partial x}{\partial r}. \tag{5.24}$$

Similarly, to transform in the other direction we would need

$$\Lambda^r{}_x = \frac{\partial r}{\partial x}. \tag{5.25}$$

The transformation matrices are exceedingly simple: just keeping track of which index is up and which is down gives the right derivative to use.

The basis one-forms are, analogously,

$$\tilde{d}\theta = \frac{\partial \theta}{\partial x}\,\tilde{d}x + \frac{\partial \theta}{\partial y}\,\tilde{d}y,$$

$$= -\frac{1}{r} \sin \theta\,\tilde{d}x + \frac{1}{r} \cos \theta\,\tilde{d}y. \tag{5.26}$$

(Notice the similarity to ordinary calculus, Eq. 5.4.) Similarly, we find

$$\tilde{d}r = \cos \theta\,\tilde{d}x + \sin \theta\,\tilde{d}y. \tag{5.27}$$

We can draw pictures of the bases at various points (Figure 5.5). Drawing the basis vectors is no problem. Drawing the basis one-forms is most easily done by drawing surfaces of constant r and θ for $\tilde{d}r$ and $\tilde{d}\theta$. These surfaces have different orientations in different places.

There is a point of great importance to note here: the bases change from point to point. For the vectors, the basis vectors at A in Figure 5.5 are not parallel to those at C. This is so because they point in the direction of increasing coordinate, which changes from point to point. Moreover, the lengths of the bases are not constant. For example, from Eq. 5.23 we find

$$|\vec{e}_\theta|^2 = = \vec{e}_\theta \cdot \vec{e}_\theta = r^2 \sin^2 \theta + r^2 \cos^2 \theta = r^2, \tag{5.28a}$$

so that $\vec{e}_\theta$ increases in magnitude as one gets further from the origin. The reason is that the basis vector $\vec{e}_\theta$, having components $(0,1)$ with respect to r and θ, has essentially a θ displacement of one unit, i.e. one radian. It needs to be longer to be able do this at large radii than at small radii. So we do not have a *unit* basis. It is easy to verify that

$$|\vec{e}_r| = 1, \quad |\tilde{d}r| = 1, \quad |\tilde{d}\theta| = r^{-1}. \tag{5.28b}$$

Again, $|\tilde{d}\theta|$ gets larger near $r = 0$ because a given vector can span a larger range of θ near the origin than farther away.

Metric tensor. The dot products above were all calculated by knowing the metric in Cartesian coordinates x, y:

$$\vec{e}_x \cdot \vec{e}_x = \vec{e}_y \cdot \vec{e}_y = 1, \qquad \vec{e}_x \cdot \vec{e}_y = 0;$$

or, put in tensor notation,

$$\mathbf{g}(\vec{e}_\alpha, \vec{e}_\beta) = \delta_{\alpha\beta} \quad \text{in Cartesian coordinates.} \qquad (5.29)$$

What are the components of $\mathbf{g}$ in polar coordinates? Simply

$$g_{\alpha'\beta'} = \mathbf{g}(\vec{e}_{\alpha'}, \vec{e}_{\beta'}) = \vec{e}_{\alpha'} \cdot \vec{e}_{\beta'} \qquad (5.30)$$

or, by Eq. 5.28,

$$g_{rr} = 1, \quad g_{\theta\theta} = r^2, \qquad (5.31\text{a})$$

and, from Eqs. 5.22 and 5.23,

$$g_{r\theta} = 0. \qquad (5.31\text{b})$$

So we can write the components of $\mathbf{g}$ as

$$(g_{\alpha\beta})_{\text{polar}} = \begin{pmatrix} 1 & 0 \\ 0 & r^2 \end{pmatrix}, \qquad (5.32)$$

A convenient way of displaying the components of $\mathbf{g}$ and at the same time showing the coordinates is the line element, which is the magnitude of an arbitrary 'infinitesimal' displacement $\vec{dl}$:

$$\vec{dl} \cdot \vec{dl} = \mathrm{d}s^2 = |\mathrm{d}r\, \vec{e}_r + \mathrm{d}\theta \vec{e}_\theta|^2$$
$$= \mathrm{d}r^2 + r^2 \mathrm{d}\theta^2. \qquad (5.33)$$

Do *not* confuse $\mathrm{d}r$ and $\mathrm{d}\theta$ here with the basis one-forms $\tilde{d}r$ and $\tilde{d}\theta$. The quantities in this equation are the components of $\vec{dl}$ in polar coordinates, and here 'd' simply means 'infinitesimal Δ'.

There is another way of deriving Eq. 5.33 which is instructive. Recall Eq. 3.26 in which a general $\binom{0}{2}$ tensor is written as a sum over basis $\binom{0}{2}$ tensors $\tilde{d}x^\alpha \otimes \tilde{d}x^\beta$. For the metric this is

$$\mathbf{g} = g_{\alpha\beta}\, \tilde{d}x^\alpha \otimes \tilde{d}x^\beta = \tilde{d}r \otimes \tilde{d}r + r^2\tilde{d}\theta \otimes \tilde{d}\theta.$$

Although this has a superficial resemblance to Eq. 5.33, it is different: it is an operator which, when supplied with the vector $\vec{dl}$, whose components are $\mathrm{d}r$ and $\mathrm{d}\theta$, gives Eq. 5.33. Unfortunately, the two expressions resemble each other rather too closely because of the confusing way notation has evolved in this subject. Most texts and research papers still use the 'old-fashioned' form in Eq. 5.33 for displaying the components of the metric, and we follow the same practice. The metric has an inverse:

$$\begin{pmatrix} 1 & 0 \\ 0 & r^2 \end{pmatrix}^{-1} = \begin{pmatrix} 1 & 0 \\ 0 & r^{-2} \end{pmatrix}. \tag{5.34}$$

So we have $g^{rr} = 1$, $g^{r\theta} = 0$, $g^{\theta\theta} = 1/r^2$. This enables us to make the mapping between one-forms and vectors. For instance, if ϕ is a scalar field and the one-form $\tilde{d}\phi$ is its gradient, then the vector $\vec{d}\phi$ has components

$$(\vec{d}\phi)^\alpha = g^{\alpha\beta}\phi_{,\beta}, \tag{5.35}$$

or

$$(\vec{d}\phi)^r = g^{r\beta}\phi_{,\beta} = g^{rr}\phi_{,r} + g^{r\theta}\phi_{,\theta}$$
$$= \partial\phi/\partial r, \tag{5.36a}$$
$$(\vec{d}\phi)^\theta = g^{\theta r}\phi_{,r} + g^{\theta\theta}\phi_{,\theta}$$
$$= \frac{1}{r^2}\frac{\partial\phi}{\partial\theta}. \tag{5.36b}$$

So, while $(\phi_{,r}, \phi_{,\theta})$ are components of a one-form, the vector gradient has components $(\phi_{,r}, \phi_{,\theta}/r^2)$. Even though we are in Euclidean space, vectors generally have different components from their associated one-forms. Cartesian coordinates are the only coordinates in which the components are the same.

5.3 Tensor calculus in polar coordinates

The fact that the basis vectors of polar coordinates are not constant everywhere leads to some problems when one tries to differentiate vectors. For instance, consider the simple vector $\vec{e}_x$, which is a constant vector field, the same at any point. In polar coordinates it has components $\vec{e}_x \rightarrow (\Lambda^r{}_x, \Lambda^\theta{}_x) = (\cos\theta, -r^{-1}\sin\theta)$. These are clearly not constant, even though $\vec{e}_x$ is constant. The reason is that they are components in a nonconstant basis. If we were to differentiate them with respect to, say, θ, we would most certainly *not* get $\partial\vec{e}_x/\partial\theta$, which must be identically zero. So, from this example, one sees that differentiating the components of a vector does not necessarily give the derivative of the vector: one must also differentiate the nonconstant basis vectors. This is the key to the understanding of curved coordinates and, indeed, of curved spaces. We shall now make these ideas systematic.

Derivatives of basis vectors. Since $\vec{e}_x$ and $\vec{e}_y$ are constant vector fields, we easily find that

$$\frac{\partial}{\partial r}\vec{e}_r = \frac{\partial}{\partial r}(\cos\theta\,\vec{e}_x + \sin\theta\,\vec{e}_y) = 0, \tag{5.37a}$$

$$\frac{\partial}{\partial\theta}\vec{e}_r = \frac{\partial}{\partial\theta}(\cos\theta\vec{e}_x + \sin\theta\,\vec{e}_y)$$

$$= -\sin\theta\,\vec{e}_x + \cos\theta\vec{e}_y = \frac{1}{r}\vec{e}_\theta. \tag{5.37b}$$

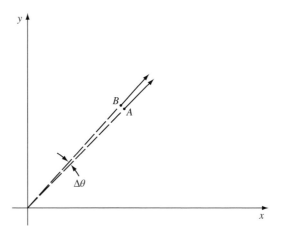

Figure 5.6 The change in $\vec{e}_r$ (given by the radial arrows) when θ changes by $\Delta\theta$.

These have a simple geometrical picture, shown in Figure 5.6. At two nearby points, A and B, $\vec{e}_r$ must point directly away from the origin, and so it points in slightly different directions at A and at B. The derivative of $\vec{e}_r$ with respect to θ is just the difference between $\vec{e}_r$ at A and B divided by $\Delta\theta$. The difference in this case is clearly a vector parallel to $\vec{e}_\theta$, which then makes Eq. 5.37b reasonable.

Similarly,

$$\frac{\partial}{\partial r}\,\vec{e}_\theta = \frac{\partial}{\partial r}\,(-r\sin\theta\,\vec{e}_x + r\cos\theta\,\vec{e}_y)$$

$$= -\sin\theta\,\vec{e}_x + \cos\theta\,\vec{e}_y = \frac{1}{r}\,\vec{e}_\theta, \tag{5.38a}$$

$$\frac{\partial}{\partial\theta}\,\vec{e}_\theta = -r\cos\theta\,\vec{e}_x - r\sin\theta\,\vec{e}_y = -r\vec{e}_r. \tag{5.38b}$$

The reader is encouraged to draw a picture similar to Figure 5.6 to explain these formulae.

Derivatives of general vectors. Let us go back to the derivative of $\vec{e}_x$. Since

$$\vec{e}_x = \cos\theta\,\vec{e}_r - \frac{1}{r}\sin\theta\,\vec{e}_\theta, \tag{5.39}$$

we have

$$\frac{\partial}{\partial\theta}\,\vec{e}_x = \frac{\partial}{\partial\theta}\,(\cos\theta)\,\vec{e}_r + \cos\theta\frac{\partial}{\partial\theta}\,(\vec{e}_r)$$

$$- \frac{\partial}{\partial\theta}\left(\frac{1}{r}\sin\theta\right)\vec{e}_\theta - \frac{1}{r}\sin\theta\frac{\partial}{\partial\theta}\,(\vec{e}_\theta) \tag{5.40}$$

$$= -\sin\theta\,\vec{e}_r + \cos\theta\left(\frac{1}{r}\vec{e}_\theta\right)$$

$$- \frac{1}{r}\cos\theta\,\vec{e}_\theta - \frac{1}{r}\sin\theta\,(-r\vec{e}_r). \tag{5.41}$$

To get this we used Eqs. 5.37 and 5.38. Simplifying gives

$$\frac{\partial}{\partial \theta}\, \vec{e}_x = 0,\tag{5.42}$$

just as one should have. Now, in Eq. 5.40 the first and third terms come from differentiating the *components* of $\vec{e}_x$ in the polar coordinate basis; the other two terms are the derivatives of the polar basis vectors themselves, and are necessary for cancelling out the derivatives of the components.

A general vector $\vec{V}$ has components (V^r, V^θ) in the polar basis. Its derivative, by analogy with Eq. 5.40, is

$$\frac{\partial \vec{V}}{\partial r} = \frac{\partial}{\partial r}(V^r\, \vec{e}_r + V^\theta\, \vec{e}_\theta)$$

$$= \frac{\partial V^r}{\partial r}\, \vec{e}_r + V^r \frac{\partial \vec{e}_r}{\partial r} + \frac{\partial V^\theta}{\partial r}\, \vec{e}_\theta + V^\theta \frac{\partial \vec{e}_\theta}{\partial r},$$

and similarly for $\partial \vec{V}/\partial \theta$. Written in index notation, this becomes

$$\frac{\partial \vec{V}}{\partial r} = \frac{\partial}{\partial r}(V^\alpha\, \vec{e}_\alpha) = \frac{\partial V^\alpha}{\partial r}\, \vec{e}_\alpha + V^\alpha \frac{\partial \vec{e}_\alpha}{\partial r}.$$

(Here α runs over r and θ, of course.)

This shows explicitly that the derivative of $\vec{V}$ is more than just the derivative of its components V^α. Now, since r is just one coordinate, we can generalize the above equation to

$$\frac{\partial \vec{V}}{\partial x^\beta} = \frac{\partial V^\alpha}{\partial x^\beta}\, \vec{e}_\alpha + V^\alpha \frac{\partial \vec{e}_\alpha}{\partial x^\beta},\tag{5.43}$$

where, now, x^β can be either r or θ, for $\beta = 1$ or 2.

The Christoffel symbols

The final term in Eq. 5.43 is obviously of great importance. Since $\partial \vec{e}_\alpha/\partial x^\beta$ is itself a vector, it can be written as a linear combination of the basis vectors; we introduce the symbol $\Gamma^\mu{}_{\alpha\beta}$ to denote the coefficients in this combination:

$$\frac{\partial \vec{e}_\alpha}{\partial x^\beta} = \Gamma^\mu{}_{\alpha\beta}\, \vec{e}_\mu.\tag{5.44}$$

The interpretation of $\Gamma^\mu{}_{\alpha\beta}$ is that it is the μth component of $\partial \vec{e}_\alpha/\partial x^\beta$. It needs three indices: one (α) gives the basis vector being differentiated; the second (β) gives the coordinate with respect to which it is being differentiated; and the third (μ) denotes the component of the resulting derivative vector. These quantities, $\Gamma^\mu{}_{\alpha\beta}$, are so useful that they have been given a name: the Christoffel symbols. The question whether they are components of tensors we postpone until much later.

We have of course already calculated the Christoffel symbols for polar coordinates. From Eqs. 5.37 and 5.38 we find

$$
\left.
\begin{array}{ll}
(1) & \partial \vec{e}_r / \partial r = 0 \Rightarrow \Gamma^\mu{}_{rr} = 0 \quad \text{for all } \mu, \\[2mm]
(2) & \partial \vec{e}_r / \partial \theta = \dfrac{1}{r}\,\vec{e}_\theta \Rightarrow \Gamma^r{}_{r\theta} = 0, \qquad \Gamma^\theta{}_{r\theta} = \dfrac{1}{r}, \\[3mm]
(3) & \partial \vec{e}_\theta / \partial r = \dfrac{1}{r}\,\vec{e}_\theta \Rightarrow \Gamma^r{}_{\theta r} = 0, \qquad \Gamma^\theta{}_{\theta r} = \dfrac{1}{r}, \\[3mm]
(4) & \partial \vec{e}_\theta / \partial \theta = -r\,\vec{e}_r \Rightarrow \Gamma^r{}_{\theta\theta} = -r, \qquad \Gamma^\theta{}_{\theta\theta} = 0.
\end{array}
\right\}
\tag{5.45}
$$

In the definition Eq. 5.44, all indices must refer to the same coordinate system. Thus, although we computed the derivatives of $\vec{e}_r$ and $\vec{e}_\theta$ by using the constancy of $\vec{e}_x$ and $\vec{e}_y$, the Cartesian bases do not, in the end, make any appearance in Eq. 5.45. The Christoffel symbols' importance is that they enable one to express these derivatives without using any other than polar coordinates.

The covariant derivative

Using the definition of the Christoffel symbols, Eq. 5.44, the derivative in Eq. 5.43 becomes

$$
\frac{\partial \vec{V}}{\partial x^\beta} = \frac{\partial V^\alpha}{\partial x^\beta}\,\vec{e}_\alpha + V^\alpha \Gamma^\mu{}_{\alpha\beta}\,\vec{e}_\mu.
\tag{5.46}
$$

In the last term there are two sums, on α and μ. Relabeling the dummy indices will help here: we change μ to α and α to μ and get

$$
\frac{\partial \vec{V}}{\partial x^\beta} = \frac{\partial V^\alpha}{\partial x^\beta}\,\vec{e}_\alpha + V^\mu \Gamma^\alpha{}_{\mu\beta}\,\vec{e}_\alpha.
\tag{5.47}
$$

The reason for the relabeling was that now $\vec{e}_\alpha$ can be factored out of both terms:

$$
\frac{\partial \vec{V}}{\partial x^\beta} = \left(\frac{\partial V^\alpha}{\partial x^\beta} + V^\mu \Gamma^\alpha{}_{\mu\beta} \right) \vec{e}_\alpha.
\tag{5.48}
$$

So, the vector field $\partial \vec{V}/\partial x^\beta$ has components

$$
\partial V^\alpha / \partial x^\beta + V^\mu \Gamma^\alpha{}_{\mu\beta}.
\tag{5.49}
$$

Recall our original notation for the partial derivative, $\partial V^\alpha / \partial x^\beta = V^\alpha{}_{,\beta}$. We keep this notation and define a *new* symbol:

$$
V^\alpha{}_{;\beta} := V^\alpha{}_{,\beta} + V^\mu \Gamma^\alpha{}_{\mu\beta}.
\tag{5.50}
$$

Then, with this shorthand semicolon notation, we have

$$
\partial \vec{V}/\partial x^\beta = V^\alpha{}_{;\beta}\,\vec{e}_\alpha,
\tag{5.51}
$$

a very compact way of writing Eq. 5.48.

Now $\partial \vec{V}/\partial x^\beta$ is a vector field if we regard β as a given fixed number. But there are two values that β *can* have, and so we can also regard $\partial \vec{V}/\partial x^\beta$ as being associated with a $\binom{1}{1}$ tensor field which maps the vector $\vec{e}_\beta$ into the vector $\partial \vec{V}/\partial x^\beta$. This tensor field is called the *covariant derivative* of $\vec{V}$, denoted, naturally enough, as $\nabla \vec{V}$. Then its components are

$$(\nabla \vec{V})^\alpha{}_\beta = (\nabla_\beta \vec{V})^\alpha = V^\alpha{}_{;\beta}. \tag{5.52}$$

On a Cartesian basis the components are just $V^\alpha{}_{,\beta}$. In a curvilinear basis, however, the derivatives of the basis vectors must be taken into account, and we find that $V^\alpha{}_{;\beta}$ are the components of $\nabla \vec{V}$ in whatever coordinate system the Christoffel symbols in Eq. 5.50 refer to. The significance of this statement should not be underrated, as it is the foundation of all our later work. There is a single $\binom{1}{1}$ tensor called $\nabla \vec{V}$. In Cartesian coordinates its components are $\partial V^\alpha/\partial x^\beta$. In general coordinates $\{x^{\mu'}\}$ its components are called $V^{\alpha'}{}_{;\beta'}$ and can be obtained in either of two equivalent ways: (i) they can be computed directly in $\{x^{\mu'}\}$ using Eq. 5.50 and a knowledge of the $\Gamma^{\alpha'}{}_{\mu'\beta'}$ coefficients in these coordinates; or (ii) they can be obtained by the usual tensor transformation laws from Cartesian coordinates to $\{x^{\mu'}\}$.

What is the covariant derivative of a scalar? The covariant derivative differs from the partial derivative with respect to the coordinates only because the basis vectors change. But a scalar does not depend on the basis vectors, so its covariant derivative is the same as its partial derivative, which is its gradient:

$$\nabla_\alpha f = \partial f/\partial x^\alpha, \qquad \nabla f = \tilde{d}f. \tag{5.53}$$

Divergence and Laplacian. Before doing any more theory, let us link this up with quantities we have seen before. In Cartesian coordinates the divergence of a vector V^α is $V^\alpha{}_{,\alpha}$. This is the scalar obtained by contracting $V^\alpha{}_{,\beta}$ on its two indices. Since contraction is a frame-invariant operation, the divergence of $\vec{V}$ can be calculated in other coordinates $\{x^{\mu'}\}$ also by contracting the components of $\nabla \vec{V}$ on their two indices. This results in a scalar with the value $V^{\alpha'}{}_{;\alpha'}$. It is important to realize that this is the *same* number as $V^\alpha{}_{,\alpha}$ in Cartesian coordinates:

$$V^\alpha{}_{,\alpha} \equiv V^{\beta'}{}_{;\beta'}, \tag{5.54}$$

where unprimed indices refer to Cartesian coordinates and primed indices to the arbitrary system.

For polar coordinates (dropping primes for convenience here)

$$V^\alpha{}_{;\alpha} = \frac{\partial V^\alpha}{\partial x^\alpha} + \Gamma^\alpha{}_{\mu\alpha} V^\mu.$$

Now, from Eq. 5.45 we can calculate

$$\left. \begin{array}{l} \Gamma^\alpha{}_{r\alpha} = \Gamma^r{}_{rr} + \Gamma^\theta{}_{r\theta} = 1/r, \\ \Gamma^\alpha{}_{\theta\alpha} = \Gamma^r{}_{\theta r} + \Gamma^\theta{}_{\theta\theta} = 0. \end{array} \right\} \tag{5.55}$$

Therefore we have

$$V^\alpha{}_{;\alpha} = \frac{\partial V^r}{\partial r} + \frac{\partial V^\theta}{\partial \theta} + \frac{1}{r}V^r,$$

$$= \frac{1}{r}\frac{\partial}{\partial r}(rV^r) + \frac{\partial}{\partial \theta}V^\theta. \tag{5.56}$$

This may be a familiar formula. What is probably more familiar is the Laplacian, which is the divergence of the gradient. But divergence is defined only for vectors, and the gradient is a one-form. Therefore we must first convert the one-form to a vector. Thus, given a scalar ϕ, we have the vector gradient (see Eq. 5.53 and the last part of § 5.2 above) with components $(\phi_{,r}, \phi_{,\theta}/r^2)$. Using these as the components of the vector in the divergence formula, Eq. 5.56 gives

$$\nabla \cdot \nabla \phi := \nabla^2 \phi = \frac{1}{r}\frac{\partial}{\partial r}\left(r\frac{\partial \phi}{\partial r}\right) + \frac{1}{r^2}\frac{\partial^2 \phi}{\partial \theta^2}. \tag{5.57}$$

This is the Laplacian in plane polar coordinates. It is, of course, identically equal to

$$\nabla^2 \phi = \frac{\partial^2 \phi}{\partial x^2} + \frac{\partial^2 \phi}{\partial y^2}. \tag{5.58}$$

Derivatives of one-forms and tensors of higher types. Since a scalar ϕ depends on no basis vectors, its derivative $\tilde{d}\phi$ is the same as its covariant derivative $\nabla\phi$. We shall almost always use the symbol $\nabla\phi$. To compute the derivative of a one-form (which as for a vector will not be simply the derivatives of its components), we use the property that the combination of a one-form and a vector give a scalar. Thus, if $\tilde{p}$ is a one-form and $\vec{V}$ is an arbitrary vector then, for fixed β, $\nabla_\beta \tilde{p}$ is also a one-form, $\nabla_\beta \vec{V}$ is a vector, and $\langle \tilde{p}, \vec{V} \rangle \equiv \phi$ is a scalar. In any (arbitrary) coordinate system this scalar is just given by

$$\phi = p_\alpha V^\alpha. \tag{5.59}$$

Therefore, by the product rule for derivatives,

$$\nabla_\beta \phi = \phi_{,\beta} = \frac{\partial p_\alpha}{\partial x^\beta} V^\alpha + p_\alpha \frac{\partial v^\alpha}{\partial x^\beta}. \tag{5.60}$$

But we can use Eq. 5.50 to replace $\partial V^\alpha/\partial x^\beta$ by $V^\alpha{}_{;\beta}$, which are the components of $\nabla_\beta \vec{V}$:

$$\nabla_\beta \phi = \frac{\partial p_\alpha}{\partial x^\beta} V^\alpha + p_\alpha V^\alpha{}_{;\beta} - p_\alpha V^\mu \Gamma^\alpha{}_{\mu\beta}. \tag{5.61}$$

Rearranging terms, and relabeling dummy indices in the term that contains the Christoffel symbol, gives

$$\nabla_\beta \phi = \left(\frac{\partial p_\alpha}{\partial x^\beta} - p_\mu \Gamma^\mu{}_{\alpha\beta}\right) V^\alpha + p_\alpha V^\alpha{}_{;\beta}. \tag{5.62}$$

Now, every term in this equation except the term in parentheses is *known* to be the component of a tensor, for an arbitrary vector field $\vec{V}$. Therefore, since the multiplication and/or addition of components always gives new tensors, it must be true that the term in parentheses is also the component of a tensor. This is the covariant derivative of $\tilde{p}$:

$$(\nabla_\beta \tilde{p})_\alpha := (\nabla \tilde{p})_{\alpha\beta} := p_{\alpha;\beta} = p_{\alpha,\beta} - p_\mu \Gamma^\mu{}_{\alpha\beta}. \tag{5.63}$$

Then Eq. 5.62 reads

$$\nabla_\beta (p_\alpha V^\alpha) = p_{\alpha;\beta} V^\alpha + p_\alpha V^\alpha + p_\alpha V^\alpha{}_{;\beta}.$$

Thus covariant differentiation obeys the same sort of product rule as Eq. 5.60. It *must* do this, since in Cartesian coordinates ∇ just gives partial differentiation of components, so the above equation reduces to Eq. 5.60.

Let us compare the two formulae Eq. 5.50 and Eq. 5.63:

$$V^\alpha{}_{;\beta} = V^\alpha{}_{,\beta} + V^\mu \Gamma^\alpha{}_{\mu\beta},$$
$$p_{\alpha;\beta} = p_{\alpha,\beta} - p_\mu \Gamma^\mu{}_{\alpha\beta}.$$

There are certain similarities and certain differences. If one remembers that the derivative index β is the *last* index on Γ, then the other indices are the only ones they can be, without raising and lowering with the metric. The only thing to watch is the sign difference. It may help to remember that $\Gamma^\alpha{}_{\mu\beta}$ is related to derivatives of the basis vectors, for then it is reasonable that $-\Gamma^\mu{}_{\alpha\beta}$ is related to derivatives of the basis one-forms. The change in sign means that the basis one-forms change in the opposite way to basis vectors, which makes sense when one remembers that the contraction $\langle \tilde{\omega}^\alpha, \vec{e}_\beta \rangle = \delta^\alpha{}_\beta$ is a *constant* whose derivative must be zero.

The same procedure that led to Eq. 5.63 would lead to the following:

$$\nabla_\beta T_{\mu\nu} = T_{\mu\nu,\beta} - T_{\alpha\nu} \Gamma^\alpha{}_{\mu\beta} - T_{\mu\alpha} \Gamma^\alpha{}_{\nu\beta}; \tag{5.64}$$
$$\nabla_\beta A^{\mu\nu} = A^{\mu\nu}{}_{,\beta} + A^{\alpha\nu} \Gamma^\mu{}_{\alpha\beta} + A^{\mu\alpha} \Gamma^\nu{}_{\alpha\beta}; \tag{5.65}$$
$$\nabla_\beta B^\mu{}_\nu = B^\mu{}_{\nu,\beta} + B^\alpha{}_\nu \Gamma^\mu{}_{\alpha\beta} - B^\mu{}_\alpha \Gamma^\alpha{}_{\nu\beta}. \tag{5.66}$$

The reader should inspect these closely: they are *very* systematic. Simply throw in one Γ term for each index; a raised index is treated like a vector and a lowered index like a one-form. The geometrical meaning of Eq. 5.64 is that $\nabla_\beta T_{\mu\nu}$ is a component of the $\binom{0}{3}$ tensor $\nabla \mathbf{T}$, where $\mathbf{T}$ is a $\binom{0}{2}$ tensor. Similarly, in Eq. 5.65, $\mathbf{A}$ is a $\binom{2}{0}$ tensor and $\nabla \mathbf{A}$ is a $\binom{2}{1}$ tensor with components $\nabla_\beta A^{\mu\nu}$.

5.4 Christoffel symbols and the metric

The formalism developed above has not used any properties of the metric tensor to derive covariant derivatives. But the metric must be involved somehow, because it can convert a vector into a one-form, and so it must have something to say about the relationship between their derivatives. In particular, in Cartesian coordinates the components of the one-form and its related vector are *equal*, and since ∇ just involves the differentiation of components, the components of the covariant derivatives of the one-form and vector must be equal. This means that if $\vec{V}$ is an arbitrary vector and $\tilde{V} = \mathbf{g}(\vec{V},\ )$ is its related one-form, then in Cartesian coordinates

$$\nabla_\beta \tilde{V} = \mathbf{g}(\nabla_\beta \vec{V},\ ). \tag{5.67}$$

But Eq. 5.67 is a tensor equation, so it must be valid in *all* coordinates. We conclude that

$$V_{\alpha;\beta} = g_{\alpha\mu} V^\mu{}_{;\beta}, \tag{5.68}$$

which is the component representation of Eq. 5.67.

In case the above argument in words needs amplification, let us go through it again in equations. Let unprimed indices $\alpha, \beta, \gamma, \ldots$ denote Cartesian coordinates and primed indices $\alpha', \beta', \gamma', \ldots$ denote *arbitrary* coordinates.

We begin with the statement

$$V_{\alpha'} = g_{\alpha'\mu'} V^{\mu'}, \tag{5.69}$$

valid in any coordinate system. But in Cartesian coordinates

$$g_{\alpha\mu} = \delta_{\alpha\mu}, \quad V_\alpha = V^\alpha.$$

Now, also in Cartesian coordinates, the Christoffel symbols vanish, so

$$V_{\alpha;\beta} = V_{\alpha,\beta} \quad \text{and} \quad V^\alpha{}_{;\beta} = V^\alpha{}_{,\beta}.$$

Therefore we conclude that

$$V_{\alpha;\beta} = V^\alpha{}_{;\beta},$$

in Cartesian coordinates only. To convert this into an equation valid in all coordinate systems, we note that in Cartesian coordinates

$$V^\alpha{}_{;\beta} = g_{\alpha\mu} V^\mu{}_{;\beta},$$

so that, again in Cartesian coordinates, we have

$$V_{\alpha;\beta} = g_{\alpha\mu} V^\mu{}_{;\beta}.$$

But this equation *is* a tensor equation, so its validity in one coordinate system implies its validity in all. This is just Eq. 5.68 again:

$$V_{\alpha';\beta'} = g_{\alpha'\mu'} V^{\mu'}{}_{;\beta'} \tag{5.70}$$

This result has far-reaching implications. If we take the β' covariant derivative of Eq. 5.69 we find

$$V_{\alpha';\beta'} = g_{\alpha'\mu';\beta'} V^{\mu'} + g_{\alpha'\mu'} V^{\mu'}{}_{;\beta'}.$$

Comparison of this with Eq. 5.70 shows (since $\vec{V}$ is an arbitrary vector) that we must have

$$g_{\alpha'\mu';\beta'} \equiv 0 \tag{5.71}$$

in all coordinate systems. This is a consequence of Eq. 5.67. In Cartesian coordinates,

$$g_{\alpha\mu;\beta} \equiv g_{\alpha\mu,\beta} = \delta_{\alpha\mu,\beta} \equiv 0$$

is a trivial identity. However, in other coordinates it is not obvious, so we shall work it out as a check on the consistency of our formalism.

Using Eq. 5.64 gives (now unprimed indices are general)

$$g_{\alpha\beta;\mu} = g_{\alpha\beta,\mu} - \Gamma^{\nu}{}_{\alpha\mu}g_{\nu\beta} - \Gamma^{\nu}{}_{\beta\mu}g_{\alpha\nu}. \tag{5.72}$$

In polar coordinates let us work out a few examples. Let $\alpha = r, \beta = r, \mu = r$:

$$g_{rr;r} = g_{rr,r} - \Gamma^{\nu}{}_{rr}g_{\nu r} - \Gamma^{\nu}{}_{rr}g_{r\nu}.$$

Since $g_{rr,r} = 0$ and $\Gamma^{\nu}{}_{rr} = 0$ for all ν, this is trivially zero. Not so trivial is the case $\alpha = \theta, \beta = \theta, \mu = r$:

$$g_{\theta\theta;r} = g_{\theta\theta,r} - \Gamma^{\nu}{}_{\theta r}g_{\nu\theta} - \Gamma^{\nu}{}_{\theta r}g_{\theta\nu}.$$

With $g_{\theta\theta} = r^2, \Gamma^{\theta}{}_{\theta r} = 1/r$ and $\Gamma^{r}{}_{\theta r} = 0$, this becomes

$$g_{\theta\theta\;;r} = (r^2)_{,r} - \frac{1}{r}(r^2) - \frac{1}{r}(r^2) = 0.$$

So it works, almost magically. But it is important to realize that it is not magic: it follows directly from the facts that $g_{\alpha\beta,\mu} = 0$ in Cartesian coordinates and that $g_{\alpha\beta;\mu}$ are the components of the *same* tensor $\nabla\mathbf{g}$ in arbitrary coordinates.

Perhaps it is useful to pause here to get some perspective on what we have just done. We introduced covariant differentiation in arbitrary coordinates by using our understanding of parallelism in Euclidean space. We then showed that the metric of Euclidean space is covariantly constant: Eq. 5.71. When we go on to curved (Riemannian) spaces we will have to discuss parallelism much more carefully, but Eq. 5.71 will *still* be true, and therefore so will all its consequences, such as those we now go on to describe.

Calculating the Christoffel symbols from the metric. The vanishing of Eq. 5.72 leads to an extremely important result. One sees that Eq. 5.72 can be used to determine $g_{\alpha\beta,\mu}$ in terms of $\Gamma^{\mu}{}_{\alpha\beta}$. It turns out that the reverse is also true, that $\Gamma^{\mu}{}_{\alpha\beta}$ can be expressed in terms of $g_{\alpha\beta,\mu}$. This gives an easy way to derive the Christoffel symbols. To show this we first prove a result of some importance in its own right: *in any coordinate system* $\Gamma^{\mu}{}_{\alpha\beta} \equiv \Gamma^{\mu}{}_{\beta\alpha}$. To prove this symmetry consider an arbitrary scalar field ϕ. Its first derivative $\nabla\phi$ is a one-form with components $\phi_{,\beta}$. Its second covariant derivative $\nabla\nabla\phi$ has components $\phi_{,\beta;\alpha}$ and is a $\binom{0}{2}$ tensor. In Cartesian coordinates these components are

$$\phi_{,\beta,\alpha} := \frac{\partial}{\partial x^{\alpha}} \frac{\partial}{\partial x^{\beta}} \phi$$

and we see that they are symmetric in α and β, since partial derivatives commute. But if a tensor is symmetric in one basis it is symmetric in all bases. Therefore

$$\phi_{,\beta;\alpha} = \phi_{,\alpha;\beta} \tag{5.73}$$

in *any* basis. Using the definition, Eq. 5.63 gives

$$\phi_{,\beta,\alpha} - \phi_{,\mu}\Gamma^{\mu}{}_{\beta\alpha} = \phi_{,\alpha,\beta} - \phi_{,\mu}\Gamma^{\mu}{}_{\alpha\beta}$$

in any coordinate system. But again we have

$$\phi_{,\alpha,\beta} = \phi_{,\beta,\alpha}$$

in *any* coordinate system, which leaves us with

$$\Gamma^{\mu}{}_{\alpha\beta}\phi_{,\mu} = \Gamma^{\mu}{}_{\beta\alpha}\phi_{,\mu}$$

for arbitrary ϕ. This proves the assertion that

$$\Gamma^{\mu}{}_{\alpha\beta} = \Gamma^{\mu}{}_{\beta\alpha} \quad \text{in any coordinate system.} \tag{5.74}$$

We will use this to invert Eq. 5.72 by some advanced index gymnastics. We write three versions of Eq. 5.72 with different permutations of indices:

$$g_{\alpha\beta,\mu} = \Gamma^{\nu}{}_{\alpha\mu}g_{\nu\beta} + \Gamma^{\nu}{}_{\beta\mu}g_{\alpha\nu},$$

$$g_{\alpha\mu,\beta} = \Gamma^{\nu}{}_{\alpha\beta}g_{\nu\mu} + \Gamma^{\nu}{}_{\mu\beta}g_{\alpha\nu},$$

$$-g_{\beta\mu,\alpha} = -\Gamma^{\nu}{}_{\beta\alpha}g_{\nu\mu} - \Gamma^{\nu}{}_{\mu\alpha}g_{\beta\nu}.$$

We add these up and group terms, using the symmetry of $\mathbf{g}$, $g_{\beta\nu} = g_{\nu\beta}$:

$$g_{\alpha\beta,\mu} + g_{\alpha\mu,\beta} - g_{\beta\mu,\alpha}$$
$$= (\Gamma^{\nu}{}_{\alpha\mu} - \Gamma^{\nu}{}_{\mu\alpha})g_{\nu\beta} + (\Gamma^{\nu}{}_{\alpha\beta} - \Gamma^{\nu}{}_{\beta\alpha})g_{\nu\mu} + (\Gamma^{\nu}{}_{\beta\mu} + \Gamma^{\nu}{}_{\mu\beta})g_{\alpha\nu}.$$

In this equation the first two terms on the right vanish by the symmetry of Γ, Eq. 5.74, and we get

$$g_{\alpha\beta,\mu} + g_{\alpha\mu,\beta} - g_{\beta\mu,\alpha} = 2g_{\alpha\nu}\Gamma^{\nu}{}_{\beta\mu}.$$

We are almost there. Dividing by 2, multiplying by $g^{\alpha\gamma}$ (with summation implied on α) and using

$$g^{\alpha\gamma}g_{\alpha\nu} \equiv \delta^{\gamma}{}_{\nu}$$

gives

$$\frac{1}{2}g^{\alpha\gamma}(g_{\alpha\beta,\mu} + g_{\alpha\mu,\beta} - g_{\beta\mu,\alpha}) = \Gamma^{\gamma}{}_{\beta\mu}. \tag{5.75}$$

This is the expression for the Christoffel symbols in terms of the partial derivatives of the components of **g**. In polar coordinates, for example,

$$\Gamma^\theta{}_{r\theta} = \tfrac{1}{2} g^{\alpha\theta}(g_{\alpha r,\theta} + g_{\alpha\theta,r} - g_{r\theta,\alpha}).$$

Since $g^{r\theta} = 0$ and $g^{\theta\theta} = r^{-2}$ we have

$$\Gamma^\theta{}_{r\theta} = \frac{1}{2r^2}(g_{\theta r,\theta} + g_{\theta\theta,r} - g_{r\theta,\alpha})$$

$$= \frac{1}{2r^2} g_{\theta\theta,r} = \frac{1}{2r^2}(r^2)_{,r} = \frac{1}{r}.$$

This is the same value for $\Gamma^\theta{}_{r\theta}$ as we derived earlier. This method of computing $\Gamma^\alpha{}_{\beta\mu}$ is so useful that it is well worth committing Eq. 5.75 to memory. It will be exactly the same in curved spaces.

The tensorial nature of $\Gamma^\alpha{}_{\beta\mu}$. Since $\vec{e}_\alpha$ is a vector, $\nabla\vec{e}_\alpha$ is a $\binom{1}{1}$ tensor whose components are $\Gamma^\mu{}_{\alpha\beta}$. Here α is fixed and μ and β are the component indices: changing α changes the tensor $\nabla\vec{e}_\alpha$, while changing μ or β changes only the component under discussion. So it is possible to regard μ and β as component indices and α as a label giving the tensor to which we are referring. There is one such tensor for each basis vector $\vec{e}_\alpha$. However, this is not terribly useful, since under a change of coordinates the basis changes and the important quantities in the new system are the *new* tensors $\nabla\vec{e}_{\beta'}$ which are obtained from the old ones $\nabla\vec{e}_\alpha$ in a complicated way: they are *different* tensors, not just different components of the same tensor. So the set $\Gamma^\mu{}_{\alpha\beta}$ in one frame is not obtained by a simple tensor transformation from the set $\Gamma^{\mu'}{}_{\alpha'\beta'}$ in another frame. The easiest example of this is Cartesian coordinates, where $\Gamma^\alpha{}_{\beta\mu} \equiv 0$, while they are not zero in other frames. In many books it is said that the $\Gamma^\mu{}_{\alpha\beta}$ are not components of tensors. As we have seen, this is not strictly true: $\Gamma^\mu{}_{\alpha\beta}$ are the (μ, β) components of a *set* of $\binom{1}{1}$ tensors $\nabla\vec{e}_\alpha$. But there is no single $\binom{1}{2}$ tensor whose components are $\Gamma^\mu{}_{\alpha\beta}$, so expressions like $\Gamma^\mu{}_{\alpha\beta}V^\alpha$ are not components of a single tensor, either. The combination

$$V^\beta{}_{,\alpha} + V^\mu\Gamma^\beta{}_{\mu\alpha}$$

is a component of a single tensor $\nabla\vec{V}$.

5.5 Noncoordinate bases

In this whole discussion we have generally assumed that nonCartesian basis vectors are generated by a coordinate transformation from (x, y) to some (ξ, η). However, as we shall show below, not every field of basis vectors can be obtained in this way, and we shall have to look carefully at our results to see which of them need modification (few actually do). We will almost never use such noncoordinate bases in this book, but one frequently encounters them in the standard reference works on curved coordinates in flat space, so we should pause to take a brief look at them now.

Polar coordinate basis. The basis vectors for our polar coordinate system were defined by

$$\vec{e}_{\alpha'} = \Lambda^\beta{}_{\alpha'}\vec{e}_\beta,$$

where primed indices refer to polar coordinates and unprimed to Cartesian. Moreover, we had

$$\Lambda^\beta{}_{\alpha'} = \partial x^\beta/\partial x^{\alpha'},$$

where we regard the Cartesian coordinates $\{x^\beta\}$ as functions of the polar coordinates $\{x^{\alpha'}\}$. We found that

$$\vec{e}_{\alpha'} \cdot \vec{e}_{\beta'} \equiv g_{\alpha'\beta'} \neq \delta_{\alpha'\beta'},$$

i.e. that these basis vectors are *not* unit vectors.

Polar unit basis. Often it is convenient to work with *unit* vectors. A simple set of unit vectors derived from the polar coordinate basis is:

$$\vec{e}_{\hat{r}} = \vec{e}_r, \quad \vec{e}_{\hat{\theta}} = \frac{1}{r}\vec{e}_\theta, \tag{5.76}$$

with a corresponding unit one-form basis

$$\tilde{\omega}^{\hat{r}} = \tilde{d}r, \quad \tilde{\omega}^{\hat{\theta}} = r\tilde{d}\theta. \tag{5.77}$$

The reader should verify that

$$\left.\begin{array}{c} \vec{e}_{\hat{\alpha}} \cdot \vec{e}_{\hat{\beta}} \equiv g_{\hat{\alpha}\hat{\beta}} = \delta_{\hat{\alpha}\hat{\beta}}, \\ \tilde{\omega}^{\hat{\alpha}} \cdot \tilde{\omega}^{\hat{\beta}} \equiv g^{\hat{\alpha}\hat{\beta}} = \delta^{\hat{\alpha}\hat{\beta}} \end{array}\right\} \tag{5.78}$$

so these constitute orthonormal bases for the vectors and one-forms. Our notation, which is fairly standard, uses a 'caret' or 'hat' above an index to denote an orthornormal basis. Now, the question arises, do there exist coordinates (ξ, η) such that

$$\vec{e}_{\hat{r}} = \vec{e}_\xi = \frac{\partial x}{\partial \xi}\vec{e}_x + \frac{\partial y}{\partial \xi}\vec{e}_y \tag{5.79a}$$

and

$$\vec{e}_{\hat{\theta}} = \vec{e}_\eta = \frac{\partial x}{\partial \eta}\vec{e}_x + \frac{\partial y}{\partial \eta}\vec{e}_y? \tag{5.79b}$$

If so, then $\{\vec{e}_{\hat{r}}, \vec{e}_{\hat{\theta}}\}$ is a basis for the coordinates (ξ, η) and so can be called a coordinate basis; if such (ξ, η) can be shown not to exist then these vectors are known as a noncoordinate basis. The question is actually more easily answered if we look at the basis one-form. Thus, we seek (ξ, η) such that

$$\left.\begin{array}{c} \tilde{\omega}^{\hat{r}} = \tilde{d}\xi = \frac{\partial \xi}{\partial x}\tilde{d}x + \frac{\partial \xi}{\partial y}\tilde{d}y, \\ \tilde{\omega}^{\hat{\theta}} = \tilde{d}\eta = \frac{\partial \eta}{\partial x}\tilde{d}x + \frac{\partial \eta}{\partial y}\tilde{d}y. \end{array}\right\} \tag{5.80}$$

Since we know $\tilde{\omega}^{\hat{r}}$ and $\tilde{\omega}^{\hat{\theta}}$ in terms of $\tilde{d}r$ and $\tilde{d}\theta$, we have, from Eqs. 5.26 and 5.27,

$$\left.\begin{array}{l} \tilde{\omega}^{\hat{r}} = \tilde{d}r = \cos\theta\, \tilde{d}x + \sin\theta\, \tilde{d}y, \\ \tilde{\omega}^{\hat{\theta}} = r\, \tilde{d}\theta = -\sin\theta\, \tilde{d}x + \cos\theta\, \tilde{d}y. \end{array}\right\} \tag{5.81}$$

(The orthonormality of $\tilde{\omega}^{\hat{r}}$ and $\tilde{\omega}^{\hat{\theta}}$ are obvious here.) Thus if (ξ, η) exist we have

$$\frac{\partial\eta}{\partial x} = -\sin\theta, \quad \frac{\partial\eta}{\partial y} = \cos\theta. \tag{5.82}$$

If this were true then the mixed derivatives would be equal:

$$\frac{\partial}{\partial y}\frac{\partial\eta}{\partial x} = \frac{\partial}{\partial x}\frac{\partial\eta}{\partial y}. \tag{5.83}$$

This would imply

$$\frac{\partial}{\partial y}(-\sin\theta) = \frac{\partial}{\partial x}(\cos\theta) \tag{5.84}$$

or

$$\frac{\partial}{\partial y}\left(\frac{y}{\sqrt{(x^2 + y^2)}}\right) + \frac{\partial}{\partial x}\left(\frac{x}{\sqrt{(x^2 + y^2)}}\right) = 0.$$

This is certainly *not* true. Therefore ξ and η do *not* exist: we have a noncoordinate basis. (If this manner of proof is surprising, try it on $\tilde{d}r$ and $\tilde{d}\theta$ themselves.)

In textbooks that deal with vector calculus in curvilinear coordinates, almost all use the unit orthonormal basis rather than the coordinate basis. Thus, for polar coordinates, if a vector has components in the *coordinate* basis PC,

$$\vec{V} \xrightarrow[PC]{} (a, b) = \{V^{\alpha'}\}, \tag{5.85}$$

then it has components in the *orthonormal* basis PO

$$\vec{V} \xrightarrow[PO]{} (a, rb) = \{V^{\hat{\alpha}}\}. \tag{5.86}$$

So if, for example, such books calculate the divergence of the vector, they obtain, instead of our Eq. 5.56,

$$\nabla \cdot V = \frac{1}{r}\frac{\partial}{\partial r}(rV^{\hat{r}}) + \frac{1}{r}\frac{\partial}{\partial\theta}V^{\hat{\theta}}. \tag{5.87}$$

The difference between Eqs. 5.56 and 5.87 is purely a matter of the basis for $\vec{V}$.

General remarks on noncoordinate bases. The principal differences between coordinate and noncoordinate bases arise as follows. Consider an arbitrary scalar field ϕ and the number $\tilde{d}\phi(\vec{e}_\mu)$, where $\vec{e}_\mu$ is a basis vector of some arbitrary basis. We have used the notation

$$\tilde{d}\phi(\vec{e}_\mu) = \phi_{,\mu}. \tag{5.88}$$

Now, if $\vec{e}_\mu$ is a member of a coordinate basis, then $\tilde{d}\phi(\vec{e}_\mu) = \partial\phi/\partial x^\mu$ and we have, as defined in an earlier chapter,

$$\phi_{,\mu} = \frac{\partial \phi}{\partial x^\mu} \quad \text{coordinate basis.} \tag{5.89}$$

But if no coordinates exist for $\{\vec{e}_\mu\}$ then Eq. 5.89 must fail. For example, if we let Eq. 5.88 *define* $\phi_{,\hat{\mu}}$ then we have

$$\phi_{,\hat{\theta}} = \frac{1}{r}\frac{\partial \phi}{\partial \theta}. \tag{5.90}$$

In general, we get

$$\nabla_{\hat{\alpha}}\phi \equiv \phi_{,\hat{\alpha}} = \Lambda^\beta{}_{\hat{\alpha}}\nabla_\beta\phi = \Lambda^\beta{}_{\hat{\alpha}}\frac{\partial \phi}{\partial x^\beta} \tag{5.91}$$

for any coordinate system $\{x^\beta\}$ and noncoordinate basis $\{\vec{e}_{\hat{\alpha}}\}$. It is thus convenient to continue with the notation in Eq. 5.88, and to make a rule that $\phi_{,\mu} = \partial\phi/\partial x^\mu$ only in a coordinate basis.

The Christoffel symbols may be defined just as before:

$$\nabla_{\hat{\beta}}\vec{e}_{\hat{\alpha}} = \Gamma^{\hat{\mu}}{}_{\hat{\alpha}\hat{\beta}}\vec{e}_{\hat{\mu}}; \tag{5.92}$$

but now

$$\nabla_{\hat{\beta}} = \Lambda^\alpha{}_{\hat{\beta}}\frac{\partial}{\partial x^\alpha}, \tag{5.93}$$

where $\{x^\alpha\}$ is any coordinate system and $\{\vec{e}_{\hat{\beta}}\}$ any basis (coordinate or not). Now, however, one *cannot* prove that $\Gamma^{\hat{\mu}}{}_{\hat{\alpha}\hat{\beta}} = \Gamma^{\hat{\mu}}{}_{\hat{\beta}\hat{\alpha}}$, since that proof used $\phi_{,\hat{\alpha},\hat{\beta}} = \phi_{,\hat{\beta},\hat{\alpha}}$, which was true in a coordinate basis (partial derivatives commute) but is not true otherwise. Hence, also, Eq. 5.75 for $\Gamma^\mu{}_{\alpha\beta}$ in terms of $g_{\alpha\beta,\gamma}$ applies only in a coordinate basis. More general expressions are worked out in Exercise 5.20.

What is the general reason for the nonexistence of coordinates for a basis? If $\{\tilde{\omega}^{\tilde{\alpha}}\}$ is a coordinate one-form basis, then its relation to another such basis $\{\tilde{d}x^\alpha\}$ is

$$\tilde{\omega}^{\tilde{\alpha}} = \Lambda^{\tilde{\alpha}}{}_\beta \tilde{d}x^\beta = \frac{\partial x^{\tilde{\alpha}}}{\partial x^\beta}dx^\beta. \tag{5.94}$$

The key point is that $\Lambda^{\tilde{\alpha}}{}_\beta$, which is generally a function of position, must actually be the partial derivative $\partial x^{\tilde{\alpha}}/\partial x^\beta$ everywhere. Thus we have

$$\frac{\partial}{\partial x^\gamma}\Lambda^{\tilde{\alpha}}{}_\beta = \frac{\partial^2 x^{\tilde{\alpha}}}{\partial x^\gamma \partial x^\beta} = \frac{\partial^2 x^{\tilde{\alpha}}}{\partial x^\beta \partial x^\gamma} = \frac{\partial}{\partial x^\beta}\Lambda^{\tilde{\alpha}}{}_\gamma. \tag{5.95}$$

These 'integrability conditions' must be satisfied by all the elements $\Lambda^{\tilde{\alpha}}{}_\beta$ in order for $\tilde{\omega}^{\tilde{\alpha}}$ to be a coordinate basis. Clearly, one can always choose a transformation matrix for which this fails, thereby generating a noncoordinate basis.

Noncoordinate bases in this book. We shall not have occasion to use such bases very often. Mainly, it is important to understand that they exist: that not every basis is derivable from a coordinate system. The algebra of coordinate bases is simpler in almost every respect. One may ask why then do standard treatments of curvilinear coordinates in vector calculus stick to orthonormal bases, which are noncoordinate? The reason is that in such a basis in Euclidean space the metric has components $\delta_{\alpha\beta}$, so the form of the dot product and the equality of vector and one-form components carry over directly from Cartesian coordinates (which have the *only* orthonormal coordinate basis!). In order to gain the simplicity of coordinate bases for vector and tensor calculus, one has to spend time learning the difference between vectors and one-forms!

5.6 Looking ahead

The work we have done in this chapter has developed almost all the notation and concepts we will need in our study of curved spaces and spacetimes. It is particularly important that the reader should understand §§ 5.2–5.4, because the mathematics of curvature will be developed by analogy with the development here. What we have to add to all this is a discussion of parallelism, of how to measure the extent to which the Euclidean parallelism axiom fails. This measure is the famous Riemann tensor.

5.7 Bibliography

The Eötvös and Pound–Rebka–Snider experiments, and other experimental fundamentals underpinning GR, are discussed in Dicke (1964), Misner *et al.* (1973), and Will (2014, 2018). See Hoffmann (1983) for a less mathematical discussion of the motivation for introducing curvature. For a review of the GPS system's use of relativity, see Ashby (2003).

The mathematics of curvilinear coordinates is developed from a variety of points of view in Abraham *et al.* (1988), Lovelock & Rund (1990), and Schutz (1980b).

Exercises

5.1 Repeat the argument that led to Eq. 5.1 under more realistic assumptions: suppose a fraction ε of the kinetic energy of the mass at the bottom can be converted into a photon and sent back up, the remaining energy staying at ground level in a useful form. Devise a perpetual motion engine if Eq. 5.1 is violated.

5.2 Explain why a *uniform* external gravitational field would raise no tides on Earth.

5.3 (a) Show that the coordinate transformation $(x, y) \to (\xi, \eta)$ with $\xi = x$ and $\eta = 1$ violates Eq. 5.6.

(b) Are the following coordinate transformations good ones? Compute the Jacobian and list any points at which the transformations fail.

(i) $\xi = (x^2 + y^2)^{1/2}, \eta = \arctan(y/x)$;
(ii) $\xi = \ln x, \eta = y$;
(iii) $\xi = \arctan(y/x), \eta = (x^2 + y^2)^{-1/2}$.

5.4   A curve is defined by $\{x = f(\lambda), \ y = g(\lambda), \ 0 \leqslant \lambda \leqslant 1\}$. Show that the tangent vector $(dx/d\lambda, dy/d\lambda)$ does actually lie tangent to the curve.

5.5   Sketch the following curves. Which have the same paths? Find also their tangent vectors where the parameter equals zero.

(a) $x = \sin \lambda, y = \cos \lambda$; (b) $x = \cos(2\pi t^2), y = \sin(2\pi t^2 + \pi)$; (c) $x = s, y = s + 4$; (d) $x = s^2, y = -(s - 2)(s + 2)$; (e) $x = \mu, y = 1$.

5.6   Justify the pictures in Figure 5.5.

5.7   Calculate all elements of the transformation matrices $\Lambda^{\alpha'}{}_\beta$ and $\Lambda^\mu{}_{\nu'}$ for the transformation from Cartesian coordinates (x, y) – the unprimed indices – to polar coordinates (r, θ) – the primed indices.

5.8   (a) (This exercise uses the result of Exercise 5.7.) Let $f = x^2 + y^2 + 2xy$, and in Cartesian coordinates let $\vec{V} \rightarrow (x^2 + 3y, y^2 + 3x)$, $\vec{W} \rightarrow (1, 1)$. Compute f as a function of r and θ, and find the components of $\vec{V}$ and $\vec{W}$ in the polar basis, expressing them as functions of r and θ.

(b) Find the components of $\tilde{d}f$ in Cartesian coordinates and obtain them in polar coordinates (i) by direct calculation in polars, and (ii) by transforming components from Cartesian.

(c) (i) Use the metric tensor in polar coordinates to find the polar components of the one-forms $\tilde{V}$ and $\tilde{W}$ associated with $\vec{V}$ and $\vec{W}$. (ii) Obtain the polar components of $\tilde{V}$ and $\tilde{W}$ by transformation of their Cartesian components.

5.9   Draw a diagram similar to Figure 5.6 to explain Eq. 5.38.

5.10   Prove that $\nabla \vec{V}$, defined in Eq. 5.52, is a $\binom{1}{1}$ tensor.

5.11   (This exercise uses the result of Exercises 5.7 and 5.8.) For the vector field $\vec{V}$ whose Cartesian components are $(x^2 + 3y, y^2 + 3x)$, compute: (a) $V^\alpha{}_{,\beta}$ in Cartesian coordinates; (b) the transformation $\Lambda^{\mu'}{}_\alpha \Lambda^\beta{}_{\nu'} V^\alpha{}_{,\beta}$ to polar coordinates; (c) the components $V^{\mu'}{}_{;\nu'}$ directly in polars using the Christoffel symbols, Eq. 5.45, in Eq. 5.50; (d) the divergence $V^\alpha{}_{,\alpha}$ using your results for (a); (e) the divergence $V^{\mu'}{}_{;\mu'}$ using your results for either (b) or (c); (f) the divergence $V^{\mu'}{}_{;\mu'}$ using Eq. 5.56 directly.

5.12   For the one-form field $\tilde{p}$ whose Cartesian components are $(x^2 + 3y, y^2 + 3x)$, compute: (a) $p_{\alpha,\beta}$ in Cartesian; (b) the transformation $\Lambda^\alpha{}_{\mu'} \Lambda^\beta{}_{\nu'} \, p_{\alpha,\beta}$ to polar coordinates; (c) the components $p_{\mu';\nu'}$ directly in polars using the Christoffel symbols, Eq. 5.45, in Eq. 5.63.

5.13   If you have done both Exercises 5.11 and 5.12, show in polar coordinates that $g_{\mu'\alpha'} V^{\alpha'}{}_{;\nu'} = p_{\mu';\nu'}$.

5.14   For the tensor whose polar components are

$$(A^{rr} = r^2, A^{r\theta} = r \sin \theta, A^{\theta r} = r \cos \theta, A^{\theta\theta} = \tan \theta),$$

compute Eq. 5.65 in polar coordinates for all possible indices.

5.15   For the vector whose polar components are $(V^r = 1, V^\theta = 0)$, compute in polars all components of the second covariant derivative $V^\alpha{}_{;\mu;\nu}$. (Hint: to find the second derivative, treat the first derivative $V^\alpha{}_{;\mu}$ as any $\binom{1}{1}$ tensor: see Eq. 5.66.)

5.16 Fill in all the missing steps leading from Eq. 5.74 to Eq. 5.75.

5.17 Discover how each expression $V^\beta{}_{,\alpha}$ and $V^\mu \Gamma^\beta{}_{\mu\alpha}$ separately transforms under a change of coordinates (for $\Gamma^\beta{}_{\mu\alpha}$, begin with Eq. 5.44). Show that neither obeys the standard tensor law, but that their *sum* does obey the standard law.

5.18 Verify Eq. 5.78.

5.19 Verify that the calculation in Eqs. 5.81–5.84, when repeated for $\tilde{d}r$ and $\tilde{d}\theta$, shows them to form a coordinate basis.

5.20 For a noncoordinate basis $\{\vec{e}_\mu\}$, define $\nabla_{\vec{e}_\mu}\vec{e}_\nu - \nabla_{\vec{e}_\nu}\vec{e}_\mu := c^\alpha{}_{\mu\nu}\vec{e}_\alpha$ and use this in place of Eq. 5.74 to generalize Eq. 5.75.

5.21 Consider the x–t plane of an inertial observer in SR. A certain uniformly accelerated observer wishes to set up an orthonormal coordinate system. By Exercise 2.21, the observer's world line is

$$t(\lambda) = a \sinh \lambda, \quad x(\lambda) = a \cosh \lambda, \tag{5.96}$$

where a is a constant and $a\lambda$ is the local proper time (the clock time on the observer's wristwatch).

(a) Show that the spacelike line described by Eq. 5.96 with a as the variable parameter and λ fixed is orthogonal to the observer's world line where they intersect. Changing λ in Eq. 5.96 then generates a *family* of such lines.

(b) Show that Eq. 5.96 defines a transformation from coordinates (t, x) to coordinates (λ, a) which form an *orthogonal* coordinate system. Draw these coordinates and show that they cover only one half of the original t–x plane. Show that the coordinates are bad on the lines $|x| = |t|$, so they really cover two disjoint quadrants.

(c) Find the metric tensor and all the Christoffel symbols in this coordinate system. This observer will do a satisfactory job by using Christoffel symbols appropriately and sticking to events in just one quadrant. In this sense, SR admits accelerated observers. The right-hand quadrant in these coordinates is sometimes called *Rindler space*, and the boundary lines $x = \pm t$ bear some resemblance to the black-hole horizons we will study later.

5.22 Show that if $U^\alpha \nabla_\alpha V^\beta = W^\beta$ then $U^\alpha \nabla_\alpha V_\beta = W_\beta$.

6 Curved manifolds

6.1 Differentiable manifolds and tensors

The mathematical concept of a curved space begins (but does not end) with the idea of a *manifold*. A manifold is essentially a continuous space which looks locally like Euclidean space. To the concept of a manifold is added the idea of curvature itself. The introduction of curvature into a manifold will be the subject of subsequent sections. First we study the idea of a manifold, which one can regard as just a fancy word for 'space'.

Manifolds. The surface of a sphere is a manifold. So is any m-dimensional 'hyperplane' in an n-dimensional Euclidean space ($m \leqslant n$). More abstractly, the set of all rigid rotations of Cartesian coordinates in three-dimensional Euclidean space will be shown below to be a manifold. Basically, a manifold is any set that can be continuously parametrized. The number of independent parameters is the *dimension* of the manifold, and the parameters themselves are the *coordinates* of the manifold. Consider the examples just mentioned. The surface of a sphere is 'parametrized' by two coordinates θ and ϕ. An m-dimensional 'hyperplane' has m Cartesian coordinates, and the set of all rotations can be parametrized by the three Euler angles, which in effect give the direction of the axis of rotation (two parameters for this) and the amount of rotation (one parameter). So, the set of rotations is a manifold: each point is a particular rotation, and the coordinates are the three parameters. It is a three-dimensional manifold.

Mathematically, the association of points with the values of their parameters can be thought of as a mapping of points of a manifold into points of the Euclidean space of the correct dimension. This is the meaning of the fact that a manifold looks locally like Euclidean space: it is 'smooth' and has a certain number of dimensions. It must be stressed that the large-scale topology of a manifold may be very different from Euclidean space: the surface of a torus is not Euclidean, even topologically. But locally the correspondence is good: a small patch of the surface of a torus can be mapped one to one into the plane tangent to it. This is the way to think of a manifold: it is a space with coordinates that locally looks Euclidean but that globally can warp, bend, and do almost anything (as long as it stays continuous).

Differential structure. In practice, we shall only consider differentiable manifolds. These are spaces that are continuous and differentiable. Roughly, this means that in the neighborhood of each point in the manifold it is possible to define a smooth map to Euclidean space that preserves derivatives of scalar functions at that point. The surface of a sphere is differentiable everywhere. That of a cone is differentiable except at its apex. Nearly all manifolds of use in physics are differentiable almost everywhere. The curved spacetimes of GR certainly are thus differentiable.

The assumption of differentiability immediately means that we can define one-forms and vectors. That is, in a certain coordinate system on the manifold, the members of the set $\{\phi_{,\alpha}\}$ are the components of the one-form $\tilde{d}\phi$; and any set of the form $\{a\phi_{,\alpha} + b\psi_{,\alpha}\}$, where a and b are functions, is also a one-form field. Similarly, every curve (with parameter, say, λ) has a tangent vector $\vec{V}$ defined as the linear function that takes the one-form $\tilde{d}\phi$ into the derivative of ϕ along the curve, $d\phi/d\lambda$:

$$\langle \tilde{d}\phi, \vec{V} \rangle = \vec{V}(\tilde{d}\phi) = \nabla_{\vec{V}}\phi = d\phi/d\lambda. \tag{6.1}$$

Any linear combination of vectors is also a vector. Using the vectors and one-forms so defined, we can build up the whole set of tensors of type $\binom{M}{N}$, just as we did in SR. Since we have not yet picked out any $\binom{0}{2}$ tensor to serve as the metric, there is not yet any correspondence between forms and vectors. Everything else, however, is exactly as we had in SR and in polar coordinates. All of this comes only from differentiability, so the set of all tensors is said to be part of the 'differential structure' of the manifold. We will not have much occasion to use that term.

Review. It is useful here to review the fundamentals of tensor algebra. We can summarize the following rules.

(1) A tensor *field* defines a tensor at every point.
(2) Vectors and one-forms are linear operators on each other, producing real numbers. The linearity means that:

$$\langle \tilde{p}, a\vec{V} + b\vec{W} \rangle = a\langle \tilde{p}, \vec{V} \rangle + b\langle \tilde{p}, \vec{W} \rangle,$$
$$\langle a\tilde{p} + b\tilde{q}, \vec{V} \rangle = a\langle \tilde{p}, \vec{V} \rangle + b\langle \tilde{q}, \vec{V} \rangle,$$

where a and b are any scalar fields.
(3) Tensors are also linear operators on one-forms and vectors, producing real numbers.
(4) If two tensors of the same type have equal components in a given basis, they have equal components in all bases and are said to be identical (or equal, or the same). Only tensors of the same type can be equal. In particular, if a tensor's components are all zero in one basis then they are zero in all bases, and the tensor is said to be zero.
(5) A number of manipulations of components of tensor fields are called 'permissible tensor operations' because they produce components of new tensors:

  (i) Multiplication by a scalar field produces components of a new tensor of the same type.
  (ii) Addition of the components of two tensors of the same type gives the components of a new tensor of the same type. (In particular, only tensors of the same type can be added.)
  (iii) Multiplication of the components of two tensors of arbitrary type gives the components of a new tensor of the sum of the types, the outer product of the two tensors.

(iv) Covariant differentiation (to be discussed later) of the components of a tensor of type $\binom{N}{M}$ gives the components of a tensor of type $\binom{N}{M+1}$.

(v) Contraction on a pair of indices of the components of a tensor of type $\binom{N}{M}$ produces the components of a tensor of type $\binom{N-1}{M-1}$. (Contraction is only defined between an upper and lower index.)

(6) If an equation is formed using components of tensors combined only by the permissible tensor operations, and if the equation is true in one basis, then it is true in any other basis. This is a very useful result. It comes from the fact that (from (5) above) such an equation is simply an equality between the components of two tensors of the same type, which, from (4), is then true in any system.

6.2 Riemannian manifolds

So far we have not introduced a metric onto the manifold. Indeed, on certain manifolds a metric would be unnecessary or inconvenient for whichever problem is being considered. But in our case the metric is absolutely fundamental, since it will carry the information about the rates at which clocks run and the distances between points, just as it does in SR. A differentiable manifold on which a symmetric $\binom{0}{2}$ tensor field $\mathbf{g}$ has been singled out to act as the metric at each point is called a Riemannian manifold. (Strictly speaking, only if the metric is positive-definite – that is, $\mathbf{g}(\vec{V}, \vec{V}) > 0$ for all $\vec{V} \neq 0$ – is it called Riemannian; indefinite metrics, as in SR and GR, are called pseudo-Riemannian. This is a distinction that we will not bother to make.)

It is important to understand that in picking out a metric we add structure to the manifold; we shall see that the metric completely defines the curvature of the manifold. Thus, by our choosing one metric $\mathbf{g}$ the manifold gets a certain curvature (perhaps that of a sphere), while a different metric $\mathbf{g}'$ would give it a different curvature (perhaps an ellipsoid of revolution). The differentiable manifold itself is 'primitive': an amorphous collection of points, arranged locally like the points of Euclidean space but not having any distance relation or shape specified. Giving the manifold a metric $\mathbf{g}$ gives it a specific shape, as we shall see. From now on we shall study Riemannian manifolds, on which a metric $\mathbf{g}$ is assumed to be defined at every point.

For completeness we should remark that it is in fact possible to define the notion of curvature on a manifold without introducing a metric (as for so-called 'affine' manifolds). Some texts actually approach the subject this way. But since the metric is essential in GR, we shall simply study those manifolds whose curvature is defined by a metric.

The metric and local flatness. The metric, of course, provides a mapping between vectors and one-forms at every point. Thus, given a vector field $\vec{V}(\mathcal{P})$ (this notation means that $\vec{V}$ depends on the position $\mathcal{P}$, where $\mathcal{P}$ is any point), there is a unique one-form field $\tilde{V}(\mathcal{P}) = \mathbf{g}(\vec{V}(\mathcal{P}), \quad)$. The mapping must be invertible, so that associated with $\tilde{V}(\mathcal{P})$ there is a unique $\vec{V}(\mathcal{P})$. The components of $\mathbf{g}$ are called $g_{\alpha\beta}$; the components of the inverse matrix are called $g^{\alpha\beta}$. The metric permits the raising and lowering of indices in the same way as in SR, which means that

$$V_\alpha = g_{\alpha\beta} V^\beta.$$

In general, the $\{g_{\alpha\beta}\}$ will be complicated functions of position, so it will not be true that there would is a simple relation between, say, V_0 and V^0 in an arbitrary coordinate system.

Since we wish to study general curved manifolds, we have to allow any coordinate system. In SR we studied only Lorentz (inertial) frames, because they are simple. But because gravity prevents such frames from being global, we shall have to allow all coordinates, and hence all coordinate transformations, that are nonsingular. (Nonsingular means, as in § 5.2, that the matrix of the transformation, $\Lambda^{\alpha'}{}_\beta \equiv \partial x^{\alpha'}/\partial x^\beta$, has an inverse.) Now, the matrix $(g_{\alpha\beta})$ is symmetric by definition. It is a well-known theorem of matrix algebra (see Exercise 6.3) that a transformation matrix can always be found that will make any symmetric matrix into a diagonal matrix with each entry on the main diagonal either $+1$, -1, or zero. The number of $+1$ entries equals the number of positive eigenvalues of $(g_{\alpha\beta})$, while the number of -1 entries is the number of negative eigenvalues. So, if we choose **g** originally to have three positive eigenvalues and one negative, then we can always find a $\Lambda^{\alpha'}{}_\beta$ to make the metric components become

$$(g_{\alpha'\beta'}) = \begin{pmatrix} -1 & 0 & 0 & 0 \\ 0 & 1 & 0 & 0 \\ 0 & 0 & 1 & 0 \\ 0 & 0 & 0 & 1 \end{pmatrix} \equiv (\eta_{\alpha\beta}). \tag{6.2}$$

From now on we will use $\eta_{\alpha\beta}$ to denote *only* the matrix in Eq. 6.2, which is of course the metric of SR.

There are two remarks that must be made here. The first is that we can write Eq. 6.2 only if we choose $(g_{\alpha\beta})$ from among the matrices that have three positive and one negative eigenvalues. The sum of the diagonal elements in Eq. 6.2 is called the *signature* of the metric. For SR and GR it is $+2$. Thus, the fact that we have previously deduced from physical arguments that one can always construct a *local* inertial frame at any event finds its mathematical representation in Eq. 6.2, which states that the metric can be transformed into $\eta_{\alpha\beta}$ at that point. This in turn implies that the metric has to have signature $+2$ if it is to describe a spacetime with gravity.

The second remark is that the matrix $\Lambda^{\alpha'}{}_\beta$ that produces Eq. 6.2 at every point may *not* be a coordinate transformation. That is, the set $\{\tilde{\omega}^{\alpha'} = \Lambda^{\alpha'}{}_\beta \, \tilde{\mathrm{d}}x^\beta\}$ may not be a coordinate basis. From our earlier discussion of noncoordinate bases, it would be a coordinate transformation only if Eq. 5.95 holds:

$$\frac{\partial \Lambda^{\alpha'}{}_\beta}{\partial x^\gamma} = \frac{\partial \Lambda^{\alpha'}{}_\gamma}{\partial x^\beta}.$$

In a general gravitational field this is impossible, because it would imply the existence of coordinates for which Eq. 6.2 is true everywhere: a global Lorentz frame. However, having found a basis at a particular point $\mathcal{P}$ for which Eq. 6.2 is true, it is possible to find coordinates such that, in the neighborhood of $\mathcal{P}$, Eq. 6.2 is 'nearly' true. This is embodied in the following theorem, whose (rather long) proof is at the end of this section. Choose any point $\mathcal{P}$ of the manifold. A coordinate system $\{x^\alpha\}$ can be found whose origin is at $\mathcal{P}$ and in which

$$g_{\alpha\beta}(x^\mu) = \eta_{\alpha\beta} + 0[(x^\mu)^2]. \tag{6.3}$$

That is, the metric near $\mathcal{P}$ is approximately that of SR, the difference being of second order in the coordinates. From now on we shall refer to such coordinate systems as 'local Lorentz frames' or 'local inertial frames'. Equation 6.3 can be rephrased in a somewhat more precise way as

$$g_{\alpha\beta}(\mathcal{P}) = \eta_{\alpha\beta} \quad \text{for all } \alpha, \beta, \tag{6.4}$$

$$\frac{\partial}{\partial x^\gamma} g_{\alpha\beta}(\mathcal{P}) = 0 \quad \text{for all } \alpha, \beta, \gamma; \tag{6.5}$$

but generally it is the case that

$$\frac{\partial^2}{\partial x^\gamma \partial x^\mu} g_{\alpha\beta}(\mathcal{P}) \neq 0$$

for at least some values of α, β, γ, and μ, if the manifold is not exactly flat.

That local Lorentz frames exist is merely the statement that any curved space has a flat space 'tangent' to it at any point. Recall that straight lines in flat spacetime are the world lines of free particles; the absence of first-derivative terms (Eq. 6.5) in the metric of a curved spacetime will mean that free particles are moving on lines that are locally straight in this coordinate system. This makes such coordinates very useful for us, since the equations of physics will be nearly as simple in them as in flat spacetime, and if constructed by the rules of § 6.1 will be valid in any coordinate system. The proof of this theorem is at the end of this section, and is worth studying.

Lengths and volumes

The metric of course gives a way to define lengths of curves. Let $d\vec{x}$ be a small vector displacement on some curve. Then $d\vec{x}$ has squared length $ds^2 = g_{\alpha\beta} \, dx^\alpha \, dx^\beta$. (Recall that we call this the *line element* of the metric.) If we take the absolute value of this and find its square root, we get a measure of length: $dl \equiv |g_{\alpha\beta} \, dx^\alpha \, dx^\beta|^{1/2}$. Then, integrating it gives

$$l = \int_{\substack{\text{along} \\ \text{curve}}} |g_{\alpha\beta} \, dx^\alpha \, dx^\beta|^{1/2} \tag{6.6}$$

$$= \int_{\lambda_0}^{\lambda_1} \left| g_{\alpha\beta} \frac{dx^\alpha}{d\lambda} \frac{dx^\beta}{d\lambda} \right|^{1/2} d\lambda, \tag{6.7}$$

where λ is the parameter of the curve (whose endpoints are λ_0 and λ_1). But since the tangent vector $\vec{V}$ has components $V^\alpha = dx^\alpha/d\lambda$, we finally have

$$l = \int_{\lambda_0}^{\lambda_1} |\vec{V} \cdot \vec{V}|^{1/2} \, d\lambda \tag{6.8}$$

as the length of the arbitrary curve.

The computation of volumes is very important for integration in spacetime. Here, we mean by 'volume' the four-dimensional volume element that we used for integration in Gauss' law in § 4.9. Let us go to a local Lorentz frame, where we know that a small four-dimensional region has four-volume $dx^0 \, dx^1 \, dx^2 \, dx^3$, where $\{x^\alpha\}$ are the coordinates which at this point give the nearly Lorentz metric, Eq. 6.3. In *any* other coordinate system $\{x^{\alpha'}\}$ it is a well-known result of the calculus of several variables that:

$$dx^0 \, dx^1 \, dx^2 \, dx^3 = \frac{\partial(x^0, x^1, x^2, x^3)}{\partial(x^{0'}, x^{1'}, x^{2'}, x^{3'})} dx^{0'} \, dx^{1'} \, dx^{2'} \, dx^{3'}, \tag{6.9}$$

where the factor $\partial(\ )/\partial(\ )$ is the Jacobian of the transformation from $\{x^{\alpha'}\}$ to $\{x^\alpha\}$, as defined in § 5.2:

$$\frac{\partial(x^0, x^1, x^2, x^3)}{\partial(x^{0'}, x^{1'}, x^{2'}, x^{3'})} = \det \begin{pmatrix} \partial x^0/\partial x^{0'} & \partial x^0/\partial x^{1'} & \cdots \\ \partial x^1/\partial x^{0'} & & \\ \vdots & & \end{pmatrix}$$

$$= \det(\Lambda^\alpha{}_{\beta'}). \tag{6.10}$$

This would be a rather tedious way to calculate the Jacobian, but there is an easier way using the metric. In matrix terminology, the transformation of the metric components is

$$(g) = (\Lambda)(\eta)(\Lambda)^T, \tag{6.11}$$

where (g) is the matrix of $g_{\alpha\beta}$, (η) of $\eta_{\alpha\beta}$, etc., and where 'T' denotes a transpose. It follows that the determinants satisfy

$$\det(g) = \det(\Lambda) \det(\eta) \det(\Lambda^T). \tag{6.12}$$

But, for any matrix,

$$\det(\Lambda) = \det(\Lambda^T), \tag{6.13}$$

and we can easily see from Eq. 6.2 that

$$\det(\eta) = -1. \tag{6.14}$$

Therefore we get

$$\det(g) = -[\det(\Lambda)]^2. \tag{6.15}$$

Now we introduce the notation

$$g := \det(g_{\alpha'\beta'}), \tag{6.16}$$

which enables us to conclude from Eq. 6.15 that

$$\det(\Lambda^\alpha{}_{\beta'}) = (-g)^{1/2}. \tag{6.17}$$

Thus, from Eq. 6.9 we get

$$dx^0 \, dx^1 \, dx^2 \, dx^3 = [-\det(g_{\alpha'\beta'})]^{1/2} dx^{0'} \, dx^{1'} \, dx^{2'} \, dx^{3'}$$
$$= (-g)^{1/2} dx^{0'} \, dx^{1'} \, dx^{2'} \, dx^{3'}. \qquad (6.18)$$

This is a very useful result. It is also conceptually important because it is the first example of a kind of argument we will frequently employ, an argument that uses locally flat coordinates to generalize our flat-space concepts to analogous concepts in curved space. In this case we began with $dx^0 \, dx^1 \, dx^2 \, dx^3 = d^4x$ in a locally flat coordinate system. We argued that this volume element at $\mathcal{P}$ must be the volume physically measured by rods and clocks, since the space is the same as Minkowski space in this small region. We then found that the value of this expression in arbitrary coordinates $\{x^{\mu'}\}$ is Eq. 6.18, $(-g)^{1/2} \, d^4x'$, which is thus the expression for the true volume in a curved space at any point in any coordinates. We call this the *proper volume element*.

It should not be surprising that the metric comes into it, of course, since the metric measures lengths. One only need remember that in any coordinates the square root of the negative of the determinant of $(g_{\alpha\beta})$ is the quantity that multiplies by d^4x to get the proper volume element.

Perhaps it would be helpful to quote an example from three dimensions. Here the proper volume is $(g)^{1/2}$, since the metric is positive-definite (Eq. 6.14 would have a + sign). In spherical coordinates the line element is $dl^2 = dr^2 + r^2 d\theta^2 + r^2 \sin^2 \theta \, d\phi^2$, so the metric is

$$(g_{ij}) = \begin{pmatrix} 1 & 0 & 0 \\ 0 & r^2 & 0 \\ 0 & 0 & r^2 \sin^2 \theta \end{pmatrix}. \qquad (6.19)$$

Its determinant is $r^4 \sin^2 \theta$, so $(g)^{1/2} \, d^3x'$ is

$$r^2 \sin \theta \, dr \, d\theta \, d\phi, \qquad (6.20)$$

which we know is the correct volume element in these coordinates.

Local flatness and curvature

Local flatness fails over larger distances because of curvature. To understand the relationship between flatness and curvature, we begin by proving the local-flatness theorem quoted earlier. Let $\{x^\alpha\}$ be an arbitrary given coordinate system and $\{x^{\alpha'}\}$ the desired system: it reduces to the inertial system at a certain fixed point $\mathcal{P}$. (A point in this four-dimensional manifold is, of course, an event.) Then there is some relation

$$x^\alpha = x^\alpha(x^{\mu'}), \qquad (6.21)$$
$$\Lambda^\alpha{}_{\mu'} = \partial x^\alpha / \partial x^{\mu'}. \qquad (6.22)$$

Expanding $\Lambda^{\alpha}{}_{\mu'}$ in a Taylor series about $\mathcal{P}$ (whose coordinates are $x_0^{\mu'}$) gives the transformation at an arbitrary point $\vec{x}$ near $\mathcal{P}$:

$$\Lambda^{\alpha}{}_{\mu'}(\vec{x}) = \Lambda^{\alpha}{}_{\mu'}(\mathcal{P}) + (x^{\gamma'} - x_0^{\gamma'})\frac{\partial \Lambda^{\alpha}{}_{\mu'}}{\partial x^{\gamma'}}(\mathcal{P})$$

$$+ \frac{1}{2}(x^{\gamma'} - x_0^{\gamma'})(x^{\lambda'} - x_0^{\lambda'})\frac{\partial^2 \Lambda^{\alpha}{}_{\mu'}}{\partial x^{\lambda'}\partial x^{\gamma'}}(\mathcal{P}) + \cdots,$$

$$= \Lambda^{\alpha}{}_{\mu'}\Big|_{\mathcal{P}} + (x^{\gamma'} - x_0^{\gamma'})\frac{\partial^2 x^{\alpha}}{\partial x^{\gamma'}\partial x^{\mu'}}\Big|_{\mathcal{P}}$$

$$+ \frac{1}{2}(x^{\gamma'} - x_0^{\gamma'})(x^{\lambda'} - x_0^{\lambda'})\frac{\partial^3 x^{\alpha}}{\partial x^{\lambda'}\partial x^{\gamma'}\partial x^{\mu'}}\Big|_{\mathcal{P}} + \cdots. \tag{6.23}$$

Expanding the metric in the same way gives

$$g_{\alpha\beta}(\vec{x}) = g_{\alpha\beta}\Big|_{\mathcal{P}} + (x^{\gamma'} - x_0^{\gamma'})\frac{\partial g_{\alpha\beta}}{\partial x^{\gamma'}}\Big|_{\mathcal{P}}$$

$$+ \frac{1}{2}(x^{\gamma'} - x_0^{\gamma'})(x^{\lambda'} - x_0^{\lambda'})\frac{\partial^2 g_{\alpha\beta}}{\partial x^{\lambda'}\partial x^{\gamma'}}\Big|_{\mathcal{P}} + \cdots. \tag{6.24}$$

We put these into the transformation

$$g_{u'v'} = \Lambda^{\alpha}{}_{\mu'}\Lambda^{\beta}{}_{v'}g_{\alpha\beta} \tag{6.25}$$

to obtain

$$g_{\mu'v'}(\vec{x}) = \Lambda^{\alpha}{}_{\mu'}|_{\mathcal{P}}\,\Lambda^{\beta}{}_{v'}|_{\mathcal{P}}\,g_{\alpha\beta}|_{\mathcal{P}}$$

$$+ (x^{\gamma'} - x_0^{\gamma'})[\Lambda^{\alpha}{}_{\mu'}|_{\mathcal{P}}\,\Lambda^{\beta}{}_{v'}|_{\mathcal{P}}\,g_{\alpha\beta,\gamma'}|_{\mathcal{P}}$$

$$+ \Lambda^{\alpha}{}_{\mu'}|_{\mathcal{P}}\,g_{\alpha\beta}|_{\mathcal{P}}\,\partial^2 x^{\beta}/\partial x^{\gamma'}\partial x^{v'}|_{\mathcal{P}}$$

$$+ \Lambda^{\beta}{}_{v'}|_{\mathcal{P}}\,g_{\alpha\beta}|_{\mathcal{P}}\,\partial^2 x^{\alpha}/\partial x^{\gamma'}\partial x^{\mu'}|_{\mathcal{P}}]$$

$$+ \frac{1}{2}(x^{\gamma'} - x_0^{\gamma'})(x^{\lambda'} - x_0^{\lambda'})[\cdots]. \tag{6.26}$$

Now, we do not know the transformation Eq. 6.21, but we can define it by its Taylor expansion. Let us count the number of free variables we have for this purpose. The matrix $\Lambda^{\alpha}{}_{\mu'}|_{\mathcal{P}}$ has 16 numbers, all of which are freely specifiable. The array $\{\partial^2 x^{\alpha}/\partial x^{\gamma'}\partial x^{\mu'}|_{\mathcal{P}}\}$ has $4 \times 10 = 40$ free numbers (not $4 \times 4 \times 4$, since it is *symmetric* in γ' and μ'). The array $\{\partial^3 x^{\alpha}/\partial x^{\lambda'}\partial x^{\gamma'}\partial x^{\mu'}|_{\mathcal{P}}\}$ has $4 \times 20 = 80$ free variables, since symmetry on *all* rearrangements of λ', γ', and μ' gives only 20 independent arrangements (the general expression for three indices is $n(n + 1)(n + 2)/3!$, where n is the number of values each index can take, four in our case). On the other hand $g_{\alpha\beta}|_{\mathcal{P}}$, $g_{\alpha\beta,\gamma'}|_{\mathcal{P}}$, and $g_{\alpha\beta,\gamma'\mu'}|_{\mathcal{P}}$ are all given initially. They have, respectively, 10, $10 \times 4 = 40$, and $10 \times 10 = 100$ independent numbers for a fully general metric. The first question is, can we satisfy Eq. 6.4,

$$g_{\mu'v'}|_{\mathcal{P}} = \eta_{\mu'v'}? \tag{6.27}$$

This can be written as

$$\eta_{\mu'v'} = \Lambda^{\alpha}{}_{\mu'}|_{\mathcal{P}}\,\Lambda^{\beta}{}_{v'}|_{\mathcal{P}}\,g_{\alpha\beta}|_{\mathcal{P}}. \tag{6.28}$$

By symmetry these are ten equations, which for general matrices are independent. To satisfy them we have 16 free values in $\Lambda^\alpha{}_{\mu'}|_{\mathcal{P}}$. The equations can indeed, therefore, be satisfied, leaving six elements of $\Lambda^\alpha{}_{\mu'}|_{\mathcal{P}}$ unspecified. These six correspond to the six degrees of freedom in the Lorentz transformations that preserve the form of the metric $\eta_{\mu'\nu'}$. That is, one can boost by a velocity $\boldsymbol{v}$ (three free parameters) or rotate by an angle θ around a direction defined by two other angles. These add up to six degrees of freedom in $\Lambda^\alpha{}_{\mu'}|_{\mathcal{P}}$ that leave the local inertial frame inertial.

The next question is, can we choose the 40 free numbers $\partial \Lambda^\alpha{}_{\mu'}/\partial x^{\gamma'}|_{\mathcal{P}}$ in Eq. 6.26 in such a way as to satisfy the 40 independent equations, Eq. 6.5,

$$g_{\alpha'\beta',\gamma'}|_{\mathcal{P}} = 0? \tag{6.29}$$

Since 40 equals 40, the answer is yes, just barely. Given the matrix $\Lambda^\alpha{}_{\mu'}|_{\mathcal{P}}$, there is one and only one way to arrange the coordinates near $\mathcal{P}$ in such a way that $\Lambda^\alpha{}_{\mu',\gamma'}|_{\mathcal{P}}$ has the right values to make $g_{\alpha'\beta',\gamma'}|_{\mathcal{P}} = 0$. So there is no extra freedom other than that with which to make local Lorentz transformations.

The final question is, can we make this work at higher order? Can we find 80 numbers $\Lambda^\alpha{}_{\mu',\gamma'\lambda'}|_{\mathcal{P}}$ which can make the 100 numbers $g_{\alpha'\beta',\gamma'\lambda'}|_{\mathcal{P}} = 0$? The answer, since $80 < 100$, is no. There are, for a general metric, $100-80 = 20$ 'degrees of freedom' among the second derivatives $g_{\alpha'\beta',\mu'\lambda'}|_{\mathcal{P}}$ that must represent real aspects of the geometry.

Therefore we see that a general metric is characterized at any point $\mathcal{P}$ not so much by its value at $\mathcal{P}$ (which can always be made to equal $\eta_{\alpha\beta}$), nor by its first derivatives there (which can be made zero), but by the 20 second derivatives there, which in general cannot be made to vanish. These 20 numbers will be seen to be the independent components of a tensor which represents the curvature; this we shall show later. In a *flat* space, of course, all 20 vanish. In a general space they do not.

6.3 Covariant differentiation on a general manifold

We now look at the subject of differentiation. By definition, the derivative of a vector field involves the difference between vectors at two different points (in the limit as the points come together). In a curved space the notion of the difference between vectors at different points must be handled with care, since in between the points the space is curved and the idea that vectors at the two points might point in the 'same' direction is fuzzy. However, the local flatness of the Riemannian manifold helps us out. We need to compare vectors only in the limit as they get infinitesimally close together, and we know that we can construct a coordinate system at any point which is as close to being flat as we would like in this same limit. So, in a small region the manifold looks flat, and it is then natural to say that the derivative of a vector whose components are constant in this coordinate system is zero at that point. That is, we say that the derivatives of the basis vectors of the locally inertial coordinate system are zero at $\mathcal{P}$.

Let us emphasize that this is a *definition* of the covariant derivative. For us, its justification is in the physics: the local inertial frame is a frame in which everything is locally like SR, and in SR the derivatives of these basis vectors are zero. This definition immediately leads to the fact that in these coordinates and at this point, the covariant derivative of a vector has components given by the partial derivatives of the vector's components (that is, the Christoffel symbols vanish):

$$V^{\alpha}{}_{;\beta} = V^{\alpha}{}_{,\beta} \quad \text{at } \mathcal{P} \text{ in this frame.} \tag{6.30}$$

This is of course also true for any other tensor, including the metric:

$$g_{\alpha\beta;\gamma} = g_{\alpha\beta,\gamma} = 0 \quad \text{at } \mathcal{P}.$$

(The second equality is just Eq. 6.5.) Now, the equation $g_{\alpha\beta;\gamma} = 0$ is true in one frame (the locally inertial frame), and is a valid tensor equation; therefore it is true in *any* basis:

$$g_{\alpha\beta;\gamma} = 0 \quad \text{in any basis.} \tag{6.31}$$

This is a very important result, and comes directly from our definition of the covariant derivative. Recalling § 5.4, we see that *if* we have $\Gamma^{\mu}{}_{\alpha\beta} = \Gamma^{\mu}{}_{\beta\alpha}$ then Eq. 6.31 leads to Eq. 5.75 for *any* metric:

$$\Gamma^{\alpha}{}_{\mu\nu} = \frac{1}{2} g^{\alpha\beta} (g_{\beta\mu,\nu} + g_{\beta\nu,\mu} - g_{\mu\nu,\beta}). \tag{6.32}$$

It is left to Exercise 6.5 to demonstrate, by repeating the flat-space argument but now in the locally inertial frame, that $\Gamma^{\mu}{}_{\beta\alpha}$ is indeed symmetric in any coordinate system, so that Eq. 6.32 is correct in any coordinates. We assumed at the start that at $\mathcal{P}$ in a locally inertial frame, $\Gamma^{\alpha}{}_{\mu\nu} = 0$. But, importantly, the derivatives of $\Gamma^{\alpha}{}_{\mu\nu}$ at $\mathcal{P}$ in this frame are not all zero generally, since they involve $g_{\alpha\beta,\gamma\mu}$. This means that even though coordinates can be found in which $\Gamma^{\alpha}{}_{\mu\nu} = 0$ at a point, the Christoffel symbols do not generally vanish elsewhere. This differs from flat space, where a coordinate system exists in which $\Gamma^{\alpha}{}_{\mu\nu} = 0$ everywhere. So we can see that, at any given point, the difference between a general manifold and a flat one manifests itself in the derivatives of the Christoffel symbols.

From Eq. 6.32, given $g_{\alpha\beta}$ one can calculate $\Gamma^{\alpha}{}_{\mu\nu}$ everywhere. One can therefore calculate all covariant derivatives, given **g**. To review the formulae:

$$V^{\alpha}{}_{;\beta} = V^{\alpha}{}_{,\beta} + \Gamma^{\alpha}{}_{\mu\beta} V^{\mu}, \tag{6.33}$$

$$P_{\alpha;\beta} = P_{\alpha,\beta} - \Gamma^{\mu}{}_{\alpha\beta} P_{\mu}, \tag{6.34}$$

$$T^{\alpha\beta}{}_{;\gamma} = T^{\alpha\beta}{}_{,\gamma} + \Gamma^{\alpha}{}_{\mu\gamma} T^{\mu\beta} + \Gamma^{\beta}{}_{\mu\gamma} T^{\alpha\mu}. \tag{6.35}$$

Divergence and Gauss' law

Quite often one deals with the divergence of vectors. Given an arbitrary vector field V^α, its divergence is defined by Eq. 5.50,

$$V^\alpha{}_{;\alpha} = V^\alpha{}_{,\alpha} + \Gamma^\alpha{}_{\mu\alpha} V^\mu. \tag{6.36}$$

This formula involves a sum in the Christoffel symbol, which, from Eq. 6.32, is given by

$$\Gamma^\alpha{}_{\mu\alpha} = \frac{1}{2} g^{\alpha\beta} (g_{\beta\mu,\alpha} + g_{\beta\alpha,\mu} - g_{\mu\alpha,\beta})$$

$$= \frac{1}{2} g^{\alpha\beta} (g_{\beta\mu,\alpha} - g_{\mu\alpha,\beta}) + \frac{1}{2} g^{\alpha\beta} g_{\alpha\beta,\mu}. \tag{6.37}$$

The terms have been rearranged to simplify the formula: notice that the expression in parentheses is antisymmetric in α and β, but is contracted on α and β with $g^{\alpha\beta}$, which is symmetric. The first term therefore vanishes (see Exercise 3.26(a)) and we find that

$$\Gamma^\alpha{}_{\mu\alpha} = \frac{1}{2} g^{\alpha\beta} g_{\alpha\beta,\mu}. \tag{6.38}$$

Since $(g^{\alpha\beta})$ is the inverse matrix of $(g_{\alpha\beta})$, it can be shown (see Exercise 6.7) that the derivative of the determinant g of the matrix $(g_{\alpha\beta})$ is

$$g_{,\mu} = g g^{\alpha\beta} g_{\beta\alpha,\mu}. \tag{6.39}$$

Using this in Eq. 6.38, one finds

$$\Gamma^\alpha{}_{\mu\alpha} = (\sqrt{-g})_{,\mu} / \sqrt{-g}. \tag{6.40}$$

Then we can write the divergence, Eq. 6.36, as

$$V^\alpha{}_{;\alpha} = V^\alpha{}_{,\alpha} + \frac{1}{\sqrt{-g}} V^\alpha (\sqrt{-g})_{,\alpha} \tag{6.41}$$

or

$$V^\alpha{}_{;\alpha} = \frac{1}{\sqrt{-g}} (\sqrt{-g}\, V^\alpha)_{,\alpha}. \tag{6.42}$$

This is a very much easier formula to use than Eq. 6.36. It is also important for Gauss' law, where we integrate the divergence over a volume (using, of course, the proper volume element):

$$\int V^\alpha{}_{;\alpha} \sqrt{-g}\, \mathrm{d}^4 x = \int (\sqrt{-g}\, V^\alpha)_{,\alpha}\, \mathrm{d}^4 x. \tag{6.43}$$

Since the right-hand side of (6.43) involves just simple partial derivatives, the mathematics of Gauss' law applies to it, just as in SR (§ 4.9):

$$\int (\sqrt{-g}\, V^{\alpha})_{,\alpha}\, d^4 x = \oint V^{\alpha} n_{\alpha} \sqrt{-g}\, d^3 S. \tag{6.44}$$

This means that

$$\int V^{\alpha}_{\;;\alpha} \sqrt{-g}\, d^4 x = \oint V^{\alpha} n_{\alpha} \sqrt{-g}\, d^3 S. \tag{6.45}$$

So Gauss' law does apply on a curved manifold, in the form given by Eq. 6.45. One needs to integrate the divergence over the proper volume and to use the *proper surface element*, $n_{\alpha} \sqrt{-g}\, d^3 S$, in the surface integral.

6.4 Parallel transport, geodesics, and curvature

Until now, we have used the local-flatness theorem to develop as much mathematics on curved manifolds as possible without considering the curvature explicitly. Indeed, we have yet to give a precise mathematical definition of curvature. It is important to distinguish two different kinds of curvature: intrinsic and extrinsic. Consider, for example, a cylinder. Since a cylinder is round in one direction, one thinks of it as curved. This is its *extrinsic* curvature: the curvature it has in relation to the flat three-dimensional space of which it is part. On the other hand, a cylinder can be made by rolling a flat piece of paper without tearing or crumpling it, so the *intrinsic* geometry is that of the original paper: it is flat. This means that the distance in the surface of the cylinder between any two points is the same as it was in the original paper; parallel lines remain parallel when continued; in fact, *all* of Euclid's axioms hold for the surface of a cylinder. A two-dimensional ant confined to that surface would decide it was flat; only its global topology is strange, in that going in a certain direction in a straight line brings the ant back to where it started.

The *intrinsic* geometry of an *n*-dimensional manifold takes into account only the relationships between its points on paths that remain in the manifold (for the cylinder, on the two-dimensional surface). The *extrinsic* curvature of the cylinder comes from considering it as a surface in a space of higher dimension, and asking about the curvature of lines that stay in the surface compared with 'straight' lines that go off it. So *extrinsic* curvature relies on the notion of a higher-dimensional space. In this book, when we talk about the curvature of spacetime, we are talking about its *intrinsic* curvature, since it is clear that all world lines are confined to remain in spacetime. Whether there is a higher-dimensional space in which our four-dimensional space exists is an open question that is becoming more and more a subject of discussion within the framework of string theory. The only thing of interest to us at present, however, is the intrinsic geometry of spacetime.

The cylinder, as we have just seen, is intrinsically flat; a sphere, on the other hand, has an intrinsically curved surface. To see this, consider Figure 6.1, in which two neighboring

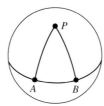

Figure 6.1   A spherical triangle *APB*.

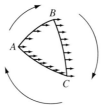

Figure 6.2   A 'triangle' made of curved lines in flat space.

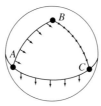

Figure 6.3   Parallel transport around a spherical triangle.

lines begin at *A* and *B* perpendicular to the equator, and hence are parallel. When continued as locally straight lines they follow the arc of great circles, and the two lines meet at the pole *P*. Parallel lines, when continued, do not remain parallel, so the space is not flat.

There is an even more striking illustration of the curvature of the sphere. Consider, first, flat space. In Figure 6.2 a closed path in flat space is drawn, and, starting at *A*, at each point a vector is drawn parallel to the vector at the previous point. This construction is carried around the loop from *A* to *B* to *C* and back to *A*. The vector finally drawn at *A* is, of course, parallel to the original one. A completely different thing happens on a sphere! Consider the path shown in Figure 6.3. Remember, we are drawing the vector as it is seen to a two-dimensional ant on the sphere, so it must always be tangent to the sphere. Aside from that, each vector is drawn as parallel as possible to the previous one. In this loop, *A* and *C* are on the equator 90° apart, and *B* is at the pole. Each arc is the arc of a great circle, and each is 90° long. At *A* we choose the vector parallel to the equator. As we move up toward *B*, each new vector is therefore drawn perpendicular to the arc *AB*. When we get to *B*, the vectors are tangent to *BC*. So, going from *B* to *C*, we keep drawing tangents to *BC*. These are perpendicular to the equator at *C*, and so, from *C* to *A* the new vectors remain perpendicular to the equator. Thus the vector field has rotated 90° in this construction! Despite the fact that each vector is drawn parallel to its neighbor, the closed

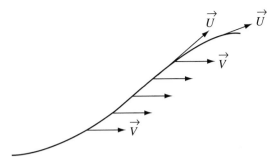

Figure 6.4   Parallel transport of $\vec{V}$ along $\vec{U}$.

loop has caused a discrepancy. Since this doesn't happen in flat space, it must be an effect of the sphere's curvature.

This result has radical implications: on a curved manifold it simply is not possible to define globally parallel vector fields. One can still define local parallelism, for instance how to move a vector from one point to another, keeping it parallel and of the same length. But the result of such 'parallel transport' from point A to point B depends on the path taken. One therefore cannot assert that a vector at A is or is not parallel to (or the same as) a certain vector at B.

Parallel transport. The construction we have just made on the sphere is called parallel transport. Suppose a vector field $\vec{V}$ is defined on the sphere and we examine how it changes along a curve, as in Figure 6.4. If the vectors $\vec{V}$ at infinitesimally close points of the curve are parallel and of equal length, then $\vec{V}$ is said to be parallel-transported along the curve. It is easy to write down an equation for this. If $\vec{U} = \mathrm{d}\vec{x}/\mathrm{d}\lambda$ is the tangent to the curve (λ being the parameter along it; $\vec{U}$ is not necessarily normalized), then in a locally inertial coordinate system at a point $\mathcal{P}$ the components of $\vec{V}$ must be constant along the curve at $\mathcal{P}$:

$$\frac{\mathrm{d}V^\alpha}{\mathrm{d}\lambda} = 0 \quad \text{at } \mathcal{P}. \tag{6.46}$$

This can be written as

$$\frac{\mathrm{d}V^\alpha}{\mathrm{d}\lambda} = U^\beta V^\alpha{}_{,\beta} = U^\beta V^\alpha{}_{;\beta} = 0 \quad \text{at } \mathcal{P}. \tag{6.47}$$

The first equality is the definition of the derivative of a function (in this case V^α) along the curve; the second equality comes from the fact that $\Gamma^\alpha{}_{\mu\nu} = 0$ at $\mathcal{P}$ in these coordinates. But the third equality is a frame-invariant expression and holds in any basis, so it can be taken as a frame-invariant *definition of the parallel transport of $\vec{V}$ along $\vec{U}$:*

$$U^\beta V^\alpha{}_{;\beta} = 0 \quad \Leftrightarrow \quad \frac{\mathrm{d}}{\mathrm{d}\lambda}\vec{V} = \nabla_{\vec{U}}\vec{V} = 0. \tag{6.48}$$

The last step uses the notation for the derivative along $\vec{U}$ introduced in Eq. 3.67.

Geodesics. The most important curves in flat space are straight lines. One of Euclid's axioms is that two straight lines that are initially parallel remain parallel when extended.

What does he mean by 'extended'? He *does not* mean 'continued in such a way that the distance between them remains constant', because even then they could both bend. What he means is that each line keeps going in the direction it has been going in. More precisely, the tangent to the curve at one point is parallel to the tangent at the previous point. In fact, a straight line in Euclidean space is the *only* curve that parallel-transports its own tangent vector! In a curved space, we can also draw lines that are 'as nearly straight as possible' by demanding parallel transport of the tangent vector. These are called *geodesics*:

$$\{\vec{U} \text{ is tangent to a geodesic}\} \quad \Leftrightarrow \quad \nabla_{\vec{U}} \vec{U} = 0. \tag{6.49}$$

(Note that in a locally inertial system these lines *are* straight.) In component notation,

$$U^\beta U^\alpha{}_{;\beta} = U^\beta U^\alpha{}_{,\beta} + \Gamma^\alpha{}_{\mu\beta} U^\mu U^\beta = 0. \tag{6.50}$$

Now, if we let λ be the parameter of the curve then $U^\alpha = \mathrm{d}x^\alpha/\mathrm{d}\lambda$ and $U^\beta \partial/\partial x^\beta = \mathrm{d}/\mathrm{d}\lambda$:

$$\frac{\mathrm{d}}{\mathrm{d}\lambda}\left(\frac{\mathrm{d}x^\alpha}{\mathrm{d}\lambda}\right) + \Gamma^\alpha{}_{\mu\beta} \frac{\mathrm{d}x^\mu}{\mathrm{d}\lambda}\frac{\mathrm{d}x^\beta}{\mathrm{d}\lambda} = 0. \tag{6.51}$$

Since the Christoffel symbols $\Gamma^\alpha{}_{\mu\beta}$ are known functions of the coordinates $\{x^\alpha\}$, this is a nonlinear (but quasi-linear) second-order differential equation for $x^\alpha(\lambda)$. It has a unique solution when initial conditions at $\lambda = \lambda_0$ are given: $x_0^\alpha = x^\alpha(\lambda_0)$ and $U_0^\alpha = (\mathrm{d}x^\alpha/\mathrm{d}\lambda)_{\lambda_0}$. So, by giving an initial position (x_0^α) and an initial direction (U_0^α), one obtains a unique geodesic.

Recall that if we change parameter then we change, mathematically speaking, the curve (though not the points it passes through). Now, if λ is a parameter of a geodesic (so that Eq. 6.51 is satisfied), and if we define a new parameter

$$\phi = a\lambda + b, \tag{6.52}$$

where a and b are *constants* (not depending on their position on the curve), then ϕ is also a parameter for which Eq. 6.51 is satisfied:

$$\frac{\mathrm{d}^2 x^\alpha}{\mathrm{d}\phi^2} + \Gamma^\alpha{}_{\mu\beta} \frac{\mathrm{d}x^\mu}{\mathrm{d}\phi}\frac{\mathrm{d}x^\beta}{\mathrm{d}\phi} = 0.$$

Generally speaking, *only* linear transformations of λ like Eq. 6.52 will give new parameters for which the geodesic equation is satisfied. A parameter like λ and ϕ above is called an *affine* parameter. A curve having the same path as a geodesic but parametrized by a nonaffine parameter is, strictly speaking, not a geodesic curve.

A geodesic is also a curve of *extremal length* between any two points: its length is unchanged to first order in small changes in the curve. You are urged to prove this by using Eq. 6.7, finding the Euler–Lagrange equations for it to be an extremal for fixed λ_0 and λ_1, and showing that these reduce to Eq. 6.51 when Eq. 6.32 is used. This is a very instructive exercise. One can also show that proper distance along the geodesic is itself an affine parameter. (See Exercises 6.13–6.15.)

6.5 The curvature tensor

At last we are in a position to give a mathematical description of the intrinsic curvature of a manifold. We go back to the curious example of the parallel transport of a vector around a closed loop, and use it for our *definition* of curvature. Let us imagine in our manifold a very small closed loop (Figure 6.5) whose four sides are the coordinate lines $x^1 = a$, $x^1 = a + \delta a$, $x^2 = b$, and $x^2 = b + \delta b$. A vector $\vec{V}$ defined at A is parallel-transported to B. The parallel transport law $\nabla_{\vec{e}_1} \vec{V} = 0$ has component form

$$\frac{\partial V^\alpha}{\partial x^1} = -\Gamma^\alpha{}_{\beta 1} V^\beta. \tag{6.53}$$

Integrating this from A to B gives

$$V^\alpha(B) = V^\alpha(A) + \int_A^B \frac{\partial V^\alpha}{\partial x^1} \, dx^1$$
$$= V^\alpha(A) - \int_{x^2=b} \Gamma^\alpha{}_{\beta 1} V^\beta dx^1, \tag{6.54}$$

where the notation '$x^2 = b$' under the integral sign denotes the path AB. Similar transport from B to C to D gives

$$V^\alpha(C) = V^\alpha(B) - \int_{x^1=a+\delta a} \Gamma^\alpha{}_{\beta 2} V^\beta \, dx^2, \tag{6.55}$$

$$V^\alpha(D) = V^\alpha(C) + \int_{x^2=b+\delta b} \Gamma^\alpha{}_{\beta 1} V^\beta \, dx^1. \tag{6.56}$$

The integral in the last equation has a different sign because the direction of transport from C to D is in the negative x^1 direction. Similarly, the completion of the loop gives

$$V^\alpha(A_{\text{final}}) = V^\alpha(D) + \int_{x^1=a} \Gamma^\alpha{}_{\beta 2} V^\beta \, dx^2. \tag{6.57}$$

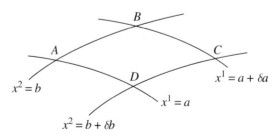

Small section of a coordinate grid.

The net change in $V^\alpha(A)$ is a vector δV^α, found by adding up Eqs. 6.54–6.57:

$$\delta V^\alpha = V^\alpha(A_{\text{final}}) - V^\alpha(A_{\text{initial}})$$

$$= \int_{x^1=a} \Gamma^\alpha{}_{\beta 2} V^\beta dx^2 - \int_{x^1=a+\delta a} \Gamma^\alpha{}_{\beta 2} V^\beta dx^2$$

$$+ \int_{x^2=b+\delta b} \Gamma^\alpha{}_{\beta 1} V^\beta dx^1 - \int_{x^2=b} \Gamma^\alpha{}_{\beta 1} V^\beta dx^1. \tag{6.58}$$

Notice that these would cancel in pairs if $\Gamma^\alpha{}_{\beta v}$ and V^β were constants on the loop, as they would be in flat space. But in curved space they are not, so if we combine the integrals over similar integration variables and work to first order in the separation in the paths, we get, to lowest order,

$$\delta V^\alpha \simeq - \int_b^{b+\delta b} \delta a \frac{\partial}{\partial x^1}(\Gamma^\alpha{}_{\beta 2} V^\beta) \, dx^2$$

$$+ \int_a^{a+\delta a} \delta b \frac{\partial}{\partial x^2}(\Gamma^\alpha{}_{\beta 1} V^\beta) \, dx^1 \tag{6.59}$$

$$\approx \delta a \, \delta b \left[-\frac{\partial}{\partial x^1}(\Gamma^\alpha{}_{\beta 2} V^\beta) + \frac{\partial}{\partial x^2}(\Gamma^\alpha{}_{\beta 1} V^\beta) \right]. \tag{6.60}$$

This involves derivatives of Christoffel symbols and of V^α. The derivatives V^α can be eliminated using Eq. 6.53 and its equivalent with 1 replaced by 2. Then Eq. 6.60 becomes

$$\delta V^\alpha = \delta a \, \delta b \left[\Gamma^\alpha{}_{\beta 1,2} - \Gamma^\alpha{}_{\beta 2,1} + \Gamma^\alpha{}_{v2}\Gamma^v{}_{\beta 1} - \Gamma^\alpha{}_{v1}\Gamma^v{}_{\beta 2} \right] V^\beta. \tag{6.61}$$

(To obtain this, one needs to relabel dummy indices in the terms quadratic in Γs.) Notice that this turns out to be just a number times V^β, summed on β. Now, the indices 1 and 2 appear because the path was chosen to go along those coordinates. Notice that it is antisymmetric in 1 and 2. The reason is that the change δV^α would have to have the opposite sign if one went around the loop in the opposite direction (that is, interchanging the roles of 1 and 2). If one used general coordinate lines x^v and x^μ, one would find

$$\delta V^\alpha = \text{change in } V^\alpha \text{ due to transport}$$

$$= \text{first } \delta a \, \vec{e}_v, \text{ then } \delta b \, \vec{e}_\mu, \text{ then } -\delta a \, \vec{a}_v, \text{ and finally } -\delta b \, \vec{e}_\mu$$

$$= \delta a \, \delta b \left[\Gamma^\alpha{}_{\beta v,\mu} - \Gamma^\alpha{}_{\beta\mu,v} + \Gamma^\alpha{}_{\sigma\mu}\Gamma^\sigma{}_{\beta v} - \Gamma^\alpha{}_{\sigma v}\Gamma^\sigma{}_{\beta\mu} \right] V^\beta. \tag{6.62}$$

Now, δV^α depends on $\delta a \delta b$, the coordinate 'area' of the loop. So it is clear that if the length of the loop in one direction is doubled, δV^α is doubled. This means that δV^α depends *linearly* on $\delta a \, \vec{e}_v$ and $\delta b \, \vec{e}_\mu$. Moreover, it certainly also depends linearly, in Eq. 6.62, on V^α itself and on $\tilde{\omega}^\alpha$, which is the basis one-form that gives δV^α from the vector $\delta \vec{V}$. Hence we have the following result: if we define

$$R^\alpha{}_{\beta\mu v} := \Gamma^\alpha{}_{\beta v,\mu} - \Gamma^\alpha{}_{\beta\mu,v} + \Gamma^\alpha{}_{\sigma\mu}\Gamma^\sigma{}_{\beta v} - \Gamma^\alpha{}_{\sigma v}\Gamma^\sigma{}_{\beta\mu}, \tag{6.63}$$

then the $R^\alpha{}_{\beta\mu\nu}$ must be components of the $\binom{1}{3}$ tensor which, when supplied with arguments $\tilde{\omega}^\alpha$, $\vec{V}$, $\epsilon_1\vec{e}_\nu$, $\epsilon_2\vec{e}_\mu$, gives δV^α, the component of the change in $\vec{V}$ after parallel-transport around a very small loop given by steps first of $\epsilon_1\vec{e}_\nu$, second of $\epsilon_2\vec{e}_\mu$, and then back by $-\epsilon_1\vec{e}_\nu$ followed by $-\epsilon_2\vec{e}_\mu$. This tensor is called the *Riemann curvature tensor* **R**.[1]

It should be clear from the derivation that the change given by Eq. 6.62 is accurate only if the loop is small enough that second derivatives of the Christoffel symbols are not needed in the expansions along the edges of the loop. The loop shown in Figure 6.3, for example, is much too large for this formula to apply.

It is useful to look at the components of **R** in a locally inertial frame at a point $\mathcal{P}$. We have $\Gamma^\alpha{}_{\mu\nu} = 0$ at $\mathcal{P}$, but we can find its derivative from Eq. 6.32:

$$\Gamma^\alpha{}_{\beta\nu,\mu} = \tfrac{1}{2}g^{\alpha\sigma}\left(g_{\sigma\beta,\nu\mu} + g_{\sigma\nu,\beta\mu} - g_{\beta\nu,\sigma\mu}\right). \tag{6.64}$$

Since the second derivatives of $g_{\alpha\beta}$ do not vanish, we get at $\mathcal{P}$

$$R^\alpha{}_{\beta\mu\nu} = \tfrac{1}{2}g^{\alpha\sigma}(g_{\sigma\beta,\nu\mu} + g_{\sigma\nu,\beta\mu} - g_{\beta\nu,\sigma\mu} - g_{\sigma\beta,\mu\nu} - g_{\sigma\mu,\beta\nu} + g_{\beta\mu,\sigma\nu}). \tag{6.65}$$

Using the symmetry of $g_{\alpha\beta}$ and the fact that

$$g_{\alpha\beta,\mu\nu} = g_{\alpha\beta,\nu\mu}, \tag{6.66}$$

because partial derivatives always commute, we find that

$$R^\alpha{}_{\beta\mu\nu} = \tfrac{1}{2}g^{\alpha\sigma}\left(g_{\sigma\nu,\beta\mu} - g_{\sigma\mu,\beta\nu} + g_{\beta\mu,\sigma\nu} - g_{\beta\nu,\sigma\mu}\right). \tag{6.67}$$

If we lower the index α we get (in the locally flat coordinate system at its origin $\mathcal{P}$)

$$R_{\alpha\beta\mu\nu} := g_{\alpha\lambda}R^\lambda{}_{\beta\mu\nu} = \tfrac{1}{2}\left(g_{\alpha\nu,\beta\mu} - g_{\alpha\mu,\beta\nu} + g_{\beta\mu,\alpha\nu} - g_{\beta\nu,\alpha\mu}\right). \tag{6.68}$$

In this form it is easy to verify the following identities:

$$R_{\alpha\beta\mu\nu} = -R_{\beta\alpha\mu\nu} = -R_{\alpha\beta\nu\mu} = R_{\mu\nu\alpha\beta}, \tag{6.69}$$
$$R_{\alpha\beta\mu\nu} + R_{\alpha\nu\beta\mu} + R_{\alpha\mu\nu\beta} = 0. \tag{6.70}$$

Thus, $R_{\alpha\beta\mu\nu}$ is antisymmetric on the first pair and on the second pair of indices, and symmetric on exchange of the two pairs. Since Eqs. 6.69 and 6.70 are valid tensor equations true in one coordinate system, they are true in all bases. (Note that an equation like Eq. 6.67 is not a valid tensor equation, since it involves partial derivatives, not covariant derivatives. Therefore it is true only in the coordinate system in which it was derived.)

---

[1] As with other definitions introduced earlier, there is no universal agreement about the overall sign of the Riemann tensor, or even on the placement of its indices. Always check the conventions of whatever text you read.

It can be shown (Exercise 6.18) that the various identities in Eqs. 6.69 and 6.70 reduce the number of independent components of $R_{\alpha\beta\mu\nu}$ (and hence of $R^{\alpha}{}_{\beta\mu\nu}$) to 20, in four dimensions. This is, *not* coincidentally, the same number of independent $g_{\alpha\beta,\mu\nu}$ that as we found at the end of § 6.2 could not be made to vanish by a coordinate transformation. Thus $R^{\alpha}{}_{\beta\mu\nu}$ characterizes the curvature in a tensorial way.

A *flat* manifold is one which has a global definition of parallelism: a vector can be moved around parallel to itself on an arbitrary curve and will return to its starting point unchanged. This clearly means that

$$R^{\alpha}{}_{\beta\mu\nu} = 0 \iff \text{flat manifold.} \qquad (6.71)$$

(Try showing that this is true in polar coordinates for the Euclidean plane.)

Covariant derivatives do not commute. An important use of the curvature tensor comes when we examine the consequences of taking two covariant derivatives of a vector field $\vec{V}$. We found in § 6.3 that first derivatives were like flat-space ones, since we could find coordinates in which the metric was flat to first order. But second derivatives present a different story:

$$\nabla_{\alpha}\nabla_{\beta}V^{\mu} = \nabla_{\alpha}(V^{\mu}{}_{;\beta})$$
$$= (V^{\mu}{}_{;\beta})_{,\alpha} + \Gamma^{\mu}{}_{\sigma\alpha}V^{\sigma}{}_{;\beta} - \Gamma^{\sigma}{}_{\beta\alpha}V^{\mu}{}_{;\sigma}. \qquad (6.72)$$

In locally inertial coordinates whose origin is at $\mathcal{P}$, all the Γs are zero, but their partial derivatives are not. Therefore we have at $\mathcal{P}$

$$\nabla_{\alpha}\nabla_{\beta}V^{\mu} = V^{\mu}{}_{,\beta\alpha} + \Gamma^{\mu}{}_{\nu\beta,\alpha}V^{\nu}. \qquad (6.73)$$

Bear in mind that this expression is valid only in this specially chosen coordinates system, and that is true also for Eqs. 6.74–6.76 below. These coordinates make the computation easier: consider now Eq. 6.73 with α and β exchanged:

$$\nabla_{\beta}\nabla_{\alpha}V^{\mu} = V^{\mu}{}_{,\alpha\beta} + \Gamma^{\mu}{}_{\nu\alpha,\beta}V^{\nu}. \qquad (6.74)$$

If we subtract these then we get the *commutator* of the covariant derivative operators ∇_{α} and ∇_{β}, written in the same notation as we would employ in quantum mechanics:

$$[\nabla_{\alpha}, \nabla_{\beta}]V^{\mu} := \nabla_{\alpha}\nabla_{\beta}V^{\mu} - \nabla_{\beta}\nabla_{\alpha}V^{\mu}$$
$$= \left(\Gamma^{\mu}{}_{\nu\beta,\alpha} - \Gamma^{\mu}{}_{\nu\alpha,\beta}\right)V^{\nu}. \qquad (6.75)$$

The terms involving the second derivatives of V^{μ} drop out here, since

$$V^{\mu}{}_{,\alpha\beta} = V^{\mu}{}_{,\beta\alpha}. \qquad (6.76)$$

(Let us pause to recall that $V^{\mu}{}_{,\alpha}$ is the partial derivative of the component V^{μ}, so by the laws of partial differentiation the partial derivatives must commute. On the other hand, $\nabla_{\alpha}V^{\mu}$ is a component of the tensor $\nabla\vec{V}$, and $\nabla_{\alpha}\nabla_{\beta}V^{\mu}$ is a component of $\nabla\nabla\vec{V}$: there is no reason (from differential calculus) why it must be symmetric on α and β. We have proved, by showing that Eq. 6.75 is nonzero, that the double covariant derivative generally is *not* symmetric.)

Now, in this frame (where $\Gamma^\mu{}_{\alpha\beta} = 0$ at $\mathcal{P}$), we can compare Eq. 6.75 with Eq. 6.63 and see that at $\mathcal{P}$

$$[\nabla_\alpha, \nabla_\beta] V^\mu = R^\mu{}_{\nu\alpha\beta} V^\nu. \tag{6.77}$$

Now, this is a valid tensor equation so it is true in *any* coordinate system: the Riemann tensor gives the commutator of covariant derivatives. We can drop the restriction to locally inertial coordinates: they were simply a convenient way of arriving at a general tensor expression for the commutator. What this means is that, in curved spaces, one must be careful to know the order in which covariant derivatives are taken: they do not commute. This can be extended to tensors of higher rank. For example, for a $\binom{1}{1}$ tensor,

$$[\nabla_\alpha, \nabla_\beta] F^\mu{}_\nu = R^\mu{}_{\sigma\alpha\beta} F^\sigma{}_\nu + R_\nu{}^\sigma{}_{\alpha\beta} F^\mu{}_\sigma. \tag{6.78}$$

That is, *each* index lies on a Riemann tensor, and each one comes with a + sign. (They *must* all have the same sign because raising and lowering indices with **g** is unaffected by ∇_α, since $\nabla \mathbf{g} = 0$.)

Equation 6.77 is closely related to our original derivation of the Riemann tensor from parallel transport around loops, because the parallel transport problem can be thought of as computing first the change of $\vec{V}$ in one direction, and then in another, followed by subtracting changes in the reverse order: this is what commuting covariant derivatives also does.

Geodesic deviation. We have often mentioned that, in a curved space, parallel lines when extended do not remain parallel. This can now be formulated mathematically in terms of the Riemann tensor. Consider two geodesics (with tangents $\vec{V}$ and $\vec{V}'$) that begin parallel and near each other, as in Figure 6.6, at points A and A'. Let the affine parameter on the geodesics be called λ. We define a 'connecting vector' $\vec{\xi}$ which 'reaches' from one geodesic to another, connecting points at equal intervals of λ (i.e. A to A', B to B', etc.). For simplicity, let us adopt a locally inertial coordinate system at A, in which the coordinate x^0 points along the geodesics and advances at the same rate as λ there (this is just a scaling of the coordinate). Then, because $V^\alpha = dx^\alpha/d\lambda$, at A we have $V^\alpha = \delta^\alpha_0$. The equation of

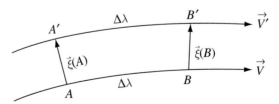

Figure 6.6   A connecting vector $\vec{\xi}$ between two geodesics connects points of the same parameter value.

the geodesic at A is

$$\left.\frac{d^2 x^\alpha}{d\lambda^2}\right|_A = 0, \tag{6.79}$$

since all Christoffel symbols vanish at A. The Christoffel symbols do not vanish at A', so the equation of the geodesic $\vec{V}'$ at A' is

$$\left.\frac{d^2 x^\alpha}{d\lambda^2}\right|_{A'} + \Gamma^\alpha{}_{00}(A') = 0, \tag{6.80}$$

where again at A' we have arranged the coordinates so that $V^\alpha = \delta^\alpha_0$. But, since A and A' are separated by $\vec{\xi}$, we have

$$\Gamma^\alpha{}_{00}(A') \cong \Gamma^\alpha{}_{00,\beta}\xi^\beta, \tag{6.81}$$

the right-hand side being evaluated at A. With Eq. 6.80 this gives

$$\left.\frac{d^2 x^\alpha}{d\lambda^2}\right|_{A'} = -\Gamma^\alpha{}_{00,\beta}\xi^\beta. \tag{6.82}$$

Now, the difference $x^\alpha(\lambda, \text{geodesic } \vec{V}') - x^\alpha(\lambda, \text{geodesic } \vec{V})$ is just the component ξ^α of the vector $\vec{\xi}$. Therefore, at A, we have

$$\frac{d^2 \xi^\alpha}{d\lambda^2} = \left.\frac{d^2 x^\alpha}{d\lambda^2}\right|_{A'} - \left.\frac{d^2 x^\alpha}{d\lambda^2}\right|_A = -\Gamma^\alpha{}_{00,\beta}\xi^\beta. \tag{6.83}$$

This then tells us how the components of $\vec{\xi}$ change. But since the coordinates are to some extent arbitrary, we want to have, not merely the second derivative of the component ξ^α, but the full second covariant derivative $\nabla_V \nabla_V \vec{\xi}$. We can use Eq. 6.48 to obtain

$$\nabla_V \nabla_V \xi^\alpha = \nabla_V (\nabla_V \xi^\alpha)$$
$$= \frac{d}{d\lambda}(\nabla_v \xi^\alpha) + \Gamma^\alpha{}_{\beta 0}(\nabla_V \xi^\beta). \tag{6.84}$$

Now, using $\Gamma^\alpha{}_{\beta\gamma} = 0$ at A, we have

$$\nabla_V \nabla_V \xi^\alpha = \frac{d}{d\lambda}\left(\frac{d}{d\lambda}\xi^\alpha + \Gamma^\alpha{}_{\beta 0}\xi^\beta\right) + 0$$
$$= \frac{d^2}{d\lambda^2}\xi^\alpha + \Gamma^\alpha{}_{\beta 0,0}\xi^\beta \tag{6.85}$$

at A. (We have also used $\xi^\beta{}_{,0} = 0$ at A, which is the condition that the curves begin parallel.) So we get

$$\nabla_V \nabla_V \xi^\alpha = \left(\Gamma^\alpha{}_{\beta 0,0} - \Gamma^\alpha{}_{00,\beta}\right)\xi^\beta$$
$$= R^\alpha{}_{00\beta}\xi^\beta = R^\alpha{}_{\mu\nu\beta}V^\mu V^\nu \xi^\beta, \tag{6.86}$$

where the second equality follows from Eq. 6.63. The final expression is frame invariant, and A is an arbitrary point, so we have, in *any* basis,

$$\nabla_V \nabla_V \xi^\alpha = R^\alpha{}_{\mu\nu\beta} V^\mu V^\nu \xi^\beta. \tag{6.87}$$

Geodesics in flat space maintain their separation; those in curved spaces do not. This is called the equation of geodesic deviation and shows mathematically that the tidal forces of a gravitational field (which cause the trajectories of neighboring particles to diverge) can be represented by the curvature of a spacetime in which particles follow geodesics.

6.6 Bianchi identities; Ricci and Einstein tensors

Let us return to Eq. 6.63 for the components of the Riemann tensor. If we differentiate it with respect to x^λ (just the partial derivative) and evaluate the result in locally inertial coordinates, we find

$$R_{\alpha\beta\mu\nu,\lambda} = \frac{1}{2} \left(g_{\alpha\nu,\beta\mu\lambda} - g_{\alpha\mu,\beta\nu\lambda} + g_{\beta\mu,\alpha\nu\lambda} - g_{\beta\nu,\alpha\mu\lambda} \right). \tag{6.88}$$

From this equation, the symmetry $g_{\alpha\beta} = g_{\beta\alpha}$ and the fact that partial derivatives commute, one can show that

$$R_{\alpha\beta\mu\nu,\lambda} + R_{\alpha\beta\lambda\mu,\nu} + R_{\alpha\beta\nu\lambda,\mu} = 0. \tag{6.89}$$

Since in our coordinates $\Gamma^\mu{}_{\alpha\beta} = 0$ at this point, this equation is *equivalent* to

$$R_{\alpha\beta\mu\nu;\lambda} + R_{\alpha\beta\lambda\mu;\nu} + R_{\alpha\beta\nu\lambda;\mu} = 0. \tag{6.90}$$

But this is a set of tensor equations, valid in *any* system, called the *Bianchi identities*, and will be very important for our work.

The Ricci tensor. Before pursuing the consequences of the Bianchi identities, we shall need to define the Ricci tensor $R_{\alpha\beta}$:

$$R_{\alpha\beta} := R^\mu{}_{\alpha\mu\beta} = R_{\beta\alpha}. \tag{6.91}$$

It is the contraction of $R^\mu{}_{\alpha\nu\beta}$ on the first and third indices. Other contractions would in principle also be possible: on the first and second, the first and fourth, etc. But because $R_{\alpha\beta\mu\nu}$ is antisymmetric on α and β and on μ and ν, all these contractions either vanish identically or reduce to $\pm R_{\alpha\beta}$. Therefore the Ricci tensor is essentially the *only* contraction of the Riemann tensor. Note that Eq. 6.69 implies that it is a *symmetric* tensor (Exercise 6.25).

Similarly, the *Ricci scalar* is defined as

$$R := g^{\mu\nu} R_{\mu\nu} = g^{\mu\nu} g^{\alpha\beta} R_{\alpha\mu\beta\nu}. \tag{6.92}$$

The Einstein tensor. Let us apply the Ricci contraction to the Bianchi identities, Eq. 6.90:

$$g^{\alpha\mu} \left[R_{\alpha\beta\mu\nu;\lambda} + R_{\alpha\beta\lambda\mu;\nu} + R_{\alpha\beta\nu\lambda;\mu} \right] = 0$$

or

$$R_{\beta\nu;\lambda} + (-R_{\beta\lambda;\nu}) + R^{\mu}{}_{\beta\nu\lambda;\mu} = 0. \tag{6.93}$$

To derive this result one needs two facts. First, by Eq. 6.31 we have

$$g_{\alpha\beta;\mu} = 0.$$

Since $g^{\alpha\mu}$ is a function only of $g_{\alpha\beta}$ it follows that

$$g^{\alpha\beta}{}_{;\mu} = 0. \tag{6.94}$$

Therefore $g^{\alpha\mu}$ and $g_{\beta\nu}$ can be taken in and out of covariant derivatives at will: index-raising and index-lowering commute with covariant differentiation. The second fact is that

$$g^{\alpha\mu} R_{\alpha\beta\lambda\mu;\nu} = -g^{\alpha\mu} R_{\alpha\beta\mu\lambda;\nu} = -R_{\beta\lambda;\nu}, \tag{6.95}$$

accounting for the second term in Eq. 6.93. The differential equations 6.93 are the contracted Bianchi identities. A more useful set of equations is obtained by contracting again on the indices β and ν:

$$g^{\beta\nu} \left[R_{\beta\nu;\lambda} - R_{\beta\lambda;\nu} + R^{\mu}{}_{\beta\nu\lambda;\mu} \right] = 0$$

or

$$R_{;\lambda} - R^{\mu}{}_{\lambda;\mu} + (-R^{\mu}{}_{\lambda;\mu}) = 0. \tag{6.96}$$

Again the antisymmetry of **R** has been used to get the correct sign in the last term. Note that, since R is a scalar, $R_{;\lambda} \equiv R_{,\lambda}$ in all coordinates. Now Eq. 6.96 can be written in the form

$$(2R^{\mu}{}_{\lambda} - \delta^{\mu}{}_{\lambda} R)_{;\mu} = 0. \tag{6.97}$$

These equations are the twice-contracted Bianchi identities, often simply also called the Bianchi identities. If we define the symmetric tensor

$$G^{\alpha\beta} \equiv R^{\alpha\beta} - \tfrac{1}{2} g^{\alpha\beta} R = G^{\beta\alpha}, \tag{6.98}$$

then we can see that Eq. 6.97 is equivalent to

$$G^{\alpha\beta}{}_{;\beta} = 0. \tag{6.99}$$

The tensor $G^{\alpha\beta}$ is constructed only from the Riemann tensor and the metric, and is automatically divergence free as an identity. It is called the Einstein tensor, since its importance for gravity was first understood by Einstein. (In fact we shall see that the Einstein field equations for GR are

$$G^{\alpha\beta} = 8\pi T^{\alpha\beta},$$

where $T^{\alpha\beta}$ is the stress–energy tensor. The Bianchi identities then imply that

$$T^{\alpha\beta}{}_{;\beta} \equiv 0,$$

which is the equation for the local conservation of energy and momentum. However, this is looking somewhat far ahead.)

6.7 Curvature in perspective

The mathematical machinery for dealing with curvature is formidable. There are many important equations in this chapter, but few of them need to be memorized. It is far more important to understand their derivation and particularly their geometrical interpretation. This interpretation is something we will build up over the next few chapters, but the material already in hand should give you some idea of what the mathematics means. Let us review the important features of curved spaces.

(1) We work on Riemannian manifolds, which are smooth spaces with a metric defined on them.
(2) The metric has signature +2, and there always exists a coordinate system in which, at a single point, one can have

$$g_{\alpha\beta} = \eta_{\alpha\beta},$$
$$g_{\alpha\beta,\gamma} = 0 \;\Rightarrow\; \Gamma^{\alpha}_{\beta\gamma} = 0.$$

(3) The element of proper volume is

$$|g|^{1/2}\mathrm{d}^4 x,$$

where g is the determinant of the matrix of components $(g_{\alpha\beta})$.
(4) The covariant derivative is simply the ordinary derivative in locally inertial coordinates. Because of curvature ($\Gamma^{\alpha}_{\beta\gamma,\sigma} \neq 0$) these derivatives do not commute.
(5) The definition of parallel transport is that the covariant derivative along the curve is zero. A geodesic parallel-transports its own tangent vector. Its affine parameter can be taken to be the proper distance itself.
(6) The Riemann tensor is the characterization of the curvature. Only if it vanishes identically is the manifold flat. It has 20 independent components (in four dimensions) and satisfies the Bianchi identities, which are differential equations. The Riemann tensor in a general coordinate system depends on $g_{\alpha\beta}$ and its first and second partial derivatives. The Ricci tensor, Ricci scalar, and Einstein tensor are contractions of the

Riemann tensor. In particular, the Einstein tensor is symmetric and of second rank, so it has ten independent components. They satisfy the four differential identities, given by Eq. 6.99.

6.8 Bibliography

The theory of differentiable manifolds is introduced in a large number of books. The following are suitable for exploring the subject further with a view toward its physical applications, particularly outside relativity: Abraham *et al.* (1988), Bishop & Goldberg (1981), Lovelock & Rund (1990), and Schutz (1980b). Standard mathematical reference classics include Kobayashi & Nomizu (2009), Schouten (2011), and Spivak (1999). A beautifully constructed and intuitive survey of differential geometry from a purely geometric point of view (that is, with an absolute minimum of algebra) is Needham (2021). If you are interested in topology, the mathematical exposition by Nahmad-Achar (2018) brings topology into the subjects we cover here and then into applications in SR and GR.

Exercises

6.1   Decide whether the following sets are manifolds and say why. If there are exceptional points at which the sets are not manifolds, give them:

   (a) the phase space of Hamiltonian mechanics, i.e. the space of the canonical coordinates and momenta p_i and q^i;
   (b) the interior of a circle of unit radius in two-dimensional Euclidean space;
   (c) the set of permutations of n objects;
   (d) the subset of Euclidean space of two dimensions (with coordinates x and y) which is a solution to $xy(x^2 + y^2 - 1) = 0$.

6.2   Of the manifolds in Exercise 6.1, on which is it customary to use a metric, and what is that metric? On which would a metric not normally be defined, and why?

6.3   It is well known that, for any symmetric matrix A (with real entries), there exists a matrix H for which the matrix $H^T A H$ is a diagonal matrix whose entries are the eigenvalues of A.

   (a) Show that there is a matrix R such that $R^T H^T A H R$ is the same matrix as $H^T A H$ except with the eigenvalues rearranged in ascending order along the main diagonal from top to bottom.
   (b) Show that there exists a third matrix N such that $N^T R^T H^T A H R N$ is a diagonal matrix whose entries on the diagonal are $-1, 0$, or $+1$.
   (c) Show that if A has an inverse, none of the diagonal elements in (b) is zero.
   (d) Show from (a)–(c) that there exists a transformation matrix Λ which produces Eq. 6.2.

6.4 Prove the following results used in the proof of the local-flatness theorem in § 6.2:

(a) The number of independent values of $\partial^2 x^\alpha / \partial x^{\gamma'} \partial x^{\mu'}|_0$ is 40.
(b) The corresponding number for $\partial^3 x^\alpha / \partial x^{\lambda'} \partial x^{\mu'} \partial x^{\nu'}|_0$ is 80.
(c) The corresponding number for $g_{\alpha\beta, \gamma' \mu'}|_0$ is 100.

6.5 (a) Prove that $\Gamma^\mu{}_{\alpha\beta} = \Gamma^\mu{}_{\beta\alpha}$ in any coordinate system in a curved Riemannian space.
(b) Use this to prove that Eq. 6.32 can be derived in the same manner as in flat space.

6.6 Prove that the first term in Eq. 6.37 vanishes.

6.7 (a) Give the definition of the determinant of a matrix A in terms of cofactors of elements.
(b) Differentiate the determinant of an arbitrary 2×2 matrix and show that it satisfies Eq. 6.39.
(c) Generalize Eq. 6.39 (by induction or otherwise) to arbitrary $n \times n$ matrices.

6.8 Fill in the missing algebra leading to Eqs. 6.40 and 6.42.

6.9 Show that Eq. 6.42 leads to Eq. 5.56. Derive the divergence formula for the metric in Eq. 6.19.

6.10 A 'straight line' on a sphere is a great circle, and it is well known that the sum of the interior angles of any triangle on a sphere whose sides are arcs of great circles exceeds $180°$. Show that the amount by which a vector is rotated by parallel transport around such a triangle (as in Figure 6.3) equals the excess of the sum of the angles over $180°$.

6.11 In this exercise we will determine the condition that a vector field $\vec{V}$ can be considered to be globally parallel on a manifold. More precisely, what guarantees that we can find a vector field $\vec{V}$ satisfying the equation

$$(\nabla \vec{V})^\alpha{}_\beta = V^\alpha{}_{;\beta} = V^\alpha{}_{,\beta} + \Gamma^\alpha{}_{\mu\beta} V^\mu = 0?$$

(a) A necessary condition, called the *integrability condition* for this equation, follows from the commuting of partial derivatives. Show that $V^\alpha{}_{,\nu\beta} = V^\alpha{}_{,\beta\nu}$ implies that

$$\left(\Gamma^\alpha{}_{\mu\beta,\nu} - \Gamma^\alpha{}_{\mu\nu,\beta} \right) V^\mu = \left(\Gamma^\alpha{}_{\mu\beta} \Gamma^\mu{}_{\sigma\nu} - \Gamma^\alpha{}_{\mu\nu} \Gamma^\mu{}_{\sigma\beta} \right) V^\sigma.$$

(b) By relabeling indices, work this into the form

$$\left(\Gamma^\alpha{}_{\mu\beta,\nu} - \Gamma^\alpha{}_{\mu\nu,\beta} + \Gamma^\alpha{}_{\sigma\nu} \Gamma^\sigma{}_{\mu\beta} - \Gamma^\alpha{}_{\sigma\beta} \Gamma^\sigma{}_{\mu\nu} \right) V^\mu = 0.$$

This turns out to be a *sufficient* condition, as well.

6.12 Prove that Eq. 6.52 defines a new affine parameter.

6.13 (a) Show that if $\vec{A}$ and $\vec{B}$ are parallel-transported along a curve, then $g(\vec{A}, \vec{B}) = \vec{A} \cdot \vec{B}$ is constant on the curve.
(b) Conclude from this that if a *geodesic* is spacelike (or timelike or null) somewhere, it is spacelike (or timelike or null) everywhere.

6.14 The proper distance along a curve whose tangent is $\vec{V}$ is given by Eq. 6.8. Show that if the curve is a geodesic, then the proper length is an affine parameter. (Use the result of Exercise 6.13.)

6.15 Use Exercises 6.13 and 6.14 to prove that the proper length of the geodesic between two points is unchanged to first order by small changes in the curve that do not change its endpoints.

6.16 (a) Derive Eqs. 6.59 and 6.60 from Eq. 6.58.
(b) Fill in the algebra needed to justify Eq. 6.61.

6.17 (a) Prove that Eq. 6.5 implies that $g^{\alpha\beta}{}_{,\mu}(P) = 0$.
(b) Use this to establish Eq. 6.64.
(c) Fill in the steps needed to establish Eq. 6.68.

6.18 (a) Derive Eqs. 6.69 and 6.70 from Eq. 6.68.
(b) Show that Eq. 6.69 reduces the number of independent components of $R_{\alpha\beta\mu\nu}$ from $4 \times 4 \times 4 \times 4 = 256$ to $6 \times 7/2 = 21$. (Hint: treat *pairs* of indices. Calculate how many independent choices of pairs there are for the first and the second pairs in $R_{\alpha\beta\mu\nu}$.)
(c) Show that Eq. 6.70 imposes only one further relation independent of Eq. 6.69 on the components, reducing the total of independent components to 20.

6.19 Prove that $R^{\alpha}{}_{\beta\mu\nu} = 0$ for polar coordinates in the Euclidean plane. Use Eq. 5.45 or equivalent results.

6.20 Fill in the algebra necessary to establish Eq. 6.73.

6.21 Consider the sentences following Eq. 6.78. Why does the argument in parentheses *not* apply to the signs in

$$V^{\alpha}{}_{;\beta} = V^{\alpha}{}_{,\beta} + \Gamma^{\alpha}{}_{\mu\beta}V^{\mu} \quad \text{and} \quad V_{\alpha;\beta} = V_{\alpha,\beta} - \Gamma^{\mu}{}_{\alpha\beta}V_{\mu}?$$

6.22 Fill in the algebra necessary to establish Eqs. 6.84–6.86.

6.23 Prove Eq. 6.88. (Be careful: one cannot simply differentiate Eq. 6.67 since it is valid only at P, not in the neighborhood of P.)

6.24 Establish Eq. 6.89 from Eq. 6.88.

6.25 (a) Prove that the Ricci tensor is the only independent contraction of $R^{\alpha}{}_{\beta\mu\nu}$: all others are multiples of it.
(b) Show that the Ricci tensor is symmetric.

6.26 Use Exercise 6.17(a) to prove Eq. 6.94.

6.27 Fill in the algebra necessary to establish Eqs. 6.95, 6.97 and 6.99.

6.28 (a) Derive Eq. 6.19 by using the usual coordinate transformation from Cartesian to spherical polars.
(b) Deduce from Eq. 6.19 that the metric of the surface of a sphere of radius r has components $g_{\theta\theta} = r^2$, $g_{\phi\phi} = r^2 \sin^2\theta$, $g_{\theta\phi} = 0$ in the usual spherical coordinates.
(c) Find the components $g^{\alpha\beta}$ for the sphere.

6.29 In polar coordinates, calculate the Riemann curvature tensor of a sphere of unit radius, whose metric is given in Exercise 6.28. (Note that in two dimensions there is only *one* independent component, by the same arguments as in Exercise 6.18(b). So calculate $R_{\theta\phi\theta\phi}$ and obtain all other components in terms of it.)

6.30 Calculate the Riemann curvature tensor of the cylinder. (Since the cylinder is flat, this should vanish. Use whatever coordinates you like, and make sure you write down the metric properly!)

6.31 Show that covariant differentiation obeys the usual product rule, e.g. $(V^{\alpha\beta}W_{\beta\gamma})_{;\mu} = V^{\alpha\beta}{}_{;\mu}W_{\beta\gamma} + V^{\alpha\beta}W_{\beta\gamma;\mu}$. (Hint: use a locally inertial frame.)

6.32 A four-dimensional manifold has coordinates (u, v, w, p) in which the metric has components $g_{uv} = g_{ww} = g_{pp} = 1$, all other independent components vanishing.

(a) Show that the manifold is flat and the signature is +2.

(b) The result in (a) implies that the manifold must be Minkowski spacetime. Find a coordinate transformation to the usual coordinates (t, x, y, z). (You may find it a useful hint to calculate $\vec{e}_v \cdot \vec{e}_v$ and $\vec{e}_u \cdot \vec{e}_u$.)

6.33 A 'three-sphere' is the three-dimensional surface in four-dimensional Euclidean space (coordinates x, y, z, w), given by the equation $x^2 + y^2 + z^2 + w^2 = r^2$, where r is the radius of the sphere.

(a) Define new coordinates (r, θ, ϕ, χ) by the equations $w = r \cos \chi$, $z = r \sin \chi \cos \theta$, $x = r \sin \chi \sin \theta \cos \phi$, $y = r \sin \chi \sin \theta \sin \phi$. Show that (θ, ϕ, χ) are coordinates for the sphere. These generalize the familiar polar coordinates.

(b) Show that the metric of the three-sphere of radius r has components in these coordinates $g_{\chi\chi} = r^2$, $g_{\theta\theta} = r^2 \sin^2 \chi$, $g_{\phi\phi} = r^2 \sin^2 \chi \sin^2 \theta$, all other components vanishing. (Use the same method as in Exercise 6.28.)

6.34 Establish the following identities for a general metric tensor in a general coordinate system. You may find Eqs. 6.39 and 6.40 useful.

(a) $\Gamma^{\mu}{}_{\mu\nu} = \frac{1}{2}(\ln |g|)_{,\nu}$;

(b) $g^{\mu\nu}\Gamma^{\alpha}{}_{\mu\nu} = -(g^{\alpha\beta}\sqrt{-g})_{,\beta}/\sqrt{-g}$;

(c) for an antisymmetric tensor $F^{\mu\nu}$, $F^{\mu\nu}{}_{;\nu} = (\sqrt{-g}\,F^{\mu\nu})_{,\nu}/\sqrt{-g}$;

(d) $g^{\alpha\beta}g_{\beta\mu,\nu} = -g^{\alpha\beta}{}_{,\nu}g_{\beta\mu}$. (Hint: what is $g^{\alpha\beta}g_{\beta\mu}$?)

(e) $g^{\mu\nu}{}_{,\alpha} = -\Gamma^{\mu}{}_{\beta\alpha}g^{\beta\nu} - \Gamma^{\nu}{}_{\beta\alpha}g^{\mu\beta}$. (Hint: use Eq. 6.31.)

6.35 Compute 20 independent components of $R_{\alpha\beta\mu\nu}$ for a manifold with line element $ds^2 = -e^{2\Phi}\,dt^2 + e^{2\Lambda}\,dr^2 + r^2(d\theta^2 + \sin^2\theta\,d\phi^2)$, where Φ and Λ are arbitrary functions of the coordinate r alone. (First, identify the coordinates and the components $g_{\alpha\beta}$; next, compute $g^{\alpha\beta}$ and the Christoffel symbols. Then decide on the indices of the 20 components of $R_{\alpha\beta\mu\nu}$ that you wish to calculate, and compute them. Remember that one can deduce the remaining 236 components from those 20.)

6.36 A four-dimensional manifold has coordinates (t, x, y, z) and line element $ds^2 = -(1 + 2\phi)\,dt^2 + (1 - 2\phi)(dx^2 + dy^2 + dz^2)$, where $|\phi(t, x, y, z)| \ll 1$ everywhere. At any point P with coordinates (t_0, x_0, y_0, z_0), find a coordinate transformation to a locally inertial coordinate system, to first order in ϕ. At what rate does such a frame accelerate with respect to the original coordinates, again to first order in ϕ?

6.37 (a) The 'proper volume' of a two-dimensional manifold is usually called the 'proper area'. Using the metric in Exercise 6.28, integrate Eq. 6.18 to find the proper area of a sphere of radius r.

(b) Do the analogous calculation for the three-sphere of Exercise 6.33.

6.38 Integrate Eq. 6.8 to find the length of a circle of constant coordinate θ on a sphere of radius r.

6.39 (a) For any two vector fields $\vec{U}$ and $\vec{V}$, their *Lie bracket* is defined to be the vector field $[\vec{U}, \vec{V}]$ with components

$$[\vec{U}, \vec{V}]^\alpha = U^\beta \nabla_\beta V^\alpha - V^\beta \nabla_\beta U^\alpha. \tag{6.100}$$

Show that

$$[\vec{U}, \vec{V}] = -[\vec{V}, \vec{U}],$$
$$[\vec{U}, \vec{V}]^\alpha = U^\beta \, \partial V^\alpha / \partial x^\beta - V^\beta \, \partial U^\alpha / \partial x^\beta.$$

This is one tensor field in which partial derivatives need not be accompanied by Christoffel symbols!

(b) Show that $[\vec{U}, \vec{V}]$ is a derivative operator on $\vec{V}$ along $\vec{U}$, i.e. show that for any scalar f,

$$[\vec{U}, f\vec{V}] = f[\vec{U}, \vec{V}] + \vec{V}(\vec{U} \cdot \nabla f). \tag{6.101}$$

This is sometimes called the *Lie derivative* with respect to $\vec{U}$ and denoted by

$$[\vec{U}, \vec{V}] := \pounds_{\vec{U}} \vec{V}, \qquad \vec{U} \cdot \nabla f := \pounds_{\vec{U}} f. \tag{6.102}$$

This would be written in the more conventional form of the Leibnitz rule for the derivative operator $\pounds \vec{U}$:

$$\pounds_{\vec{U}}(f\vec{V}) = f \pounds_{\vec{U}} \vec{V} + \vec{V} \pounds_{\vec{U}} f. \tag{6.103}$$

The result of (a) shows that this derivative operator may be defined without a connection or metric, and is therefore very fundamental. See Schutz (1980b) for an introduction.

(c) Calculate the components of the Lie derivative of a one-form field $\tilde{\omega}$ from the knowledge that, for any vector field $\vec{V}$, $\tilde{\omega}(\vec{V})$ is a scalar like f above, and from the definition that $\pounds_{\vec{U}} \tilde{\omega}$ is a one-form field:

$$\pounds_{\vec{U}}[\tilde{\omega}(\vec{V})] = (\pounds_{\vec{U}} \tilde{\omega})(\vec{V}) + \tilde{\omega}(\pounds_{\vec{U}} \vec{V}).$$

This is the analog of Eq. 6.103.

7 Physics in a curved spacetime

7.1 The transition from differential geometry to gravity

The essence of a physical theory expressed in mathematical form is the identification of the mathematical concepts with certain physically measurable quantities. This must be our first concern when we look at the relation of the concepts of geometry we have developed to the effects of gravity in the physical world. We have already discussed this to some extent. In particular, we have assumed that spacetime is a differentiable manifold, and we have shown that there do not exist global inertial frames in the presence of nonuniform gravitational fields. Behind these statements are two identifications:

> (I) Spacetime (the set of all events) is a four-dimensional manifold with a metric.
> (II) The metric is measurable by rods and clocks. The distance along a rod between two nearby points is $|\mathrm{d}\vec{x} \cdot \mathrm{d}\vec{x}|^{1/2}$ and the time measured by a clock that experiences two events closely separated in time is $|-\mathrm{d}\vec{x} \cdot \mathrm{d}\vec{x}|^{1/2}$.

So, there do not generally exist coordinates in which $\mathrm{d}\vec{x} \cdot \mathrm{d}\vec{x} = -(\mathrm{d}x^0)^2 + (\mathrm{d}x^1)^2 + (\mathrm{d}x^2)^2 + (\mathrm{d}x^3)^2$ everywhere. On the other hand, we have also argued that such frames *do* exist locally. This clearly suggests a curved manifold, in which coordinates can be found which make the dot product at a particular point look like it does in a Minkowski spacetime.

Therefore we make a further requirement:

> (III) The metric of spacetime can be put into the Lorentz form $\eta_{\alpha\beta}$ at any particular event by an appropriate choice of coordinates.

Having chosen this way of representing spacetime, we must do two more things to get a complete theory. First, we must specify how physical objects (particles, electric fields, fluids) behave in a curved spacetime and, second, we need to say how the curvature is generated or determined by the objects in the spacetime.

Let us consider Newtonian gravity as an example of a physical theory. For Newton, spacetime consisted of three-dimensional Euclidean space, repeated endlessly in time. (Mathematically, this is called $R^3 \times R$.) There was no metric on spacetime as a whole manifold, but the Euclidean space had its usual metric and time was measured by a universal clock. Observers with different velocities were all equally valid: this form of relativity was built into Galilean mechanics. Therefore there was no universal standard of rest, and different observers would have different definitions of whether two events occurring at different times happened at the same location. But all observers would agree on simultaneity, on whether two events happened in the same time-slice or not. Thus the 'separation in time' between two events meant the time elapsed between the two Euclidean slices containing the two events. This was independent of the spatial locations of the events, so in Newtonian gravity there was a universal notion of time: all observers, regardless of position or motion, would agree on the elapsed time between two given events. Similarly, the 'separation in space' between two events meant the Euclidean distance between them. If the events were simultaneous, occurring in the same Euclidean time-slice, then this was simple to compute using the metric of that slice, and all observers would agree on it. If the events happened at different times, each observer would take the location of the events in their respective space-slices and compute the Euclidean distance between them. The coordinates of the locations would differ for different observers, but again the distance between them would be the same for all observers.

However, in Newtonian theory there was no way to combine the time and distance measures: there was no invariant measure of the length of a general curve that changed position and time as it went along. Without an invariant way of converting times to distances, this was not possible. What Einstein brought to relativity was the invariance of the speed of light, which then permits a unification of time and space measures. Einstein's four-dimensional spacetime has a much simpler structure than Newton's!

Now, within this model of spacetime, Newton gave a law for the behavior of objects that experienced gravitational forces: $F = ma$, where $F = -m\nabla\phi$ for a given gravitational field ϕ. And he also gave a law determining how ϕ is generated: $\nabla^2\Phi = 4\pi G\rho$. These two laws are the ones for which we must now find analogs in our relativistic point of view on spacetime. The second law will be dealt with in the next chapter. In this chapter, we ask only how a given metric affects bodies in spacetime.

We have already discussed this for the simple case of particle motion. Since we know that the 'acceleration' of a particle in a gravitational field is independent of its mass, we can go to a freely falling frame in which nearby particles have no acceleration. This is what we have identified as a locally inertial frame. Since freely falling particles have no acceleration in that frame, they follow straight lines, at least locally. But straight lines in a local inertial frame are, of course, the *definition* of geodesics in the full curved manifold. So we have our first postulate on the way in which particles are affected by the metric:

> (IV) *Weak equivalence principle.* Freely falling particles move on timelike geodesics of the spacetime.[1]

By 'freely falling' we mean particles unaffected by other forces, such as electric fields, etc. All other known forces in physics are distinguished from gravity by the fact that there

---

[1] It is more common to define the WEP without reference to a curved spacetime, but just to say that all particles fall at the same rate in a gravitational field, independent of their mass and composition. But the Einstein equivalence principle (Postulate IV′) is normally taken to imply that gravity can be represented by spacetime curvature, so we shall simply start with the assumption that we have a curved spacetime.

are particles unaffected by them. So the weak equivalence principle (Postulate IV) is a very strong statement, capable of experimental test. And it has been tested, and continues to be tested, to high accuracy. Experiments typically compare the rate of fall of objects that are composed of different materials; current experimental limits bound the fractional differences in acceleration to a few parts in 10^{13}. The WEP is therefore one of the most precisely tested laws in all of physics. There are even proposals to test it up to the level of parts in 10^{18} using satellite-borne experiments.

But the WEP refers only to particles. How are, say, fluids affected by a non-flat metric? We need a generalization of (IV):

(IV′) *Einstein equivalence principle.* Any local physical experiment not involving gravity will have the same result if performed in a freely falling inertial frame as if it were performed in the flat spacetime of special relativity.

In this case 'local' means that the experiment does not involve fields, such as electric fields, that may extend over large regions and therefore extend outside the domain of validity of the local inertial frame. All of *local* physics is the same in a freely falling inertial frame as it is in special relativity. Gravity introduces nothing new *locally*. All the effects of gravity are felt over extended regions of spacetime. This, too, has been tested strongly (Will 2006).

This may seem strange to someone used to blaming gravity for making it hard to climb stairs or mountains, or even to get out of bed! But these local effects of gravity are, in Einstein's point of view, really the effects of our being pushed around by the Earth and objects on it. Our 'weight' is caused by the solid Earth exerting forces on us that prevent us from falling freely on a geodesic (weightlessly, through the floor). This is a very reasonable point of view. Consider astronauts orbiting the Earth. At an altitude of some 300 km, they are hardly any further from the center of the Earth than you or I, so the strength of the Newtonian gravitational force on them is almost the same as on us. But they are weightless, as long as their orbit prevents them from encountering the solid Earth. Once we acknowledge that spacetime has natural curves, the geodesics, and that when we fall on them we are in free fall and feel no gravity, then we can dispose of the Newtonian concept of a gravitational force altogether. We are only following the natural spacetime curve.

The true measure of gravity on the Earth is its tides. These are nonlocal effects, because they arise from the difference of the Moon's Newtonian gravitational acceleration across the Earth, or in other words from the geodesic deviation near the Earth. If the Earth were permanently cloudy, an Earthling would not know about the Moon from its overall gravitational acceleration, since the Earth falls freely: we don't feel the Moon locally. But Earthlings could in principle discover the Moon even without seeing it, by observing and understanding the tides. Tidal forces are the only measurable aspect of gravity.

Mathematically, what the Einstein equivalence principle means, roughly speaking, is that if we have a local law of physics that is expressed in tensor notation in SR then its mathematical form should be the same in a locally inertial frame of a curved spacetime.

This principle is often called the 'comma-goes-to-semicolon rule', because if a law contains derivatives in its special-relativistic form ('commas'), then it has these same derivatives in the local inertial frame. To convert the law into an expression valid in *any* coordinate frame, one simply makes the derivatives covariant ('semicolons'). It is an extremely simple way to generalize the physical laws. In particular, it forbids 'curvature coupling': it is conceivable that the correct form of, say, thermodynamics in a curved spacetime would involve the Riemann tensor somehow, which would vanish in SR. Postulate (IV′) would not allow any Riemann-tensor terms in the equations.

As an example of how (IV′) translates into mathematics, we discuss fluid dynamics, which will be our main interest in this course. The law of conservation of particles in SR is expressed as

$$(n\, U^{\alpha})_{,\alpha} = 0, \tag{7.1}$$

where n is the density of particles in the momentarily comoving reference frame (MCRF), and where U^{α} is the four-velocity of a fluid element. In a curved spacetime, at any event, one can find a locally inertial frame comoving momentarily with the fluid element at that event, and define n in exactly the same way. Similarly one can define $\vec{U}$ to be the time basis vector of that frame, just as in SR. Then, according to the Einstein equivalence principle (see Chapter 5), the law of conservation of particles in the locally inertial frame is *exactly* Eq. 7.1. But, because the Christoffel symbols are zero at the given event because it is the origin of the locally inertial frame, this is equivalent to

$$(n\, U^{\alpha})_{;\alpha} = 0. \tag{7.2}$$

This form of the law is valid in *all* frames and so allows us to compute the conservation law in any frame and be sure that it is the one implied by the Einstein equivalence principle. We have therefore generalized the law of particle conservation to a curved spacetime. We will follow this method for other laws of physics as we need them.

Is this just a game with tensors, or is there physical content in what we have done? Is it possible that in a curved spacetime the conservation law would actually be something other than Eq. 7.2? The answer is yes: consider postulating the equation

$$(n\, U^{\alpha})_{;\alpha} = qR^{2}, \tag{7.3}$$

where R is the Ricci scalar, defined in Eq. 6.92 as the double trace of the Riemann tensor, and where q is a constant. This would also reduce to Eq. 7.1 in SR, since in a flat spacetime the Riemann tensor vanishes. But in curved spacetime, this equation predicts something very different: the curvature would either create or destroy particles, according to the sign of the constant q. Thus, both the previous equations are consistent with the laws of physics in SR. The Einstein equivalence principle asserts that we should generalize Eq. 7.1 in the simplest possible manner, that is, to Eq. 7.2. It is of course a matter for experiment, or astronomical observation, to decide whether Eq. 7.2 or Eq. 7.3 is correct. General relativity, however, is founded on the assumption that the Einstein equivalence principle is correct. There is no observational evidence to the contrary.

Similarly, the law of conservation of entropy in SR is

$$U^{\alpha} S_{,\alpha} = 0. \tag{7.4}$$

Since there are no Christoffel symbols in the covariant derivative of a scalar like S, this law is *unchanged* in a curved spacetime. Finally, the conservation of four-momentum is

$$T^{\mu\nu}{}_{,\nu} = 0. \tag{7.5}$$

The generalization is

$$T^{\mu\nu}{}_{;\nu} = 0, \tag{7.6}$$

with the definition

$$T^{\mu\nu} = (\rho + p)U^\mu U^\nu + pg^{\mu\nu}, \tag{7.7}$$

exactly as before. (Notice that $g^{\mu\nu}$ is the tensor whose components in the local inertial frame equal the flat-space metric tensor $\eta^{\mu\nu}$.)

7.2 Physics in slightly curved spacetimes

To see the implications of (IV′) for the motion of a particle or fluid, one must know the metric on the manifold. Since we have not yet studied the way in which a metric is generated, we will at this stage have to be content with assuming a form for the metric which we shall derive later. We will see later that for *weak* gravitational fields (where, in Newtonian language, the gravitational potential energy of a particle is much less than its rest-mass energy) the ordinary Newtonian potential ϕ completely determines the metric, which has the form

$$ds^2 = -(1 + 2\phi)dt^2 + (1 - 2\phi)(dx^2 + dy^2 + dz^2). \tag{7.8}$$

(The sign of ϕ is chosen to be negative so that, far from a source of mass M, we have $\phi = -GM/r$.) Now, the condition above that the field be weak means that $|m\phi| \ll m$, so that $|\phi| \ll 1$. The metric Eq. 7.8 is really only correct to first order in ϕ, so we shall work to this order from now on.

Let us compute the motion of a freely falling particle. We denote its four-momentum by $\vec{p}$. For all except massless particles, this is $m\vec{U}$, where $\vec{U} = \mathrm{d}\vec{x}/\mathrm{d}\tau$. Now, by (IV), the particle's path is a geodesic, and we know that proper time is an affine parameter on such a path. Therefore $\vec{U}$ must satisfy the geodesic equation,

$$\nabla_{\vec{U}} \vec{U} = 0. \tag{7.9}$$

For convenience later, however, we note that any constant times the proper time is an affine parameter, in particular τ/m. Then $\mathrm{d}\vec{x}/\mathrm{d}(\tau/m)$ is also a vector satisfying the geodesic equation. This vector is just $m\mathrm{d}\vec{x}/\mathrm{d}\tau = \vec{p}$. So we can also write the equation of motion of the particle as

$$\nabla_{\vec{p}} \vec{p} = 0. \tag{7.10}$$

This equation can also be used for photons, which have a well-defined $\vec{p}$ but no $\vec{U}$ since $m = 0$.

If the particle has a nonrelativistic velocity in the coordinates of Eq. 7.8, we can find an approximate form for Eq. 7.10. First let us consider the 0 component of the equation, noting that the ordinary derivative along $\vec{p}$ is m times the ordinary derivative along $\vec{U}$, or in other words $m \, \mathrm{d}/\mathrm{d}\tau$:

$$m \frac{\mathrm{d}}{\mathrm{d}\tau} p^0 + \Gamma^0{}_{\alpha\beta} p^\alpha p^\beta = 0. \tag{7.11}$$

Because the particle has a nonrelativistic velocity we have $p^0 \gg p^1$, so Eq. 7.11 is approximately

$$m \frac{\mathrm{d}}{\mathrm{d}\tau} p^0 + \Gamma^0{}_{00} (p^0)^2 = 0. \tag{7.12}$$

We need to compute $\Gamma^0{}_{00}$:

$$\Gamma^0{}_{00} = \tfrac{1}{2} g^{0\alpha} (g_{\alpha 0,0} + g_{\alpha 0,0} - g_{00,\alpha}). \tag{7.13}$$

Now because $[g_{\alpha\beta}]$ is diagonal, $[g^{\alpha\beta}]$ is also diagonal and its elements are the reciprocals of those of $[g_{\alpha\beta}]$. Therefore $g^{0\alpha}$ is nonzero only when $\alpha = 0$, so Eq. 7.13 becomes

$$\Gamma^0{}_{00} = \tfrac{1}{2} g^{00} g_{00,0} = \frac{1}{2} \frac{-1}{1 + 2\phi} (-2\phi)_{,0}$$

$$= \phi_{,0} + O(\phi^2). \tag{7.14}$$

To lowest order in the velocity of the particle and in ϕ, we can replace $(p^0)^2$ in the second term of Eq. 7.12 by m^2, obtaining

$$\frac{\mathrm{d}}{\mathrm{d}\tau} p^0 = -m \frac{\partial \phi}{\partial \tau}. \tag{7.15}$$

Since p^0 is the energy of the particle in this frame, this means that the energy is conserved unless the gravitational field depends on time. This result is true also in Newtonian theory. Here, however, we must note that p^0 is the energy of the particle with respect to this frame only.

The spatial components of the geodesic equation give the counterparts of the Newtonian law $\mathbf{F} = m\mathbf{a}$. They are

$$p^\alpha p^i{}_{,\alpha} + \Gamma^i{}_{\alpha\beta} p^\alpha p^\beta = 0, \tag{7.16}$$

or, to lowest order in the velocity,

$$m \frac{\mathrm{d}p^i}{\mathrm{d}\tau} + \Gamma^i{}_{00} (p^0)^2 = 0. \tag{7.17}$$

Again we have neglected p^i compared to p^0 in the Γ summation. Consistently with this we can again put $(p^0)^2 = m^2$ to a first approximation and get

$$\frac{\mathrm{d}p^i}{\mathrm{d}\tau} = -m \Gamma^i{}_{00}. \tag{7.18}$$

We calculate the Christoffel symbol as:

$$\Gamma^i{}_{00} = \tfrac{1}{2} g^{i\alpha} (g_{\alpha 0,0} + g_{\alpha 0,0} - g_{00,\alpha}). \tag{7.19}$$

Now, since $[g^{\alpha\beta}]$ is diagonal, we can write

$$g^{i\alpha} = (1 - 2\phi)^{-1} \delta^{i\alpha} \tag{7.20}$$

and get

$$\Gamma^i{}_{00} = \tfrac{1}{2}(1 - 2\phi)^{-1}\delta^{ij}(2g_{j0,0} - g_{00,j}), \tag{7.21}$$

where we have changed α to j because δ^{i0} is zero. Now we notice that $g_{j0} \equiv 0$ and so we get

$$\Gamma^i{}_{00} = \tfrac{1}{2}g_{00,j}\delta^{ij} + 0(\phi^2) \tag{7.22}$$

$$= -\tfrac{1}{2}(-2\phi)_{,j}\delta^{ij}. \tag{7.23}$$

With this the equation of motion, Eq. 7.17, becomes

$$\mathrm{d}p^i/\mathrm{d}\tau = -m\phi_{,j}\delta^{ij}. \tag{7.24}$$

This is the usual equation in Newtonian theory, since the force of a gravitational field is $-m\nabla\phi$. This demonstrates that general relativity predicts the Keplerian motion of the planets, at least so long as the higher-order effects neglected here are too small to measure. We shall see that this is true for most planets, but not for Mercury.

Both the energy-conservation equation and the equation of motion were derived as approximations based on two things: that the metric is nearly the Minkowski metric ($|\phi| \ll 1$) and that the particle's velocity is nonrelativistic ($p^0 \gg p^i$). These two limits are just the circumstances under which Newtonian gravity is verified, so it is reassuring – indeed, essential – that we have recovered the Newtonian equations. However, there is no magic here. It almost *had* to work, given that we know that particles fall on straight lines in freely falling frames.

One can do the same sort of calculation to verify that the Newtonian equations hold for other systems in the appropriate limit. For instance, you have an opportunity to do this for the perfect fluid in Exercise 7.5. Note that the condition that the fluid is nonrelativistic means not only that its velocity is small but also that the random velocities of its particles are nonrelativistic, which means $p \ll \rho$.

This correspondence of our relativistic point of view with the older, Newtonian theory in the appropriate limit is very important. *Any* new theory must make the same predictions as the old theory in the regime in which the old theory was known to be correct. The equivalence principle plus the form of the metric, Eq. 7.8, does this.

7.3 Curved intuition

Although in the appropriate limit our curved-spacetime picture of gravity predicts the same things as Newtonian theory predicts, it is very different from Newton's theory in concept. One must therefore work gradually toward an understanding of its new point of view.

The first difference is the absence of a preferred frame. In Newtonian physics *and* in SR, inertial frames are preferred. Since 'velocity' cannot be measured locally but 'acceleration' can be, both theories single out special classes of coordinate systems for spacetimes in which particles which have no physical acceleration (i.e. $d\vec{U}/d\tau = 0$) also have no coordinate acceleration ($d^2x^i/dt^2 = 0$). In our new picture, there is no coordinate system which is inertial everywhere, i.e. in which $d^2x^i/dt^2 = 0$ for every particle for which $d\vec{U}/d\tau = 0$. Therefore we have to allow all coordinates to be on an equal footing. By using the Christoffel symbols we correct coordinate-dependent quantities such as d^2x^i/dt^2 to obtain coordinate-independent quantities such as $d^2\vec{U}/d\tau$. Therefore, one need not, and in fact one *should not*, develop coordinate-dependent ways of thinking.

A second difference concerns energy and momentum. In Newtonian physics, SR, and our geometrical gravity theory, each particle has a definite energy and momentum, whose values depend on the frame in which they are evaluated. In the latter two theories, energy and momentum are components of a single four-vector $\vec{p}$. In SR, the total four-momentum of a system is the sum of the four-momenta of all the particles, $\sum_i \vec{p}_{(i)}$. But in a curved spacetime, one *cannot* add up vectors that are defined at different points, because one does not know how to do so: two vectors can only be said to be parallel if they are compared at the same point, and the value of a vector at a point to which it has been parallel-transported depends on the curve along which it was moved. So there is *no* invariant way of adding up all the $\vec{p}$s, and so if a system has a definable four-momentum, it is not just the simple thing that it was in SR.

It turns out that for any system whose spatial extent is bounded (i.e. an isolated system), a total energy and momentum *can* be defined, in a manner which we will discuss later. One way to see that the *total* mass–energy of a system should not be the sum of the energies of the particles is that this neglects what in Newtonian language is called its gravitational self-energy, a negative quantity which is the work one gains by assembling the system from isolated particles at infinity. This energy, if it is to be included, cannot be assigned to any particular particle but resides in the geometry itself. The notion of the gravitational potential energy, however, is itself not well defined in the new picture: it must in some sense represent the difference between the sum of the energies of the particles and the total mass of the system, but since the sum of the energies of the particles is not well defined, neither is the gravitational potential energy. Only the *total* energy–momentum of a system is, in general, definable, in addition to the four-momentum of individual particles.

7.4 Conserved quantities

The previous discussion of energy may make one wonder what one can say about conserved quantities associated with a particle or system. For a particle, one must realize that gravity, in the old viewpoint, is a 'force', so that a particle's kinetic energy and momentum need not be conserved under its action. In our new viewpoint, then, one cannot expect to find a coordinate system in which the components of $\vec{p}$ are constants along the trajectory of a particle. There is one notable exception to this, and it is important enough for us to look at it in detail.

The geodesic equation can be written for the 'lowered' components of $\vec{p}$ as follows:

$$p^\alpha p_{\beta;\alpha} = 0,\tag{7.25}$$

or

$$p^\alpha p_{\beta,\alpha} - \Gamma^\gamma{}_{\beta\alpha}p^\alpha p_\gamma = 0,$$

or

$$m\frac{\mathrm{d}p_\beta}{\mathrm{d}\tau} = \Gamma^\gamma{}_{\beta\alpha}p^\alpha p_\gamma.\tag{7.26}$$

Now, the right-hand side turns out to be simple:

$$\begin{aligned}\Gamma^\gamma{}_{\alpha\beta}p^\alpha p_\gamma &= \frac{1}{2}g^{\gamma\nu}(g_{\nu\beta,\alpha} + g_{\nu\alpha,\beta} - g_{\alpha\beta,\nu})p^\alpha p_\gamma\\ &= \frac{1}{2}(g_{\nu\beta,\alpha} + g_{\nu\alpha,\beta} - g_{\alpha\beta,\nu})g^{\gamma\nu}p_\gamma p^\alpha\\ &= \frac{1}{2}(g_{\nu\beta,\alpha} + g_{\nu\alpha,\beta} - g_{\alpha\beta,\nu})p^\nu p^\alpha.\end{aligned}\tag{7.27}$$

The product $p^\nu p^\alpha$ is symmetric in ν and α, while the first and third terms inside the parentheses are, together, antisymmetric in ν and α. Therefore they cancel, leaving only the middle term:

$$\Gamma^\gamma{}_{\beta\alpha}p^\alpha p_\gamma = \frac{1}{2}g_{\nu\alpha,\beta}p^\nu p^\alpha.\tag{7.28}$$

The geodesic equation can thus, in complete generality, be written

$$m\frac{\mathrm{d}p_\beta}{\mathrm{d}\tau} = \frac{1}{2}g_{\nu\alpha,\beta}p^\nu p^\alpha.\tag{7.29}$$

We therefore have the following important result: *if all the components $g_{\alpha\nu}$ are independent of x^β for some fixed index β, then p_β is a constant along any particle's trajectory.*

For instance, suppose we have a stationary (i.e. time-independent) gravitational field. Then a coordinate system can be found in which the metric components are time independent, and in that system p_0 is conserved. Therefore p_0 (or, really, $-p_0$) is usually called the 'energy' of the particle, *without* qualifying it with 'in this frame'. Notice that coordinates can also be found in which the same metric has time-dependent components: any time-dependent coordinate transformation from the 'nice' system will do this. In fact, most freely falling locally inertial systems are like this, since a freely falling particle sees a gravitational field that varies with its position, and therefore with time in its coordinate system. The frame in which the metric components are stationary is special, and is the usual 'laboratory frame' on Earth. Therefore p_0 in this frame is related to the usual energy defined in the lab, and includes the particle's gravitational potential energy, as we shall now show. Consider the equation

$$\begin{aligned}\vec{p}\cdot\vec{p} = -m^2 &= g_{\alpha\beta}p^\alpha p^\beta\\ &= -(1+2\phi)(p^0)^2 + (1-2\phi)[(p^x)^2 + (p^y)^2 + (p^z)^2]\end{aligned}\tag{7.30}$$

where we have used the metric in Eq. 7.8. This can be solved to give

$$(p^0)^2 = [m^2 + (1 - 2\phi)(\boldsymbol{p}^2)](1 + 2\phi)^{-1}, \tag{7.31}$$

where, as a shorthand, we denote by $\boldsymbol{p}^2$ the sum $(p^x)^2 + (p^y)^2 + (p^z)^2$. Keeping within the approximation $|\phi| \ll 1$, $|\boldsymbol{p}| \ll m$, we can simplify this to

$$(p^0)^2 \approx m^2 (1 - 2\phi + \boldsymbol{p}^2/m^2)$$

or

$$p^0 \approx m(1 - \phi + \boldsymbol{p}^2/2m^2). \tag{7.32}$$

Now we lower the index and get

$$p^0 = g_{0\alpha} p^\alpha = g_{00} p^0 = -(1 + 2\phi) p^0, \tag{7.33}$$

$$-p_0 \approx m(1 + \phi + \boldsymbol{p}^2/2m^2) = m + m\phi + \boldsymbol{p}^2/2m. \tag{7.34}$$

The first term is the rest mass of the particle. The second and third are the Newtonian parts of its energy: the gravitational potential energy and kinetic energy. This means that the constancy of p_0 along a particle's trajectory generalizes the Newtonian concept of a conserved energy.

Notice that a *general* gravitational field will not be stationary in *any* frame,[2] so no conserved energy can be defined.

In a similar manner, if a metric is axially symmetric, then coordinates can be found in which $g_{\alpha\beta}$ is independent of the angle ψ around the axis. Then p_ψ will be conserved. This is the particle's angular momentum. In the nonrelativistic limit we have

$$p_\psi = g_{\psi\psi} p^\psi \approx g_{\psi\psi} m \, \mathrm{d}\psi/\mathrm{d}t \approx m g_{\psi\psi} \Omega, \tag{7.35}$$

where Ω is the angular velocity of the particle. Now, for a nearly flat metric we have

$$g_{\psi\psi} = \vec{e}_\psi \cdot \vec{e}_\psi \approx r^2 \tag{7.36}$$

in cylindrical coordinates (r, ψ, z), so that the conserved quantity is

$$p_\psi \approx m r^2 \Omega. \tag{7.37}$$

This is the usual Newtonian definition of angular momentum.

We have discussed conservation laws of particle motion. Similar considerations apply to fluids, since they are just large collections of particles. But the situation with regard to the total mass and momentum of a self-gravitating system is more complicated. It turns out that an isolated system's mass and momentum *are* conserved, but we must postpone any discussion of this until we see how they are defined.

---

[2] It is easy to see that there is generally no coordinate system which makes a given metric time independent. The metric has ten independent components (the same number as a 4×4 symmetric matrix), while a change of coordinates only enables one to introduce four degrees of freedom to change the components (these are the four functions $x^{\tilde{\alpha}}(x^\mu)$). It is a special metric indeed if all ten components can be made time independent this way.

7.5 Bibliography

The question of how curvature and physics fit together is discussed in more detail in Geroch (1978). Conserved quantities are treated in detail in any of the advanced texts. The material in this chapter also serves as a preparation for the theory of quantum fields in a fixed curved spacetime. See Birrell & Davies (1984), Fulling (1989), and Parker & Toms (2009). This in turn leads to one of the most active areas of gravitation research today, the quantization of general relativity. While we will not treat this area in this book, readers who want to approach this subject from the starting point of classical general relativity (as contrasted with approaching it from the starting point of string theory) may wish to look at Rovelli & Vidotto (2020), Bojowald (2008), and Thiemann (2007).

Exercises

7.1　If Eq. 7.3 were the correct generalization of Eq. 7.1 to a curved spacetime, how would you interpret it? What would happen to the number of particles in a comoving volume of the fluid, as time evolves? In principle, can one distinguish experimentally between Eq. 7.2 and Eq. 7.3?

7.2　To first order in ϕ, compute $g^{\alpha\beta}$ for Eq. 7.8.

7.3　Calculate all the Christoffel symbols for the metric given by Eq. 7.8, to first order in ϕ. Assume that ϕ is a general function of t, x, y, and z.

7.4　Verify that the results in Eqs. 7.15 and 7.24 depend only on g_{00}: the form of g_{xx} does not affect them, as long as it is $1 + O(\phi)$.

7.5　(a) For a perfect fluid, verify that the spatial components of Eq. 7.6 in the Newtonian limit reduce to

$$v_{,t} + (v \cdot \nabla)v + \nabla p/\rho + \nabla\phi = 0 \tag{7.38}$$

for the metric given in Eq. 7.8. This is known as Euler's equation for nonrelativistic fluid flow in a gravitational field. You will need to use Eq. 7.2 to get this result.

(b) Examine the time component of Eq. 7.6 under the same assumptions, and interpret each term.

(c) Equation 7.38 implies that a static fluid ($v = 0$) in a static Newtonian gravitational field obeys the equation of hydrostatic equilibrium:

$$\nabla p + \rho\nabla\phi = 0. \tag{7.39}$$

A metric tensor is said to be static if there exist coordinates in which $\vec{e}_0$ is timelike, $g_{i0} = 0$, and $g_{\alpha\beta,0} = 0$. Deduce from Eq. 7.6 that a static fluid ($U^i = 0$, $p_{,0} = 0$, etc.) obeys the relativistic equation of hydrostatic equilibrium:

$$p_{,i} + (\rho + p)\left[\tfrac{1}{2}\ln(-g_{00})\right]_{,i} = 0. \tag{7.40}$$

(d) This suggests that, at least for static situations, there is a close relation between g_{00} and $-\exp(2\phi)$, where ϕ is the Newtonian potential for a similar physical situation. Show that Eq. 7.8 and Exercise 7.4 are consistent with this.

7.6 Deduce Eq. 7.25 from Eq. 7.10.

7.7 Consider the following four different metrics, as given by their line elements:

(i) $ds^2 = -dt^2 + dx^2 + dy^2 + dz^2$;

(ii) $ds^2 = -(1 - 2M/r)\,dt^2 + (1 - 2M/r)^{-1}\,dr^2 + r^2(d\theta^2 + \sin^2\theta\,d\phi^2)$, where M is a constant;

(iii)

$$ds^2 = -\frac{\Delta - a^2\sin^2\theta}{\rho^2}\,dt^2 - 2a\frac{2Mr\sin^2\theta}{\rho^2}\,dt\,d\phi$$
$$+ \frac{(r^2 + a^2)^2 - a^2\Delta\sin^2\theta}{\rho^2}\sin^2\theta\,d\phi^2 + \frac{\rho^2}{\Delta}dr^2 + \rho^2\,d\theta^2,$$

where M and a are constants and we have introduced the shorthand notation $\Delta = r^2 - 2Mr + a^2$, $\rho^2 = r^2 + a^2\cos^2\theta$;

(iv) $ds^2 = -dt^2 + R^2(t)\left[(1 - kr^2)^{-1}dr^2 + r^2(d\theta^2 + \sin^2\theta\,d\phi^2)\right]$, where k is a constant and $R(t)$ is an arbitrary function of t alone.

The first metric should be familiar by now. We shall encounter the other three in later chapters. They are, respectively, the Schwarzschild, Kerr, and Robertson–Walker metrics.

(a) For each metric find as many conserved components p_α of a freely falling particle's four-momentum as possible.

(b) Use the result of Exercise 6.28 to put (i) in the form

$$(i')\ ds^2 = -dt^2 + dr^2 + r^2(d\theta^2 + \sin^2\theta\,d\phi^2).$$

From this, argue that (ii) and (iv) are spherically symmetric. Does this increase the number of conserved components p_α?

(c) It can be shown that for (i') and (ii)–(iv), a geodesic which begins with $\theta = \pi/2$ and $p^\theta = 0$ – i.e. one which begins as a tangent to the equatorial plane – always has $\theta = \pi/2$ and $p^\theta = 0$. For cases (i'), (ii) and (iii), use the equation $\vec{p}\cdot\vec{p} = -m^2$ to solve for p^r in terms of m, other conserved quantities, and known functions of position.

(d) For (iv), spherical symmetry implies that if a geodesic begins with $p^\theta = p^\phi = 0$, these remain zero. Use this to show from Eq. 7.29 that when $k = 0$, p_r is a conserved quantity.

7.8 Suppose that in some coordinate system the components of the metric $g_{\alpha\beta}$ are independent of some coordinate x^μ.

(a) Show that the conservation law $T^\nu{}_{\mu;\nu} = 0$ for *any* stress–energy tensor becomes

$$\frac{1}{\sqrt{-g}}(\sqrt{-g}\,T^\nu{}_\mu){}_{,\nu} = 0. \tag{7.41}$$

(b) Suppose that in these coordinates $T^{\alpha\beta} \neq 0$ only in some bounded region of each spacelike hypersurface $x^0 = $ const. Show that Eq. 7.41 implies that

$$\int_{x^0=\text{const.}} T^\nu{}_\mu \sqrt{-g}\, n_\nu \, \mathrm{d}^3 x$$

is independent of x^0, if n_ν is the unit normal to the hypersurface. This is the generalization to continuous media of the conservation law stated after Eq. 7.29.

(c) Consider flat Minkowski space in a global inertial frame with spherical polar coordinates (t, r, θ, ϕ). Show from (b) that

$$J = \int_{t=\text{const.}} T^0{}_\phi r^2 \sin\theta \, \mathrm{d}r \, \mathrm{d}\theta \, \mathrm{d}\phi \tag{7.42}$$

is independent of t. This is the total angular momentum of the system.

(d) Express the integral in (c) in terms of the components of $T^{\alpha\beta}$ on the Cartesian basis (t, x, y, z), showing that

$$J = \int (x T^{y0} - y T^{x0}) \mathrm{d}x \, \mathrm{d}y \, \mathrm{d}z. \tag{7.43}$$

This is the continuum version of the nonrelativistic expression $(\mathbf{r} \times \mathbf{p})_z$ for a particle's angular momentum about the z axis.

7.9 (a) Find the components of the Riemann tensor $R_{\alpha\beta\mu\nu}$ for the metric given in Eq. 7.8, to first order in ϕ.

(b) Show that the equation of geodesic deviation, Eq. 6.87, implies that (to lowest order in ϕ and the velocities)

$$\frac{\mathrm{d}^2 \xi^i}{\mathrm{d}t^2} = -\phi_{,ij} \xi^j. \tag{7.44}$$

(c) Interpret this equation when the geodesics are world lines of freely falling particles which begin from rest at nearby points in a Newtonian gravitational field.

7.10 (a) Show that if a vector field ξ^α satisfies *Killing's equation*,

$$\nabla_\alpha \xi_\beta + \nabla_\beta \xi_\alpha = 0, \tag{7.45}$$

then, along a geodesic, $p^\alpha \xi_\alpha = $ const. This is a coordinate-invariant way of characterizing the conservation law we deduced from Eq. 7.29. One only has to know whether a metric admits Killing fields.

(b) Find ten Killing fields of Minkowski spacetime.

(c) Show that if $\vec{\xi}$ and $\vec{\eta}$ are Killing fields, then so is $\alpha\vec{\xi} + \beta\vec{\eta}$ for *constant* α and β.

(d) Show that Lorentz transformations of the fields in (b) simply produce linear combinations as in (c).

(e) Use the results of Exercise 7.7(a) to find the Killing vectors of the metrics (ii)–(iv).

8 The Einstein field equations

8.1 Purpose and justification of the field equations

Having decided upon a description of gravity, and its action on matter, that is based on the idea of a curved manifold with a metric, we must now complete the theory by postulating a law which shows how the sources of the gravitational field determine the metric. The Newtonian analog is

$$\nabla^2 \phi = 4\pi G\rho, \tag{8.1}$$

where ρ is the mass density. Its solution for a point particle of mass m is (see Exercise 8.1).

$$\phi = -\frac{Gm}{r}, \tag{8.2}$$

which is dimensionless in units where $c = 1$.

The source of the gravitational field in Newton's theory is the mass density. In our relativistic theory of gravity the source must be related to this, but it must be a relativistically meaningful concept, which 'mass' alone is not. An obvious relativistic generalization is the total energy, including rest mass. In the MCRF of a fluid element, we denoted the density of total energy by ρ in Chapter 4. So, we might be tempted to use this ρ as the source of the relativistic gravitational field. This would not be very satisfactory, however, because ρ is the energy density as measured by only one observer, the MCRF. Other observers measure the energy density to be the component T^{00} in their own reference frames. If we were to use ρ as the source of the field, we would be saying that one class of observers is preferred above all others, namely those for whom ρ is the energy density. This point of view is at variance with the approach we adopted in the previous chapter, where we stressed that one must allow *all* coordinate systems to be on an equal footing. So we shall reject ρ as the source and instead insist that the generalization of Newton's mass density should be T^{00}. But again, if T^{00} alone were the source, one would have to specify a frame in which T^{00} was evaluated. An invariant theory can avoid introducing preferred coordinate systems by using the *whole* of the stress–energy tensor **T** as the source of the gravitational field. The generalization of Eq. 8.1 to relativity would then have the form

$$\mathbf{O}(\mathbf{g}) = k\mathbf{T}, \tag{8.3}$$

where k is a constant (as yet undetermined) and **O** is a differential operator acting on the metric tensor **g**, which we have already seen in Eq. 7.8 is the generalization of ϕ. There will thus be 10 differential equations, one for each independent component of Eq. 8.3, in place of the single one, Eq. 8.1. (Recall that **T** is symmetric, so it has only 10 independent components, not 16.)

By analogy with Eq. 8.1, we should look for a second-order differential operator **O** that produces a tensor of rank $\binom{2}{0}$, since in Eq. 8.3 it is equated to the $\binom{2}{0}$ tensor **T**. In other words, $\{O^{\alpha\beta}\}$ must be the components of a $\binom{2}{0}$ tensor and must be combinations of $g_{\mu\nu,\lambda\sigma}$, $g_{\mu\nu,\lambda}$, and $g_{\mu\nu}$. It should be clear from Chapter 6 that the Ricci tensor $R^{\alpha\beta}$ satisfies these conditions. In fact, *any* tensor of the form

$$O^{\alpha\beta} = R^{\alpha\beta} + \mu g^{\alpha\beta} R + \Lambda g^{\alpha\beta} \tag{8.4}$$

satisfies these conditions, if μ and Λ are constants. To determine μ we employ a property of $T^{\alpha\beta}$ which we have not yet used, namely that the *Einstein equivalence principle* demands the local conservation of energy and momentum (Eq. 7.6):

$$T^{\alpha\beta}{}_{;\beta} = 0.$$

This equation must be true for *any* metric tensor. Then Eq. 8.3 implies that

$$O^{\alpha\beta}{}_{;\beta} = 0, \tag{8.5}$$

which again must be true for any metric tensor. Since $g^{\alpha\beta}{}_{;\mu} = 0$, we now find, from Eq. 8.4, that

$$(R^{\alpha\beta} + \mu g^{\alpha\beta} R)_{;\beta} = 0. \tag{8.6}$$

By comparing this with Eq. 6.98, we see that we must have $\mu = -\frac{1}{2}$ if Eq. 8.6 is to be an identity for arbitrary $g_{\alpha\beta}$. So we are led by this chain of argument to the equation

$$G^{\alpha\beta} + \Lambda g^{\alpha\beta} = k T^{\alpha\beta}, \tag{8.7}$$

with undetermined constants Λ and k. In index-free form this is

$$\mathbf{G} + \Lambda \mathbf{g} = k\mathbf{T}. \tag{8.8}$$

These are called the field equations of GR, or Einstein's field equations. We shall see below that we can determine the constant k by demanding that Newton's gravitational field equation comes out right, but that Λ remains arbitrary.

But first let us summarize the discussion so far. We were led to Eq. 8.7 by asking for equations that (i) resemble but generalize Eq. 8.1; (ii) introduce no preferred coordinate system; and (iii) guarantee the local conservation of energy–momentum for any metric tensor. Equation 8.7 is not the only equation which satisfies (i)–(iii). Many alternatives have been proposed, beginning even before Einstein arrived at equations like Eq. 8.7.

In recent years, when technology has made it possible to test Einstein's equations fairly precisely, even in the weak gravity of the solar system, many new alternative theories have been proposed. Some have even been designed to agree with Einstein's predictions at the precision of foreseeable solar-system experiments, differing only for much stronger fields. The competitors of GR are, however, invariably more complicated than Einstein's equations, and simply on aesthetic grounds are unlikely to attract much attention from physicists unless Einstein's equations are eventually found to conflict with some experiment. A number of the competing theories and the increasingly accurate experimental tests which have been used to eliminate them since the 1960s are discussed in Misner *et al.* (1973), Will (2014), and Will (2018). (We will study two classical tests in Chapter 11.) Einstein's equations have stood up extremely well to these tests, so we will not discuss any alternative theories in this book. In this we are in the good company of the Nobel-Prize-winning astrophysicist S. Chandrasekhar (1980):

> The element of controversy and doubt, that [has] continued to shroud the general theory of relativity to this day, derives precisely from this fact, namely that in the formulation of his theory Einstein incorporates aesthetic criteria; and every critic feels that he is entitled to his own differing aesthetic and philosophic criteria. Let me simpy say that I do not share these doubts; and I shall leave it at that.

Although Einstein's theory is essentially unchallenged at the moment, there are still reasons for expecting that it is not the last word, and therefore for continuing to probe it experimentally. Einstein's theory is, of course, not a quantum theory, and strong theoretical efforts are currently being made to formulate a consistent quantum theory of gravity. One expects that, at some level of experimental precision, there will be measurable quantum corrections to the theory, which might for example come in the form of extra fields coupled to the metric. The source of such a field might violate the Einstein equivalence principle. The field itself might carry an additional form of gravitational waves. In principle any of the predictions of general relativity might be violated in some such theory. Precision experiments on gravitation could some day provide the essential clue needed to guide the theoretical development of a quantum theory of gravity. However, interesting as they might be, such considerations are outside the scope of this introduction. For the purposes of this book we will not consider alternative theories any further.

Geometrized units. We have not determined the value of the constant k in Eq. 8.7, which plays the same role as $4\pi G$ in Eq. 8.1. Before discussing it below we will establish a more convenient set of units, namely those in which $G = 1$. Just as in SR we found it convenient to choose units in which the fundamental constant c was set to unity, so in studies of gravity it is more natural to work in units where G also has the value unity. A convenient conversion factor from SI units to these geometrized units (where $c = G = 1$) is

$$1 = G/c^2 = 7.425 \times 10^{-28} \, \text{m} \, \text{kg}^{-1}. \tag{8.9}$$

We shall use this to eliminate kg as a unit, measuring mass in meters. We list in Table 8.1 the values of certain useful constants in SI and geometrized units. Exercise 8.2 should help you to become accustomed to these units.

Table 8.1 Comparison of SI and geometrized values of fundamental constants

Constant	SI value	Geometrized value
c	$2.998 \times 10^8 \, \mathrm{ms}^{-1}$	1
G	$6.674 \times 10^{-11} \, \mathrm{m^3 \, kg^{-1} \, s^{-2}}$	1
$\hbar$	$1.055 \times 10^{-34} \, \mathrm{kg \, m^2 \, s^{-1}}$	$2.612 \times 10^{-70} \, \mathrm{m^2}$
m_e	$9.109 \times 10^{-31} \, \mathrm{kg}$	$6.764 \times 10^{-58} \, \mathrm{m}$
m_p	$1.673 \times 10^{-27} \, \mathrm{kg}$	$1.242 \times 10^{-54} \, \mathrm{m}$
$M_\odot$	$1.988 \times 10^{30} \, \mathrm{kg}$	$1.476 \times 10^3 \, \mathrm{m}$
$M_\oplus$	$5.972 \times 10^{24} \, \mathrm{kg}$	$4.434 \times 10^{-3} \, \mathrm{m}$
$L_\odot$	$3.85 \times 10^{26} \, \mathrm{kg \, m^2 \, s^{-3}}$	1.06×10^{-26}

Notes: The symbols m_e and m_p stand respectively for the rest masses of the electron and proton; $M_\odot$ and $M_\oplus$ denote, respectively, the masses of the Sun and Earth; and $L_\odot$ is the Sun's luminosity (the SI unit is equivalent to joules per second). Values are rounded to at most four figures even when known more accurately. Data from Yao (2006).

An illustration of the fundamental nature of geometrized units in gravitational problems is provided by the uncertainties in the two values given for $M_\oplus$. Earth's mass is measured by examining satellite orbits and using Kepler's laws. This measures the Newtonian potential, which involves the product $GM_\oplus$, c^2 times the *geometrized* value of the mass. This number is known to nine significant figures, from the laser tracking of satellites orbiting the Earth. Moreover, the speed of light c now has a *defined* value, so there is no uncertainty in it. Thus, the geometrized value of $M_\oplus$ is known to nine significant figures. The value of G, however, is measured in laboratory experiments, where the weakness of gravity introduces large uncertainty. The conversion factor G/c^2 is uncertain by one part in 10^4, so that is also the accuracy of the SI value of $M_\oplus$. Similarly, the Sun's geometrized mass is known to nine figures by precise radar tracking of the planets. Again, its mass in kilograms is far more uncertain.

8.2 Einstein's equations

In component notation, Einstein's equations, Eq. 8.7, take the following form if we specialize to $\Lambda = 0$ (a simplification at present, but one we will drop later), and if we take $k = 8\pi$:

$$G^{\alpha\beta} = 8\pi T^{\alpha\beta}. \tag{8.10}$$

The constant Λ is called the *cosmological constant*, and was originally not present in Einstein's equations; he inserted it many years later in order to obtain static cosmological solutions – solutions for the large-scale behavior of the Universe – which he felt at the time were desirable. Observations of the expansion of the Universe subsequently made him reject the term and regret he had ever invented it. However, astronomical observations today strongly suggest that it is small but not zero. We shall return to the discussion of Λ in Chapter 13, but for the moment we shall set $\Lambda = 0$. The justification for doing this, and the possible danger of it, are discussed in Exercise 8.18.

The value $k = 8\pi$ is obtained by demanding that Einstein's equations predict the correct behavior of planets in the solar system. This is the Newtonian limit, in which we require that the predictions of GR agree with those of Newton's theory when the latter are well tested by observation. We saw in the last chapter that Newtonian motions are produced when the metric has the form Eq. 7.8. One of our tasks in this chapter is to show that Einstein's equations, Eq. 8.10, do indeed have Eq. 7.8 as a solution when we assume that gravity is weak (see Exercise 8.3). We could, of course, keep k arbitrary until then, adjusting its value to whatever is required to obtain the solution, Eq. 7.8. It is more convenient, however, for our subsequent use of the equations of this chapter if we simply set k to 8π at the outset and verify at the appropriate time that this value is correct.

Equation 8.10 should be regarded as a system of ten coupled differential equations (not 16, since $T^{\alpha\beta}$ and $G^{\alpha\beta}$ are symmetric). They are to be solved for the ten components $g_{\alpha\beta}$ when the source $T^{\alpha\beta}$ is given. The equations are nonlinear, but they have a well-posed initial-value structure – that is, they determine future values of $g_{\alpha\beta}$ from given initial data. However, one point must be made: since $\{g_{\alpha\beta}\}$ are the components of a tensor in some coordinate system, a change in coordinates induces a change in them. In particular, there are four coordinates, so there are four arbitrary functional degrees of freedom among the ten $g_{\alpha\beta}$. It should be impossible, therefore, to determine all ten $g_{\alpha\beta}$ from any initial data, since the coordinates to the future of the initial moment can be changed arbitrarily. In fact, Einstein's equations have exactly this property: the Bianchi identities

$$G^{\alpha\beta}{}_{;\beta} = 0 \qquad\qquad (8.11)$$

mean that there are four differential identities (one for each value of α above) among the ten $G^{\alpha\beta}$. These ten, then, are not independent, and the ten Einstein equations are really only six independent differential equations for the six functions among the ten $g_{\alpha\beta}$ that characterize the geometry independently of the coordinates.

These considerations are of key importance if one wants to solve Einstein's equations in order to watch systems evolve in time from some initial state. In this book we will do this in a limited way for weak gravitational waves in Chapter 9. Because of the complexity of Einstein's equations, dynamical situations are usually studied numerically. The field of numerical relativity has evolved a well-defined approach to the problem of separating the coordinate freedom in $g_{\alpha\beta}$ from the true geometric and dynamical freedom. This is described in more advanced texts, for instance Misner *et al.* (1973), or Hawking & Ellis (1973). See also Choquet–Bruhat & York (1980), Cook (2000), or the recent review by Carlotto (2021). It will suffice here simply to note that there are really only six equations for six quantities among the $g_{\alpha\beta}$, and that Einstein's equations permit complete freedom in choosing the coordinate system.

8.3 Einstein's equations for weak gravitational fields

Nearly-Lorentz coordinate systems. Since the absence of gravity leaves spacetime flat, a weak gravitational field is one in which spacetime is 'nearly' flat. This is defined as a manifold on which coordinates exist in which the metric has components

$$g_{\alpha\beta} = \eta_{\alpha\beta} + h_{\alpha\beta}, \tag{8.12}$$

where

$$|h_{\alpha\beta}| \ll 1 \tag{8.13}$$

everywhere in spacetime. Such coordinates are called nearly-Lorentz coordinates. It is important to say 'there exist coordinates' rather than 'for all coordinates', since one can find coordinates even in Minkowski space in which $g_{\alpha\beta}$ is not close to the simple diagonal $(-1, +1, +1, +1)$ form of $\eta_{\alpha\beta}$. On the other hand, if one coordinate system exists in which Eqs. 8.12 and 8.13 are true, then there must be many such coordinate systems. Two fundamental types of coordinate transformations that take one nearly-Lorentz coordinate system into another will be discussed below: background Lorentz transformations and gauge transformations.

But why should we specialize to nearly-Lorentz coordinates at all? Haven't we just said that Einstein's equations allow complete coordinate freedom, so shouldn't their physical predictions be the same in any coordinates? Of course the answer is yes, the physical predictions will be the same. On the other hand, the amount of work we would have to do to arrive at the physical predictions could be enormous in a poorly chosen coordinate system. (For example, try to solve Newton's equation of motion for a particle free of all forces in spherical polar coordinates, or try to solve Poisson's equation in a coordinate system in which it does not separate!) Perhaps even more serious is the possibility that in a crazy coordinate system we may not have sufficient creativity and insight into the physics to know what calculations to make in order to arrive at interesting physical predictions. Therefore it is extremely important that the first step in the solution of any problem in GR must be an attempt to construct coordinates which will make the calculation simplest. Precisely because Einstein's equations have complete coordinate freedom, we should use this freedom intelligently. The construction of helpful coordinate systems is an art, and it is often rather difficult. In the present problem, however, it should be clear that $\eta_{\alpha\beta}$ is the simplest form for the flat-space metric, so that Eqs. 8.12 and 8.13 give the simplest and most natural 'nearly flat' metric components.

Background Lorentz transformations. The matrix of a Lorentz transformation in SR is

$$(\Lambda^{\bar{\alpha}}{}_{\beta}) = \begin{pmatrix} \gamma & -v\gamma & 0 & 0 \\ -v\gamma & \gamma & 0 & 0 \\ 0 & 0 & 1 & 0 \\ 0 & 0 & 0 & 1 \end{pmatrix}, \quad \gamma = (1 - v^2)^{-1/2} \tag{8.14}$$

(for a boost of velocity v in the x direction). For weak gravitational fields we define a 'background Lorentz transformation' to be one which has the form

$$x^{\bar{\alpha}} = \Lambda^{\bar{\alpha}}{}_\beta x^\beta, \tag{8.15}$$

in which $\Lambda^{\bar{\alpha}}{}_\beta$ is identical to a Lorentz transformation in SR, whose matrix elements are constant everywhere. Of course, we are not in SR, so this is only one class of transformations out of all possible classes. But it has a particularly nice feature, which we discover by transforming the metric tensor given in Eq. 8.12:

$$g_{\bar{\alpha}\bar{\beta}} = \Lambda^\mu{}_{\bar{\alpha}}\Lambda^\nu{}_{\bar{\beta}} g_{\mu\nu} = \Lambda^\mu{}_{\bar{\alpha}}\Lambda^\nu{}_{\bar{\beta}} \eta_{\mu\nu} + \Lambda^\mu{}_{\bar{\alpha}}\Lambda^\nu{}_{\bar{\beta}} h_{\mu\nu}. \tag{8.16}$$

Now, the Lorentz transformation is designed so that

$$\Lambda^\mu{}_{\bar{\alpha}}\Lambda^\nu{}_{\bar{\beta}} \eta_{\mu\nu} = \eta_{\bar{\alpha}\bar{\beta}}, \tag{8.17}$$

so we get

$$g_{\bar{\alpha}\bar{\beta}} = \eta_{\bar{\alpha}\bar{\beta}} + h_{\bar{\alpha}\bar{\beta}} \tag{8.18}$$

with the *definition*

$$h_{\bar{\alpha}\bar{\beta}} := \Lambda^\mu{}_{\bar{\alpha}}\Lambda^\nu{}_{\bar{\beta}} h_{\mu\nu}. \tag{8.19}$$

We see that, under a background Lorentz transformation, $h_{\mu\nu}$ transforms *as if* it were a tensor in SR all by itself! It is, of course, not a tensor, but just part of $g_{\alpha\beta}$. But this restricted transformation property leads to a convenient fiction: we can think of a slightly curved spacetime as a *flat* spacetime with a 'tensor' $h_{\mu\nu}$ defined on it. Then all physical fields – like $R_{\mu\nu\alpha\beta}$ – will be defined in terms of $h_{\mu\nu}$, and they will 'look like' fields superimposed on a flat background spacetime. It is important to bear in mind, however, that spacetime is really curved, that this fiction results from considering only one type of coordinate transformation. We shall find this fiction to be useful, however, in our calculations below.

Gauge transformations. There is another very important kind of coordinate change which leaves Eqs. 8.12 and 8.13 unchanged: a very small change in coordinates, of the form

$$x^{\alpha'} = x^\alpha + \xi^\alpha(x^\beta),$$

generated by a 'vector' ξ^α whose components are functions of position. If we demand that ξ^α be small, in the sense that $|\xi^\alpha{}_{,\beta}| \ll 1$, then we have

$$\Lambda^{\alpha'}{}_\beta = \frac{\partial x^{\alpha'}}{\partial x^\beta} = \delta^\alpha{}_\beta + \xi^\alpha{}_{,\beta}, \tag{8.20}$$

$$\Lambda^\alpha{}_{\beta'} = \delta^\alpha{}_\beta - \xi^\alpha{}_{,\beta} + O(|\xi^\alpha{}_{,\beta}|^2). \tag{8.21}$$

From these relations, one can easily verify that, to first order in small quantities,

$$g_{\alpha'\beta'} = \eta_{\alpha\beta} + h_{\alpha\beta} - \xi_{\alpha,\beta} - \xi_{\beta,\alpha}, \tag{8.22}$$

where we *define*

$$\xi_\alpha := \eta_{\alpha\beta}\xi^\beta. \tag{8.23}$$

This means that the effect of the coordinate change is to change $h_{\alpha\beta}$:

$$h_{\alpha\beta} \rightarrow h_{\alpha\beta} - \xi_{\alpha,\beta} - \xi_{\beta,\alpha}. \tag{8.24}$$

If all $|\xi^\alpha{}_{,\beta}|$ are small then the new $h_{\alpha\beta}$ is still small, and we are still in an acceptable coordinate system. This change is called a *gauge transformation*, a term used because of the strong analogies between Eq. 8.24 and the gauge transformations of electromagnetism. This analogy is explored in Exercise 8.11. The coordinate freedom of Einstein's equations means that we are free to choose an arbitrary (small) 'vector' ξ^α in Eq. 8.24. We will use this freedom below to simplify our equations enormously.

A word about the role of indices like α' and β' in Eqs. 8.21 and 8.22 may be helpful here, as beginning students are often uncertain on this point. A prime or bar on an index is an indication that it refers to a particular coordinate system, e.g. that $g_{\alpha'\beta'}$ is a component of **g** in the $\{x^{\nu'}\}$ coordinates. But the index still takes the same values (0, 1, 2, 3). On the right-hand side of Eq. 8.22 there are no primes because all quantities are defined in the unprimed system. Thus, if $\alpha = \beta = 0$ we read Eq. 8.22 as: 'The 0–0 component of **g** in the primed coordinate system is a function whose value at any point is the value of the 0–0 component of η plus the value of the 0–0 'component' of $h_{\alpha\beta}$ in the unprimed coordinates at that point minus twice the derivative of the function ξ_0 – defined by Eq. 8.23 – with respect to the unprimed coordinate x^0 there.' Equation 8.22 may look strange because – unlike, say, Eq. 8.15 – its indices do not 'match up'. But that is acceptable, since Eq. 8.22 is *not* what we have called a valid tensor equation. It expresses the relation between components of a tensor in two specific coordinates; it is not intended to be a general coordinate-invariant expression.

Riemann tensor. It is easy to show that, to first order in $h_{\mu\nu}$,

$$\mathbf{R}_{\alpha\beta\mu\nu} = \frac{1}{2}(h_{\alpha\nu,\beta\mu} + h_{\beta\mu,\alpha\nu} - h_{\alpha\mu,\beta\nu} - h_{\beta\nu,\alpha\mu}). \tag{8.25}$$

As demonstrated in Exercise 8.5, these components are *independent* of the gauge; they are unaffected by Eq. 8.24. The reason for this is that a coordinate transformation transforms the components of **R** into linear combinations of one another. A small coordinate transformation – a gauge transformation – changes the components by a small amount; but since they are already small, this change is of second order, and so the first-order expression, Eq. 8.25, remains unchanged.

Weak-field Einstein equations. We shall now consistently adopt the point of view mentioned earlier, the fiction that $h_{\alpha\beta}$ is a tensor on a 'background' Minkowski spacetime, i.e. a tensor in SR. Then all our equations will be expected to be valid tensor equations when interpreted in SR, but not necessarily valid under more general coordinate transformations. Gauge transformations will be allowed, of course, but we will not regard them as coordinate transformations. Rather, they define equivalence classes among all symmetric tensors $h_{\alpha\beta}$: any two related by Eq. 8.24 for some ξ_α will produce equivalent physical effects. Consistently with this point of view we can *define* index-raised quantities:

$$h^\mu{}_\beta := \eta^{\mu\alpha} h_{\alpha\beta}, \tag{8.26}$$

$$h^{\mu\nu} := \eta^{\nu\beta} h^\mu{}_\beta, \tag{8.27}$$

the trace

$$h := h^\alpha{}_\alpha, \tag{8.28}$$

and a 'tensor' called the 'trace reverse' of $h_{\alpha\beta}$,

$$\bar{h}^{\alpha\beta} := h^{\alpha\beta} - \frac{1}{2}\eta^{\alpha\beta} h. \tag{8.29}$$

It has this name because

$$\bar{h} := \bar{h}^\alpha{}_\alpha = -h. \tag{8.30}$$

Moreover, one can show that the inverse of Eq. 8.29 is the same:

$$h^{\alpha\beta} = \bar{h}^{\alpha\beta} - \frac{1}{2}\eta^{\alpha\beta} \bar{h}. \tag{8.31}$$

With these definitions it is not difficult to show, beginning with Eq. 8.25, that the Einstein tensor is

$$G_{\alpha\beta} = -\tfrac{1}{2}[\bar{h}_{\alpha\beta,\mu}{}^{,\mu} + \eta_{\alpha\beta}\bar{h}_{\mu\nu}{}^{,\mu\nu} - \bar{h}_{\alpha\mu,\beta}{}^{,\mu}$$
$$- \bar{h}_{\beta\mu,\alpha}{}^{,\mu} + O(h_{\alpha\beta}^2)]. \tag{8.32}$$

(Recall that for any function f,

$$f^{,\mu} := \eta^{\mu\nu} f_{,\nu}.)$$

It is clear that Eq. 8.32 would simplify considerably if we could require

$$\bar{h}^{\mu\nu}{}_{,\nu} = 0. \tag{8.33}$$

These are four equations, and since we have four free gauge functions ξ^α, we might expect to be able to find a gauge in which Eq. 8.33 is true. We shall show that this expectation is correct: it is always possible to choose a gauge to satisfy Eq. 8.33. Thus, we refer to it as a gauge condition and, specifically, as the *Lorentz gauge* condition. If we have an $h_{\mu\nu}$ satisfying this, we say we are working in the Lorentz gauge. Again, the gauge has this name by analogy with electromagnetism (see Exercise 8.11). Other names one encounters in the literature for the same gauge include the harmonic gauge and the de Donder gauge.

That this gauge exists can be shown as follows. Suppose we have some arbitrary $\bar{h}^{(\text{old})}_{\mu\nu}$ for which $\bar{h}^{(\text{old})\mu\nu}{}_{,\nu} \neq 0$. Then under the gauge change Eq. 8.24, one can show (Exercise 8.12) that $\bar{h}_{\mu\nu}$ changes to

$$\bar{h}^{(\text{new})}_{\mu\nu} = \bar{h}^{(\text{old})}_{\mu\nu} - \xi_{\mu,\nu} - \xi_{\nu,\mu} + \eta_{\mu\nu}\xi^\alpha{}_{,\alpha}. \tag{8.34}$$

Then the divergence is

$$\bar{h}^{(\text{new})\mu\nu}{}_{,\nu} = \bar{h}^{(\text{old})\mu\nu}{}_{,\nu} - \xi^{\mu,\nu}{}_{,\nu}. \tag{8.35}$$

If we want a gauge in which $\bar{h}^{(\text{new})\mu\nu}{}_{,\nu} = 0$ then ξ^{μ} is determined by the equation

$$\Box \, \xi^{\mu} = \xi^{\mu,\nu}{}_{,\nu} = \bar{h}^{(\text{old})\mu\nu}{}_{,\nu} \tag{8.36}$$

where the symbol $\Box$ is used for the four-dimensional Laplacian:

$$\Box f = f^{,\mu}{}_{,\mu} = \eta^{\mu\nu} f_{,\mu\nu} = \left(-\frac{\partial^2}{\partial t^2} + \boldsymbol{\nabla}^2 \right) f. \tag{8.37}$$

This operator is also called the *D'Alembertian* or wave operator, and is sometimes denoted by Δ. The equation

$$\Box f = g \tag{8.38}$$

is the three-dimensional inhomogeneous wave equation, and it always has a solution for any (sufficiently well-behaved) g (see Choquet–Bruhat *et al.*, 1977), so there always exists some ξ^{μ} which will transform from an arbitrary $h_{\mu\nu}$ to the Lorentz gauge. In fact, this ξ^{μ} is not unique, since any vector η^{μ} satisfying the homogeneous wave equation

$$\Box \, \eta^{\mu} = 0 \tag{8.39}$$

can be added to ξ^{μ} and the result will still obey

$$\Box(\xi^{\mu} + \eta^{\mu}) = \bar{h}^{(\text{old})\mu\nu}{}_{,\nu} \tag{8.40}$$

and so will still give a Lorentz gauge. Thus, the Lorentz gauge is really a class of gauges.
In this gauge, Eq. 8.32 becomes (see Exercise 8.10)

$$G^{\alpha\beta} = -\tfrac{1}{2}\Box \, \bar{h}^{\alpha\beta}. \tag{8.41}$$

Then the weak-field Einstein equations are

$$\Box \, \bar{h}^{\mu\nu} = -16\,\pi\,T^{\mu\nu}. \tag{8.42}$$

These are called the field equations of *linearized theory*, since they result from keeping terms linear in $h_{\alpha\beta}$.

8.4 Newtonian gravitational fields

Newtonian limit

Newtonian gravity is known to be valid when gravitational fields are too weak to produce velocities near the speed of light: $|\phi| \ll 1, |v| \ll 1$. In such situations, GR must make the same predictions as Newtonian gravity. The fact that velocities are small means that the components $T^{\alpha\beta}$ typically obey the inequalities $|T^{00}| \gg |T^{0i}| \gg |T^{ij}|$. We can only say 'typically' because in special cases T^{0i} might vanish, say for a spherical star, and the second inequality would not hold. But in a strongly rotating Newtonian star T^{0i} would greatly exceed any of the components of T^{ij}. Now, these inequalities should be expected to transfer to $\bar{h}_{\alpha\beta}$ because of Eq. 8.42: $|\bar{h}^{00}| \gg |\bar{h}^{0i}| \gg |\bar{h}^{ij}|$. Of course, one must again be careful about making too broad a statement: one can add in any solution to the homogeneous form of Eq. 8.42, where the right-hand side is set to zero. In such a solution the sizes of the components would not be controlled by the sizes of the components of $T^{\alpha\beta}$. These homogeneous solutions are what we call gravitational waves, as we shall see in the next chapter. So the ordering given here on the components of $\bar{h}_{\alpha\beta}$ holds only in the absence of significant gravitational radiation. Newtonian gravity, of course, has no gravitational waves, so the ordering is just what we need if we want to reproduce Newtonian gravity in general relativity. Thus, we can expect that the dominant 'Newtonian' gravitational field comes from the dominant field equation

$$\Box \, \bar{h}^{00} = -16\,\pi\rho, \tag{8.43}$$

where we use the fact that $T^{00} = \rho + O(\rho v^2)$. For fields that change only because the sources move with velocity v, we have that $\partial/\partial t$ is of the same order as $v\partial/\partial x$, so that

$$\Box = \nabla^2 + O(v^2\nabla^2). \tag{8.44}$$

Thus, Eq. 8.43 is, to lowest order,

$$\nabla^2 \bar{h}^{00} = -16\pi\rho. \tag{8.45}$$

Comparing this with the Newtonian equation, Eq. 8.1, i.e.

$$\nabla^2 \phi = 4\pi\rho$$

(with $G = 1$), we see that we must make the identification

$$\bar{h}^{00} = -4\phi. \tag{8.46}$$

Since all other components of $\bar{h}^{\alpha\beta}$ are negligible at this order, we have

$$h = h^{\alpha}{}_{\alpha} = -\bar{h}^{\alpha}{}_{\alpha} = \bar{h}^{00}, \tag{8.47}$$

and this implies that

$$h^{00} = -2\phi, \tag{8.48}$$
$$h^{xx} = h^{yy} = h^{zz} = -2\phi, \tag{8.49}$$

or

$$ds^2 = -(1 + 2\phi)dt^2 + (1 - 2\phi)(dx^2 + dy^2 + dz^2). \tag{8.50}$$

This is identical to the metric given in Eq. 7.8. We saw there that this metric gives the correct Newtonian laws of motion, so the demonstration here that it follows from Einstein's equations completes the proof that Newtonian gravity is a limiting case of GR. Importantly it also confirms that the constant 8π in Einstein's equations is the correct value of k.

Most astronomical systems are well described by Newtonian gravity as a first approximation. But there are many systems for which it is important to compute the corrections beyond Newtonian theory. These are called post-Newtonian effects, and in Exercises 8.19 and 8.20, we encounter two of them. Post-Newtonian effects in the Solar System include the famous fundamental tests of general relativity, such as the precession of the perihelion of Mercury and the bending of light by the Sun, both of which we will meet in Chapter 11. Outside the Solar System one of the most important post-Newtonian tests is the shrinking of the orbit of the Hulse–Taylor binary pulsar, which confirms general relativity's predictions concerning gravitational radiation (see Chapter 9). Post-Newtonian effects therefore lead to important high-precision tests of general relativity, and the theory of these effects is very well developed. The approximation has been carried to very high orders (Blanchet 2014).

The opening of the Universe to observation by gravitational wave detectors has led to many more opportunities for testing general relativity. The signals from binary black holes and binary neutron stars as they spiral together and merge – a process that inevitably faces the stars in the Hulse–Taylor binary pulsar system some 10^8 years from now – are observed over many orbits, and again this can be compared with not only the post-Newtonian description of the orbits but also fully general-relativistic numerical simulations of the mergers. General relativity agrees with all the observations so far. And these gravitational wave tests go beyond testing orbital mechanics; they can compare the speed of the electromagnetic and gravitational waves as they travel to us from their sources. This is a subject taken up in Chapter 12.

The far field of stationary relativistic sources

For any source in the full Einstein equations which is confined within a limited region of space (a 'localized' source), one can always go far enough away from it that its gravitational field becomes so weak that linearized theory applies in that region. We say that such a spacetime is *asymptotically flat*: spacetime becomes flat asymptotically at large distances from the source. One might be tempted, then, to carry our previous discussion over to this case and say that Eq. 8.50 describes the far field of the source, with ϕ the Newtonian potential. This procedure would be wrong, for two related reasons. First, the derivation of Eq. 8.50 assumed that gravity was weak everywhere, including inside the source, because a crucial step was the identification of Eq. 8.45 with Eq. 8.1 inside the source. In the present discussion we wish to make no assumptions about the weakness of gravity in the source. The second reason why the method would be wrong is that in any case we do not know how to define the Newtonian potential ϕ of a highly relativistic source, so Eq. 8.50 would not make sense.

So we shall work from the linearized field equations directly. Since at first we will assume that the source of the field $T^{\mu\nu}$ is stationary (i.e. independent of time), we can assume that far away from it $h_{\mu\nu}$ is independent of time. (Later we will relax this assumption.) Then Eq. 8.42 becomes

$$\nabla^2 \bar{h}^{\mu\nu} = 0, \tag{8.51}$$

far from the source. This has the solution

$$\bar{h}^{\mu\nu} = A^{\mu\nu}/r + O(r^{-2}),$$

(8.52)

where $A^{\mu\nu}$ is constant. In addition, we must demand that the gauge condition, Eq. 8.33, be satisfied:

$$0 = \bar{h}^{\mu\nu}{}_{,\nu} = \bar{h}^{\mu j}{}_{,j} = -A^{\mu j}n_j/r^2 + O(r^{-3}),$$

(8.53)

where the sum on ν reduces to a sum on the spatial index j because $\bar{h}^{\mu\nu}$ is time independent, and where n_j is the unit radial normal,

$$n_j = x_j/r.$$

(8.54)

The consequence of Eq. 8.53 for all x^i is that

$$A^{\mu j} = 0,$$

(8.55)

for all μ and j. This means that *only* $\bar{h}^{00}$ survives or, in other words, that far from the source we have

$$|\bar{h}^{00}| \gg |\bar{h}^{ij}|, |\bar{h}^{00}| \gg |\bar{h}^{0j}|.$$

(8.56)

These conditions guarantee that the gravitational field does indeed behave like a Newtonian field out there, so we can reverse the identification that led to Eq. 8.46 and *define* the 'Newtonian potential' for the far field of any stationary source to be

$$(\phi)_{\text{relativistic far field}} := -\tfrac{1}{4}(\bar{h}^{00})_{\text{far field}}.$$

(8.57)

With this identification, Eq. 8.50 now does make sense for our problem, and it describes the far field of our source.

Definition of the mass of a relativistic body

Now, far from a Newtonian source the potential is

$$(\phi)_{\text{Newtonian far field}} = -M/r + O(r^{-2}),$$

(8.58)

where M is the mass of the source (with $G = 1$). Thus, if in Eq. 8.52 we rename the constant A^{00} as $4M$, the identification in Eq. 8.57 says that

$$(\phi)_{\text{relativistic far field}} = -M/r.$$

(8.59)

Any small body, for example a planet, that falls freely in the relativistic source's gravitational field but stays far away from it will follow the geodesics of the metric in Eq. 8.50, with ϕ given by Eq. 8.59. In Chapter 7 we saw that these geodesics obey Kepler's laws for the gravitational field of a body of mass M. We therefore *define* this constant M to be the *total mass* of the relativistic source.

Notice that this definition is not an integral over the source: we do not add up the masses of its constituent particles. Instead, we simply measure its mass – 'weigh it' – by the orbits it produces in test bodies far away: this is how astronomers determine the masses of the

Earth, Sun, and planets. This definition enables us to write Eq. 8.50 in the following form far from *any* stationary source:

$$ds^2 = - [1 - 2M/r + 0(r^{-2})]dt^2$$
$$+ [1 + 2M/r + 0(r^{-2})](dx^2 + dy^2 + dz^2). \tag{8.60}$$

It is important to understand that, because we are not integrating over the source, the source of the far-field constant M could be quite different from our expectations from Newtonian theory. As an example of this, see the discussion of the *active gravitational mass* in Exercise 8.20.

The assumption that the source is stationary was necessary to reduce the wave equation, Eq. 8.42, to Laplace's equation, Eq. 8.51. A source which changes with time can emit gravitational waves, and these, as we shall see in the next chapter, travel out from it at the speed of light and do not obey the inequalities Eq. 8.56, so they cannot be regarded as Newtonian fields. Nevertheless, there are situations in which the definition of the mass we have just given may be used with confidence: the waves may be very weak, so that the stationary part of $\bar{h}^{00}$ dominates the wave part; or the source may have been stationary in the distant past, so that one can choose r large enough that any waves have not yet had time to reach that r. The definition of the mass of a time-dependent source is discussed in greater detail in more advanced texts, such as Misner *et al.* (1973) or Wald (1984).

8.5 Bibliography

There is a wide variety of ways to 'derive' (really, to justify) Einstein's field equations, and a selection of them may be found in the texts listed below. The weak-field or linearized equations are useful for many investigations where the full equations are too difficult to solve. We shall use them frequently in subsequent chapters, and most texts discuss them. Our extraction of the Newtonian limit was very heuristic, but there are more rigorous approaches which reveal the geometric nature of Newton's equations (Misner *et al.* 1973, Cartan 1923) and the asymptotic nature of the limit (Damour 1987, Futamase & Schutz 1983). Post-Newtonian theory is described in review articles by Blanchet (2014) and Futamase & Itoh (2007). Einstein's own road to the field equations was much less direct than the one we took. As scholars have studied his notebooks, many previous assumptions about his thinking have proved wrong. See the monumental set edited by Renn (2007).

It seems appropriate here to list a sample of widely available textbooks on GR. They differ in the background and sophistication that they assume of the reader. Some excellent texts expect little background: Hartle (2003), Rindler (2006). Some might be classed as first-year graduate texts: Carroll (2019), Glendenning (2007), Gron & Hervik (2007), parts of Misner *et al.* (1973), Møller (1972), Stephani (2004), Weinberg (2008), and Woodhouse (2007). Some make heavy demands of the student: Hawking & Ellis (1973), Landau & Lifshitz (1980), much of Misner *et al.* (1973), Synge (1960), and Wald (1984).

The material in the present text ought to be sufficient preparation, in most cases, for supplementary reading in even the most advanced texts.

Solving problems is an essential ingredient of learning a theory, and the book of problems by Lightman *et al.* (2017) is an excellent supplement to those in the present book.

Exercises

8.1   Show that Eq. 8.2 is a solution of Eq. 8.1 by the following method. Assume the point particle to be at the origin, $r = 0$, and to produce a spherically symmetric field. Then use Gauss' law on a sphere of radius r to conclude

$$\frac{d\phi}{dr} = \frac{Gm}{r^2}.$$

Deduce Eq. 8.2 from this. (Consider the behavior at infinity.)

8.2   (a) Derive the following important conversion factors from the SI values of G and c:

$$G/c^2 = 7.425 \times 10^{-28} \, \text{m} \, \text{kg}^{-1} = 1,$$
$$c^5/G = 3.629 \times 10^{52} \, \text{J} \, \text{s}^{-1} = 1.$$

(b) Derive the values in geometrized units of the constants in Table 8.1 from their given values in SI units.

(c) Express the following quantities in geometrized units:

   (i) a density (typical of neutron stars) $\rho = 10^{17} \, \text{kg} \, \text{m}^{-3}$;
   (ii) a pressure (also typical of neutron stars) $p = 10^{33} \, \text{kg} \, \text{s}^{-2} \, \text{m}^{-1}$;
   (iii) the acceleration of gravity on Earth's surface $g = 9.80 \, \text{m} \, \text{s}^{-2}$;
   (iv) the luminosity of a supernova $L = 10^{41} \, \text{J} \, \text{s}^{-1}$.

(d) Three dimensioned constants in nature are regarded as fundamental: $c, G,$ and $\hbar$. With $c = G = 1, \hbar$ has units m^2, so $\hbar^{1/2}$ defines a fundamental unit of length called the Planck length. From Table 8.1, we calculate $\hbar^{1/2} = 1.616 \times 10^{-35}$ m. Since this number involves the fundamental constants of relativity, gravitation, and quantum theory, many physicists feel that this length will play an important role in quantum gravity. Express the length in terms of the SI values of $c, G,$ and $\hbar$. Similarly, use the conversion factors to calculate the Planck mass and Planck time, fundamental numbers formed from $c, G,$ and $\hbar$ that have the units of mass and time respectively. Compare these fundamental numbers with characteristic masses, lengths, and timescales that are known from elementary particle theory.

8.3   Calculate in geometrized units:

(a) the Newtonian potential ϕ of the Sun at the Sun's surface, radius 6.960×10^8 m;
(b) the Newtonian potential ϕ of the Sun at the radius of Earth's orbit, $r = 1 \, \text{AU} = 1.496 \times 10^{11}$ m;
(c) the Newtonian potential ϕ of Earth at its surface, radius $= 6.371 \times 10^6$ m;

(d) the velocity of Earth in its orbit around the Sun.

(e) You should have found that your answer to (b) was larger than to (c). Why, then, do we on Earth feel Earth's gravitational pull much more than the Sun's?

(f) Show that a circular orbit around a body of mass M has an orbital velocity, in Newtonian theory, of $v^2 = -\phi$, where ϕ is the Newtonian potential.

8.4 (a) Let A be an $n \times n$ matrix whose entries are all very small, $|A_{ij}| \ll 1/n$, and let I be the unit matrix. Show that

$$(I + A)^{-1} = I - A + A^2 - A^3 + A^4 - \cdots.$$

by proving that (i) the series on the right-hand side converges absolutely for each of the n^2 entries, and (ii) $(I + A)$ times the right-hand side equals I.

(b) Use (a) to establish Eq. 8.21 from Eq. 8.20.

8.5 (a) Show that if $h_{\alpha\beta} = \xi_{\alpha,\beta} + \xi_{\beta,\alpha}$ then Eq. 8.25 vanishes.

(b) Argue from this that Eq. 8.25 is gauge invariant.

(c) Relate this to Exercise 7.10.

8.6 Weak-field theory assumes that $g_{\mu\nu} = \eta_{\mu\nu} + h_{\mu\nu}$, with $|h_{\mu\nu}| \ll 1$. Similarly, $g^{\mu\nu}$ must be close to $\eta^{\mu\nu}$, say $g^{\mu\nu} = \eta^{\mu\nu} + \delta g^{\mu\nu}$. Show from Exercise 8.4(a) that $\delta g^{\mu\nu} = -h^{\mu\nu} + O(h^2)$. Thus, $\eta^{\mu\alpha}\eta^{\nu\beta}h_{\alpha\beta}$ is *not* the deviation of $g^{\mu\nu}$ from flatness.

8.7 (a) Prove that $\bar{h}^{\alpha}{}_{\alpha} = -h^{\alpha}{}_{\alpha}$.

(b) Prove Eq. 8.31.

8.8 Derive Eq. 8.32 in the following manner.

(a) Show that $R^{\alpha}{}_{\beta\mu\nu} = \eta^{\alpha\sigma}R_{\sigma\beta\mu\nu} + O(h^2_{\alpha\beta})$.

(b) From this calculate $R_{\alpha\beta}$ to first order in $h_{\mu\nu}$.

(c) Show that $g_{\alpha\beta}R = \eta_{\alpha\beta}\eta^{\mu\nu}R_{\mu\nu} + O(h^2_{\alpha\beta})$.

(d) From this conclude that

$$G_{\alpha\beta} = R_{\alpha\beta} - \tfrac{1}{2}\eta_{\alpha\beta}R,$$

i.e. that the linearized $G_{\alpha\beta}$ is the *trace reverse* of the linearized $R_{\alpha\beta}$, in the sense of Eq. 8.29.

(e) Use this to simplify somewhat the calculation of Eq. 8.32.

8.9 (a) Show from Eq. 8.32 that G_{00} and G_{0i} do not contain second time derivatives of any $\bar{h}_{\alpha\beta}$. Thus only the *six* equations $G_{ij} = 8\pi T_{ij}$ are true dynamical equations. Relate this to the discussion at the end of § 8.2. The equations $G_{0\alpha} = 8\pi T_{0\alpha}$ are called *constraint* equations because they are relations among the initial data for the other six equations which prevent one choosing all these data freely.

(b) Equation 8.42 contains second time derivatives even when μ or ν is zero. Does this contradict (a)? Why?

8.10 Use the Lorentz gauge condition, Eq. 8.33, to simplify $G_{\alpha\beta}$ to Eq. 8.41.

8.11 When one writes Maxwell's equations in special-relativistic form, one identifies the scalar potential ϕ and three-vector potential A_i (with signs defined by $E_i = -\phi_{,i} - A_{i,0}$) as components of a one-form defined by the following two relations: $A_0 = -\phi$; A_i (one-form) $= A_i$ (three-vector). A gauge transformation is the replacement

$\phi \rightarrow \phi - \partial f/\partial t, A_i \rightarrow A_i + f_{,i}$. This leaves the electric and magnetic fields unchanged. The Lorentz gauge is a gauge in which $\partial\phi/\partial t + \nabla_i A^i = 0$. Write both the gauge transformation and the Lorentz gauge condition in four-tensor notation. Draw the analogy with similar equations in linearized gravity.

8.12 Prove Eq. 8.34.

8.13 The inequalities $|T^{00}| \gg |T^{0i}| \gg |T^{ij}|$ for a Newtonian system are illustrated in Exercise 8.2(c) and Exercise 8.3(d)–(e). Devise physical arguments to justify them in general.

8.14 From Eq. 8.46 and the inequalities among the components $h_{\alpha\beta}$, derive all of Eqs. 8.47–8.50.

8.15 We have argued that we should use convenient coordinates to solve the weak-field problem (or any other!), but that any physical results should be expressible in coordinate-free language. From this point of view our demonstration of the Newtonian limit is as yet incomplete, since in Chapter 7 we merely showed that the metric Eq. 7.8 led to Newton's law $dp/dt = -m\nabla\phi$.

    However, this is a coordinate-dependent equation, involving coordinate time and position. It is certainly not a valid four-dimensional tensor equation. Fill in this gap in our reasoning by showing that one can make physical measurements to verify that the relativistic predictions match the Newtonian ones. (For example, what is the relation between the proper time for one orbit and the proper circumference of the orbit?)

8.16 Redo the derivation of the Newtonian limit by replacing 8π in Eq. 8.10 by k and following through the changes this makes in subsequent equations. Verify that one recovers Eq. 8.50 only if $k = 8\pi$.

8.17 (a) A small planet orbits a static neutron star in a circular orbit whose proper circumference is 6×10^{11} m. The orbital period takes 200 days of the planet's proper time. Estimate the mass M of the star.

    (b) Five satellites are placed into circular orbits around a static black hole. The proper circumferences and proper periods of their orbits are given in the table below. Use the method of (a) to estimate the black hole's mass. Explain the trend of the results you get for the satellites.

Proper circumference	2.5×10^6 m	6.3×10^6 m	6.3×10^7 m	3.1×10^8 m	6.3×10^9 m
Proper period	8.4×10^{-3} s	0.055 s	2.1 s	23 s	2.1×10^3 s

8.18 Consider the field equations with cosmological constant, Eq. 8.7, with Λ arbitrary and $k = 8\pi$.

(a) Find the Newtonian limit and show that one recovers the motion of the planets only if $|\Lambda|$ is very small. Given that the radius of Pluto's orbit is 5.9×10^{12} m, set an upper bound on $|\Lambda|$ from Solar-System measurements.

(b) By bringing Λ over to the right-hand side of Eq. 8.7, one can regard $-\Lambda g^{\mu\nu}/8\pi$ as the stress–energy tensor of 'empty space'. Given that the observed mass of the region of the Universe near our Galaxy would have a density of about 10^{-27} kg m^{-3} if it were uniformly distributed, do you think that a value of $|\Lambda|$

near the limit that you established in (a) could have observable consequences for cosmology? Conversely, if Λ is comparable to the mass density of the Universe, do we need to include it in the equations when we discuss the Solar System?

8.19   In this exercise we shall compute the first correction to the Newtonian solution caused by a source that rotates. In Newtonian gravity, the angular momentum of the source does not affect the field: two sources with the same $\rho(x^i)$ but different angular momenta have the same field. Not so in relativity, since *all* components of $T^{\mu\nu}$ generate the field. This is our first example of a *post-Newtonian effect*, an effect that introduces an aspect of general relativity that is not present in Newtonian gravity.

(a) Suppose that a spherical body of uniform density ρ and radius R rotates rigidly about the x^3 axis with constant angular velocity Ω. Write down the components $T^{0\nu}$ in a Lorentz frame at rest with respect to the center of mass of the body, assuming ρ, Ω, and R are independent of time. For each component, work to the lowest nonvanishing order in ΩR.

(b) The general solution to the equation $\nabla^2 f = g$ which vanishes at infinity is the generalization of Eq. 8.2,

$$f(x) = -\frac{1}{4\pi} \int \frac{g(y)}{|x - y|} d^3 y,$$

which reduces to Eq. 8.2 when g is nonzero in a very small region. Use this to solve Eq. 8.42 for $\bar{h}^{00}$ and $\bar{h}^{0j}$ for the source described in (a). Obtain the solutions only outside the body, and only to lowest nonvanishing order in r^{-1}, where r is the distance from the body's center. Express the result for $\bar{h}^{0j}$ in terms of the body's angular momentum. Find the metric tensor within this approximation, and transform it to spherical coordinates.

(c) Because the metric is independent of t and the azimuthal angle ϕ, particles orbiting this body will have p_0 and p_ϕ constant along their trajectories (see § 7.4). Consider a particle of nonzero rest mass in a circular orbit of radius r in the equatorial plane. To lowest order in Ω, calculate the difference between its orbital period in the positive sense (i.e., when rotating in the sense of the central body's rotation) and in the negative sense. (Define the period to be the coordinate time taken for one orbit of $\Delta\phi = 2\pi$.)

(d) From this devise an experiment to measure the angular momentum J of the central body. Take the central body to be the Sun ($M = 2 \times 10^{30}$ kg, $R = 7 \times 10^8$ m, $\Omega = 3 \times 10^{-6}$ s^{-1}) and the orbiting particle to be Earth ($r = 1.5 \times 10^{11}$ m). What would be the difference in the length of the year between positive and negative orbits?

8.20   This exercise introduces the concept of the active gravitational mass. After deriving the weak-field Einstein equations in Eq. 8.42, we immediately specialized to the low-velocity Newtonian limit. Here we go a few steps further without making the assumption that velocities are small or that pressures are weak compared with densities.

(a) Perform a trace-reverse operation on Eq. 8.42 to get

$$\Box\, h^{\mu\nu} = -16\,\pi\,\left(T^{\mu\nu} - \tfrac{1}{2}T^{\alpha}{}_{\alpha}\eta^{\mu\nu}\right). \qquad (8.61)$$

(b) If the system is isolated and stationary then its gravitational field far away will be dominated by h^{00}, as in the argument leading up to Eq. 8.50. If the system has weak internal gravity but strong pressure, show that $h^{00} = -2\Phi$ where Φ satisfies a Newtonian-like Poisson equation,

$$\nabla^2\Phi = 4\pi\left(\rho + T^{k}{}_{k}\right). \qquad (8.62)$$

For a perfect fluid, this source term is just $\rho + 3p$ and is called the *active gravitational mass* in general relativity. If the system is Newtonian then $p \ll \rho$ and we have the usual Newtonian limit. This is another example of a *post-Newtonian effect*.

Fundamentals of gravitational radiation

9.1 The role of general relativity in the physical Universe

When Einstein introduced his theory of general relativity in 1915, he felt its primary importance was to remedy the logical conflict between special relativity and Newton's theory of gravity. The two early tests of the theory that convinced the world of science that it was a triumph were the small unexplained part of the precession of Mercury's orbit and the bending of light by the Sun, both of which we will describe in Chapter 11. These effects involve astronomy, but they are tiny.

Einstein apparently remained convinced for his whole life that general relativity would not make a serious impact on astronomy except to describe the overall behavior of the Universe, the subject we call cosmology and treat in Chapter 13. Karl Schwarzschild had discovered the solution of Einstein's equations that we now call the black hole – Chapter 11 – in 1916, but Einstein never accepted that this solution would describe a real physical object. Similarly, Einstein himself in 1916 worked out the basic description of gravitational waves, which we introduce in this chapter, but right away he realized how weak they would be, and believed they would never be detectable. Later in his life he even doubted that they were real, worrying that they might just be an effect of wavy coordinates rather than wavy geometry.

If Einstein's doubts had been right, then the rest of this book would be quite short. But the fact is that black holes are commonplace in the Universe; neutron stars are even more so; and gravitational waves are indeed real, have been detected, and are now being used to investigate parts of the physical Universe that we never had information about before. In fact, the science of gravitational radiation has so far been awarded the Nobel Prize for Physics twice: to Joseph Taylor and Richard Hulse in 1993 for their discovery in 1974 of a binary pulsar system (two neutron stars orbiting one another) that, among other things, clearly shows that the stars' orbits are shrinking because of the loss of energy to gravitational radiation; and to Rainer Weiss, Kip Thorne, and Barry Barish in 2017 for their leading roles in creating the LIGO gravitational wave detectors, which made the first direct detection of gravitational waves in 2015.

This chapter begins our study of the central role that general relativity now plays in our understanding of the Universe we observe around us. We begin here with the theory of gravitational radiation, which is the simplest possible deviation from special relativity predicted by Einstein's equations. Then we go on with separate chapters on relativistic stars (Chapter 10) and on black holes (Chapter 11). Having gained an understanding of these primary sources of detectable gravitational waves, we take up the subject (Chapter 12) of what we are already learning about the Universe from gravitational wave observations, and where the field is going in the future. Finally (Chapter 13) we broaden our viewpoint to embrace the biggest subject of all, cosmology.

9.2 The propagation of gravitational waves

It may happen that in a region of spacetime the gravitational field is weak but not stationary. This can happen far from a fully relativistic source undergoing rapid changes that took place long enough ago for the disturbances produced by the changes to reach the distant region under consideration. We shall study this problem by using the weak-field equations developed in the last chapter, but first we study the solutions of the homogeneous system of equations that we excluded from the Newtonian treatment in § 8.4.

The Einstein equations, Eq. 8.42, in vacuum ($T^{\mu\nu} = 0$) and far outside the source of the field are

$$\left(-\frac{\partial^2}{\partial t^2} + \nabla^2\right)\bar{h}^{\alpha\beta} = 0. \tag{9.1}$$

In this chapter we do not neglect $\partial^2/\partial t^2$. Equation 9.1 is called the three-dimensional wave equation. We shall show that it has a (complex) solution of the form

$$\bar{h}^{\alpha\beta} = A^{\alpha\beta}\exp{(ik_\alpha x^\alpha)}, \tag{9.2}$$

where $\{k_\alpha\}$ are the (real) constant components of some one-form and $\{A^{\alpha\beta}\}$ the (complex) constant components of some tensor.[1]

Equation 9.1 can be written as

$$\eta^{\mu\nu}\bar{h}^{\alpha\beta}{}_{,\mu\nu} = 0, \tag{9.3}$$

and, from Eq. 9.2, we have

$$\bar{h}^{\alpha\beta}{}_{,\mu} = ik_\mu\bar{h}^{\alpha\beta}. \tag{9.4}$$

Therefore Eq. 9.3 becomes

$$\eta^{\mu\nu}\bar{h}^{\alpha\beta}{}_{,\mu\nu} = -\eta^{\mu\nu}k_\mu k_\nu\bar{h}^{\alpha\beta} = 0.$$

This can vanish only if

$$\eta^{\mu\nu}k_\mu k_\nu = k^\nu k_\nu = 0. \tag{9.5}$$

So Eq. 9.2 gives a solution to Eq. 9.1 if k_α is a *null* one-form or, equivalently, if the associated four-vector k^α is *null*, i.e. tangent to the world line of a photon. (Recall that we raise and lower indices with the flat-space metric tensor $\eta^{\mu\nu}$, so k^α is a Minkowski

---

[1] Eventually we shall take the real part of any complex solutions; this is a standard technique to keep oscillator-type equations simpler to write down. It avoids our having to keep track of sine and cosine separately.

null vector.) Equation 9.2 describes a *wavelike* solution. The value of $\bar{h}^{\alpha\beta}$ is constant on a hypersurface on which $k_{\alpha}x^{\alpha}$ is constant:

$$k_{\alpha}x^{\alpha} = k_0 t + \boldsymbol{k} \cdot \boldsymbol{x} = \text{const.,} \tag{9.6}$$

where $\boldsymbol{k}$ refers to $\{k^i\}$. It is conventional to refer to k^0 as ω, the frequency of the wave:

$$\vec{k} \rightarrow (\omega, \boldsymbol{k}) \tag{9.7}$$

is the time–space decomposition of $\vec{k}$.

Imagine a photon moving in the direction of the null vector $\vec{k}$. It travels on a curve

$$x^{\mu}(\lambda) = k^{\mu}\lambda + l^{\mu}, \tag{9.8}$$

where λ is a parameter and l^{μ} is a constant vector (the photon's position at $\lambda = 0$). From Eqs. 9.8 and 9.5 we find

$$k_{\mu}x^{\mu}(\lambda) = k_{\mu}l^{\mu} = \text{const.} \tag{9.9}$$

Comparing this with Eq. 9.6, we see that the photon travels with the gravitational wave, staying forever at the same phase as the gravitational wave. We express this by saying that the wave itself travels at the speed of light, and $\vec{k}$ is its direction of travel. The nullity of $\vec{k}$ implies

$$\omega^2 = |\boldsymbol{k}|^2, \tag{9.10}$$

which is referred to as the dispersion relation for the wave. Readers familiar with wave theory will immediately see from Eq. 9.10 that the wave's phase velocity is 1, as is its group velocity, and that there is no dispersion.

The Einstein equations assume the simple form given by Eq. 9.1 only if we impose the gauge condition

$$\bar{h}^{\alpha\beta}{}_{,\beta} = 0, \tag{9.11}$$

whose consequences we must therefore consider. From Eq. 9.4 we find that

$$A^{\alpha\beta}k_{\beta} = 0, \tag{9.12}$$

which is a restriction on $A^{\alpha\beta}$: it must be orthogonal to $\vec{k}$.

The solution $A^{\alpha\beta}\exp(ik_{\mu}x^{\mu})$ is a *plane wave*. (Of course, in physical applications, one uses only the real part of this expression, allowing $A^{\alpha\beta}$ to be complex.) By the theorems of Fourier analysis, *any* solution of Eqs. 9.1 and 9.11 is a superposition of plane wave solutions. (See Exercise 9.3.)

The transverse–traceless gauge

We so far have only one constraint, Eq. 9.12, on the amplitude $A^{\alpha\beta}$, but we can use our gauge freedom to restrict it further. Recall from Eq. 8.38 that we can change the gauge while remaining within the Lorentz class of gauges, using any vector solving

$$\left(-\frac{\partial^2}{\partial t^2} + \nabla^2\right)\xi_{\alpha} = 0. \tag{9.13}$$

Let us choose a solution

$$\xi_\alpha = B_\alpha \exp(ik_\mu x^\mu), \tag{9.14}$$

where B_α is a constant and k^μ is the same null vector as for our wave solution. This produces a change in $h^{\alpha\beta}$, given by Eq. 8.24,

$$h_{\alpha\beta}^{\text{(NEW)}} = h_{\alpha\beta}^{\text{(OLD)}} - \xi_{\alpha,\beta} - \xi_{\beta,\alpha} \tag{9.15}$$

and a consequent change in $\bar{h}_{\alpha\beta}$, given by Eq. 8.34,

$$\bar{h}_{\alpha\beta}^{\text{(NEW)}} = \bar{h}_{\alpha\beta}^{\text{(OLD)}} - \xi_{\alpha,\beta} - \xi_{\beta,\alpha} + \eta_{\alpha\beta}\xi^\mu_{,\mu}. \tag{9.16}$$

Using Eq. 9.14 and dividing out the exponential factor common to all terms gives

$$A_{\alpha\beta}^{\text{(NEW)}} = A_{\alpha\beta}^{\text{(OLD)}} - i\,B_\alpha k_\beta - i\,B_\beta k_\alpha + i\,\eta_{\alpha\beta}B^\mu k_\mu. \tag{9.17}$$

In Exercise 9.5 it is shown that B_α can be chosen to impose two further restrictions on $A_{\alpha\beta}^{\text{(NEW)}}$:

$$A^\alpha{}_\alpha = 0 \tag{9.18}$$

and

$$A_{\alpha\beta}U^\beta = 0, \tag{9.19}$$

where $\vec{U}$ is some fixed four-velocity, i.e. any constant timelike unit vector that we choose. Equations 9.12, 9.18, and 9.19 together are called the *transverse–traceless* (TT) gauge conditions. (The word 'traceless' refers to Eq. 9.18; 'transverse' will be explained below.) We have now used up all our gauge freedom, so any remaining independent components of $A_{\alpha\beta}$ must be physically important. Notice, by the way, that the trace condition, Eq. 9.18, implies (see Eq. 8.29)

$$\bar{h}_{\alpha\beta}^{\text{TT}} = h_{\alpha\beta}^{\text{TT}}. \tag{9.20}$$

Let us go to a Lorentz frame for the background Minkowski spacetime (i.e. make a background Lorentz transformation), in which the vector $\vec{U}$ upon which we have based the TT gauge is the time basis vector $U^\beta = \delta^\beta{}_0$. Then Eq. 9.19 implies $A_{\alpha 0} = 0$ for all α. In this frame, let us orient our spatial coordinate axes so that the wave is traveling in the z direction, $\vec{k} \rightarrow (\omega, 0, 0, \omega)$. Then, with Eq. 9.19, Eq. 9.12 implies $A_{\alpha z} = 0$ for all α. (This is the origin of the adjective 'transverse' for the gauge: $A_{\mu\nu}$ is 'across' the direction of propagation $\vec{e}_z$.) These two restrictions mean that only A_{xx}, A_{yy}, and $A_{xy} = A_{yx}$ are nonzero. Moreover, the trace condition, Eq. 9.18, implies $A_{xx} = -A_{yy}$. In matrix form, we therefore have in this specially chosen frame

$$(A_{\alpha\beta}^{\mathrm{TT}}) = \begin{pmatrix} 0 & 0 & 0 & 0 \\ 0 & A_{xx} & A_{xy} & 0 \\ 0 & A_{xy} & -A_{xx} & 0 \\ 0 & 0 & 0 & 0 \end{pmatrix} \qquad (9.21)$$

There are only *two* independent constants, A_{xx}^{TT} and A_{xy}^{TT}. What is their physical significance?

The effect of waves on free particles

As we remarked earlier, any wave is a superposition of plane waves; if the wave travels in the z direction we can put all the plane waves into the form Eq. 9.21, so that any wave has only the two independent components h_{xx}^{TT} and h_{xy}^{TT}. Consider a situation in which a particle initially in a wave-free region of spacetime encounters a gravitational wave. Choose a background Lorentz frame in which the particle is initially at rest, and choose the TT gauge referred to this frame (i.e. the four-velocity U^α in Eq. 9.19 is the initial four-velocity of the particle). A free particle obeys the geodesic equation, Eq. 7.9,

$$\frac{\mathrm{d}}{\mathrm{d}\tau}U^\alpha + \Gamma^\alpha{}_{\mu\nu}U^\mu U^\nu = 0. \qquad (9.22)$$

Since the particle is initially at rest, the initial value of its acceleration is

$$(\mathrm{d}U^\alpha/\mathrm{d}\tau)_0 = -\Gamma^\alpha{}_{00} = -\tfrac{1}{2}\eta^{\alpha\beta}(h_{\beta 0,0} + h_{0\beta,0} - h_{00,\beta}). \qquad (9.23)$$

But, by Eq. 9.21, $h_{\beta 0}^{\mathrm{TT}}$ vanishes so initially the acceleration vanishes. This means the particle will still be at rest a moment later, and then, by the same argument, the acceleration will still be zero a moment later. The result is that the particle remains at rest forever, regardless of the wave! However, being 'at rest' simply means remaining at a constant coordinate position, so we should not be too hasty in its interpretation. All we have discovered is that by choosing the TT gauge – which means making a particular adjustment in the 'wiggles' of our coordinates – we have found a coordinate system that stays attached to individual particles. This in itself has no invariant geometrical meaning.

To get a better measure of the effect of the wave, let us consider two nearby particles, one at the origin and another at $x = \varepsilon$, $y = z = 0$, both beginning at rest. Both then remain at these coordinate positions, and the proper distance between them is

$$\Delta l \equiv \int |\mathrm{d}s^2|^{1/2} = \int |g_{\alpha\beta}\,\mathrm{d}x^\alpha\,\mathrm{d}x^\beta|^{1/2}$$

$$= \int_0^\varepsilon |g_{xx}|^{1/2}\,\mathrm{d}x \approx |g_{xx}(x=0)|^{1/2}\varepsilon$$

$$\approx [1 + \tfrac{1}{2}h_{xx}^{\mathrm{TT}}(x=0)]\varepsilon. \qquad (9.24)$$

Now, since h_{xx}^{TT} is not generally zero, the *proper* distance (as opposed to the coordinate distance) does change with time. This is an illustration of the difference between computing a coordinate-dependent number (the position of a particle) and a coordinate-independent number (the proper distance between two particles). The effect of the wave is unambiguously seen in the coordinate-independent number. The physical effects of gravitational fields always show up in measurables.

The proper distance between two particles can be measured, and this is in fact what all modern gravitational wave detectors do. We will discuss two ways of measuring it in the subsection on 'Measuring the stretching of space' below, and later in the chapter we discuss how both ways might be used to detect real gravitational waves.

Equation 9.24 tells us a lot. First, the change in the proper distance between two particles is proportional to their initial separation ε: gravitational waves create a bigger distance change if the original distance is bigger. This is the reason why modern gravitational wave detectors, which we discuss below, are designed and built on huge scales, measuring changes in separation over many kilometers (for ground-based detectors) or millions of kilometers (as planned in space). The second thing we learn from Eq. 9.24 is that the effect is small, proportional to h_{ij}^{TT}. We will see when we study the generation of waves below that these dimensionless components are typically 10^{-21} or smaller. So gravitational wave detectors have to sense relative distance changes smaller than one part in 10^{21}. This experimental milestone was achieved for the first time in 2005. By 2015 improvements in sensitivity had allowed the detectors to reach nearly to 10^{-22}, and the first detection followed almost immediately.

Tidal accelerations: gravitational wave forces

Another approach to the same question of how gravitational waves affect free particles involves the equation of geodesic deviation, Eq. 6.87. This will lead us, in the following subsection, to a way of understanding the action of gravitational waves as a tidal force on particles, whether they are free or not.

Consider again two freely falling particles, and set up the connecting vector ξ^α between them. If we were to work in a TT-coordinate system, as in the previous subsection, then the fact that the particles remain at rest in the coordinates means that the components of $\vec{\xi}$ would remain constant; although correct, this would not be a helpful result since we have not associated the components of $\vec{\xi}$ in TT-coordinates with the result of any measurement. Instead we shall work in a different coordinate system, one closely associated with measurements: the local inertial frame at the point of the first geodesic, where $\vec{\xi}$ originates. In this frame, coordinate distances are proper distances, as long as we can neglect quadratic terms in the coordinates. That means that, in these coordinates, the components of $\vec{\xi}$ do indeed correspond to measurable proper distances if the geodesics are near enough to one another.

What is more, in this frame the second covariant derivative in Eq. 6.87 simplifies. It starts out as $\nabla_U \nabla_U \xi^\alpha$, where we are calling the tangent to the geodesic $\vec{U}$ here instead of $\vec{V}$. Now, the first derivative acting on $\vec{\xi}$ just gives $\mathrm{d}\xi^\alpha/\mathrm{d}\tau$. But the second derivative is a covariant one, and should contain not just $\mathrm{d}/\mathrm{d}\tau$ but also a term with a Christoffel symbol. But in this local inertial frame the Christoffel symbols all vanish at this point, so the second derivative is just an ordinary second derivative with respect to τ. The result is, again in the locally inertial frame,

$$\frac{\mathrm{d}^2}{\mathrm{d}\tau^2}\xi^\alpha = R^\alpha{}_{\mu\nu\beta}\,U^\mu U^\nu \xi^\beta, \tag{9.25}$$

where $\vec{U} = \mathrm{d}\vec{x}/\mathrm{d}\tau$ is the four-velocity of the two particles. In these coordinates the components of $\vec{U}$ are needed only to lowest (i.e. flat-space) order, since any corrections to U^α that depend on $h_{\mu\nu}$ will give terms second order in $h_{\mu\nu}$ in the above equation (because $R^\alpha{}_{\mu\nu\beta}$ is already first order in $h_{\mu\nu}$). Therefore $\vec{U} \to (1,\ 0,\ 0,\ 0)$ and, initially, $\vec{\xi} \to (0, \varepsilon, 0, 0)$. Then, to first order in $h_{\mu\nu}$, Eq. 9.25 reduces to

$$\frac{\mathrm{d}^2}{\mathrm{d}\tau^2}\xi^\alpha = \frac{\partial^2}{\partial t^2}\xi^\alpha = \varepsilon R^\alpha{}_{00x} = -\varepsilon R^\alpha{}_{0x0}. \tag{9.26}$$

This is the fundamental result, which shows that the Riemann tensor is locally measurable by simply watching the proper-distance changes between nearby geodesics.

Now, the Riemann tensor is itself gauge invariant, so its components do not depend on the choice between a local inertial frame and the TT coordinates. It follows also that the left-hand side of Eq. 9.26 must have an interpretation that is independent of the coordinate gauge. We identify ξ^α as the *proper lengths* of the components of the connecting vector $\vec{\xi}$, in other words the proper distances along the four coordinate directions over the coordinate intervals spanned by the vector. With this interpretation, we free ourselves from the choice of gauge and arrive at a gauge-invariant interpretation of the whole of Eq. 9.26.

Just to emphasize that we have restored gauge freedom to this equation, let us write the Riemann tensor components in terms of the components of the metric in TT gauge. This is possible, since the Riemann components are gauge invariant, and it is desirable, since these components are particularly simple in the TT gauge. It is not hard to use Eq. 8.25 to show that, for a wave traveling in the z direction, the components are

$$\left.\begin{aligned} R^x{}_{0x0} &= R_{x0x0} = -\tfrac{1}{2}h^{\mathrm{TT}}_{xx,00}, \\[2pt] R^y{}_{0x0} &= R_{y0x0} = -\tfrac{1}{2}h^{\mathrm{TT}}_{xy,00}, \\[2pt] R^y{}_{0y0} &= R_{y0y0} = -\tfrac{1}{2}h^{\mathrm{TT}}_{yy,00} = -R^x{}_{0x0}, \end{aligned}\right\} \tag{9.27}$$

with all other independent components vanishing. This means, for example, that two particles initially separated in the x direction have a separation vector $\vec{\xi}$ whose components' proper lengths obey

$$\frac{\partial^2}{\partial t^2}\xi^x = \tfrac{1}{2}\varepsilon\frac{\partial^2}{\partial t^2}h^{\mathrm{TT}}_{xx}, \qquad \frac{\partial^2}{\partial t^2}\xi^y = \tfrac{1}{2}\varepsilon\frac{\partial^2}{\partial t^2}h^{\mathrm{TT}}_{xy}. \tag{9.28a}$$

This is clearly consistent with Eq. 9.24. Similarly, two particles initially separated by ε in the y direction obey

$$\frac{\partial^2}{\partial t^2}\xi^y = \tfrac{1}{2}\varepsilon\frac{\partial^2}{\partial t^2}h^{\mathrm{TT}}_{yy} = -\tfrac{1}{2}\varepsilon\frac{\partial^2}{\partial t^2}h^{\mathrm{TT}}_{xx},$$

$$\frac{\partial^2}{\partial t^2}\xi^x = \tfrac{1}{2}\varepsilon\frac{\partial^2}{\partial t^2}h^{\mathrm{TT}}_{xy}. \tag{9.28b}$$

Remember, from Eq. 9.21, that $h^{\mathrm{TT}}_{yy} = -h^{\mathrm{TT}}_{xx}$.

Measuring the stretching of space

The action of gravitational waves is sometimes characterized as a stretching of space. Equation 9.24 makes it clear what this means: as the wave passes through, the proper separations of free objects that are simply sitting at rest change with time. However, students of general relativity sometimes find this concept confusing. A frequent question is, if space is stretched, why is a ruler (which consists, after all, mostly of empty space with a few electrons and nuclei scattered through it) not also stretched, so that the stretching is not measurable by the ruler? The answer to this question lies in considering the additional electromagnetic forces that keep the ruler from stretching. Mathematically, we can understand this not from Eq. 9.24 but instead by using the geodesic deviation equation, Eq. 9.26.

Although the two formulations of the action of a gravitational wave, Eq. 9.24 and Eq. 9.26, are essentially equivalent for free particles, the second is far more useful and instructive when we consider the behavior of particles that have other forces acting on them as well. The first formulation is the complete solution for the relative motion of particles that are freely falling, but it gives no way of including other forces. The second formulation is not a *solution* but a *differential equation*. It shows the acceleration of one particle (let's call it B), induced by the wave, as measured in a freely falling local inertial frame that initially coincides with the motion of the other particle (let's call it A). It says that, in this local frame, the wave acts just like a force pushing on B. This is called the *tidal force* associated with the wave. This force depends on the separation $\vec{\xi}$. From Eq. 9.25 it is clear that the acceleration resulting from this effective force is

$$\frac{\partial^2}{\partial t^2} \xi^i = -R^i{}_{0j0} \xi^j. \tag{9.29}$$

Now, if particle B has another force on it as well, say $\vec{F}_B$, then to get the complete motion of B we must simply solve Newton's second law with all forces included. This means solving the differential equation

$$\frac{\partial^2}{\partial t^2} \xi^i = -R^i{}_{0j0} \xi^j + \frac{1}{m_B} F^i_B, \tag{9.30}$$

where m_B is the mass of particle B. Indeed, if particle A also has a force $\vec{F}_A$ on it, then it will not remain at rest in the local inertial frame. But we can still make measurements in that frame, and in this case the separation of the two particles will obey

$$\frac{\partial^2}{\partial t^2} \xi^i = -R^i{}_{0j0} \xi^j + \frac{1}{m_B} F^i_B - \frac{1}{m_A} F^i_A. \tag{9.31}$$

This equation allows one to treat material systems acted on by gravitational waves. It allows us now to answer the question of what happens to a ruler when the wave hits. Since the atoms in the ruler are not free, but instead are acted upon by electric forces from nearby atoms, the ruler will stretch by an amount that depends on how strong the tidal gravitational forces are compared to the internal binding forces. Now, gravitational forces are very weak indeed compared to electric forces, so in practice the ruler does not stretch

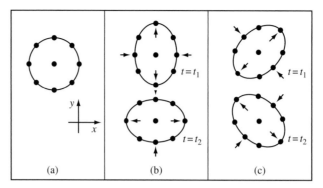

Figure 9.1 (a) A circle of free particles before a wave traveling in the *z* direction reaches them. (b) Distortions of the circle produced by a wave with the '+' polarization. The two pictures represent the same wave at phases separated by 180°. The particles are positioned according to their proper distances from one another. (c) As (b) but for the '×' polarization.

at all. This is one way of measuring the 'stretching of space': use a solid ruler to record the tidal displacements of nearby free particles.

There is, however, an even better way to measure the stretching: send photons back and forth between the free particles and measure the changes in the light-travel time between them. This is the principle of the laser-interferometer gravitational wave detector, and we will discuss it in some detail below. The continuum version of Eq. 9.31 can be used to understand how 'bar' detectors – the first instruments that were built, unsuccessfully, to detect gravitational waves – react to incident waves. We will discuss this also below.

Polarization of gravitational waves

Equations 9.28a and 9.28b define the *polarization* of a gravitational wave. Consider a ring of particles initially at rest in the x–y plane, as in Figure 9.1(a). Suppose that a wave has $h_{xx}^{TT} \neq 0, h_{xy}^{TT} = 0$. Then the particles will be moved (in terms of proper distance relative to the particle in the center) in the way shown in Figure 9.1(b), as the wave oscillates and $h_{xx}^{TT} = -h_{yy}^{TT}$ changes sign. If, instead, the wave had $h_{xy}^{TT} \neq 0$ but $h_{xx}^{TT} = h_{yy}^{TT} = 0$, then the picture would distort as in Figure 9.1(c). Since h_{xy}^{TT} and h_{xx}^{TT} are independent, (b) and (c) provide a pictorial representation for two different linear polarizations. Notice that the two states are simply rotated 45° relative to one another. This contrasts with the two polarization states of an electromagnetic wave, which are at 90° to each other. As Exercise 9.16 shows, this pattern of polarization is due to the fact that gravity is represented by the second-rank symmetric tensor $h_{\mu\nu}$. (By contrast, electromagnetism is represented by the vector potential A^μ of Exercise 8.11.)

An exact plane wave

Although all the gravitational waves that we can expect to detect on Earth are so weak that linearized theory ought to describe them very accurately, it is interesting to see whether the linear plane wave corresponds to some exact solution of the nonlinear equations that has similar properties. We shall briefly derive such a solution.

Suppose the wave is traveling in the z direction. By analogy with Eq. 9.2 we might hope to find a solution that depends only on

$$u := t - z.$$

This suggests using u as a coordinate, as we did in Exercise 3.34 (with x replaced by z). In flat space it is natural, then, to define a complementary null coordinate

$$v = t + z,$$

so that the line element of flat spacetime becomes

$$ds^2 = -du\,dv + dx^2 + dy^2.$$

Now, we have seen that the linear wave affects only proper distances perpendicular to its motion, i.e. in the x–y coordinate plane. So let us look for a nonlinear generalization of this, i.e. for a solution with metric

$$ds^2 = -dudv + f^2(u)dx^2 + g^2(u)dy^2,$$

where f and g are functions to be determined by Einstein's equations. It is a straightforward calculation to discover that the only nonvanishing Christoffel symbols and Riemann tensor components are

$$\Gamma^x{}_{xu} = \dot{f}/f, \qquad \Gamma^y{}_{yu} = \dot{g}/g,$$

$$\Gamma^v{}_{xx} = 2\dot{f}/f, \qquad \Gamma^v{}_{yy} = 2\dot{g}/g,$$

$$R^x{}_{uxu} = -\ddot{f}/f, \qquad R^y{}_{uyu} = -\ddot{g}/g,$$

and others obtainable by symmetry. Here, the dots denote derivatives with respect to u. The only vacuum field equation then becomes

$$\ddot{f}/f + \ddot{g}/g = 0. \tag{9.32}$$

We can therefore prescribe an arbitrary function $g(u)$ and solve this equation for $f(u)$. This is the same freedom as we had in the linear case, where Eq. 9.2 can be multiplied by an arbitrary $f(k_z)$ and integrated over k_z to give the Fourier representation of an arbitrary function of $(z - t)$. In fact, if g is nearly 1,

$$g \approx 1 + \varepsilon(u),$$

so that we are near the linear case; then Eq. 9.32 has a solution

$$f \approx 1 - \varepsilon(a).$$

This is just the linear wave in Eq. 9.21 having plane polarization with $h_{xy} = 0$, i.e. the polarization shown in Figure 9.1(b). Moreover, it is easy to see that the geodesic equation implies, in the nonlinear case, that a particle initially at rest in our coordinates remains at rest. We have, therefore, a simple nonlinear solution corresponding to our approximate linear solution.

This solution is one of a class called plane-fronted waves with parallel rays. See Ehlers & Kundt (1962), § 2.5, and Stephani (2004), § 24.5.

Geometrical optics: waves in a curved spacetime

In this chapter we have made the simplifying assumption that our gravitational waves are the only gravitational field and that they are perturbations of flat spacetime, starting from Eq. 8.12. We found that they behave like a wave field moving at the speed of light in special relativity. But the real Universe contains other gravitational fields, and gravitational waves have to make their way to our detectors through the fields of stars, galaxies, and the Universe as a whole. How do they move?

The answer comes from studying waves as perturbations of a curved metric, of the form $g_{\alpha\beta} + h_{\alpha\beta}$, where $\mathbf{g}$ could be the metric created by any combination of sources of gravity. The computation of the dynamical equation governing $\mathbf{h}$ is very similar to the one we went through at the beginning of this chapter, but the mathematics of curved spacetime must be used. We will not go into the details here, but it is important to understand qualitatively that the results are very similar to the results one would get for electromagnetic waves traveling through complicated media.

If the waves have short wavelength then they basically follow a null geodesic, and they parallel-transport their polarization tensor. This is exactly the same as for electromagnetic waves, so that photons and gravitational waves leaving the same source at the same time will continue to travel together through the Universe, provided they move through a vacuum. For this *geometrical optics* approximation to hold, the wavelength has to be short in two ways. It must be short compared with the typical curvature scale, so that the wave is merely a ripple on a smoothly curved background spacetime; and its period must be short compared to the timescale on which the background gravitational fields might change. If nearby null geodesics converge, then gravitational and electromagnetic waves traveling on them will be focused and become stronger. This is called *gravitational lensing*, and we will see an example of it in Chapter 11.

Gravitational waves do not always keep step with their electromagnetic counterparts. Electromagnetic waves are strongly affected by ordinary matter, so that if their null geodesic passes through matter then they can suffer additional lensing, scattering, or absorption. Gravitational waves are hardly disturbed by matter at all. They follow the null geodesics even through matter. The reason is the weakness of their interaction with matter, as we saw in Eq. 9.24. If the wave amplitudes h_{ij}^{TT} are small then their effect on any matter they pass through is also small, and the back-effect of the matter on them will be of the same order of smallness. Gravitational waves are therefore highly prized carriers of information from distant regions of the Universe: we can in principle use them to 'see' into the centers of supernova explosions, through obscuring dust clouds, or right back to the first fractions of a second after the Big Bang.

9.3 The detection of gravitational waves

General considerations

The great progress that astronomy has made since about 1960 is due largely to the fact that technology has permitted astronomers to begin to observe in many different parts of the electromagnetic spectrum. Because they were restricted to observing visible light, the astronomers of the 1940s could have had no inkling of such diverse and exciting phenomena as quasars, pulsars, black holes in X-ray binaries, giant black holes in galactic centers, gamma-ray bursts, and the cosmic microwave background radiation. As technology has progressed, each new wavelength region has revealed unexpected and important information. Most regions of the electromagnetic spectrum have now been explored at some level of sensitivity, but there is a completely different spectrum which, at the time of writing, has only just begun to be explored: the gravitational wave spectrum.

As we shall see in Chapter 12, nearly all astrophysical phenomena emit gravitational waves, and the most violent ones (which are of course among the most interesting) give off radiation in copious amounts. In some situations, gravitational radiation carries information that no electromagnetic radiation can give us. For example, gravitational waves come to us direct from the heart of supernova explosions; the electromagnetic radiation from the same region is scattered countless times by the dense material surrounding the explosion, taking days to eventually make its way out, and in the process losing most of the detailed information it might carry about the explosion. As another example, detectable gravitational waves from the Big Bang originated when the Universe was perhaps only 10^{-25} s old; they are our earliest messengers from the beginning of our Universe, and they should carry the imprint of unknown physics in the early Universe at energies far higher than anything we can hope to reach in accelerators on the Earth.

Beyond what we can predict, we can be virtually certain that the gravitational wave spectrum has surprises for us, clues to phenomena we never suspected. Astronomers know that only 4% of the mass–energy of the Universe is made up of ordinary matter, which can emit or receive electromagnetic waves; the remaining 96% cannot radiate electromagnetically but it nevertheless couples to gravity, and some of it could turn out to radiate gravitational waves.

The technical difficulties involved in the detection of gravitational radiation are enormous, because the amplitudes of the metric perturbations $h_{\mu\nu}$ that can be expected from distant sources are so small (see §§ 9.5 and 12.2 below). The successful detections have spurred further plans for more sensitive detectors and detectors in other parts of the gravitational wave spectrum. We outline the current and likely future developments here, but to get updated you should consult the scientific literature referred to in the bibliographies of this chapter and of Chapter 12.

Ground-based detectors aim to detect waves with frequencies higher than 1 Hz. Below this there are disturbances in the Newtonian gravitational field of Earth (such as from moving vehicles, moving air masses, seismic waves, ...) that could excite detectors more than the expected astronomical gravitational waves. There have been two basic designs: bars and interferometers.

- *Resonant mass detectors.* Also known as 'bar detectors', these are solid masses that respond to incident gravitational waves by going into vibration. The first purpose-built detectors were of this kind (Weber 1961), and in the 1980s there were up to six detectors operating as a network. We shall study them below, because they can teach us a lot about how gravitational waves interact with matter, and we shall use them later to help us derive the energy carried by a gravitational wave. But bar detectors in the end were unable to reach sufficient sensitivity to make detections, and they have now been phased out.
- *Laser interferometers.* This is the beam detector design of the LIGO instruments that made the first direct detection in 2015. Interferometers use highly stable laser light to monitor the proper distances between masses that are free to move in response to a gravitational wave; when a wave comes past, these proper distances will change, as we saw earlier. The principle is illustrated by the distance monitor described in Exercise 9.26. Very-large-scale interferometers (Bond *et al.* 2016) are now being operated by two laboratories: LIGO in the USA (two 4-km detectors, one in the state of Washington, the other in Louisiana) and Virgo in Italy (3 km from Pisa). Further such detectors are planned on the ground and in space, as we will discuss when we examine below how interferometers work.

To observe waves at frequencies below 1 Hz one must use methods that are not sensitive to Earth gravity disturbances. A space-based interferometer called LISA is under development by the European Space Agency for launch in the 2030s, and we discuss it below. Other ways of detecting gravitational waves are also being pursued.

- *Spacecraft tracking.* This principle has been used to search for gravitational waves using the communication data between Earth and interplanetary spacecraft (Armstrong 2006). This uses radio waves to monitor the distance between a spacecraft and Earth, the second method of measuring tidal distortion that we mentioned above. The sensitivity of these searches is not very high, however, because they are limited by the stability of the atomic clocks that are used for the timing and by delays caused by the plasma in the solar wind. Return-travel times to spacecraft are typically a few hours, so this technique is most sensitive to waves with frequencies below 0.1 mHz.
- *Pulsar timing.* Radio astronomers search for small irregularities in the times of arrival of signals from pulsars. Pulsars are spinning neutron stars that emit strong directed beams of radio waves, apparently because they have ultra-strong magnetic fields that can be strongly misaligned to the axis of rotation. Each time a magnetic pole happens to point toward Earth, the beamed emission is observed as a pulse of radio waves. As we shall see in the next chapter, neutron stars can rotate very rapidly, even hundreds of times per second. Because the pulses are tied to the rotation rate, many pulsars are intrinsically very good clocks, and may accordingly be used for gravitational wave detection.

    Special-purpose pulsar timing arrays are currently searching for gravitational waves, by looking for correlated timing irregularities that could be caused by gravitational waves passing the radio array (Manchester 2015). When the planned Square Kilometer

Array (SKA) radio telescope facility is built (perhaps by 2025), radio astronomers will have a superb tool for monitoring hundreds of very stable pulsars and digging deep for gravitational wave signals. Because pulses come from trapped plasma in the polar regions of neutron stars, individual pulses can be quite irregular but, when averaged over months or more, they are very good clocks, potentially better than manmade ones (Cordes *et al.* 2004). Such searches are accordingly sensitive to gravitational waves below frequencies of 100 nHz. Pulsar astronomers have organized themselves into the International Pulsar Timing Array to pursue the goal of making a first detection in this frequency band (Verbiest *et al.* 2016).

- *The polarization of cosmic microwave background radiation anisotropies.* Cosmologists study the fluctuations in the cosmic microwave background temperature distribution on the sky for telltale signatures of gravitational waves left over from the Big Bang. Waves present at the time the radiation was emitted can affect its polarization, which is now being measured by several purpose-built radio instruments. The waves will still be present today as a random wave field called the cosmological gravitational wave background. Its spectrum is likely to be rather flat across all gravitational wavelengths, and the wave amplitudes today are likely to be very small.

  Detecting these cosmological waves either in the cosmic background or directly would tell us much about the physics of the early Universe. A positive measurement would be a momentous discovery. We will discuss the physics of how gravitational waves polarize the cosmic microwave background, and the strategies for detecting this important background, in Chapter 13.

Interferometers, spacecraft tracking, and pulsar timing all share a common principle: they monitor electromagnetic radiation to look for the effects of gravitational waves on the arrival times of the beams. These are all examples of the general class of *beam detectors*.

It is important to bear in mind that these different approaches are suitable for different parts of the gravitational wave spectrum. Just as for the electromagnetic spectrum, gravitational waves at different frequencies carry different kinds of information. While ground-based detectors (bars and interferometers) typically observe between 10 Hz and a few kHz (although some prototypes have been built for MHz frequencies), space-based interferometers will explore the mHz part of the spectrum. Pulsar timing is suitable only at nHz frequencies, because one has to average over short-time fluctuations in the arrival times of pulses. Furthermore, the cosmic microwave background was affected by gravitational waves that had extremely long wavelengths at the time when the Universe was only a few hundred thousand years old (Chapter 13), so the frequencies of those waves today are of order 10^{-16} Hz!

A resonant detector

Resonant bar detectors are a good case study for the interaction of gravitational waves with continuous matter. To understand how they work in principle, we consider the following idealized detector, depicted in Figure 9.2. Two point particles, each of mass m, are connected by a massless spring with spring constant k, damping constant ν, and

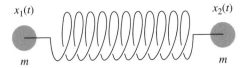

$x_1(t)$ $x_2(t)$

m m

Figure 9.2 A spring with two identical masses as a bar detector of gravitational waves.

unstretched length l_0. The system lies on the x axis of our TT coordinate system, with the masses at coordinate positions x_1 and x_2. In flat space time, the masses would obey the equations

$$mx_{1,00} = -k(x_1 - x_2 + l_0) - v(x_1 - x_2)_{,0} \qquad (9.33)$$

and

$$mx_{2,00} = -k(x_2 - x_1 - l_0) - v(x_2 - x_1)_{,0}. \qquad (9.34)$$

If we define the stretch ξ, resonant frequency ω_0, and damping rate γ by

$$\xi = x_2 - x_1 - l_0, \qquad \omega_0^2 = 2\,k/m, \qquad \gamma = v/m, \qquad (9.35)$$

then we can combine Eqs. 9.33 and 9.34 to give

$$\xi_{,00} + 2\gamma\xi_{,0} + \omega_0^2\xi = 0, \qquad (9.36)$$

the usual damped-harmonic-oscillator equation.

What is the situation as a gravitational wave passes? We shall analyze the problem in three steps.

(1) A *free* particle remains at rest in the TT coordinates. This means that a local inertial frame at rest at, say, x_1, before the wave arrives remains at rest there after the wave hits. Let its coordinates be $\{x^{\alpha'}\}$. Suppose that the only motions in the system are those produced by the wave, i.e. that $\xi = 0(l_0|h_{\mu\nu}|) \ll l_0$. Then the masses' velocities will be small as well, and Newton's equations for the masses will apply in the local inertial frame:

$$mx^{j'}_{,0'0'} = F^{j'}, \qquad (9.37)$$

where $\{F^{j'}\}$ are the components of any nongravitational forces on the masses. Because $\{x^{\alpha'}\}$ can differ from our TT coordinates $\{x^{\alpha}\}$ only by terms of order $h_{\mu\nu}$, and because x_1, $x_{1,0}$, and $x_{1,00}$ are all of order $h_{\mu\nu}$, we can use the TT coordinates in Eq. 9.37 with negligible error:

$$mx^{j}_{,00} = F^{j} + O(|h_{\mu\nu}|^2). \qquad (9.38)$$

(2) The only nongravitational force on each mass is that due to the spring. Since all the motions are slow, the spring will exert a force proportional to its instantaneous *proper* extension, as measured using the metric. If the proper length of the spring is l, and if the gravitational wave travels in the z direction, then

$$l(t) = \int_{x_1(t)}^{x_2(t)} [1 + h^{TT}_{xx}(t)]^{1/2} \, dx = [x_2(t) - x_1(t)][1 + \tfrac{1}{2}h^{TT}_{xx}(t)] + O(|h_{\mu\nu}|^2), \qquad (9.39)$$

and Eq. 9.38 for our system gives

$$mx_{1,00} = -k(l_0 - l) - v(l_0 - l)_{,0,} \tag{9.40}$$

$$mx_{2,00} = -k(l - l_0) - v(l - l_0)_{,0,} \tag{9.41}$$

(3) Let us write the physical stretch ξ as

$$\xi = l - l_0. \tag{9.42}$$

We substitute Eq. 9.39 into this:

$$\xi = x_2 - x_1 - l_0 + \tfrac{1}{2}(x_2 - x_1)h_{xx}^{TT} + O(|h_{\mu\nu}|^2). \tag{9.43}$$

Noting that the factor $(x_2 - x_1)$ multiplying h_{xx}^{TT} can be replaced by l_0 without changing the equation, to the required order of accuracy, we can solve this to give

$$x_2 - x_1 = l_0 + \xi - \tfrac{1}{2}h_{xx}^{TT}l_0 + O(|h_{\mu\nu}|^2). \tag{9.44}$$

If we use this in the difference between Eqs. 9.41 and 9.40, we obtain

$$\xi_{,00} + 2\gamma\xi_{,0} + \omega_0^2\xi = \tfrac{1}{2}l_0 h_{xx,00}^{TT,} \tag{9.45}$$

correct to first order in h_{xx}^{TT}. This is the fundamental equation governing the response of the detector to the gravitational wave. It has the simple form of a forced damped harmonic oscillator. The forcing term is the tidal acceleration produced by the gravitational wave, as given in Eq. 9.28a, although our derivation started with the proper length computation in Eq. 9.24. This shows again the self-consistency of the two approaches to understanding the action of a gravitational wave on matter. An alternative derivation of this result using the equation of geodesic deviation may be found in Exercise 9.22. The generalization to waves incident from other directions is dealt with in Exercise 9.21.

One might use a detector of this sort as a resonant detector for sources of gravitational radiation of a fixed frequency (e.g. pulsars or close binary stars). (It can also be used to detect bursts – short wave packets of broad-spectrum radiation – but we will not discuss this here.) Suppose that the incident wave has the form

$$h_{xx}^{TT} = A \cos \Omega t. \tag{9.46}$$

Then the steady solution of Eq. 9.45 for ξ is

$$\xi = R\cos(\Omega t + \phi), \tag{9.47}$$

with

$$R = \tfrac{1}{2}l_0\Omega^2 A/[(\omega_0^2 - \Omega^2)^2 + 4\Omega^2\gamma^2]^{1/2}, \tag{9.48}$$

$$\tan \phi = 2\gamma\Omega/(\omega_0^2 - \Omega^2). \tag{9.49}$$

(Of course, the general initial-value solution for ξ will also contain transients, which damp away on a timescale $1/\gamma$.) The energy of oscillation of the detector is, to lowest order in h_{xx}^{TT},

$$E = \tfrac{1}{2}m(x_{1,0})^2 + \tfrac{1}{2}m(x_{2,0})^2 + \tfrac{1}{2}k\xi^2. \tag{9.50}$$

For a detector which was at rest before the wave arrived, we have $x_{1,0} = -x_{2,0} = -\xi_{,0}/2$ (see Exercise 9.23), so that

$$E = \tfrac{1}{4}m[(\xi_{,0})^2 + \omega_0^2\xi^2] \tag{9.51}$$

$$= \tfrac{1}{4}mR^2[\Omega^2\sin^2(\Omega t + \phi) + \omega_0^2\cos^2(\Omega t + \phi)]. \tag{9.52}$$

The mean value of this is its average over one period, $2\pi/\Omega$:

$$\langle E \rangle = \tfrac{1}{8}mR^2(\omega_0^2 + \Omega^2). \tag{9.53}$$

We shall always use angle brackets $\langle \; \rangle$ to denote time averages.

If we wish to detect a specific source whose frequency Ω is known, then we should adjust ω_0 to equal Ω for maximum response (resonance), as we see from Eq. 9.48. In this case the amplitude of the response will be

$$R_{\text{resonant}} = \tfrac{1}{4}l_0 A(\Omega/\gamma) \tag{9.54}$$

and the energy of vibration is

$$E_{\text{resonant}} = \tfrac{1}{64}ml_0^2\Omega^2 A^2(\Omega/\gamma)^2. \tag{9.55}$$

The ratio Ω/γ is related to the quality factor Q of an oscillator, where $1/Q$ is defined as the average fraction of the energy of the undriven oscillator lost to friction in one radian of oscillation (see Exercise 9.25):

$$Q = \omega_0/2\gamma. \tag{9.56}$$

In the resonant case we have

$$E_{\text{resonant}} = \tfrac{1}{16}ml_0^2\Omega^2 A^2 Q^2. \tag{9.57}$$

What numbers are realistic for laboratory detectors? Most such detectors were massive cylindrical bars, in which the 'spring' was the elasticity of the bar when stretched along its axis. When waves hit the bar broadside, they would excite its longitudinal modes of vibration. The first detectors, built by Joseph Weber of the University of Maryland in the 1960s, were aluminum bars of mass 1.4×10^3 kg, length $l_0 = 1.5$ m, resonant frequency $\omega_0 = 10^4\,\text{s}^{-1}$, and Q about 10^5. This means that a strong resonant gravitational wave of $A = 10^{-20}$ (see § 9.4 below) will excite the bar to an energy of the order of 10^{-20} J. The resonant amplitude given by Eq. 9.54 is only about 10^{-15} m, roughly the diameter of an atomic nucleus! Many realistic gravitational waves will have amplitudes many orders of magnitude smaller than this, and will last for much too short a time to bring the bar to its full resonant amplitude.

Notice that Eq. 9.57 implies that gravitational waves can put energy into material systems that they encounter. This suggests that gravitational waves themselves carry energy. In § 9.5 below, we will start with this calculation in order to derive the expression for the energy flux of a gravitational wave.

Limitations of bar detectors

The measurement of the response to a gravitational wave has to compete with other disturbances that could masquerade as signals. Here, we shall consider how this affected bar detectors, because interferometers have to deal with many of the same problems.

When trying to measure such tiny effects, there are in general two ways to improve things: one way is to increase the size of the effect and the other is to reduce any extraneous disturbances that might obscure the measurement. And then one has to determine how best to make the measurement. The size of the effect is controlled by the amplitude of the wave, the length of the bar, and the Q-value of the material. We cannot control the wave's amplitude, and unfortunately extending the length is not an option: realistic bars may be as long as 3 m, but longer bars would be much harder to isolate from external disturbances. In order to achieve high values of Q some bars have actually been made of single sapphire crystals, but it is hard to do better than that. Novel designs, such as spherical detectors that respond efficiently to waves from any direction, can increase the signal somewhat, but the difficulty of making bars intrinsically more sensitive is the main reason why they have been retired. Beam detectors, on the other hand, can be made (in principle) as long as we want, so they can amplify the response by many orders of magnitude over bars.

The other issue in gravitational wave detection, whether by bars or beams, is to reduce the instrumental and environmental noise. For example, thermal noise in any oscillator induces random vibrations with a mean energy of $k_B T$, where T is the absolute temperature and k_B is Boltzmann's constant,

$$k_B = 1.38 \times 10^{-23} \text{ J/K}.$$

In our example, this will be comparable with the energy of excitation of a bar detector if T is room temperature (~ 300 K). But we chose a very optimistic wave amplitude. To detect reliably a wave with an amplitude ten times smaller would require a temperature 100 times smaller. For this reason, bar detectors in the 1980s began to change from room-temperature operation to cryogenic operation at around 3 K. The coldest operating temperature was reached by the Auriga bar, which went below 100 mK. (See the bibliography at the end of the chapter.) Interferometers today are similarly beginning to use cryogenically cooled mirrors in order to reduce thermal vibrations, as we shall discuss below.

Other sources of noise, such as vibrations from passing vehicles and everyday seismic disturbances, could be considerably larger than thermal noise, so a bar had to be very carefully isolated. This was done by hanging the bar from a support so that it formed a pendulum with a low resonant frequency, say 1 Hz. Vibrations from the ground may move the top attachment point of the pendulum, but little of this is transmitted through to the bar at frequencies above the pendulum frequency: pendulums are good low-pass mechanical filters. In practice, several sequential pendulums may be used, and the hanging frame is further isolated from vibration by using absorbing mounts. This technology is now being used in interferometers, to isolate the mirrors from ground vibrations. The isolation has been further improved with active vibration cancellation at the mounts: external seismometers measure vibrations and they feed back to activators that move the supports in the other way. The principle is much like that of noise-cancelling audio headphones.

How did resonant detectors attempt to measure the tiny stretching created by a gravitational wave? Such a measuring apparatus is called a transducer: it converts the mechanical disturbance into an electrical output. Weber's original aluminum bar was instrumented with strain detectors at its waist, where the stretching of the metal is maximum. Other groups tried to extract the energy of vibration from the bar into a transducer of very small mass that was resonant at the same frequency; if the energy extraction was efficient then the transducer's amplitude of oscillation would be much larger. The most sensitive readout schemes involved ultra-low-noise low-temperature superconducting devices called SQUIDs.

We have confined our discussion to on-resonance detection of a continuous wave, in the case when there are no motions in the detector. If the wave comes in as a burst with a wide range of frequencies, where the excitation amplitude might be smaller than the broadband noise level, then one has to do a more careful analysis of the detector's sensitivity, but the general picture does not change. One difficulty bars encountered with broadband signals is that it was difficult for them in practice (although not impossible in principle) to measure the frequency components of a waveform very far from their resonant frequencies, which normally lie above 600 Hz. Since, as we now know, most strong gravitational waves are at lower frequencies, this was a serious problem, with which, however, broadband beam detectors have less difficulty.

A second difficulty was that, to reach a sensitivity to bursts of amplitude around 10^{-21} (which is the level that interferometers reached in 2005), bars would have had to conquer the so-called quantum limit. At these small excitations, the energy put into the vibrations of the bar by the wave is below one quantum (one phonon) of excitation of the resonant mode being used to detect them. The theory of how to detect below the quantum limit – of how to manipulate the Heisenberg uncertainty relation in a macroscopic object like a bar – is fascinating. For more details, see Misner *et al.* (1973), Smarr (1979), or Blair (1991). Bars never reached the quantum limit. By contrast, interferometric detectors have succeeded in reducing their noise level below their version of the standard quantum limit, using squeezed light; see Abadie *et al.* (2011).

Measuring distances with light

One of the most convenient ways of measuring the range to a distant object is by radar: send out a pulse of electromagnetic radiation, measure how long it takes to return after reflecting from the distant object, divide that by two and multiply by c, and that is the distance. Remarkably, because light occupies such a privileged position in the theory of relativity, this method is also an excellent way of measuring proper distances, even in curved spacetime. It is the foundation of laser interferometry for gravitational wave detection.

We shall compute how to use light to measure the distance between two freely falling objects. Because the objects are freely falling, and because we make no assumption that they are close to one another, we shall use the TT coordinate system. Let us first consider, for simplicity, a wave traveling in the z direction with pure + polarization, so that the metric is given by

$$ds^2 = -dt^2 + [1 + h_+(z - t)]dx^2 + [1 - h_+(t - z)]dy^2 + dz^2. \qquad (9.58)$$

Suppose, again for simplicity, that the two objects lie on the x axis, one at the origin $x = 0$ and the other at coordinate location $x = L$. In TT coordinates, they remain at these coordinate locations all the time. To make our measurement, the object at the origin sends a photon along the x axis toward the other object, which reflects it back. The first object measures the amount of proper time that has elapsed since first emitting the photon. How is this related to the distance between the objects and to the metric of the gravitational wave?

Note that a photon traveling along the x axis moves along a null world line $(ds^2 = 0)$ with $dy = dz = 0$. That means that it has an effective speed

$$\left(\frac{dx}{dt}\right)^2 = \frac{1}{1 + h_+}. \qquad (9.59)$$

Although this is not equal to 1, it is just a *coordinate speed*, so it does not contradict relativity. A photon emitted at time t_{start} from the origin reaches any coordinate location x in a time $t(x)$; this is essentially what we are trying to solve for. The photon reaches the other end, at the fixed coordinate position $x = L$, at the coordinate time given by integrating the effective speed of light from Eq. 9.59:

$$t_{\text{far}} = t_{\text{start}} + \int_0^L [1 + h_+(t(x))]^{1/2} \, dx. \qquad (9.60)$$

This is an implicit equation since the function we want to find, $t(x)$, is inside the integral. However, in linearized theory we can solve this by using the fact that h_+ is small. Where $t(x)$ appears in the argument of h_+ we can use its flat-spacetime value, since corrections due to h_+ will only bring in terms of order h_+^2 in Eq. 9.60. So, we set $t(x) = t_{\text{start}} + x$ inside the integral and expand the square root. The result is the explicit integration

$$t_{\text{far}} = t_{\text{start}} + L + \frac{1}{2}\int_0^L h_+(t_{\text{start}} + x)dx.$$

The light is reflected back, and a similar argument gives the total time for the return trip:

$$t_{\text{return}} = t_{\text{start}} + 2L + \frac{1}{2}\int_0^L h_+(t_{\text{start}} + x)dx + \frac{1}{2}\int_0^L h_+(t_{\text{start}} + L + x)dx. \qquad (9.61)$$

Note that the coordinate time t in the TT coordinates is the proper time, so that this equation gives a value that can be measured.

What does Eq. 9.61 tell us? First, suppose that L is actually rather small compared with the gravitational wavelength, or equivalently that the return time is small compared with the period of the wave, so that h_+ is effectively constant during the flight of the photon. Then the return time is just proportional to the proper distance to L as measured by this metric. This should not be surprising: for a small separation one could set up a local inertial frame in free fall with the particles, and in this frame all experiments should come out as they do in special relativity. In SR one knows that radar ranging gives the correct proper distance, so it must do so here as well.

More generally, how do we use this equation to measure the metric of the wave? The simplest way to use it is to differentiate t_{return} with respect to t_{start}, i.e. to monitor the rate of change of the return time as the wave passes. Since the only way in which t_{start} enters the integrals in Eq. 9.61 is as the argument $t_{\text{start}} + x$ of h_+, a derivative of the integral acts only as a derivative of h_+ with respect to its argument. Then the integration with respect to x is an integral of the derivative of h_+ over its argument, which simply produces h_+ again. The result of this is the vastly simpler expression

$$\frac{dt_{\text{return}}}{dt_{\text{start}}} = 1 + \tfrac{1}{2}\left[h_+(t_{\text{start}} + 2L) - h_+(t_{\text{start}})\right]. \tag{9.62}$$

This is rather a remarkable result: the rate of change of the return time depends only on the metric of the wave at the times when the wave was emitted and when it was received back at the origin. In particular, the wave amplitude when the photon was reflected off the distant object plays no role.

Now, if the signal sent out from the origin is not a single photon but a continuous electromagnetic wave with some frequency ν, then each 'crest' of the wave can be thought of as another null ray or another photon being sent out and reflected back. The derivative of the time it takes these rays to return is nothing more than the change in frequency of the electromagnetic wave:

$$\frac{dt_{\text{return}}}{dt_{\text{start}}} = \frac{\nu_{\text{return}}}{\nu_{\text{start}}}.$$

So, if we monitor changes in the redshift of the returning wave, we can relate that directly to changes in the amplitude of the gravitational waves.

So far we have used a rather special arrangement of objects and wave: the wave was traveling in a direction perpendicular to the separation of the objects. If instead the wave is traveling at an angle θ to the z axis in the x–z plane, the return time derivative does involve the wave amplitude at the reflection time:

$$\frac{dt_{\text{return}}}{dt_{\text{start}}} = 1 + \tfrac{1}{2}\{(1 - \sin\theta)h_+(t_{\text{start}} + 2L) - (1 + \sin\theta)h_+(t_{\text{start}})$$

$$+ 2\sin\theta\, h_+[t_{\text{start}} + (1 - \sin\theta)L]\}. \tag{9.63}$$

This three-term relation is the starting point for analyzing the response of beam detectors, as we shall see next. For its derivation see Exercise 9.27.

Beam detectors

The simplest beam detector uses spacecraft tracking (Armstrong 2006). Interplanetary spacecraft carry transponders, which are radio receivers that amplify and return the signals they receive from the ground tracking station. A measurement of the return time of the signal tells the space agency how far away the spacecraft is. If the measurement is accurate enough, small changes in the return time might be caused by gravitational waves. In practice this is a difficult measurement to make, because fluctuations might also be caused by changes in the index of refraction of the thin interplanetary plasma or of the ionosphere, through both of which the signals must pass. These effects can be discriminated from a true gravitational wave by using the three-term relation, Eq. 9.63. The detected waveform has to appear in three different places in the data, once with the opposite sign. Random fluctuations are unlikely to do this.

Plasma fluctuations can also be suppressed by using multiple transponding frequencies (allowing them to be measured) or by using higher-frequency transponding, even employing infrared laser light, which is hardly affected by plasma at all. But even then, there will be another limit on the accuracy: the stability of the clock at the tracing station that measures the elapsed time t_{return}. Even the best clocks are, at present, not stable at the 10^{-19} level. It follows that a beam detector of this type could not expect to measure gravitational waves with amplitudes of 10^{-20} or below. Unfortunately, as we shall see below, this is where almost all amplitudes lie.

The remedy is to use an interferometer. In an interferometer, light from a stable laser passes through a beam splitter, which sends half the light down one arm and the other half down a perpendicular arm. The two beams of light have correlated phases. When they return after reflecting off mirrors at the ends of the arms, they are brought back into interference. (See Figure 9.3.) The interference measures the difference in the arm-lengths of the two arms. If this difference changes, say because a gravitational wave passes through, then the interference pattern changes and the wave can in principle be detected.

Now, an interferometer can usefully be thought of as two beam detectors laid perpendicular to one another. The two beams are correlated in phase: any given 'crest' of the light wave starts out down both arms at the same time. If the arms have the same proper length, then their beams will return in phase, interfering constructively. If the arms differ, say, by half the wavelength of light, then they will return out of phase and they will destructively interfere. An interferometer solves the problem that a beam detector needs a stable clock. The 'clock' in this case is one of the arms. The travel time of the light in one arm essentially provides a reference for the return time of the light in the other arm. The technology of lasers, mirrors, and other systems is such that this reference can be used to measure changes in the return travel time of the other arm much more stably than the best atomic clocks would be able to do.

Interferometers are well suited to registering gravitational waves: to see how this works, look at the polarization diagram in Figure 9.1. Imagine putting an interferometer in the circle in panel (a), with the beamsplitter at the center and the ends of the arms where the circle intersects the x and y directions. Then, when a wave with + polarization arrives, as in panel (b), it will stretch one arm and at the same time compress the other. This doubles the effective arm-length difference that the interference pattern measures. But even if the wave arrives from, say, the x direction, then its transverse action will still compress and expand the y-arm, giving a signal half as large as the maximum. However, a wave with $\times$ polarization, as in panel (c), arriving perpendicular to both arms, will not stretch the interferometer arms at all, and so will not be detected. The interferometer is therefore a linearly polarized detector that responds to signals arriving from almost all directions.

Because interferometers bring light beams together and interfere them, it is often mistakenly thought that they measure the changes in their arm-lengths against a standard wavelength, and this in turn sometimes leads students to ask whether the wavelength of light is affected by the gravitational wave, thereby invalidating the measurement. Our discussion of beam detectors should make clear that the wavelength of the light is not the relevant quantity: interferometers basically compare two *return times*, and as long as light travels up and down the arms on null world lines, it does not matter at all what is happening to the wavelength along the way.[2] For a good and accessible discussion of this point, see Saulson (1997).

Interferometric detectors

The LIGO interferometers made the first direct detection of gravitational waves on 14 September 2015, observing the merger of two unexpectedly massive black holes. They made a number of further detections over the following two years. Virgo was added in August 2017, joining in two detections, including the landmark first detection of a binary neutron star merger on 17 August of that year. We discuss this ensemble of detections in detail in Chapter 12 below. Here we discuss how the extraordinary sensitivity of these detectors was achieved.

Interferometers have two major advantages over bar detectors: by increasing their arm-lengths, their sensitivity can be increased considerably before one runs into fundamental problems of materials or physics; and (because they do not depend on any resonant vibration) they operate over a broad spectrum of frequencies. The largest earth-bound detectors are the two 4-km LIGO detectors in the USA, closely followed by the 3-km Virgo detector in Italy. Physical length is only the first step in increasing the size of the response to a gravitational wave. The light beams inside the arms could be 'folded' over multiple passes up and down, so that the residence time of light in each arm is longer; this further increases the difference in return times when there is a gravitational wave. These large ground-based interferometers do something that is equivalent, which is to make the arms into light cavities that hold photons for many times the light travel time along the arm. They are known as Fabry–Pérot interferometers. A simplified sketch of the optical design of a Fabry–Pérot-type detector is shown in Figure 9.3.

However, even with long arms, the signal can be masked by a variety of instrumental sources of noise (Bond *et al.* 2016, Saulson 2017). To filter out vibrations from ground disturbances, interferometers use the same strategy as bar detectors: the optical components are suspended, as discussed earlier. Thermal vibrations of mirrors and their suspensions are, as for bars, a serious problem, but cooling the mirrors is challenging because of heat absorption from the laser beam. Because of this, LIGO and Virgo operate at room temperature, but the KAGRA detector in Japan (just beginning operation as this is being written) will ultimately cool its mirrors to 20 K (Akutsu *et al.* 2019). KAGRA is ambitious in its vibration-isolation as well, because it resides in underground tunnels. Seismic disturbances are strongest at the ground surface, so underground tunnels provide a quieter environment.

---

[2] In fact, the wavelength and frequency of light depend in any case on the observer, so the question cannot be posed in a frame-invariant way. This is another reason not to introduce them into discussions of how interferometers measure gravitational waves!

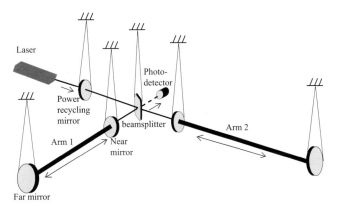

Figure 9.3 The configuration of a Fabry–Pérot interferometer (optical cavities for arms) such as LIGO or Virgo. Light of a single frequency is generated by the laser and passes through a power recycling mirror (see below) to the beamsplitter, where it is divided between the two arms. The arms form cavities that hold each photon for typically much longer than one round-trip time, because the near mirrors are almost fully reflecting. This power buildup increases the sensitivity. The light exiting the cavities returns to the beamsplitter. A beam with destructive interference (cancellation of amplitudes) goes toward the photodetector; it should be dark unless a gravitational wave is present. The other return beam from the beamsplitter is the constructive interference beam returning toward the laser. Almost all the light goes here, and in order not to waste it, the power recycling mirror returns it to the interferometer in phase with the new incoming laser light. All the mirrors, including the beamsplitter, are suspended (as illustrated), in order to filter out mechanical vibration noise. Other refinements of the optical design, such as mode cleaners and signal recycling mirrors, are not illustrated. The diagram is not to scale: the arms are 3 to 4 km long, while the central area (the near mirrors, beamsplitter, laser, photodetector) is contained in a single building. Adapted from www.ligo.caltech.edu/page/what-is-interferometer

In all the interferometers, the disturbance by thermal noise is reduced by using ultra-high-Q materials for the mirrors and suspension wires; this confines the thermal noise to narrow bands around the resonant frequencies of the mirror vibrations and pendulum modes, and these are designed to be well outside the observation band of the instruments. This approach, which includes using optical fiber as the suspension wire and bonding it to mirrors without glue, was pioneered in the GEO600 detector in Germany. The values of Q achieved so far by this technique (in current LIGO suspensions) are extraordinary: 2×10^9 (Cumming *et al.* 2020). By pioneering this technology and other sophisticated optical techniques, GEO600 had sufficient sensitivity that, despite its smaller arm-length (600 m), it could participate in the initial gravitational wave searches, along with LIGO and Virgo, starting in 2005. After the big detectors were upgraded with the same technologies, operating as Advanced LIGO (from 2015) and Advanced Virgo (from 2017), GEO600 was no longer competitive as an observatory, but it continues to test new technologies.

The final source of noise that we discuss here is what physicists call shot noise, which is the random fluctuations of intensity in the interference of the two beams; it comes from the fact that the beams are composed of discrete photons and not continuous classical electro-magnetic radiation. Shot noise is therefore a quantum noise, and it is the major limitation on sensitivity in interferometers at frequencies above about 200 Hz. It can be reduced, and hence the sensitivity increased, by increasing the amount of light stored in the arm cavities, because with more photons the fractional amplitude of the power fluctuations goes down.

The envelope of the different noise sources provides an observing band for the advanced ground-based detectors that runs from about 20 Hz up to around 2 kHz. At low frequencies, the dominant noise is ground vibration. In the middle of this range, the sensitivity limit is set by thermal noise from suspensions and mirror vibrations. At higher frequencies the limit comes from shot noise.

The detection of gravitational waves involves more than building and operating sensitive detectors. It also requires appropriate data analysis, because the expected signals will typically be below the broadband noise and must be extracted by intelligent computer-based filtering of the data. This means that different kinds of signals have different smallest-amplitude detection limits. A short burst of gravitational waves at, say, 200 Hz with an amplitude of 10^{-22} would be around the noise level of the current LIGO detectors, so we often say that LIGO is a 10^{-22} detector. However, a long wavetrain, such as from a binary neutron star merger (see below), contains much more information and can be detected with a good signal-to-noise ratio even when its amplitude is somewhat below 10^{-22}. These topics will be examined in more detail in Chapter 12.

Networks of interferometers

Because the detectors are so complex, there is always the possibility that random internal disturbances will masquerade as gravitational wave signals, so in practice signals need to be seen in more than one detector at the same time in order for the scientists to have confidence. For this reason, the USA built two LIGO detectors, separated by about 3,000 km. But there are good reasons for building more than two.

A single interferometer produces a single output, its response as a function of time. This contains much information about the time-evolution of the source, but it cannot reveal important aspects about the signal, such as where it is coming from or what its polarization is. A network of detectors is required in order to extract a complete description of the wave.

Gravitational wave events that are accompanied by the emission of electromagnetic radiation (as in the case of a merger of two neutron stars, as we will see in Chapter 12) may have their direction accurately determined if telescopes can identify a counterpart.

If no counterpart is identified (this is generally expected to be the case for mergers of black holes), the direction to the source can still be inferred if the event is observed by a network of at least three gravitational wave detectors, using the relative delays in the arrival times of the signals. If two detectors are separated by a distance d, the signal must arrive within the light-travel-time window $\pm d/c$. The measured delay within this window defines a circle on the sky, on which the source must lie. More accurately, the measurement uncertainty broadens this circle into a band on the sky. With another detector, a further time delay defines another band, which will intersect the first in two places, and the source must lie within one of them.

The inclination of the binary's orbital plane to the line of sight can be measured from the polarization. As we will see in Eqs. 9.98 and 9.99, the radiation emitted perpendicular to the orbital plane is circularly polarized, and that emitted in the orbital plane is linearly polarized. In the general case it will be elliptically polarized, and by measuring the degree of elliptical polarization we can measure the inclination. As we discussed earlier, our detectors are linearly polarized, with different sensitivities in different directions. They measure just one component of the polarization. If we know the location of the source on the sky, we require just two (nonaligned) detectors to determine the polarization and hence the inclination of the orbit.

For these reasons, as well as for the confidence that comes from detecting a weak gravitational wave independently in more than one detector, groups around the world work together to build and operate a global network of detectors, and to analyze their pooled data. In addition to the two 4-km LIGO detectors in the USA (see Figure 9.4), the 3-km Virgo detector in Italy, and the 3-km KAGRA detector in Japan, a fifth detector is under construction in India: a 4-km instrument called LIGO-India that will be the third instrument in the LIGO network. It is hoped that this will come into operation around 2025.

Figure 9.4   The LIGO gravitational wave observatory at Hanford, Washington. One of the 4-km arms stretches into the distance, the other leaves the photo off to the left. The laser light in the arms is contained inside 4-km-long stainless-steel vacuum tubes 1.6 m in diameter. For an aerial view of this site go to a geographic display engine, such as Google Earth, and type in the longitude and latitude (46.4553, −119.4060). The other LIGO detector is at Livingston, Louisiana, at (30.5632, −90.7744). The Virgo detector near Pisa, Italy, is at (43.6312, 10.5045), and the smaller GEO600 detector near Hanover, Germany, is at (52.25, 9.81). In Japan, the KAGRA detector is in underground tunnels at (36.4119, 137.3058). The LIGO-India detector, which is just beginning construction as this is being written, is near the town of Aundha in the state of Maharashtra, at (19.54, 77.04). (Photograph courtesy of the LIGO Laboratory.)

Future detector plans

Although LIGO has demonstrated that gravitational wave detection is possible, its detection range covers only a small fraction of the observable Universe. Upgrades are anticipated, but it is expected that the length of the arms will become a limiting factor by the mid- to late-2020s. Plans are therefore being formulated for what are called 'third-generation' detectors, with sensitivities ten times better than Advanced LIGO. These would observe binary mergers involving neutron stars and stellar black holes essentially everywhere they happen. Their arms would be much longer than LIGO's and they would need a number of new technologies that are currently under development. The two main projects that are currently being developed (mid-2021) are the Einstein Telescope in Europe and Cosmic Explorer in the USA. The current expectations of their potential performance and scientific returns are laid out in a white paper (Bailes *et al.* 2020) written by the Gravitational Wave International Committee (GWIC), a coordinating body among the various projects. If approved for construction, these ambitious detectors would probably come into operation sometime after 2035.

Even more ambitious than the ground-based detector projects is the LISA mission of the European Space Agency (ESA), with participation from the US space agency NASA. This is currently planned for launch in the mid-2030s. Going into space is necessary if one wants to observe at frequencies below about 1 Hz. At these low frequencies, the Earth's Newtonian gravitational field is too noisy: detectors register any change in gravity, and even the tiny changes in gravity associated with the density changes of seismic waves and weather systems are larger than the expected amplitudes of gravitational waves over the time periods associated with these low frequencies. So detectors in the mHz band need to operate far from the Earth.

LISA will consist of three spacecraft in an equilateral triangle, all orbiting the Sun at a distance of 1 AU, the same as the Earth, and trailing the Earth by $20°$. Their separations – the lengths of the interferometer arms – are planned to be 2.5×10^6 km, well-matched to detecting gravitational waves in the millihertz frequency range.

The three arms can be combined in various ways to form three different two-armed interferometers, which allows LISA to measure both polarizations of an incoming wave and to sweep the sky with a fairly uniform antenna pattern. Because the opening angle of each of these interferometer combinations is only $60°$, the sensitivity is lower than that of a right-angle interferometer by a factor of $\sin(60°) = 0.866$. But this lower sensitivity can be compensated by making the arms longer, and in any case it is a price worth paying if it allows the measurement of the polarization of the wave.

Each of LISA's three arms has two different laser beams, from one end of the arm to the other end and back. These beams are not reflected, as in ground-based detectors (see Figure 9.3), but, rather, at each end the incoming beam is received and then the outgoing beam is amplified and locked in phase to the received beam, simulating a reflection and compensating for the losses caused by the fact that by the time the beam reaches the other end of the arm, it is much wider than the receiving mirror of the spacecraft. The interferometry is, therefore, not done by recombining a beam that was split at a beamsplitter. Instead, signal extraction is done by a sophisticated combination of measurements of the phases of all the beams at different times, a technique called time-delay interferometry, or TDI (see the review by Tinto & Dhurandhar 2014).

Most of LISA's sources will be observable for months or longer: even merging black holes will take months to move through its frequency band. LISA will measure the positions of sources by the frequency modulation produced by its orbit around the Sun during the time the signal is observed. We discuss this further in Chapter 12.

As with ground-based instruments, LISA must contend with noise. Thermal noise is not an issue because its large arm-length means that the signal that it is measuring – the time difference between arms – is much larger than the signal that would be induced by the vibrations of materials. But external disturbances, caused by the Sun's radiation pressure and the solar wind, are significant, and so the LISA spacecraft must be designed to fall freely to high accuracy despite these applied forces. Each spacecraft is planned to contain two free masses (called proof masses) that are freely falling within the spacecraft, and which are the reference points that define the ends of the two arms that terminate at the spacecraft. The spacecraft acts as a shield, preventing external forces from acting on the proof masses. It senses the positions of the masses and uses very weak jets to adjust its position so it does not disturb them.

This technique is called drag-free operation, and it was demonstrated very successfully by ESA's LISA Pathfinder mission. Dedicated to testing this critical technology, Pathfinder was launched at the end of 2015 to the L1 Lagrange point of the Earth–Moon system, where during the following two years it demonstrated drag-free performance at a level significantly better than that required for the LISA mission (Armano *et al.* 2018).

Communicating with 2-W lasers, and equipped with the LISA Pathfinder isolation system, LISA is expected to achieve a remarkable sensitivity, sufficient to see mergers of very massive black holes – masses in the range from 10^4 to $10^7 \, M_\odot$ will be in its frequency band – anywhere in the Universe.

But there is a gap between the masses detectable from the ground (up to perhaps $100 \, M_\odot$) and those radiating in the LISA band. These are the intermediate-mass black holes, and they are best detected in the decihertz frequency band, between say 0.01 Hz and 10 Hz. At present (mid-2021) there are very few proposals for such detectors in space, the best developed of which is the DECIGO project (Sato *et al.* 2017). But the astrophysics in this band is rich (Sedda *et al.* 2019), so this is likely to be a prime candidate for a mission after LISA.

We will return in Chapter 12 to the kinds of source that might be detected by space- and ground-based detectors. First we consider how gravitational waves are generated in general relativity.

Cosmic sound

We finish this section by drawing what should be a helpful analogy between gravitational waves and sound, i.e. between gravitational wave detectors and microphones. Most of astronomy uses telescopes that are directed into narrow fields of view, capturing photons and recording their energy. Except in radio astronomy and in optical interferometry, the phase of the photons is not measured and is not useful, since they arrive incoherently, having been emitted independently from different parts of the source (a star, say, or a galaxy). Gravitational wave detection is different in every respect.

Gravitational wave detectors record not just the amplitude (a proxy for energy) but also the phase of the waves. This is possible because the waves have frequencies in the acoustic range (kHz) or below, very low compared with most detectable electromagnetic waves. Since the corresponding wavelengths are much larger than the sizes of individual detectors, detectors cannot focus on a single narrow cone of directions. They are essentially omnidirectional, with quadrupolar antenna patterns, which we will discuss in § 12.3. These properties are analogous to sound: microphones record the phase oscillations of incident pressure waves and, because the waves generally have long wavelengths, it is difficult to focus a microphone in any particular direction. If we want to locate the source of a gravitational wave on the sky we need a network where we combine the detector outputs coherently, judging direction from phase delays. This is analogous, for example, to underwater acoustic detection networks, for military or geophysical use.

As we shall see in the next section, gravitational waves are generated by the overall motions and mass distributions in their sources, rather than coming from quantum transitions in individual atoms within the source. The reason is that gravity is a field generated by a 'charge' of only one sign: masses in different locations are all positive, and they reinforce one another in generating a wave. In electromagnetism there are two charges, which in normal matter never get separated by more than something like a nanometer. This ensures that the radiation generated by one atom is not correlated with that from another. But gravity is similar to sound in this respect as well: sound is generated by local pressure changes, all of which combine coherently to create the radiated sound wave.

The propagation of gravitational waves is therefore a kind of cosmic sound, a stress in spacetime that rockets across the Universe at the speed of light. When a detector receives this wave, its one-dimensional time series output can straightforwardly be converted into sound that our ears can hear. Gravitational wave physicists have come to use the language of sound to describe their work: we *listen* to gravitational waves, the signals from binaries (see the next section) *chirp* upwards in frequency, radiating binary systems are *standard sirens* (§ 9.6). In Chapter 12 we shall give links to some websites where one can listen to detected waves. Do follow them and listen to sounds that have come to us from extraordinary events far away across the Universe.

9.4 The generation of gravitational waves

Simple estimates. It is easy to see that the amplitude of any gravitational waves incident on Earth should be small. A 'strong' gravitational wave would have $h_{\mu\nu} = O(1)$, and we should expect amplitudes like this only very near a highly relativistic source, where the Newtonian potential (if it has any meaning in such a situation) would be of order 1. For a source of mass M, this would be at distances of order M from it. As with all radiation fields, the amplitude of the gravitational waves falls off as r^{-1} far from the source. (Readers who are not familiar with solutions of the wave equation will find demonstrations of this in the following sections.) So, if Earth is a distance R from a source of mass M, the largest-amplitude waves we should expect are of order M/R.

Let us consider two examples. For the formation of a $10\,M_\odot$ black hole in a supernova explosion in a nearby galaxy 10^{23} m away, the amplitude is about 10^{-17}. This is of course an upper limit; the waves could have much smaller amplitudes than this. As another example, consider the first system detected by LIGO, a binary consisting of two black holes of mass about $30\,M_\odot$ at a distance of perhaps 10^{25} m. Here our upper amplitude limit would be 10^{-20}. The actual observed peak amplitude was 10^{-21}, so the distortions of geometry near those two black holes as they merged may have been of order 10%.

An approximate calculation of wave generation

Our object is to solve Eq. 8.42,

$$\left(-\frac{\partial^2}{\partial t^2} + \nabla^2\right)\bar{h}_{\mu\nu} = -16\pi T_{\mu\nu}. \tag{9.64}$$

We will find the exact solution in a later section. Here we will make some simplifying – but realistic – assumptions. We assume that the time-dependent part of $T_{\mu\nu}$ is in sinusoidal oscillation with frequency Ω, i.e. that it is the real part of

$$T_{\mu\nu} = S_{\mu\nu}(x^i)\,e^{-i\Omega t}, \tag{9.65}$$

and that the region of space in which $S_{\mu\nu} \neq 0$ is small compared with $2\pi/\Omega$, the wavelength of a gravitational wave of frequency Ω. The first assumption is not much of a restriction, since a general time dependence can be reduced to a sum over sinusoidal motions by Fourier analysis. Besides, many interesting astrophysical sources *are* roughly periodic: pulsating stars, pulsars, binary systems. The second assumption is called the slow-motion assumption, since it implies that the typical velocity inside the source region, which is Ω multiplied by the size of that region, should be much less than 1. All but the most powerful sources of gravitational waves probably satisfy this condition.

Let us look for a solution for $\bar{h}_{\mu\nu}$ of the form

$$\bar{h}_{\mu\nu} = B_{\mu\nu}(x^i)\,e^{-i\Omega t}. \tag{9.66}$$

(In the end we must take the real part of this for our answer.) Putting this and Eq. 9.65 into Eq. 9.64 gives

$$(\nabla^2 + \Omega^2)B_{\mu\nu} = -16\pi S_{\mu\nu}. \tag{9.67}$$

It is important to bear in mind as we proceed that the indices on $\bar{h}_{\mu\nu}$ in Eq. 9.64 play almost no role. We shall regard each component $\bar{h}_{\mu\nu}$ as simply a function on Minkowski space satisfying the wave equation. All our steps would be the same if we were solving the scalar equation $(-\partial^2/\partial t^2 + \nabla^2)f = g$, until we come to Eq. 9.75 below.

Outside the source (i.e. where $S_{\mu\nu} = 0$) we want a solution $B_{\mu\nu}$ of Eq. 9.67 which represents the outgoing radiation far away; and of all such possible solutions we want the one which dominates in the slow-motion limit. Let us define r to be the spherical polar radial coordinate whose origin is chosen to be inside the source. We show in Exercise 9.29 that the solution we seek is the simplest of all the solutions of Eq. 9.67 outside the source,

$$B_{\mu\nu} = \frac{A_{\mu\nu}}{r} e^{i\Omega r} + \frac{Z_{\mu\nu}}{r} e^{-i\Omega r}, \tag{9.68}$$

where $A_{\mu\nu}$ and $Z_{\mu\nu}$ are constants. The term in $e^{-i\Omega r}$ represents a wave traveling toward the origin $r = 0$ (an ingoing wave), while the other term is outgoing (see Exercise 9.28). We want waves emitted by the source, so we choose $Z_{\mu\nu} = 0$.

Our problem is to determine $A_{\mu\nu}$ in terms of the source. Here we make our approximation that the source is nonzero only inside a sphere of radius $\varepsilon \ll 2\pi/\Omega$. Let us integrate Eq. 9.67 over the interior of this sphere. One term that results is

$$\int \Omega^2 B_{\mu\nu} \, d^3x \leqslant \Omega^2 |B_{\mu\nu}|_{\max} 4\pi\varepsilon^3/3, \tag{9.69}$$

where $|B_{\mu\nu}|_{\max}$ is the maximum value $B_{\mu\nu}$ takes inside the source. We will see that this term is negligible. The other term that comes from integrating the left-hand side of Eq. 9.67 is

$$\int \nabla^2 B_{\mu\nu} \, d^3x = \oint \boldsymbol{n} \cdot \nabla B_{\mu\nu} \, dS, \tag{9.70}$$

by Gauss' theorem. But the surface integral is outside the source, where $B_{\mu\nu}$ is given by Eq. 9.68, which is spherically symmetric:

$$\oint \boldsymbol{n} \cdot \nabla B_{\mu\nu} \, dS = 4\pi\varepsilon^2 \left(\frac{d}{dr} B_{\mu\nu} \right)_{r=\varepsilon} \approx -4\pi A_{\mu\nu}, \tag{9.71}$$

again with the approximation $\varepsilon \ll 2\pi/\Omega$. Finally, we define the integral of the right-hand side of Eq. 9.67 to be

$$J_{\mu\nu} = \int S_{\mu\nu} \, d^3x. \tag{9.72}$$

Combining these results in the limit $\varepsilon \to 0$ gives

$$A_{\mu\nu} = 4J_{\mu\nu}, \tag{9.73}$$
$$\bar{h}_{\mu\nu} = 4J_{\mu\nu} e^{i\Omega(r-t)}/r. \tag{9.74}$$

These are the expressions for the gravitational waves generated by the source, neglecting terms of order r^{-2} and any r^{-1} terms that are higher order in $\varepsilon\Omega$.

It is possible to simplify these expressions considerably. Here we begin to use the fact that $\{\bar{h}_{\mu\nu}\}$ are components of a single tensor, not the unrelated functions we have solved for in Eq. 9.74. From Eq. 9.72 we learn that

$$J_{\mu\nu} e^{-i\Omega t} = \int T_{\mu\nu} \, d^3x, \tag{9.75}$$

one consequence of which is

$$-i\Omega J^{\mu 0} e^{-i\Omega t} = \int T^{\mu 0}{}_{,0} \, d^3x. \tag{9.76}$$

Now, the conservation law for $T_{\mu\nu}$,

$$T^{\mu\nu}{}_{,\nu} = 0, \tag{9.77}$$

implies that

$$T^{\mu 0}{}_{,0} = -T^{\mu k}{}_{,k} \tag{9.78}$$

and hence that

$$i\Omega J^{\mu 0} e^{-i\Omega t} = \int T^{\mu k}{}_{,k} \, \mathrm{d}^3 x = \oint T^{\mu k} n_k \, \mathrm{d}S, \tag{9.79}$$

the last step being the application of Gauss' theorem to any volume completely containing the source. This means that $T^{\mu\nu} = 0$ on the surface bounding this volume, so that the right-hand side of Eq. 9.79 vanishes. This means that if $\Omega \neq 0$ we have

$$J^{\mu 0} = 0, \qquad \bar{h}^{\mu 0} = 0. \tag{9.80}$$

These conditions basically embody the laws of conservation of total energy and momentum for the oscillating source.

The expression for J_{ij} can be rewritten in an instructive way, by using the result of Exercise 9.24.

$$\frac{\mathrm{d}^2}{\mathrm{d}t^2} \int T^{00} x^l x^m \, \mathrm{d}^3 x = 2 \int T^{lm} \, \mathrm{d}^3 x. \tag{9.81}$$

For a source in slow motion, we saw in Chapter 7 that $T^{00} \approx \rho$, the Newtonian mass density. It follows that the integral on the left-hand side of Eq. 9.81 is what is often referred to as the quadrupole-moment tensor of the mass distribution,

$$I^{lm} := \int T^{00} x^l x^m \, \mathrm{d}^3 x \tag{9.82a}$$

$$= D^{lm} e^{-i\Omega t}. \tag{9.82b}$$

(Conventions for defining the quadrupole moment vary from one text to another. We follow Misner *et al.* (1973).) In terms of Eq. 9.82b we have

$$\bar{h}_{jk} = -2\Omega^2 D_{jk} e^{-i\Omega(r-t)}/r. \tag{9.83}$$

It is important to remember that Eq. 9.83 is an approximation which neglects not merely all terms of order r^{-2} but also r^{-1} terms that are not dominant in the slow-motion approximation. In particular, $\bar{h}_{jk}{}^{,k}$ is of higher order, and this guarantees that the gauge condition $\bar{h}^{\mu\nu}{}_{,\nu} = 0$ is satisfied by Eqs. 9.83 and 9.80 at the lowest order in r^{-1} and Ω. Because of Eq. 9.83, this approximation is often called the *quadrupole* approximation for gravitational radiation.

As in the case of the plane waves we studied earlier, we have here the freedom to make a further restriction of the gauge. The obvious choice is to try to find a TT gauge, transverse to the direction of motion of the wave (the radial direction), whose unit vector is $n^j = x^j/r$. Exercise 9.32 shows that this is possible, so that in the TT gauge we have the simplest form of the wave. If we choose our axes so that at the point where we measure the wave it is traveling in the z direction, then we can supplement Eq. 9.80 by

$$\bar{h}^{\mathrm{TT}}_{zi} = 0, \tag{9.84}$$

$$\bar{h}^{\mathrm{TT}}_{xx} = -\bar{h}^{\mathrm{TT}}_{yy} = -\Omega^2 (I\!\!\!\!-_{xx} - I\!\!\!\!-_{yy})e^{i\Omega r}/r, \tag{9.85}$$

$$\bar{h}^{\mathrm{TT}}_{xy} = -2\Omega^2 I\!\!\!\!-_{xy} e^{i\Omega r}/r, \tag{9.86}$$

where

$$I\!\!\!\!-_{jk} = I_{jk} - \tfrac{1}{3}\delta_{jk} I^l_l \tag{9.87}$$

is called the trace-free or reduced quadrupole-moment tensor.

Example of a radiating system: an oscillator. Let us consider the waves emitted by a simple oscillator like the one we used as a detector in § 9.3. If both masses oscillate with angular frequency ω and amplitude A about mean equilibrium positions that are a distance l_0 apart then, by Exercise 9.30, the quadrupole tensor has only one nonzero component,

$$
\begin{aligned}
I_{xx} &= m[(x_1)^2 + (x_2)^2] \\
&= [(-\tfrac{1}{2}l_0 - A \cos \omega t)^2 + (\tfrac{1}{2}l_0 + A \cos \omega t)^2] \\
&= \text{const.} + mA^2 \cos 2\omega t + 2ml_0 A \cos \omega t.
\end{aligned} \tag{9.88}
$$

Recall that only the sinusoidal part of I_{xx} should be used in the formulae developed in the previous paragraph. In this case there are two such parts, with frequencies ω and 2ω. Since the wave equation Eq. 9.64 is linear, we can treat each term separately and simply add the results later. The ω term in I_{xx} is the real part of $2ml_0 A \exp(-i\omega t)$. The trace-free quadrupole tensor has components

$$
\left.
\begin{aligned}
I\!\!\!\!-_{xx} &= I_{xx} - \tfrac{1}{3}I^j_j = \tfrac{2}{3}I_{xx} = \tfrac{4}{3}ml_0 A e^{-i\omega t}, \\
I\!\!\!\!-_{yy} &= I\!\!\!\!-_{zz} - \tfrac{1}{3}I_{xx} = -\tfrac{2}{3}ml_0 A e^{-i\omega t};
\end{aligned}
\right\} \tag{9.89}
$$

all off-diagonal components vanish. If we consider the radiation traveling in the z direction, we get, from Eqs. 9.84–9.86,

$$\bar{h}^{\mathrm{TT}}_{xx} = -\bar{h}^{\mathrm{TT}}_{yy} = -2\, m\omega^2 l_0 A e^{i\omega(r-t)}/r, \qquad \bar{h}^{\mathrm{TT}}_{xy} = 0. \tag{9.90}$$

The radiation is linearly polarized, with an orientation such that the ellipse in Figure 9.1 is aligned with the line joining two masses. The same is true for the radiation going in the y direction, by symmetry. But for the radiation traveling in the x direction (i.e. along the line joining the masses), we need to make the permutation $z \to x, x \to y, y \to z$ in Eqs. 9.85–9.86, and we find

$$\bar{h}^{\mathrm{TT}}_{ij} = 0. \tag{9.91}$$

There is *no* radiation in the x direction. In Exercise 9.36 we will fill in this radiation pattern by calculating the amount of radiation and its polarization in arbitrary directions.

A similar calculation for the 2ω piece of I_{xx} gives the same radiation pattern, replacing Eq. 9.90 by

$$\bar{h}_{xx}^{\mathrm{TT}} = -\bar{h}_{yy}^{\mathrm{TT}} = -4\,m\omega^2 A \mathrm{e}^{2i\omega(r-t)}/r, \qquad \bar{h}_{xy}^{\mathrm{TT}} = 0. \qquad (9.92)$$

The total radiation field is the real part of the sum of Eqs. 9.90 and 9.92, e.g.

$$\bar{h}_{xx}^{\mathrm{TT}} = -[2\,m\,\omega^2 l_0 A \cos \omega(r-t) + 4\,m\,\omega^2 A^2 \cos 2\,\omega(r-t)]/r. \qquad (9.93)$$

Let us estimate the radiation from a laboratory-sized generator of this type. If we take $m = 10^3$ kg $= 7 \times 10^{-25}$ m, $l_0 = 1$ m, $A = 10^{-4}$ m, and $\omega = 10^4\,\mathrm{s}^{-1} = 3 \times 10^{-5}\,\mathrm{m}^{-1}$, then the 2ω contribution is negligible and we find that the amplitude is about $10^{-34}/r$, where r is measured in meters. This shows that laboratory generators are unlikely to produce useful gravitational waves in the near future!

Example of a radiating system: binary stars. A more interesting example of a gravitational wave source is a binary star system. Strictly speaking, our derivation applies only to sources whose motions result from nongravitational forces (this is the content of Eq. 9.77), but our final result, Eqs. 9.84–9.87, makes use only of the motions produced, not of the forces. It is perhaps not so surprising, then, that one can show that Eqs. 9.84–9.87 are a good first approximation for systems dominated by Newtonian gravitational forces. (See the bibliography in § 9.7 for references.)

Let us suppose, then, that we have two stars of mass m, idealized as points in circular orbit about one another, separated by a distance l_0 (i.e. moving in a circle of radius $\frac{1}{2}l_0$). Their orbit equation (gravitational force = 'centrifugal force') is

$$\frac{m^2}{l_0^2} = m\omega^2 \left(\frac{l_0}{2}\right) \Rightarrow \omega = \left(\frac{2m}{l_0^3}\right)^{1/2}, \qquad (9.94)$$

where ω is the angular velocity of the orbit. Then, with an appropriate choice of coordinates, the masses move on the curves

$$\begin{aligned} x_1(t) &= \tfrac{1}{2}l_0 \cos \omega t, \quad y_1(t) = \tfrac{1}{2}l_0 \sin \omega t, \\ x_2(t) &= -x_1(t), \qquad\quad y_2(t) = -y_1(t), \end{aligned} \right\} \qquad (9.95)$$

where the subscripts 1 and 2 refer to the respective stars. These equations give

$$\begin{aligned} I_{xx} &= \tfrac{1}{4}ml_0^2 \cos 2\omega t + \text{const.}, \\ I_{yy} &= -\tfrac{1}{4}ml_0^2 \cos 2\omega t + \text{const.}, \\ I_{xy} &= \tfrac{1}{4}ml_0^2 \sin 2\omega t. \end{aligned} \right\} \qquad (9.96)$$

The only nonvanishing components of the reduced quadrupole tensor are, in complex notation and omitting time-independent terms,

$$\begin{aligned} \text{\textcentoldstyle}_{xx} &= -\text{\textcentoldstyle}_{yy} = \tfrac{1}{4}ml_0^2 \mathrm{e}^{-2i\omega t}, \\ \text{\textcentoldstyle}_{xy} &= \tfrac{1}{4}iml_0^2 \mathrm{e}^{-2i\omega t}. \end{aligned} \right\} \qquad (9.97)$$

All the radiation comes out with frequency $\Omega = 2\,\omega$. The radiation along the z direction (perpendicular to the plane of the orbit) is, by Eqs. 9.84–9.86,

$$\left.\begin{aligned}\bar{h}_{xx} &= -\bar{h}_{yy} = -2ml_0^2\,\omega^2 e^{2i\omega(r-t)}/r,\\[4pt]\bar{h}_{xy} &= -2iml_0^2\,\omega^2 e^{2i\omega(r-t)}/r.\end{aligned}\right\}\tag{9.98}$$

This is *circularly polarized* radiation: the factor i in $\bar{h}_{xy}$ ensures that the two polarization amplitudes are out of phase; if, after taking the real part, one of them follows a cos $2\omega t$ time dependence then the other will follow sin $2\omega t$. (See Exercise 9.15). The radiation in the plane of the orbit, say in the x direction, is found in the same manner used to derive Eq. 9.91. This gives

$$\bar{h}_{yy}^{\mathrm{TT}} = -\bar{h}_{zz}^{\mathrm{TT}} = ml_0^2\omega^2 e^{2i\omega(r-t)}/r,\tag{9.99}$$

all other components vanishing. This shows linear polarization aligned with the orbital plane.

For the general case, if the normal vector to the plane of the orbit is inclined to the line of sight at an angle ι, so that a fully face-on orbit has $\iota = 0$, then the general formulae for the amplitudes is

$$\left.\begin{aligned}\bar{h}_{xx} &= -\bar{h}_{yy} = -\,ml_0^2\,\omega^2[1+\cos^2(\iota)]\,e^{2i\omega(r-t)}/r,\\[4pt]\bar{h}_{xy} &= -2iml_0^2\,\omega^2\,\cos(\iota)\,e^{2i\omega(r-t)}/r.\end{aligned}\right\}\tag{9.100}$$

Note that for $\iota = 0$ one gets Eq. 9.98 while for $\iota = \pi/2$ one gets Eq. 9.99. The convention on x and y is that the x axis in the plane perpendicular to the observer's line of sight is aligned in such a way that when $\iota = \pi/2$ the orbit is in the x–z plane. This is derived in Exercise 9.38 and the calculation is generalized to unequal masses in elliptical orbits in Exercise 9.46.

The amplitude of the radiation is of order $ml_0^2\omega^2/r$, which, by Eq. 9.94, is $\sim (m\omega)^{2/3}m/r$. Binary systems are very important sources of gravitational waves. Binary systems containing pulsars previously provided strong indirect evidence for gravitational radiation (Stairs 2003, Lorimer 2008). The first, and still the most important, of these systems is the one containing the pulsar PSR B1913+16, the discovery of which by Hulse and Taylor (1975) led to their being awarded the Nobel Prize for Physics in 1993.

This system consists of two neutron stars (see Chapter 10) orbiting each other closely. The orbital period, inferred from the Doppler shift of the pulsar's period, is 7 h 45 min 7 s (27907 s or 8.3721×10^{12} m), and both stars have masses approximately equal to $1.4\,M_\odot$ (2.07 km) (Taylor & Weisberg 1982). The system is about 8 kpc $= 2.4 \times 10^{17}$ m away,[3] so its radiation will have the approximate amplitude 10^{-20} at Earth. The frequency of the radiation, which at 72 µHz is twice the orbital frequency, is below the observation bands

---

[3] Astronomers measure distances on the large scale with a unit called a parsec, which is 3.0857×10^{16} m. Distances between stars in our Galaxy are typically a few parsecs, or pc. The sizes of galaxies are on the scale of kiloparsecs, kpc. Distances between galaxies are typically on scales of megaparsecs, Mpc. The unit derives from the way in which astronomers first began measuring distances outside the Solar System, using parallax. One "parsec" is the distance a star would be from Earth if it produced a 1 arcsec parallax angle.

of ground-based detectors (which observe above 1 Hz) and of the future space-based LISA detector (which will observe above about 300 μHz).

We will calculate the effect of the emission of this radiation, and the consequent loss of energy and angular momentum, on the binary orbit itself later in this chapter. In Chapter 10 we will discuss the dynamics of the system, including how the masses are measured.

Order-of-magnitude estimates

Although our simple approach does not enable us to write down solutions for an $\bar{h}_{\mu\nu}$ generated by more complicated, nonperiodic, motions, we can use Eq. 9.83 to obtain some order-of-magnitude estimates. Since D_{jk} is of order MR^2, for a system of mass M and size R, the radiation will have amplitude about $M(\Omega R)^2/r \approx v^2(M/r)$, where v is a typical internal velocity in the source. This applies directly to Eq. 9.99; note that in Eq. 9.93 the first term uses, instead of R^2, the product $l_0 A$ of the two characteristic lengths in the problem. If we are dealing with, say, a collapsing mass moving under its own gravitational forces then, by the virial theorem, $v^2 \sim \phi_0$, the typical Newtonian potential in the source, while $M/r \sim \phi_r$, the Newtonian potential of the source at the observer's distance r. Then we get the simple upper limit

$$h \sim \phi_0 \phi_r. \tag{9.101}$$

So the wave's amplitude is always less than, or of the order of, the Newtonian potential ϕ_r. Why then can we detect h but not ϕ_r itself? Why can we hope to find waves from a supernova in a distant galaxy without being able to detect its presence gravitationally before the explosion? The answers lie in the forces involved. The Newtonian tidal gravitational force on a detector of size l_0 at a distance r is about $\phi_r l_0/r^2$, while the wave force is $h l_0 \omega^2$ (see Eq. 9.45). The wave force is thus a factor $\phi_0(\omega r)^2 \sim (\phi_0 r/R)^2$ larger than the Newtonian force. For a relativistic system ($\phi_0 \sim 0.1$) of size 1 AU ($\sim 10^{11}$ m), observed by a detector a distance 10^{23} m away, this factor is 10^{22}. This estimate, incidentally, gives the largest distance r at which we may still approximate the gravitational field of a dynamical system as Newtonian (i.e. neglecting wave effects): $r = R/\phi_0$, where R is the size of the system.

The estimate in Eq. 9.101 is really an optimistic upper limit, because it assumes that all the mass motions contribute to D_{jk}. In realistic situations this could be a serious overestimate because of the following fundamental fact: *spherically symmetric motions do not radiate*. The rigorous proof of this is discussed in Chapter 10, but in Exercise 9.47 we derive it from linearized theory, Eq. 9.102 below. It also seems to follow from Eq. 9.82a: if T^{00} is spherically symmetric, then I^{lm} is proportional to δ^{lm} and $\textyen^{lm}$ vanishes. But this argument has to be treated with care, since Eq. 9.82a is part of an approximation designed to give only the dominant radiation. One would have to show that spherically symmetric motions would not contribute to terms of higher order in the approximation if they were present. This is in fact true, and it is interesting to ask what eliminates them. The answer is

Eq. 9.77: the conservation of energy eliminates 'monopole' radiation in a linearized theory, just as the conservation of charge eliminates monopole radiation in electromagnetism.

The danger of using Eq. 9.82a too glibly is illustrated in Exercise 9.31: four equal masses at the corners of a rotating square give no time-dependent I^{lm} and hence no radiation in this approximation. But they *would* emit radiation at a higher order of approximation.

Exact solution of the wave equation

Readers who have studied the wave equation, Eq. 9.64, will know that its outgoing-wave solution for arbitrary $T_{\mu\nu}$ is given by the retarded integral

$$\bar{h}_{\mu\nu}(t, x^i) = 4 \int \frac{T_{\mu\nu}(t - R, y^i)}{R} \, d^3 y,$$

$$R = |x^i - y^i|,$$

(9.102)

where the integral is over the past light cone of the event (t, x^i) at which $\bar{h}_{\mu\nu}$ is evaluated. We let the origin be inside the source and we suppose that the field point x^i is far away,

$$|x^i| := r \gg |y^i| := y,$$

(9.103)

and that the time derivatives of $T_{\mu\nu}$ are small. Then, inside the integral in Eq. 9.102, the dominant contribution comes from replacing R by r:

$$\bar{h}_{\mu\nu}(t, x^i) \approx \frac{4}{r} \int T_{\mu\nu}(t - r, y^i) \, d^3 y.$$

(9.104)

This is the generalization of Eq. 9.74. Now, by virtue of the conservation laws in Eq. 9.77,

$$T^{\mu\nu}{}_{,\nu} = 0,$$

we have

$$\int T_{0\mu} \, d^3 y = \text{const.},$$

(9.105)

i.e. the total energy and momentum are conserved. It follows that the $1/r$ part of $\bar{h}_{0\mu}$ is time independent, so it will not contribute to any wave field. This generalizes Eq. 9.80. Then, using Eq. 9.81, we get the generalization of Eq. 9.83:

$$\bar{h}_{jk}(t, x^i) = -\frac{2}{r} I_{jk,00}(t - r).$$

(9.106)

As before, we can adopt the TT gauge to get

$$\bar{h}^{TT}_{xx} = \frac{1}{r}[I_{xx,00}(t - r) - I_{yy,00}(t - r)],$$

$$\bar{h}^{TT}_{xx} = \frac{2}{r}[I_{xx,00}(t - r).$$

(9.107)

9.5 The energy carried away by gravitational waves

We have seen that gravitational waves can transfer energy into objects through which they pass. This is how bar detectors work. It stands to reason, then, that they also carry energy away from their sources. This is a very important aspect of gravitational wave theory because, as we shall see later in this chapter, there are some circumstances in which the effects of this loss of energy on the wave source (the back-reaction of the waves) can be observed, even when the gravitational waves themselves cannot be detected.

In the first decades after Einstein formulated general relativity, there was considerable debate and confusion about whether gravitational waves could carry energy, and Einstein's own doubts about this contributed to his doubting the reality of gravitational waves, as we mentioned at the beginning of this chapter. The doubts were not, in fact, retired until the late 1950s, not long after Einstein's death in 1955. An important argument that contributed considerable clarity to the discussion was the distinguished particle physicist Richard Feynman's 'sticky bead' thought experiment. (See the bibliography in § 9.7 for a reference.) Feynman argued that, if the tidal forces of gravitational waves could cause relative motions between test particles – which was by that time generally accepted – then one could arrange for the test particles to be beads on a rod with some friction, so that their induced motion would generate heat. Surely, Feynman argued, this heat energy has to come from somewhere, and that can only be from the energy carried by the gravitational waves.

We shall elaborate Feynman's argument in this section, showing how to use it to calculate the energy carried by the waves. We replace his beads with our bar detectors, but they are simply a device that in the end entirely drops out of the calculation, yielding a simple expression for the energy flux just in terms of the properties of the gravitational wave itself. The key point is that we *assume* energy is conserved; if it is here, as in all the rest of physics, then our result is the only possible expression for the energy flux.

In our discussion of the harmonic oscillator as a detector of waves in § 9.3, we implicitly assumed that the detector was a kind of 'test body', whose influence on the gravitational wave field is negligible. But this is, strictly speaking, inconsistent. If the detector acquires energy from the waves, then surely the waves must be weaker after passing through the detector. That is, 'downstream' of the detector they should have slightly lower amplitude than 'upstream'. It is easy to see how this comes about once we realize that in § 9.3 we ignored the fact that the oscillator, once set in motion by the waves, will radiate waves itself.

We in fact solved for this radiation in § 9.3 and found, in Eq. 9.88, that waves of two frequencies will be emitted. Consider the emitted waves with frequency Ω, the same as the incident wave. The part which is emitted exactly downstream has the same frequency as the incident wave, so the *total* downstream wave field has an amplitude which is the sum of the two. We will see below that the two interfere destructively, producing a net decrease in the downstream amplitude (see Figure 9.5). (In other directions, there is no net interference: the waves simply pass through each other.) By assuming that this amplitude change signals a change in the energy actually carried by the waves, and by equating this

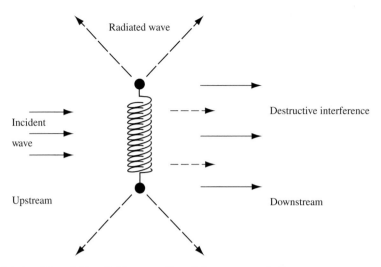

Figure 9.5   When the detector of Figure 9.2 is excited by a wave, it reradiates some waves itself.

energy change to the energy extracted from the waves by the detector, we shall arrive at a simple expression for the energy carried by the waves as a function of their amplitude. We will then be able to calculate the energy lost by bodies which radiate arbitrarily, since we know from § 9.3 what waves they produce.

The energy flux of a gravitational wave

What we require is the energy flux, the energy carried by a wave across a surface per unit area per unit time. It is more convenient, therefore, to consider not just one oscillator but an array of them filling the plane $z = 0$. We suppose they are very close together, so we may regard them as a nearly continuous distribution of oscillators, σ oscillators per unit area (Figure 9.6). If the incident wave has amplitude, in the TT gauge, given by

$$\begin{aligned} \bar{h}_{xx}^{\mathrm{TT}} &= A \cos \Omega(z - t), \\ \bar{h}_{yy}^{\mathrm{TT}} &= -\bar{h}_{xx}^{\mathrm{TT}}, \end{aligned} \tag{9.108}$$

all other components vanishing, then in § 9.3 we saw that each oscillator responds with a steady oscillation (after transients have died out) of the form

$$\xi = R \cos(\Omega t + \phi), \tag{9.109}$$

where R and ϕ are given by Eqs. 9.48 and 9.49, respectively. This motion is steady because the energy dissipated by friction in the oscillators is compensated by the work done on the spring by the tidal gravitational forces of the wave. It follows that the wave supplies an energy to each oscillator equal to

$$dE/dt = \nu(d\xi/dt)^2 = m\gamma(d\xi/dt)^2. \tag{9.110}$$

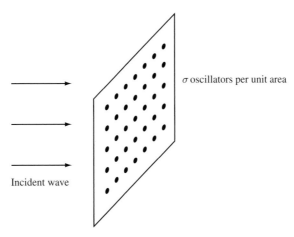

σ oscillators per unit area

Incident wave

The situation when a set of detectors like those of Figure 9.5 are arranged in the plane $z = 0$ at a density of σ per unit area.

Averaging this over one period of oscillation, $2\pi/\Omega$, in order to get a steady energy loss, gives (angle brackets denote the average)

$$\langle dE/dt \rangle = \frac{1}{2\pi/\Omega} \int_0^{2\pi/\Omega} m\gamma\Omega^2 R^2 \sin^2(\Omega t + \phi) \, dt$$

$$= \tfrac{1}{2} m\gamma\Omega^2 R^2. \tag{9.111}$$

This is the energy supplied to each oscillator per unit time. With σ oscillators per unit area, the net energy flux F of the wave must decrease on passing through the $z = 0$ plane by

$$\delta F = -\tfrac{1}{2}\sigma m\gamma\Omega^2 R^2. \tag{9.112}$$

We can calculate the change in the amplitude downstream independently of the calculation that led to Eq. 9.112. Each oscillator has a quadrupole tensor given by Eq. 9.88, with ωt replaced by $\Omega t + \phi$ and A replaced by $R/2$. (Each mass moves with an amplitude A, one-half of the total stretching of the spring R.) Since in our case R is tiny compared to l_0 ($R = 0(h_{xx}^{TT} l_0)$), the 2Ω term in Eq. 9.88 is negligible compared with the Ω term. So each oscillator has

$$I_{xx} = m l_0 R \cos(\Omega t + \phi). \tag{9.113}$$

By Eq. 9.93, each oscillator produces a wave amplitude

$$\delta \bar{h}_{xx} = -2\Omega^2 m l_0 R \cos[\Omega(r - t) - \phi]/r \tag{9.114}$$

at any point a distance r away. (We call the amplitude $\delta \bar{h}_{xx}$ to indicate that it is small compared with the incident wave.) It is a simple matter to get the total radiated field by adding up the contributions due to all the oscillators. In Figure 9.7, consider a point P that is a distance z downstream from the plane of oscillators. Set up polar coordinates $(\tilde{\omega}, \phi)$ in the plane, centered at Q beneath P. A typical oscillator O at a distance $\tilde{\omega}$ from Q contributes

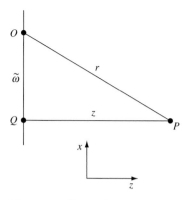

Figure 9.7 Geometry for calculating the field at P due to an oscillator at O.

a field, Eq. 9.114, at P, with $r = (\tilde{\omega}^2 + z^2)^{1/2}$. Since the number of such oscillators between $\tilde{\omega}$ and $\tilde{\omega} + \mathrm{d}\tilde{\omega}$ is $2\pi\sigma\tilde{\omega}\,\mathrm{d}\tilde{\omega}$, the total oscillator-produced field at P is

$$\delta\bar{h}_{xx}^{\text{total}} = -2m\Omega^2 l_0 R 2\pi \int_0^\infty \sigma\cos[\Omega(r-t)-\phi]\frac{\tilde{w}\,\mathrm{d}\tilde{w}}{r}.$$

But we may change the integration variable to r,

$$\tilde{w}\mathrm{d}\tilde{w} = r\mathrm{d}r,$$

obtaining

$$\delta\bar{h}_{xx}^{\text{total}} = -2m\Omega^2 l_0 R 2\pi \int_z^\infty \sigma\cos[\Omega(r-t)-\phi]\,\mathrm{d}r. \qquad (9.115)$$

If σ were constant then this would be trivial to integrate, but its value would be undefined at $r = \infty$. Physically, we should expect that the more distant oscillators play no real role, so we adopt the device of assuming that σ is proportional to $\exp(-\varepsilon r)$ and allowing ε to tend to zero after the integration. The result is

$$\delta\bar{h}_{xx}^{\text{total}} = 4\pi\sigma m\Omega l_0 R \sin[\Omega(z-t)-\phi]. \qquad (9.116)$$

So, the plane of oscillators sends out a net plane wave.[4] To compare this with the incident wave we must put Eq. 9.116 into the same TT gauge (recall that Eq. 9.83 is *not* in the TT gauge), with the result (Exercise 9.49)

$$\delta\bar{h}_{xx}^{\text{TT}} = -\delta\bar{h}_{yy}^{\text{TT}} = 2\pi\sigma m\Omega l_0 R \sin[\Omega(z-t)-\phi]. \qquad (9.117)$$

If we now add this to the incident wave, Eq. 9.108, we get the net result, to first order in R,

$$\bar{h}_{xx}^{\text{net}} = \bar{h}_{xx}^{\text{TT}} + \delta\bar{h}_{xx}^{\text{TT}}$$
$$= (A - 2\pi\sigma m\Omega l_0 R \sin\phi)\cos[\Omega(z-t)-\psi], \qquad (9.118)$$

---

[4] This explains why, at the beginning of this discussion, we did not consider the waves radiated by an individual oscillator in directions other than downstream. Those waves destructively interfere with similar waves from the other oscillators, so that the net radiation of the planar distribution of oscillators goes entirely in the direction perpendicular to the plane.

where

$$\tan \psi = \frac{2\pi \sigma m \Omega l_0 R}{A} \cos \phi. \tag{9.119}$$

Apart from a small phase shift ψ, the net effect is a reduction in the amplitude A by

$$\delta A = -2\pi \sigma m \Omega l_0 R \sin \phi. \tag{9.120}$$

This reduction must be responsible for the decrease in flux F downstream. Dividing Eq. 9.112 by Eq. 9.120 and using Eqs. 9.48 and 9.49 to eliminate R and ϕ gives the remarkably simple result

$$\frac{\delta F}{\delta A} = \frac{1}{16\pi} \Omega^2 A. \tag{9.121}$$

This is our key result. It says that a change δA in the amplitude A of a wave of frequency Ω changes its flux F (averaged over one period) by an amount depending only on Ω, A, and δA. The oscillators helped us to derive this result from the conservation of energy, but they have dropped out completely! Equation 9.121 is a property of the wave itself. We can 'integrate' Eq. 9.121 to get the total flux of a wave of frequency Ω and amplitude A:

$$F = \frac{1}{32\pi} \Omega^2 A^2. \tag{9.122}$$

Since the average of the square of the wave, Eq. 9.108, is

$$\langle (\bar{h}_{xx}^{\text{TT}})^2 \rangle = \tfrac{1}{2} A^2$$

(again, angle brackets denote an average over one period), and since there are only two nonvanishing components of $\bar{h}_{\mu\nu}^{\text{TT}}$, we can also write Eq. 9.122 as

$$F = \frac{1}{32\pi} \Omega^2 \langle \bar{h}_{\mu\nu}^{\text{TT}} \bar{h}^{\text{TT}\mu\nu} \rangle. \tag{9.123}$$

This form is invariant under background Lorentz transformations, but not under gauge changes. Since one polarization can be transformed into another by a background Lorentz transformation (a rotation), Eq. 9.123 applies to all polarizations and hence to arbitrary plane waves of frequency Ω. In fact, since it gives an energy rate per unit area, it applies to *any* wavefront, either plane waves or the spherical expanding waves of § 9.3: one can always look at a small enough area that the curvature of the wavefront is not noticeable. The generalization to arbitrary waves (with no single frequency) is in Exercise 9.50.

The reader who remembers the discussion of energy in § 7.3 may object that this whole derivation is suspect because of the difficulty of defining energy in GR. Indeed, we have not *proved* that energy is conserved, that the energy put into the oscillators must equal the decrease in flux; we have simply assumed this in order to derive the flux. Our proof may be turned around, however, to argue that the flux we have constructed is the only acceptable definition of energy for the waves, since our calculation shows it is conserved when added to other energies, to lowest order in $h_{\mu\nu}$. The qualification 'to lowest order' is important, since it is precisely because we are almost in flat spacetime that, at lowest order, we can construct conserved quantities. At higher order, away from linearized theory, local energy cannot be so easily defined, because the time dependence of the true metric becomes important. These questions are among the most fundamental in relativity, and are discussed in detail in any of the advanced texts. Our equations should be used *only* in linearized theory.

Energy lost by a radiating system

Consider a general isolated system, radiating according to Eqs. 9.82–9.87. By integrating Eq. 9.123 over a sphere surrounding the system, we can calculate its net energy loss rate. For example, at a distance r along the z axis, Eq. 9.123 is

$$F = \frac{\Omega^6}{32\pi r^2} \langle 2(I_{xx} - I_{yy})^2 + 8I^{xy} \rangle. \tag{9.124}$$

Use of the identity

$$I^i_i = I_{xx} + I_{yy} + I_{zz} = 0 \tag{9.125}$$

(which follows from Eq. 9.87) gives, after some algebra,

$$F = \frac{\Omega^6}{16\pi r^2} \langle 2I_{ij}I^{ij} - 4I_{zj}I_z{}^j + I_{zz}^2 \rangle. \tag{9.126}$$

Now, the index z appears here only because it is the distance from the center of the coordinates, where the radiation comes from. It is the only part of F which depends on the location on the sphere of radius r about the source, since all the components of I_{ij} depend on time but not position. Therefore we can generalize Eq. 9.126 to arbitrary locations on the sphere by using the unit vector normal to the sphere,

$$n^j = x^j/r. \tag{9.127}$$

We get for F

$$F = \frac{\Omega^6}{16\pi r^2} \langle 2I_{ij}I^{ij} - 4n^j n^k I_{ji}I_k^i + n^i n^j n^k n^l I_{ij}I_{kl} \rangle. \tag{9.128}$$

The total luminosity of the source is the integral of this over the sphere of radius r. In Exercise 9.52, we prove the following integrals:

$$\int n^j n^k \sin\theta \, d\theta \, d\phi = \frac{4\pi}{3} \delta^{jk}, \tag{9.129}$$

$$\int n^i n^j n^k n^l \sin\theta \, d\theta \, d\phi = \frac{4\pi}{15} (\delta^{ij}\delta^{kl} + \delta^{ik}\delta^{jl} + \delta^{il}\delta^{jk}). \tag{9.130}$$

It then follows that the luminosity L of a source of gravitational waves is

$$L = \int F r^2 \sin\theta \, d\theta d\phi$$

$$= \frac{1}{4}\Omega^6 \langle 2I_{ij}I^{ij} - \frac{4}{3}I_{ij}I^{ij} + \frac{1}{15}(I^i_i I^k_k + I^{ij}I_{ij} + I^{ij}I_{ij}) \rangle, \tag{9.131}$$

$$L = \frac{1}{5}\Omega^6 \langle I_{ij}I^{ij} \rangle. \tag{9.132}$$

The generalization to cases where f_{ij} has a more general time dependence is

$$L = \tfrac{1}{5}\langle \dddot{\mathit{f}}_{ij}\dddot{\mathit{f}}^{\,ij}\rangle, \tag{9.133}$$

where the dots denote time derivatives.

It must be stressed that Eqs. 9.132 and 9.133 are accurate only for weak gravitational fields and slow velocities. At best they can give only order-of-magnitude results for highly relativistic sources of gravitational waves. But, in the spirit of our derivation and discussion of the order-of-magnitude estimate of h_{ij} in Eq. 9.70, we can still learn something about strong sources from Eq. 9.132. Since I_{jk} is of order MR^2, Eq. 9.132 tells us that $L \sim M^2R^4\Omega^6 \sim (M/R)^2(R\Omega)^6 \sim \phi_0^2 v^6$. The luminosity is a very sensitive function of the velocity. The largest velocities that one should expect are of the order of the velocity of free fall, $v^2 \sim \phi_0$, so we should expect that

$$L \lesssim (\phi_0)^5. \tag{9.134}$$

Since $\phi_0 \lesssim 1$, the luminosity in geometrized units should never substantially exceed one. In ordinary units this is

$$L \lesssim 1 = c^5/G \approx 3.6 \times 10^{52}\ \text{W}. \tag{9.135}$$

We can understand why this particular luminosity is an upper limit by the following simple argument. The radiation field inside a source of size R and luminosity L has energy density $\gtrsim L/R^2$ (because $|T^{0i}| \sim |v^i|T^{00} = cT^{00} = T^{00}$), which is the flux across its surface. The total radiation energy is therefore $\gtrsim LR$. The Newtonian potential alone of the radiation is therefore $\gtrsim L$. We shall see in the next chapter that anything whose Newtonian potential substantially exceeds one must form a black hole: its gravitational field will be so strong that no radiation will escape at all. Therefore $L \sim 1$ is the largest luminosity that any source can have. This argument applies equally well to all forms of radiation, electromagnetic as well as gravitational. Bright quasars and gamma-ray bursts, which are the most luminous classes of object so far observed, have a (geometrized) luminosity $\lesssim 10^{-10}$. By contrast, black hole mergers (Chapter 11) have been shown by numerical simulations to reach peak luminosities $\sim 10^{-3}$, all the radiation of course being emitted in gravitational waves.

An example: the Hulse–Taylor binary pulsar

In § 9.3 we calculated f_{ij} for a binary system consisting of two stars of equal mass M in circular orbits a distance l_0 apart. If we use the real part of Eq. 9.97 in Eq. 9.133, we get

$$L = \frac{8}{5}M^2 l_0^4 \omega^6. \tag{9.136}$$

Eliminating l_0 in favor of M and ω, we get

$$L = \frac{32}{5\sqrt[3]{4}}(M\omega)^{10/3} \approx 4.0(M\omega)^{10/3}. \tag{9.137}$$

This expression illustrates two things: first, that L is dimensionless in geometrized units and, second, that it is almost always easier to compute in geometrized units, and then convert back at the end. The conversion is

$$L \text{ (SI units)} = \frac{c^5}{G} L \text{ (geometrized)}$$
$$= 3.63 \times 10^{52} \text{ J s}^{-1} \times L \text{ (geometrized)}. \tag{9.138}$$

So, for the binary pulsar system described in § 9.3, if its orbit were circular, we would have $\omega = 2\pi/P = 7.5049 \times 10^{-13} \text{ m}^{-1}$ and

$$L = 1.71 \times 10^{-29} \tag{9.139}$$

in geometrized units. We can, of course, convert this to watts, but a more meaningful procedure is to compare this with the Newtonian energy of the system, which is (defining the orbital radius $r = \frac{1}{2}l_0$) given by

$$E = \frac{1}{2}M\omega^2 r^2 + \frac{1}{2}M\omega^2 r^2 - \frac{M^2}{2r} \tag{9.140}$$
$$= \frac{M}{r}(\omega^2 r^3 - \frac{1}{2}M) = -\frac{M^2}{4r}$$
$$= -4^{-2/3}M^{5/3}\omega^{2/3} \approx -0.40M^{5/3}\omega^{2/3} \tag{9.141}$$
$$= -1.11 \times 10^{-3} \text{ m.} \tag{9.142}$$

The physical question is, how long does it take to change this? Put differently, the energy radiated in waves must change the orbit by decreasing the energy of the orbit, which makes $|E|$ larger and hence ω larger and the period smaller. What change in the period $P = 2\pi/\omega$ do we expect in, say, one year?

From Eq. 9.141, by taking logarithms and differentiating we get

$$\frac{1}{E}\frac{dE}{dt} = \frac{2}{3}\frac{1}{\omega}\frac{d\omega}{dt} = -\frac{2}{3}\frac{1}{P}\frac{dP}{dt}. \tag{9.143}$$

Since dE/dt is just $-L$, we can solve for dP/dt:

$$dP/dt = (3PL)/(2E) \approx -15\,PM^{-1}(M\Omega)^{8/3}$$
$$= -2.0 \times 10^{-13}, \tag{9.144}$$

which is dimensionless in any system of units. It can be reexpressed in seconds per year:

$$dP/dt = -6.0 \times 10^{-6} \text{ s yr}^{-1}. \tag{9.145}$$

This estimate needs to be revised to allow for the eccentricity of the orbit, which is considerable: $e = 0.617$. The correct formula is derived in Exercise 9.56. The result is that the true rate of energy loss is some 12 times our estimate in Eq. 9.139. The reason that this is such a large factor is that the stars' maximum angular velocity (when they are closest) is larger than the mean value we have used for Ω, and since L depends on the angular velocity to a high power, a small change in the angular velocity accounts for this rather large factor of 12. Timing of the orbit on timescales too short to show the changes due to radiation emission determines the masses of the two stars very accurately, enabling a precise calculation of the relativistic prediction. The most recent results (Weisberg & Huang 2016) give the prediction of general relativity for the rate of change of the period as

$$dP/dt = -2.40263 \times 10^{-12}, \tag{9.146}$$

$$= -7.2 \times 10^{-15} \, \text{s yr}^{-1}, \tag{9.147}$$

and the observed value is

$$dP/dt = -(2.398 \pm 0.004) \times 10^{-12}, \tag{9.148}$$

They agree to within 0.17%.

9.6 Standard sirens

In this section we learn that distance is encoded in the waveform of binaries, because gravitation theory is what physicists call a scale-free theory: there is no intrinsic or preferred mass, length, or timescale in the theory. In order to understand what scale-free means, we first generalize the earlier discussion of binaries to include systems where the component masses are not necessarily equal to one another.

Realistic binaries with unequal masses

We have considered so far only binaries with equal masses, in order to keep the algebra simpler and focus the discussion on the concepts. But no real binary will have exactly equal masses, so observational gravitational wave astronomy (which we treat in Chapter 12) needs to take that into account. Moreover, and even more interesting from a theoretical point of view, when we look at the equations for the radiation from unequal-mass binaries in the right way, we will discover that the signals from binaries whose orbits shrink during an observation contain enough measurables to determine the distance to the binary. This is a very unusual property in astronomy, where determining accurate distances is a major challenge. Such binaries are called *standard sirens*, and we treat them in detail in the next subsection. Here we lay the foundations of that treatment by looking at the radiation from unequal-mass binaries.

The radiation amplitudes for such a binary are obtained in Exercise 9.39 in a straightforward way from those we calculated for the equal-mass case. Given two bodies of mass m_1 and m_2, we express the amplitudes in terms of the total mass $M = m_1 + m_2$ and the Newtonian reduced mass $\mu = m_1 m_2 / M$. The circularly polarized radiation going up the z axis (perpendicular to the orbital plane) is the following generalization of Eq. 9.98:

$$\left. \begin{array}{l} \bar{h}_{xx} = -\bar{h}_{yy} = -4\mu M^{2/3} \, \omega^{2/3} e^{2i\omega(r-t)}/r, \\ \bar{h}_{xy} = -4i\mu M^{2/3} \, \omega^{2/3} e^{2i\omega(r-t)}/r. \end{array} \right\} \tag{9.149}$$

Similarly, the linearly polarized radiation going out the x axis, in the orbital plane, is the generalization of Eq. 9.99:

$$\bar{h}^{\mathrm{TT}}_{yy} = -\bar{h}^{\mathrm{TT}}_{zz} = 2\mu M^{2/3}\, \omega^{2/3} \mathrm{e}^{2\mathrm{i}\omega(r-t)}/r. \tag{9.150}$$

These equations reveal a remarkable simplification. Although the two component masses m_1 and m_2 can be chosen arbitrarily, the radiation amplitudes depend only on a single combination of these masses, $\mu M^{2/3}$. This has led to the definition of a new mass associated with the system, called (for reasons that will become clear below) the *chirp mass* $\mathcal{M}$, defined by

$$\mathcal{M} := \mu^{3/5} M^{2/5} = m_1^{3/5} m_2^{3/5}/M^{1/5}. \tag{9.151}$$

Using the chirp mass explicitly, and taking into account Eq. 9.100, we can write down the general formula for the amplitudes for *any* orbital inclination ι to the line of sight:

$$\left.\begin{aligned}
\bar{h}_{xx} &= -\bar{h}_{yy} = -2\frac{\mathcal{M}}{r}(\mathcal{M}\omega)^{2/3}\, [1+\cos^2(\iota)]\, \mathrm{e}^{2\mathrm{i}(\mathcal{M}\omega)(r-t)/\mathcal{M}}, \\[2mm]
\bar{h}_{xy} &= -4\mathrm{i}\frac{\mathcal{M}}{r}(\mathcal{M}\omega)^{2/3}\, \cos(\iota)\, \mathrm{e}^{2\mathrm{i}(\mathcal{M}\omega)(r-t)/\mathcal{M}}.
\end{aligned}\right\} \tag{9.152}$$

These still need to be put into the TT gauge for the direction along the angle ι, which means they must be projected perpendicularly to the line of sight and made trace-free.

The way in which we have written the dependence on $\mathcal{M}$ in Eq. 9.152 emphasizes the fact that the mass $\mathcal{M}$ provides the only scale in these equations: a system with twice as large a chirp mass and half the orbital frequency will have the same gravitational wave amplitudes at twice the distance. Even the phase of the wave will be the same at this distance if we double the time.

To complete the calculation we generalize the gravitational wave luminosity and the binary system's total Newtonian energy, both of which also exhibit this scaling property with $\mathcal{M}$. This is done in Exercise 9.40, and interested readers can find a more extensive discussion in Andersson (2019). We obtain for the luminosity

$$L = \frac{32}{5}(\mathcal{M}\omega)^{10/3}, \tag{9.153}$$

and for the energy

$$\frac{E}{\mathcal{M}} = -\frac{1}{2}(\mathcal{M}\omega)^{2/3}. \tag{9.154}$$

It is important to bear in mind that, in all these equations, ω is the *orbital* angular velocity, not the angular frequency of the emitted gravitational waves, which will be twice that.

Notice that the expression for the energy E in Eq. 9.154 is purely Newtonian, so the chirp mass is not an exclusively relativistic parameter. Despite this, the chirp mass does not seem to have been noted in studies of Newtonian binaries, where of course it would have been nothing more than a curiosity. Only when one considers the gravitational wave equations does it become the key scaling parameter. The gravitational wave signal allows the measurement of this scale. This is so because, as we noted above for the Hulse–Taylor binary pulsar, the luminosity is the rate of decrease of the binary's energy:

$$\frac{dE}{dt} = -L. \tag{9.155}$$

As is computed in Exercise 9.41, this equation allows one to solve for the chirp mass in terms of the frequency and rate of change of the frequency, obtaining

$$\mathcal{M} = \left(\frac{5}{96}\right)^{3/5} \dot{\omega}^{3/5} \omega^{-11/5}. \tag{9.156}$$

This shows how to obtain $\mathcal{M}$ from a detected signal: at any time, measure both ω and $\dot{\omega}$.

Scaling relations for binaries

Because the fundamental expressions for the radiation and the energy, Eqs. 9.152–9.154, all scale with $\mathcal{M}$, so will the important derived quantities associated with the evolution of a binary. We list the most important relations here and leave the (straightforward) algebra to Exercise 9.43. The characteristic timescale for the shrinking of the orbit when it has frequency ω is

$$\tau_{\text{chirp}} := \frac{\omega}{\dot{\omega}} = \frac{5}{96}\mathcal{M}(\mathcal{M}\omega)^{-8/3}. \tag{9.157}$$

This is called the 'chirp time' of the system, because it measures the timescale on which the frequency of the gravitational wave signal rises; this rising frequency is called a *chirp*, by analogy with the calls of some birds.

While the chirp time gives the instantaneous timescale for changes in the orbit, it is also of interest to know the full remaining lifetime of the system. This is, within our approximation of point masses, the time when the point masses merge, i.e. when they reach $l_0 = 0$ or $\omega = \infty$. A binary with orbital frequency ω has a remaining lifetime of

$$\tau_{\text{merger}} = \frac{5}{256}\mathcal{M}(\mathcal{M}\omega)^{-8/3}. \tag{9.158}$$

Notice that this is just 3/8 of the chirp time given in the previous equation. This shows the effect of the rapid speeding up of the orbital decay as the objects get closer and closer.

A particularly interesting quantity is the *orbital phase* $\psi(t)$. This is related to the orbital frequency by $\omega = d\psi/dt$. It is interesting since it is dimensionless, so it can depend only on the dimensionless quantity $\mathcal{M}\omega$. The amount of orbital phase remaining when the orbital frequency is ω is given by

$$\psi_{\text{merger}} = \frac{1}{32}(\mathcal{M}\omega)^{-5/3}. \tag{9.159}$$

The number of orbits remaining before merger is just $\psi_{\text{merger}}/2\pi$. Because the binary system's luminosity also depends only on $\mathcal{M}\omega$, we can express it instead as a function of the remaining orbital phase:

$$L = 1/\left(160\,\psi_{\text{merger}}^2\right). \tag{9.160}$$

This is a truly remarkable result: the gravitational wave luminosity of a circular binary depends *only* on the number of orbits remaining before coalescence: if we know the number of remaining orbits, we know the luminosity, regardless of the masses. A dramatic consequence of the scale-free nature of gravity, it was first derived, along with many of the other key scaling results given above, by Forward & Berman (1967). This prescient paper appeared even before the discovery of pulsars, and several years before the discovery of the first relativistic binary system, the Hulse–Taylor binary pulsar.

Of course, real binary stars or black holes are not point particles, so they coalesce a bit sooner than one would predict for point particles. Neutron stars will begin tidally interacting and deviating from the point-particle approximation when they are separated by a few times the neutron-star radius of around 10–15 km. (See Exercise 9.44.) Black holes, by contrast, are so compact that they merge when their corresponding point particles would have less than one orbit remaining. That means that, at the end of a black-hole merger signal, the luminosity of the system may approach 0.01, which in conventional units would be 1% of the maximum luminosity given in Eq. 9.134. This is a far higher luminosity than that of any other known class of systems in astronomy. In fact, as shown in Exercise 9.45, it is greater than the electromagnetic luminosity of the entire observable Universe!

Because Eq. 9.160 is independent of the mass, this huge final luminosity will be characteristic of any pair of black holes, no matter how tiny. The difference between small- and large-mass systems is in the duration, since this peak luminosity holds only for the last orbit. That also means that small systems radiate proportionately less total energy than large systems, ensuring that they radiate away the same fraction of their initial mass–energy: all dimensionless quantities depend only on the remaining number of orbits.

The calculation we have done is of course limited by the assumption that the orbit equations are perfectly Newtonian, and the only effect of general relativity is in the loss of energy and angular momentum to gravitational waves. Post-Newtonian corrections to the orbit will affect the waveform more and more strongly as the two components draw closer together. From such corrections it is possible to break the mass degeneracy and measure both m_1 and m_2. (See Blanchet 2014, Poisson & Will 2014, Andersson 2019.) Because these are small corrections, the accuracy with which the individual masses can be measured is poorer than the accuracy of $\mathcal{M}$.

Measuring distances using gravitational waves

The fact that the luminosity depends only on the observable $\mathcal{M}\omega$ means that a gravitational wave observation of a circular binary can track the changing luminosity of the system as it evolves. Astronomers value systems whose luminosity is known, because they can be used to measure distances. Such systems are rare.

A typical astronomical observation results in a measurement of the apparent brightness of an object, which is a measurement of the energy flux F at the telescope. For gravitational waves, this is done by measuring the wave amplitude h and computing the flux from it. If the object has a known intrinsic luminosity L, then this can unlock its distance r. The reason is that, for a source that radiates energy isotropically,

$$L = 4\pi r^2 F. \tag{9.161}$$

Now, binaries do not radiate isotropically but, if we know the inclination angle ι in Eq. 9.152, then the measured flux in the direction of our detectors can be converted into the luminosity once $\mathcal{M}$ is known.

With a single detector, there is not enough information to measure ι. But, as Eq. 9.152 shows, the inclination is measurable from the degree of circular polarization in the signal. Therefore, as we saw in § 9.3, in general we need a network of three detectors in order to make the polarization measurement and thereby to determine the distance to a binary. Two detectors not aligned with each other would suffice if the location of the binary on the sky is known from other information, for example from the identification of the explosion following the merger of two neutron stars. But if we rely on gravitational wave information alone, we require three.

Astronomers call a source of known luminosity a 'standard candle'. As mentioned earlier, gravitational wave astronomers have dubbed binary systems *standard sirens,* to emphasize the analogy between gravitational wave detection and measurement of sound waves.

If we absorb all the numerical constants along with all the angular factors involved in inferring the true flux of a gravitational wave as it passes through Earth into a factor that we call α, which itself ought to be O(1), then Eq. 9.152 shows that the distance is given by

$$r = \alpha \mathcal{M}(\mathcal{M}\omega)^{2/3}/h, \tag{9.162}$$

where h is the measured amplitude in one of the detectors. In Chapter 12 we shall discuss in more detail how α can be measured by a network of detectors. All gravitational wave signals detected to date (mid-2021) have been from binary standard sirens, and LIGO and Virgo assign distances to them using the methods described here.

This expression for the distance to a standard siren was first written down in Schutz (1986), but of course it is implicit in the work of Forward & Berman (1967) referred to earlier. They noted that Eq. 9.160 allowed the luminosity of the wave to be measured, provided that the detector was sufficiently broadband to be able to follow the remainder of the signal and estimate the remaining phase-to-merger ψ_{merger}. (Their work was done in the era of narrowband bar detectors.) They estimated that the maximum range at which then-current bar detectors might detect a binary neutron-star system was about 1 kpc, which is only in the near neighborhood of the Sun in our galaxy. For comparison, LIGO and Virgo have at the time of writing (mid-2021) detected such systems at distances exceeding 100 Mpc. The short range of bars and the fact that neutron stars had not yet been observed when they wrote their paper explains why the Forward–Berman paper fell into obscurity, and many of its results were subsequently independently rediscovered.

Astronomers prize standard candles because, among other things, they allow a measurement of the expansion rate of the Universe, a fundamental parameter we call the Hubble–Lemaître constant. Standard sirens, similarly, have the potential to contribute to this measurement, as was first pointed out in Schutz (1986). We will discuss how gravitational waves are assisting with this measurement in Chapters 12 and 13.

Notice that all the information contained in a binary signal is obtained by time measurements: the frequency is obvious, the chirp mass is simply a function of the frequency and its rate of change, and even the amplitude is measured in an interferometer from the time difference in the arrival of laser phases from the two arms. The amplitude measurement is entirely a local time measurement. But the signal frequency and its rate of change involve the dynamics at the source. Therefore, if the source has a velocity away from us, the observed frequency and its time derivative will be redshifted when we receive them. This will feed through into the chirp-mass measurement. It is easy to see from Eq. 9.156 that the observed chirp mass $\mathcal{M}_{obs}$ and the true chirp mass $\mathcal{M}_{true}$ as measured in the rest frame of the binary are related by

$$\mathcal{M}_{obs} = (1 + z)\mathcal{M}_{true}, \qquad (9.163)$$

where z is the redshift.

This will apply also to individual component masses m_1 and m_2, which are inferred from post-Newtonian effects observable in the orbital velocities. The inferred distance in Eq. 9.162 will also be affected by the redshift, being multiplied by $1 + z$. This is a familiar effect in special relativity, where the distance to a receding object that is inferred from its observed luminosity is called the luminosity distance.

Now, the typical velocities of stars and binaries in galaxies are small enough that Eq. 9.163 is normally not a significant correction. But it becomes important when we consider the cosmological expansion. We discuss this expansion in Chapter 13, but the recession velocities of galaxies from our position produce frequency redshifts in the spectrum of light emitted by the galaxies, and this redshift will affect the observed binary signals in the same way. It follows, as pointed out by Krolak & Schutz (1987), that the observed frequencies and masses will contain redshift factors, and the inferred standard-candle distance will be the cosmological luminosity distance, which includes not only redshift effects but effects due to the curvature of space between the source and us. We will discuss the cosmological luminosity distance in Chapter 13. The detections of binary gravitational waves by LIGO and Virgo, which we will discuss in Chapter 12, all apply the correction in Eq. 9.163 to the observed masses.

We have not discussed here the most general binary systems, which typically would have nonzero eccentricity. The full problem of radiation and orbit decay with eccentricity was first studied by Peters & Mathews (1963) and Peters (1964). As mentioned earlier, this solution is explored in Exercises 9.46 and 9.56. Importantly, these equations imply that the eccentricity of an orbit is radiated away faster than the shrinking of the orbital radius. To see why, compare, in the results of Exercise 9.56, the equations for the rates of change of a (semi-major axis) and e (eccentricity). The factor e multiplying the entire right-hand side of Eq. 9.187 means that eccentricity decays quasi-exponentially, and the e-folding time decreases with the $1/a^4$ factor on the right-hand side. This contrasts with the dominant dependence on a in Eq. 9.186, where the factor a^{-3} on the right-hand side means that the rate of decrease of a is roughly a power law, $1/t^{1/4}$, where t is the time remaining until the coalescence of the two components. While this goes rapidly toward the end, the exponential decrease of eccentricity is generally faster over the whole orbit evolution. Andersson (2019) provides a more extensive discussion, showing that the relation between *e* and *a* can be found explicitly and that, when the Hulse–Taylor binary pulsar enters the LIGO frequency band (after which it will have only one or two minutes left before coalescence), its eccentricity will have reduced to below 10^{-5}.

This discussion implies that only gravitational wave observations of relatively young systems or of systems that are not highly relativistic will need to take eccentricity into account. Ground-based detectors observe binaries only in their final few minutes of lifetime, so eccentricity can usually be ignored. LISA, on the other hand, is expected to observe some systems that may be highly eccentric, as will be discussed in Chapter 12. Eccentric systems can still be used for distance determination, because their eccentricity can be measured from their more complex signals: just as for Mercury, the orbit will precess, and this modulates the projected ellipticity of the orbit, in general modulating the received amplitude of the signal and the degree and orientation of elliptical polarization.

9.7 Bibliography

Joseph Weber's early thinking about detectors is to be found in Weber (1961). He became embroiled in controversy when he claimed in the late 1960s that his bars were actually detecting gravitational waves, a claim that was not supported by any of the other bar detectors built at the time to verify Weber's claim. This put the field of gravitational wave detection under a cloud for a time, but in fact most of the later bar and interferometric detectors grew out of the teams and projects that invalidated Weber's claims. For an interesting history of this episode, see Collins (2004). The full history of scientists' uncertainties about the reality of gravitational waves, starting with Einstein, is examined in the highly readable book by Kennefick (2007). The paper by Blum *et al.* (2018) investigates the reasons why it took such a long time for scientists to come to grips with the reality of gravitational waves.

The limitations on the sensitivity of bar detectors generated important research on the quantum theory of detection, leading to techniques which can measure with unexpectedly high precision by manipulating the Heisenberg uncertainty principle in such a way that the measurement does not disturb the quantity being measured. In bar detectors this was called 'quantum nondemolition' detection. See the early work by Thorne *et al.* (1979), Caves *et al.* (1980), and Braginsky (1980). The principle was implemented in laser interferometers with the name 'squeezed light', and is now being used in LIGO and Virgo (Aasi *et al.* 2013).

A full discussion of the wave equation is beyond our scope here but is amply treated in many texts on electromagnetism, such as Jackson (1998). A simplified discussion of gravitational waves can be found in Schutz (1984).

We shall address the detections of gravitational waves that have been achieved so far, and their astrophysical implications, in Chapter 12. A student who wants to go to more depth in the science of detection has a number of options. A good starting point would be the open-access review journal *Living Reviews in Relativity*, whose articles are kept up to date. These include, for example: Blanchet (2014) and Schäfer & Jaranowski (2018) for post-Newtonian treatments of radiation from binary systems like the Hulse–Taylor system and especially from binaries that coalesce; Hough & Rowan (2000) and Bond *et al.* (2016) for a more in-depth look at how interferometers work; and Sathyaprakash & Schutz (2009) concerning how information about real sources is obtained from networks of detectors. A classic reference on detection, including antenna patterns, detector noise sources, and data analysis fundamentals, is Thorne (1987).

A number of books now treat gravitational wave detection science in great depth. Saulson (2017) is an up-to-date introduction to how detectors work by one of the LIGO pioneers. In-depth treatments of the physics and astrophysics of gravitational wave detection can be found in Creighton & Anderson (2011) and in Blair *et al.* (2012). The two-volume set *Gravitational Waves* by M. Maggiore (2007, 2018) deals in depth with detector technology (first volume) and the astrophysics of the first detections (second volume). A wide-ranging treatment of detection, including not just ground-based detectors but also LISA, pulsar timing, and future technologies, is Auger & Plagnol (2017). A broad and in-depth treatment of the subject, including the mathematics of gravitational waves and of binary systems, data analysis, and the modeling of sources for both ground-based and space-based detectors is Andersson (2019). The most authoritative book on detectors for some time to come is likely to be Reitze, Saulson & Grote (2019). It is a two-volume set covering in detail all the important current and future detector technologies. The editors are senior LIGO experimenters, and the contributors are experts in their respective technologies, drawn from the LIGO–Virgo–GEO–KAGRA community.

Much information about detection science is available on the web. The LIGO website is a rich source of information on the latest detections:
`www.ligo.caltech.edu/`.
Websites of other detectors and projects also provide information about plans and specific technologies, often in languages other than English. These include:
GEO600: `http://geo600.org/`,
Virgo: `www.virgo-gw.eu/`,
KAGRA: `https://gwcenter.icrr.u-tokyo.ac.jp/en/`,
LIGO-India: `www.gw-indigo.org`,
LISA: `www.lisamission.org`,
Einstein Telescope: `www.et-gw.eu`, and
Cosmic Explorer: `https://cosmicexplorer.org`.
The Gravitational Wave International Committee (GWIC) maintains its own website at `https://gwic.ligo.org`.

Educational articles about gravitational waves and other applications of general relativity can be found on the Einstein Online website:
`www.einstein-online.info/`.
The Scienceface project has dozens of short videos and interviews with leading scientists in the field, most designed for a secondary-school audience. The videos may be viewed on the project's YouTube channel: `www.youtube.com/user/scienceface/featured/`.

For young people, the award-winning book by Andersson (2017) guides readers from Einstein's early work through to the detection of gravitational waves. An entertaining book on gravitational waves aimed at even younger children is Lundgren *et al.* (2016).

Our derivation of the energy flux of gravitational waves in § 9.5 rests on Feynman's sticky-bead argument, which he presented at the famous 1957 'Chapel Hill' conference, whose proceedings are now available as an open-access book: Rickles & DeWitt (2011). It can be freely downloaded from the website:[5]

`www.edition-open-sources.org/sources/5/index.html`.

Others at the Chapel Hill conference took similar points of view, notably Hermann Bondi and Felix Pirani. Joseph Weber attended this meeting and was so convinced by the arguments that gravitational waves were a real phenomenon that he was inspired to pioneer the field of detection.

Feynman's argument is the most directly physical method of getting the flux, and it does not rely extensively on the techniques of classical field theory. There are other ways of deriving the flux, starting for example with variational principles for the Einstein field equations, a topic we do not deal with in this book. See Misner *et al.* (1973) for details. Our approach also stays within linear theory, while of course we then use it to calculate the energy radiated by extremely relativistic sources, where gravity is nonlinear. Computing the energy loss of fully relativistic systems, and proving that at lowest order it is satisfactory to use the quadrupole formula as if in linear theory, was a problem that was not fully solved until long after the Chapel Hill meeting. See, for example, Damour (1987), Futamase (1983), and especially Blanchet (2014).

An elementary review of standard sirens is given in Holz *et al.* (2018).

Exercises

9.1   A function $f(s)$ has derivative $f'(s) = \mathrm{d}f/\mathrm{d}s$. Prove that $\partial f(k_\mu x^\mu)/\partial x^\nu = k_\nu f'(k_\mu x^\mu)$. Use this to prove Eq. 9.4 and the equation following it.

9.2   Show that the real and imaginary parts of Eq. 9.2 at a fixed spatial position $\{x^i\}$ oscillate sinusoidally in time with frequency $\omega = k^0$.

9.3   Let $\bar{h}^{\alpha\beta}(t, x^i)$ be any solution of Eq. 9.1 which has the property $\int \mathrm{d}x^\alpha |\bar{h}^{\mu\nu}|^2 < \infty$ for the integral over any particular x^α, holding other coordinates fixed. Define the Fourier transform of $\bar{h}^{\alpha\beta}(t, x^i)$ as

$$\bar{H}^{\alpha\beta}(\omega, k^i) = \int \bar{h}^{\alpha\beta}(t, x^i) \exp(\mathrm{i}\omega t - \mathrm{i}k_j x^j) \, \mathrm{d}t \mathrm{d}^3 x.$$

Show, by transforming Eq. 9.1, that $\bar{H}^{\alpha\beta}(\omega, k^i)$ is zero except for those values of ω and k^i that satisfy Eq. 9.10. By applying the inverse transform, write $\bar{h}^{\alpha\beta}(t, x^i)$ as a superposition of plane waves.

---

[5] This volume provides a fascinating insight into the conceptual problems that physicists had with general relativity at the time. Modern textbooks, such as the present one, benefit from the work that physicists did in the following two decades to retire the confusions and set the theory on firm physical foundations. See the discussion in Blum *et al.* (2018) for more on why these confusions persisted and how they were ultimately addressed.

9.4 Derive Eqs. 9.16 and 9.17.

9.5 (a) Show that $A_{\alpha\beta}^{(NEW)}$, given by Eq. 9.17, satisfies the gauge condition $A^{\alpha\beta}k_\beta = 0$ if $A_{\alpha\beta}^{(OLD)}$ does.

(b) Use Eq. 9.18 for $A_{\alpha\beta}^{(NEW)}$ to constrain B^μ.

(c) Show that Eq. 9.19 for $A_{\alpha\beta}^{(NEW)}$ imposes only three constraints on B^μ, not the four that one might expect from the fact that the free index α can take any value from 0 to 3. Do this by showing that the particular linear combination $k^\alpha(A_{\alpha\beta}U^\beta)$ vanishes for any B^μ.

(d) Using (b) and (c), solve for B^μ as a function of k^μ, $A_{\alpha\beta}^{(OLD)}$, and U^μ. These determine B^μ: there is no further gauge freedom.

(e) Show that it is possible to choose ξ^β in Eq. 9.15 to make any superposition of plane waves satisfy Eqs. 9.18 and 9.19, so that these are generally applicable to gravitational waves of any sort.

(f) Show that one cannot achieve Eqs. 9.18 and 9.19 for a static solution, i.e. one for which $\omega = 0$.

9.6 Fill in all the algebra implicit in the paragraph leading to Eq. 9.21.

9.7 Give a more rigorous proof that Eqs. 9.22 and 9.23 imply that a free particle initially at rest in the TT gauge remains at rest.

9.8 Does the free particle of the discussion following Eq. 9.23 *feel* any acceleration? For example, if the particle is a bowl of soup (whose diameter is much less than a wavelength), does the soup slosh about in the bowl as the wave passes?

9.9 Does the free particle of the discussion following Eq. 9.23 *see* any acceleration? To answer this, consider the two particles whose relative proper distance is calculated in Eq. 9.24. Let the particle at the origin send a beam of light towards the other, and let it be reflected by the other and received back at the origin. Calculate the amount of proper time elapsed at the origin between the emission and reception of the light (you may assume that the particles' separation is much less than a wavelength of the gravitational wave). By monitoring changes in this time, the particle at the origin can 'see' the relative acceleration of the two particles.

9.10 (a) We have seen that

$$h_{yz} = A \sin \omega(t - x), \quad \text{all other } h_{\mu\nu} = 0,$$

with A and ω constants, $|A| \ll 1$, is a solution to Eqs. 9.1 and 9.11. For this metric tensor, compute all the components of $R_{\alpha\beta\mu\nu}$ and show that some are not zero, so that the spacetime is not flat.

(b) Another metric is given by

$$h_{yz} = A \sin \omega(t - x), \quad h_{tt} = 2B(x - t),$$
$$h_{tx} = -B(x - t), \quad \text{all other } h_{\mu\nu} = 0.$$

Show that this also satisfies the field equations and the gauge conditions.

(c) For the metric in (b) compute $R_{\alpha\beta\mu\nu}$. Show that it is the *same* as that in (a).

(d) From (c) we conclude that the geometries are identical, and that the difference in the metrics is due to a small coordinate change. Find a ξ^μ such that

$$h_{\mu\nu}(\text{part (a)}) - h_{\mu\nu}(\text{part (b)}) = -\xi_{\mu,\nu} - \xi_{\nu,\mu}.$$

9.11 (a) Derive Eq. 9.27.

(b) Solve Eqs. 9.28a and 9.28b for the motion of the test particles in the polarization rings shown in Figure 9.1.

9.12 Do calculations analogous to those leading to Eqs. 9.28 and 9.32 to show that the separation of particles in the z direction (the direction of travel of the wave) is unaffected.

9.13 One kind of background Lorentz transformation is a simple $45°$ rotation of the x and y axes in the x–y plane. Show that, under such a rotation from (x, y) to (x', y'), we have $h^{\text{TT}}_{x'y'} = -h^{\text{TT}}_{xx}, h^{\text{TT}}_{x'x'} = h^{\text{TT}}_{xy}$. This is consistent with Figure 9.1.

9.14 (a) Show that a plane wave with $A_{xy} = 0$ in Eq. 9.21 has the metric

$$ds^2 = -dt^2 + (1 + h_+)dx^2 + (1 - h_+)dy^2 + dz^2, \qquad (9.164)$$

where $h_+ = A_{xx} \sin[\omega(t - z)]$.

(b) Show that this wave does not change the proper separations of free particles if they are aligned along a line bisecting the angle between the x and y axes.

(c) Show that a plane wave with $A_{xx} = 0$ in Eq. 9.21 has the metric

$$ds^2 = -dt^2 + dx^2 + 2h_+ dx\, dy + dy^2 + dz^2 \qquad (9.165)$$

where $h_+ = A_{xx} \sin[\omega(t - z)]$.

(d) Show that the wave in (c) does not change the proper separations of free particles if they are aligned along the coordinate axes.

(e) Show that the wave in (c) produces an elliptical distortion of the circle that is rotated by $45°$ to that of the wave in (a).

9.15 (a) A wave is said to be circularly polarized in the x–y plane if $h^{\text{TT}}_{yy} = -h^{\text{TT}}_{xx}$ and $h^{\text{TT}}_{xy} = \pm i h^{\text{TT}}_{xx}$. Show that, for such a wave, the ellipse in Figure 9.1 rotates without changing shape.

(b) A wave is said to be elliptically polarized with principal axes x and y if $h^{\text{TT}}_{xy} = \pm i a h^{\text{TT}}_{xx}$, where a is some real number, and $h^{\text{TT}}_{yy} = -h^{\text{TT}}_{xx}$. Show that if $h^{\text{TT}}_{xy} = \alpha h^{\text{TT}}_{xx}$, where α is a complex number (the general case for a plane wave), new axes x' and y' can be found for which the wave is elliptically polarized with principal axes x' and y'. Show that circular and linear polarization are special cases of elliptical polarization.

9.16 Two plane waves with TT amplitudes, $A^{\mu\nu}$ and $B^{\mu\nu}$, are said to have orthogonal polarizations if $(A^{\mu\nu})^* B_{\mu\nu} = 0$, where $(A^{\mu\nu})^*$ is the complex conjugate of $A^{\mu\nu}$. Show that if $A^{\mu\nu}$ and $B^{\mu\nu}$ are orthogonal polarizations, a $45°$ rotation of $B^{\mu\nu}$ makes it proportional to $A^{\mu\nu}$.

9.17 Find the transformation from the coordinates (t, x, y, z) of Eqs. 9.33–9.36 to the local inertial frame of Eq. 9.37. Use this to verify Eq. 9.38.

9.18 Prove Eq. 9.39.

9.19 Use the sum of Eqs. 9.40 and 9.41 to show that the center of mass of the spring remains at rest as the wave passes.

9.20 Derive Eq. 9.44 from Eq. 9.43, and then prove Eq. 9.45.

9.21 Generalize Eq. 9.45 to the case of a plane wave with arbitrary elliptical polarization (Exercise 9.15) traveling in an arbitrary direction relative to the separation of the masses.

9.22 Consider the equation of geodesic deviation, Eq. 6.87, from the point of view of the geodesic at the center of mass of the detector of Eq. 9.45. Show that the vector ξ as we have defined it in Eq. 9.42 is twice the connecting vector from the center of mass to one of the masses, as defined in Eq. 6.83. Show that the tidal force as measured by the center of mass leads directly to Eq. 9.45.

9.23 Derive Eqs. 9.48 and 9.49, and derive the general solution of Eq. 9.45 for arbitrary initial data at $t = 0$, given Eq. 9.46.

9.24 Prove Eq. 9.53.

9.25 Derive Eq. 9.56 from the given definition of Q.

9.26 (a) Use the metric for a plane wave with '+' polarization, Eq. 9.58, to show that the square of the coordinate speed (in the TT coordinate system) of a photon moving in the x direction is

$$\left(\frac{dx}{dt}\right)^2 = \frac{1}{1 + h_+}.$$

This is not identically 1. Does this violate relativity? Why or why not?

(b) Imagine that an experimenter at the center of the circle of particles in Figure 9.1 sends a photon to the particle on the circle at coordinate location $x = L$ on the positive-x axis, and that the photon is reflected when it reaches the particle and returns to the experimenter. Suppose further that this takes such a short time that h_+ does not change significantly during the experiment. To first order in h_+, show that the experimenter's proper time that elapses between sending out the photon and receiving it back is $(2 + h_+)L$.

(c) The experimenter says that this proves that the proper distance between herself and the particle is $(1 + h_+/2)L$. Is this a correct interpretation of her experiment? If the experimenter uses an alternative measuring process for proper distance, such as laying out a number of standard meter sticks between her location and the particle, would that produce the same answer? Why or why not?

(d) Show that, if the experimenter simultaneously does the same experiment with a particle on the y axis at $y = L$, that photon will return after a proper time $(2 - h_+)L$.

(e) The difference in these return times is $2h_+L$ and can be used to measure the wave's amplitude. Does this result depend on our use of the TT gauge, i.e. would we have obtained the same answer had we used a different coordinate system?

9.27 (a) Derive the full three-term return relation, Eq. 9.63, for the rate of change of the return time for a gravitational wave beam traveling through a plane light wave h_+ along the x direction, when the gravitational wave is moving at an angle θ to the z axis in the x–z plane.

(b) Show that, in the limit where L is small compared to a wavelength of the gravitational wave, the derivative of the return time is the derivative of $t + \delta L$, where $\delta L = L \cos^2 \theta \, h(t)$ is the excess proper distance for small L. Interpret the factor of $\cos^2 \theta$: why should we expect it to be there?

(c) Examine the limit of the three-term formula in (a) when the gravitational wave is traveling along the x axis too ($\theta = \pm\pi/2$); i.e. what happens to light going parallel to a gravitational wave?

9.28 (a) Reconstruct $\bar{h}_{\mu\nu}$ as in Eq. 9.66, using Eq. 9.68, and show that surfaces of constant phase of the wave move outwards for the $A_{\mu\nu}$ term and inwards for $Z_{\mu\nu}$.

(b) Fill in the missing algebra in Eqs. 9.69–9.71.

9.29 Equation 9.67 in the vacuum region outside the source – i.e., where $S_{\mu\nu} = 0$ – can be solved by the separation of variables. Assume a solution for $\bar{h}_{\mu\nu}$ has the form $\sum_{lm} A^{lm}_{\mu\nu} f_l(r) Y_{lm}(\theta, \phi)/\sqrt{r}$, where Y_{lm} is a spherical harmonic.

(a) Show that $f_l(r)$ satisfies the equation

$$f_l'' + \frac{1}{r} f_l' + \left[\Omega^2 - \frac{(l + \frac{1}{2})^2}{r^2}\right] f_l = 0.$$

(b) Show that the most general spherically symmetric solution is given by Eq. 9.68.

(c) Substitute the variable $s = \Omega r$ to show that f_l satisfies the equation

$$s^2 \frac{d^2 f_l}{ds^2} + s \frac{df_l}{ds} + [s^2 - (l + \tfrac{1}{2})^2] f_l = 0. \tag{9.166}$$

This is known as Bessel's equation, whose solutions are called Bessel functions of order $l + \frac{1}{2}$. Their properties are explored in most textbooks on mathematical physics.

(d) Show, by substitution into Eq. 9.166, that the function $f_l/\sqrt{s}$ is a linear combination of the spherical Bessel and spherical Neumann functions,

$$j_l(s) = (-1)^l s^l \left(\frac{1}{s} \frac{d}{ds}\right)^l \left(\frac{\sin s}{s}\right), \tag{9.167}$$

$$n_l(s) = (-1)^{l+1} s^l \left(\frac{1}{s} \frac{d}{ds}\right)^l \left(\frac{\cos s}{s}\right). \tag{9.168}$$

(e) Use Eqs. 9.167 and 9.168 to show that, for $s \gg l$, the dominant behavior of j_l and n_l is given by

$$j_l(s) \sim \frac{1}{s} \sin\left(s - \frac{l\pi}{2}\right), \tag{9.169}$$

$$n_l(s) \sim -\frac{1}{s} \cos\left(s - \frac{l\pi}{2}\right). \tag{9.170}$$

(f) Similarly, show that for $s \ll l$ the dominant behavior is

$$j_l(s) \sim s^l / (2l + 1)!!, \tag{9.171}$$

$$n_l(s) \sim -(2l - 1)!!/s^{l+1}, \tag{9.172}$$

where we use the standard double factorial notation

$$(m)!! = m(m-2)(m-4)\cdots 3 \cdot 1 \tag{9.173}$$

for odd m.

(g) Show from (e) that the outgoing-wave vacuum solution of Eq. 9.67 for any fixed l and m is

$$(\bar{h}_{\mu\nu})_{lm} = A_{\mu\nu}^{lm} h_l^{(1)}(\Omega r) e^{-i\Omega t} Y_{lm}(\theta,\phi), \tag{9.174}$$

where $h_l^{(1)}(\Omega r)$ is the spherical Hankel function of the first kind,

$$h_l^{(1)}(\Omega r) = j_l(\Omega r) + i n_l(\Omega r). \tag{9.175}$$

(h) Repeat the calculation of Eqs. 9.69–ch09:eqn9.65, only this time multiply Eq. 9.67 by $j_l(r\Omega) Y_{lm}^*(\theta,\phi)$ before performing the integrals. Show that the left-hand side of Eq. 9.67 becomes, when so integrated, exactly

$$\varepsilon^2 \left(j_l(\Omega\varepsilon)\frac{\mathrm{d}}{\mathrm{dr}} B_{\mu\nu}(\varepsilon) - B_{\mu\nu}(\varepsilon)\frac{\mathrm{d}}{\mathrm{dr}} j_l(\Omega\varepsilon) \right),$$

and that when $\Omega\varepsilon \ll l$ this becomes (with the help of Eq. 9.174 and Eqs. 9.171–9.172 above, since we assume $r = \varepsilon$ is outside the source) simply $i A_{\mu\nu}^{lm}/\Omega$. Similarly, show that the right-hand side of Eq. 9.67 integrates to $-16\pi\Omega^l \int T_{\mu\nu} r^l Y_{lm}^*(\theta,\phi) \mathrm{d}^3 x/(2l+1)!!$ in the same approximation.

(i) Show, then, that the solution is Eq. 9.174, with

$$A_{\mu\nu}^{lm} = 16\pi i\Omega^{l+1} J_{\mu\nu}^{lm}/(2l+1)!!, \tag{9.176}$$

where

$$J_{\mu\nu}^{lm} = \int T_{\mu\nu} r^l Y_{lm}^*(\theta,\phi)\, \mathrm{d}^3 x. \tag{9.177}$$

(j) Let $l = 0$ and deduce Eqs. 9.73 and 9.74.

(k) Show that if $J_{\mu\nu}^{lm} \neq 0$ for some l, then the terms neglected in Eq. 9.176 because of the approximation $\Omega\varepsilon \ll 1$ are of the same order as the dominant terms in Eq. 9.176 for $l+1$. In particular, this means that if $J_{\mu\nu} \neq 0$ in Eq. 9.72, any attempt to get a more accurate answer than Eq. 9.74 must take into account not only the terms for $l > 0$ but also the neglected terms in the derivation of Eq. 9.74, such as Eq. 9.69.

9.30 Rewrite Eq. 9.82a for a set of N discrete point particles, whose masses are $\{m_{(A)}, A = 1, \ldots, N\}$ and whose positions are $\{x_{(A)}^i\}$.

9.31 Calculate the quadrupole tensor I_{jk} and its traceless counterpart $\bar{I}_{jk}$ (Eq. 9.87) for the following mass distributions.

(a) A spherical star whose density is $\rho(r,t)$. Take the origin of the coordinates in Eq. 9.82 to be the center of the star.

(b) The star in (a), but with the origin of the coordinates at an arbitrary point.

(c) An ellipsoid of uniform density ρ and semiaxes of length a, b, c oriented along the x, y, and z axes respectively. Take the origin to be at the center of the ellipsoid.

(d) The ellipsoid in (c), but rotating about the z axis with angular velocity ω.

(e) Four masses m located respectively at the points $(a, 0, 0)$, $(0, a, 0)$, $(-a, 0, 0)$, $(0, -a, 0)$.

(f) The masses as in (e), but all moving counterclockwise about the z axis on a circle of radius a with angular velocity ω.

(g) Two masses m connected by a massless spring, each oscillating on the x axis with angular frequency ω and amplitude A about mean equilibrium positions a distance l_0 apart, keeping their center of mass fixed.

(h) Unequal masses m and M connected by a spring of spring constant k and equilibrium length l_0, oscillating (with their center of mass fixed) at the natural frequency of the system, with amplitude $2A$ (this is the total stretching of the spring). Their separation is along the x axis.

9.32 This exercise develops the TT gauge for spherical waves.

(a) In order to transform Eq. 9.83 to the TT gauge, use a gauge transformation generated by a vector $\xi^\alpha = B^\alpha(x^\mu)e^{i\Omega(r-t)}/r$, where B^α is a slowly varying function of x^μ. Find the general transformation law to order $1/r$.

(b) Demand that the new $\bar{h}_{\alpha\beta}$ satisfy three conditions to order $1/r$: $\bar{h}_{0\mu} = 0$, $\bar{h}^\alpha_{\ \alpha} = 0$, and $\bar{h}_{\mu j}n^j = 0$, where $n^j := x^j/r$ is the unit vector in the radial direction. Show that it is possible to find functions B^α which accomplish such a transformation *and* which satisfy $\Box\, \xi^\alpha = 0$ to order $1/r$.

(c) Show that Eqs. 9.84–9.87 hold in the TT gauge.

9.33 (a) Let n^j be a unit vector in three-dimensional Euclidean space. Show that $P^j_{\ k} = \delta^j_{\ k} - n^j n_k$ is the projection tensor orthogonal to n^j, i.e. show that, for any vector V^j, (i) $P^j_{\ k}V^k$ is orthogonal to n^j and (ii) $P^j_{\ k}P^k_{\ l}V^l = P^j_{\ k}V^k$.

(b) Show that the TT gauge $\bar{h}^{TT}_{ij}$ of Eqs. 9.84–9.86 is related to the original $\bar{h}_{kl}$ of Eq. 9.83 by

$$\bar{h}^{TT}_{ij} = P^k_{\ i}P^l_{\ j}\bar{h}_{kl} - \tfrac{1}{2}P_{ij}(P^{kl}\bar{h}_{kl}), \qquad (9.178)$$

where n^j points in the z direction.

9.34 Show that $\textnormal{+}_{jk}$ is trace-free, i.e. $\textnormal{+}^l_{\ l} = 0$.

9.35 For the systems described in Exercise 9.31, calculate the TT quadrupole radiation field, Eqs. 9.85 and 9.86 or Eq. 9.178, along the x, y, and z axes. In Eqs. 9.85 and 9.86 be sure to change the indices appropriately when doing the calculation on the x and y axes, as in the discussion leading to Eq. 9.91.

9.36 Use Eq. 9.178 or a rotation of the axes in Eqs. 9.85, 9.86 to calculate the amplitude and orientation of the polarization ellipse of the radiation from the simple oscillator, Eq. 9.88, traveling at an angle θ to the x axis.

9.37 The ω and 2ω terms in Eq. 9.93 are qualitatively different, in that the 2ω term depends only on the amplitude of oscillator A, while the ω term depends on both

A and the separation of the masses l_0. Why should l_0 be involved – the masses don't move over that distance? The answer is that stresses *are* transmitted over that distance by the spring, and stresses cause the radiation. To see this, do an analogous calculation for a similar system in which stresses are *not* passed over large distances. Consider a system consisting of two pairs of masses. Each pair has one particle of mass m and another of mass $M \gg m$. The masses within each pair are connected by a short spring whose natural frequency is ω. The pairs' centers of mass are at rest relative to one another. The springs oscillate with equal amplitude in such a way that each mass m oscillates sinusoidally with amplitude A, and the centers of oscillation of the masses are separated by $l_0 \gg A$. The masses oscillate out of phase. Use the calculation of Exercise 9.31(h) to show that the radiation field of the system is Eq. 9.93 without the ω term. The difference between this system and that in Eq. 9.93 may be thought to be the origin of the stresses that maintain the motion of the masses m.

9.38 Do the same as Exercise 9.36 for the binary system, Eqs. 9.98, 9.99, but instead of finding the orientation of the linear polarization, find the orientation of the ellipse of elliptical polarization.

9.39 Calculate the radiation produced by a binary system in a circular orbit with (possibly) unequal masses m_1 and m_2 and separation l_0, using previous results, as follows. Let their total mass be $M = m_1 + m_2$ and their reduced mass be $\mu = m_1 m_2 / M$.

(a) Show that the center of mass of the system (which will remain at rest during their motion) is a distance $m_2 l_0/M$ from the position of mass m_1 and, in the opposite direction, a distance $m_1 l_0/M$ from the position of mass m_2. Each mass therefore orbits on a circle of the corresponding radius, remaining diametrically opposite one another at the computed distances.

(b) Repeat the calculation leading from Eqs. 9.95, 9.96 to show that the quadrupole tensor can be obtained from that of the equal-mass case by replacing m by 2μ, for fixed values of l_0 and ω.

(c) Show that the orbital frequency of the unequal-mass system is given by $\omega = (M/l_0^3)^{1/2}$.

(d) Generalize Eq. 9.98, for the circularly polarized radiation amplitudes that go out perpendicular to the orbital plane, to the unequal-mass case by replacing m by 2μ and by using the previous part of this question to eliminate l_0 (which is not a direct observable for a binary at a large distance) in favor of ω. You should obtain Eq. 9.149.

(e) In the same manner, convert Eq. 9.99, for the linearly polarized radiation that comes out in the orbital plane, into the expression valid for unequal masses given in Eq. 9.150.

9.40 Calculate the energetics of a radiating, unequal-mass binary system, as follows. Use the same notation as in Exercise 9.39, and as in that exercise eliminate the unobservable l_0 in favor of ω.

(a) Generalize the expression for the gravitational wave luminosity of a circular binary as given in Eq. 9.136 to the unequal-mass case. You should obtain Eq. 9.153.

(b) Compute the Newtonian energy of the unequal-mass circular binary system by adding up the kinetic energies of the two components with the (negative) gravitational potential energy, generalizing Eq. 9.140 and the calculation following it. You should obtain Eq. 9.154.

9.41 Use the expressions for luminosity and orbital energy given in Eqs. 9.153 and 9.154 to investigate the relationship between the chirp mass and the rate of change of the orbital frequency.

(a) Use the energy balance relation for a radiating circular binary system, given in Eq. 9.155, to show that the rate at which the orbital frequency changes, $\dot{\omega}$, is

$$\dot{\omega} = \frac{96}{5}\mathcal{M}^{-2}(\mathcal{M}\omega)^{11/3}. \tag{9.179}$$

(b) Solve this equation to find $\mathcal{M}$ in terms of ω and $\dot{\omega}$. You should obtain Eq. 9.156.

9.42 We investigate some properties of the chirp mass.

(a) In Newtonian mechanics the reduced mass of a binary system is $\mu = m_1 m_2 / M$, where $M = m_1 + m_2$. Show that

$$\mathcal{M} = \mu^{3/5} M^{2/5} \quad \text{and} \quad \mathcal{M}/M = (\mu/M)^{3/5}. \tag{9.180}$$

(b) When the ratio of the component masses is very small we define $\delta := m_2/M \ll 1$. Show in this case, working to lowest order in δ, that

$$\mathcal{M} = \delta^{3/5} M. \tag{9.181}$$

9.43 In this exercise we fill in the steps needed to get the scaling relations in Eqs. 9.157–9.160.

(a) First we address the characteristic timescale for the orbit to change. By convention this is the ratio $\tau_{\text{chirp}} = \omega/\dot{\omega}$, which has dimensions of time. Show that a Taylor expansion of ω in time about an observation time t_0, at which $\omega(t_0) = \omega_0$, $\dot{\omega}(t_0) = \dot{\omega}_0$, and $\tau_{\text{chirp:0}} = \omega_0/\dot{\omega}_0$ has the form

$$\omega(t) = \omega_0[1 + (t - t_0)/\tau_{\text{chirp:0}} + \cdots].$$

This shows precisely what is meant by a characteristic timescale: it is the timescale during which ω will change by O(1). Our definition is not unique; we could have taken the timescale over which the orbital radius would change by O(1). This would differ only by a constant factor, again of O(1).

(b) Now find τ_{chirp} from its definition by using Eq. 9.179. You should get Eq. 9.157. Notice that this decreases dramatically as ω gets larger, i.e. as the two components get closer together.

(c) To get the remaining lifetime to merger, we need to write Eq. 9.179 as a differential equation for $dt/d\omega$, and then to integrate this equation over ω for t from the present time to the time at which $\omega = \infty$. You should get Eq. 9.158.

(d) The next relation we investigate is the way in which the phase ψ increases during the orbital inspiral. The phase and frequency are related simply by $\omega = d\psi/dt$. Divide this by Eq. 9.179 to get a differential equation between ψ and ω. Integrate this for ψ to show that the phase remaining from the time when the orbit has frequency ω is given by Eq. 9.159.

(e) Finally use the luminosity relation given in Eq. 9.153 to eliminate $\mathcal{M}\omega$ from Eq. 9.159 and so obtain the fundamental relationship between the luminosity and the orbital phase, Eq. 9.160.

9.44 For each of the following systems of compact objects, calculate $\mathcal{M}$ and then use the Newtonian orbit equation to estimate their orbital angular velocity ω at a given moment. From this calculate the dimensionless parameter $\mathcal{M}\omega$ and the number of remaining orbits if the system were to follow the Newtonian point-particle approximation. Finally calculate their gravitational wave luminosity at a given moment.

(a) Two neutron stars, each with mass $1.4\,M_\odot$, in a circular orbit at the moment their separation (in the Newtonian approximation) is $l_0 = 30$ km.

(b) Two black holes of equal mass $M = 10^6\,M_\odot$ at the moment they are merging. In the Newtonian approximation their separation is $l_0 = 4M$.

(c) The same as (b) but with $M = 10^{-5}$ g.

(d) Two black holes, one with mass $m_1 = 10^6\,M_\odot$ and the other with mass $m_2 = 10\,M_\odot$. Assume that they merge when their separation l_0 is the radius of the horizon of the large hole, namely $2m_1$.

9.45 We will see in Chapter 13 that the observable Universe contains of order 10^{11} galaxies, each with of order 10^{11} stars. Our Sun is a typical star, so use its luminosity to estimate the total electromagnetic luminosity of the entire Universe, in watts. Convert this to a dimensionless value and compare it with the luminosity of a pair of merging black holes.

9.46 Let two spherical stars of mass m_1 and m_2 be in elliptical orbit about one another in the x–y plane. Let the orbit be characterized by its total energy E and its angular momentum J.

(a) Use Newtonian gravity to calculate the equation of the orbits of both masses about their center of mass. Express the orbital period P, minimum separation l_0, and eccentricity e as functions of E and J.

(b) Calculate $\mathcal{I}_{kj}$ for this system.

(c) Calculate from Eq. 9.107 the TT radiation field along the x and z axes. Show that your result reduces to Eqs. 9.98, 9.99 when $m_1 = m_2$ and the orbits are circular.

9.47 Show from Eq. 9.102 that spherically symmetric motions produce no gravitational radiation.

9.48 Derive Eq. 9.116 from Eq. 9.115 in the manner suggested in the text.

9.49 (a) Derive Eq. 9.117.

(b) Derive Eqs. 9.118 and 9.119 by superposing Eqs. 9.108 and 9.117 and assuming R is small.

(c) Derive Eq. 9.121 in the manner indicated.

9.50   Show that if we define an averaged stress–energy tensor for the waves

$$T_{\alpha\beta} = \ll \bar{h}^{TT}_{\mu\nu,\alpha} \bar{h}^{TT\,\mu\nu}{}_{,\beta} \gg /32\pi \tag{9.182}$$

(where $\ll \;\; \gg$ denotes an average over both one period of oscillation in time and one wavelength of distance in all spatial directions), then the flux F of Eq. 9.123 is the component T^{0z} for that wave. A more detailed argument shows that Eq. 9.182 can in fact be regarded as the stress–energy tensor of any wave packet, provided the averages are defined suitably. This is called the Isaacson stress–energy tensor. See Misner *et al.* (1973) for details.

9.51   (a) Derive Eq. 9.126 from Eq. 9.124.
   (b) Justify Eq. 9.128 from Eq. 9.126.
   (c) Derive Eq. 9.128 from Eq. 9.123 using Exercise 9.33(b).

9.52   (a) Consider the integral in Eq. 9.129. We shall do it by the following method. (i) Argue on grounds of symmetry that $\int n^j n^k \sin\theta \, d\theta \, d\phi$ must be proportional to δ^{jk}. (ii) Evaluate the constant of proportionality by explicitly considering the case $j = k = z$.
   (b) Follow the same method for Eq. 9.130. In (i) argue that the integral can depend only on δ^{ij}, and show that the given tensor is the only tensor constructed only from δ^{ij} which has the symmetry of being unchanged when the values of any two of its indices are exchanged.

9.53   Derive Eqs. 9.131 and 9.132, remembering Eq. 9.125 and the fact that $\dot{I}_{ij}$ is symmetric.

9.54   (a) Recall that the angular momentum of a particle is p_ϕ. It follows that the angular momentum flux of a continuous system across a surface $x^i = $ const. is $T_{i\phi}$. Use this and Exercise 9.50 to show that the total z component of angular momentum radiated by a source of gravitational waves (which is the integral over a sphere of large radius of $T_{r\phi}$ in Eq. 9.182) is

$$F_J = -\tfrac{2}{5}(\dddot{I}_{xl}\ddot{I}_y{}^l - \dddot{I}_{yl}\ddot{I}_x{}^l). \tag{9.183}$$

   (b) Show that if $\bar{h}^{TT}_{\mu\nu}$ depends on t and ϕ only as $\cos(\Omega t - m\phi)$, then the ratio of the total energy radiated and the total angular momentum radiated is Ω/m.

9.55   Calculate Eq. 9.136.

9.56   For the arbitrary binary system of Exercise 9.46, with eccentricity e:

(a) Show that the average energy loss rate over one orbit is

$$\langle dE/dt\rangle = -\frac{32}{5}\frac{\mu^2(m+M)^3}{a^5(1-e^2)^{7/2}}\left(1 + \frac{73}{24}e^2 + \frac{37}{96}e^4\right) \tag{9.184}$$

and, from the result of Exercise 9.54(a)

$$\langle dL/dt\rangle = -\frac{32}{5}\frac{\mu^2(m+M)^{5/2}}{a^{7/2}(1-e^2)^{7/2}}\left(1 + \frac{7}{8}e^2\right), \tag{9.185}$$

where $\mu = mM/(m+M)$ is the reduced mass.

(b) Show that

$$\langle da/dt \rangle = -\frac{64}{5} \frac{\mu(m+M)^2}{a^3(1-e^2)^{7/2}} \left(1 + \frac{73}{24}e^2 + \frac{37}{96}e^4\right), \tag{9.186}$$

$$\langle de/dt \rangle = -\frac{304}{15} \frac{\mu(m+M)^2 e}{a^4(1-e^2)^{5/2}} \left(1 + \frac{121}{304}e^2\right), \tag{9.187}$$

$$\langle dP/dt \rangle = -\frac{192\pi}{5} \frac{\mu(m+M)^{3/2}}{a^{5/2}(1-e^2)^{7/2}} \left(1 + \frac{73}{24}e^2 + \frac{37}{96}e^4\right). \tag{9.188}$$

(c) Verify Eq. 9.146. (Do parts (b) and (c) even if you cannot manage part (a).) These were originally derived by Peters (1964).

Spherical solutions for stars

10.1 Coordinates for spherically symmetric spacetimes

For our first study of strong gravitational fields in GR we will consider spherically symmetric systems. They are reasonably simple, yet physically very important, since very many astrophysical objects appear to be nearly spherical. We begin by choosing the coordinate system to reflect the assumed symmetry.

Flat space in spherical coordinates

By defining the usual coordinates (r, θ, ϕ), the line element of Minkowski space can be written

$$\mathrm{d}s^2 = -\mathrm{d}t^2 + \mathrm{d}r^2 + r^2(\mathrm{d}\theta^2 + \sin^2\theta\,\mathrm{d}\phi^2). \tag{10.1}$$

Each surface of constant r and t is a sphere or, more precisely, a two-sphere, a two-dimensional spherical surface. Distances along curves confined to such a sphere are given by the above equation with $\mathrm{d}t = \mathrm{d}r = 0$:

$$\mathrm{d}l^2 = r^2(\mathrm{d}\theta^2 + \sin^2\theta\,\mathrm{d}\phi^2) := r^2\,\mathrm{d}\Omega^2, \tag{10.2}$$

which defines the symbol $\mathrm{d}\Omega^2$ for the element of solid angle. We note that such a sphere has circumference $2\pi r$ and area $4\pi r^2$, i.e. 2π times the square root of the coefficient of $\mathrm{d}\Omega^2$ and 4π times the coefficient of $\mathrm{d}\Omega^2$ respectively. Conversely, any two-surface whose line element is Eq. 10.2 with r^2 independent of θ and ϕ has the intrinsic geometry of a two-sphere.

Two-spheres in a curved spacetime

The statement that a spacetime is spherically symmetric can now be made more precise: it implies that every point of spacetime is on a two-surface that is a two-sphere, i.e. whose line element is

$$\mathrm{d}l^2 = f(r', t)(\mathrm{d}\theta^2 + \sin^2\theta\,\mathrm{d}\phi^2), \tag{10.3}$$

where $f(r', t)$ is an unknown function of the other two coordinates of our manifold, r' and t. The area of each sphere is $4\pi f(r', t)$. We *define* the radial coordinate r of our spherical geometry such that $f(r', t) := r^2$. This represents a coordinate transformation

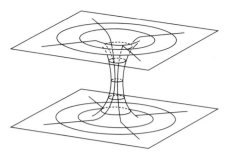

Figure 10.1 Two plane sheets connected by a circular throat: there is circular (axial) symmetry, but the center of any circle is not in the two-space.

from (r', t) to (r, t). Then any surface $r = $ const., $t = $ const. is a two-sphere of area $4\pi r^2$ and circumference $2\pi r$. This coordinate r is called the 'curvature coordinate' or 'area coordinate' because it defines the radius of curvature and area of the spheres. There is no *a priori* relation between r and the proper distance from the center of the sphere to its surface. This r is defined only by the properties of the spheres themselves. Since their 'centers' (at $r = 0$ in flat space) are not points on the spheres themselves, the statement that a spacetime is spherically symmetric does not require even that there *be* a point at the center. A simple counterexample of a two-space in which there are *circles* but no point at the center of them is shown in Figure 10.1. The space consists of two sheets which are joined by a 'throat'. The whole thing is symmetric about an axis along the middle of the throat, but the points on this axis – which are the 'centers' of the circles – are *not* part of the two-dimensional surface illustrated. Yet if ϕ is an angle about the axis, the line element on each circle is just $r^2 \mathrm{d}\phi^2$, where r is a constant labeling each circle. This r is the same sort of coordinate that we use in our spherically symmetric spacetime.

Meshing the two-spheres into a three-space

Consider the spheres at r and $r + \mathrm{d}r$. Each has a coordinate system (θ, ϕ), but up to now we have not required any relation between them. That is, one could conceive of having the pole for the sphere at r in one orientation, while that for $r + \mathrm{d}r$ was in another. The sensible thing is to say that a line of $\theta = $ const., $\phi = $ const. is *orthogonal* to the two-spheres. Such a line has, by definition, a tangent $\vec{e}_r$. Since the vectors $\vec{e}_\theta$ and $\vec{e}_\phi$ lie in the spheres, we require $\vec{e}_r \cdot \vec{e}_\theta = \vec{e}_r \cdot \vec{e}_\phi = 0$. This means that $g_{r\theta} = g_{r\phi} = 0$. (Recall Eqs. 3.3 and 3.21.) This is a definition of the coordinates allowed by spherical symmetry. We thus have restricted the metric to the form

$$\mathrm{d}s^2 = g_{00}\,\mathrm{d}t^2 + 2g_{0r}\,\mathrm{d}r\,\mathrm{d}t + 2g_{0\theta}\,\mathrm{d}\theta\,\mathrm{d}t$$
$$+ 2g_{0\phi}\,\mathrm{d}\phi\,\mathrm{d}t + g_{rr}\,\mathrm{d}r^2 + r^2\,\mathrm{d}\Omega^2. \tag{10.4}$$

Spherically symmetric spacetime. Since not only are the spaces $t = $ const. spherically symmetric, but also the whole spacetime, we must have that a line $r = $ const., $\theta = $ const., $\phi = $ const. is also orthogonal to the two-spheres. Otherwise there would be a preferred direction in space. This means that $\vec{e}_t$ is orthogonal to $\vec{e}_\theta$ and $\vec{e}_\phi$, or $g_{t\theta} = g_{t\phi} = 0$. So now we have

$$ds^2 = g_{00}\,dt^2 + 2g_{0r}\,dr\,dt + g_{rr}\,dr^2 + r^2\,d\Omega^2. \tag{10.5}$$

This is the general metric of a spherically symmetric spacetime, where g_{00}, g_{0r}, and g_{rr} are functions of r and t. We have used our coordinate freedom to reduce the metric to the simplest possible form.

10.2 Static spherically symmetric spacetimes

The metric

Clearly, the simplest physical situation we can describe is a quiescent star or black hole – a static system. We *define* a static spacetime to be one in which we can find a time coordinate t with two properties: (i) all metric components are independent of t, and (ii) the geometry is unchanged by time reversal, $t \rightarrow -t$. The second condition means that a film made of the situation looks the same when run backwards. This is not logically implied by (i), as the example of a rotating star makes clear: time reversal changes the sense of rotation, but the metric components will be constant in time. (A spacetime with property (i) but not necessarily (ii) is said to be *stationary*.)

Condition (ii) has the following implication. The coordinate transformation

$$(t, r, \theta, \phi) \rightarrow (-t, r, \theta, \phi)$$

has $\Lambda^{\bar{0}}{}_0 = -1, \Lambda^i{}_j = \delta^i{}_j$, and we find

$$\left.\begin{array}{l} g_{\bar{0}\bar{0}} = (\Lambda^0{}_{\bar{0}})^2 g_{00} = g_{00}, \\[4pt] g_{\bar{0}\bar{r}} = \Lambda^0{}_{\bar{0}}\Lambda^r{}_{\bar{r}} g_{0r} = -g_{0r}, \\[4pt] g_{\bar{r}\bar{r}} = (\Lambda^r{}_{\bar{r}})^2 g_{rr} = g_{rr}. \end{array}\right\} \tag{10.6}$$

Since the geometry must be unchanged (i.e. since $g_{\bar{\alpha}\bar{\beta}} = g_{\alpha\beta}$), we must have $g_{0r} \equiv 0$. Thus, the metric of a static, spherically symmetric spacetime is

$$ds^2 = -e^{2\Phi}\,dt^2 + e^{2\Lambda}\,dr^2 + r^2\,d\Omega^2, \tag{10.7}$$

where we have introduced $\Phi(r)$ and $\Lambda(r)$ in place of the two unknowns $g_{00}(r)$ and $g_{rr}(r)$. This replacement is acceptable provided $g_{00} < 0$ and $g_{rr} > 0$ everywhere. We shall see below that these conditions do hold inside stars, but they break down for black holes. When we study black holes in the next chapter we shall have to look carefully again at our coordinate system.

If we are interested in stars, which are bounded systems, we are entitled to demand that, far from the star, spacetime is essentially flat. This means that we can impose the following boundary conditions (or asymptotic regularity conditions) on Einstein's equations:

$$\lim_{r \to \infty} \Phi(r) = \lim_{r \to \infty} \Lambda(r) = 0. \tag{10.8}$$

With this condition we say that spacetime is *asymptotically flat*.

Physical interpretation of metric terms

Since we have constructed our coordinates to reflect the physical symmetries of the spacetime, the metric components will have a useful physical significance. The proper radial distance from any radius r_1 to another radius r_2 is

$$l_{12} = \int_{r_1}^{r_2} e^{\Lambda}\, dr, \tag{10.9}$$

since the curve is one on which $dt = d\theta = d\phi = 0$. More important is the significance of g_{00}. Since the metric is independent of t, we know from Chapter 7 that any particle following a geodesic has constant momentum component p_0, which we can define to be the constant $-E$:

$$p_0 := -E. \tag{10.10}$$

But a local *inertial* observer at rest (momentarily) at any radius r of the spacetime measures a different energy. Her four-velocity must have $U^i = d\,x^i/d\,\tau = 0$ (since she is momentarily at rest), and the condition $\vec{U} \cdot \vec{U} = 1$ implies $U^0 = e^{-\Phi}$. The energy she measures is

$$E^* = -\vec{U} \cdot \vec{p} = e^{-\Phi}E. \tag{10.11}$$

We have found therefore that a particle whose geodesic is characterized by the constant E has energy $e^{-\Phi}E$ relative to a locally inertial observer at rest in the spacetime. Since $e^{-\Phi} = 1$ far away, we see that E is the energy that a distant observer would measure if the particle moves far away. We call it the energy at infinity. Since $e^{-\Phi} > 1$ everywhere else (this will be clear later), we see that the particle has a larger energy relative to inertial observers at rest that it passes elsewhere. This extra energy is just the kinetic energy it gains by falling in a gravitational field. The energy is studied in more detail in Exercise 10.3 and § 10.8.

This is particularly significant for photons. Consider a photon emitted at radius r_1 and received very far away. If its frequency in the local inertial frame is ν_{em} (which would be determined by the process emitting it; e.g. a spectral line), then its local energy is $h\nu_{em}$ (h being Planck's constant) and its conserved constant E is $h\nu_{em} \exp[\Phi(r_1)]$. When it reaches the distant observer it is measured to have energy E, and hence frequency $E/h \equiv \nu_{rec} = \nu_{em} \exp[\Phi(r_1)]$. The *redshift* of the photon, defined by

$$z = \frac{\lambda_{rec} - \lambda_{em}}{\lambda_{em}} = \frac{\nu_{em}}{\nu_{rec}} - 1, \tag{10.12}$$

is therefore

$$z = e^{-\Phi(r_1)} - 1. \tag{10.13}$$

This important equation attaches physical significance to e^{Φ}. (Compare this calculation with the one in Chapter 2.)

The Einstein tensor

One can show that for the metric given by Eq. 10.7, the Einstein tensor has components

$$G_{00} = \frac{1}{r^2} e^{2\Phi} \frac{d}{dr}[r(1 - e^{-2\Lambda})], \tag{10.14}$$

$$G_{rr} = -\frac{1}{r^2} e^{2\Lambda}(1 - e^{-2\Lambda}) + \frac{2}{r}\Phi', \tag{10.15}$$

$$G_{\theta\theta} = r^2 e^{-2\Lambda}[\Phi'' + (\Phi')^2 + \Phi'/r - \Phi'\Lambda' - \Lambda'/r], \tag{10.16}$$

$$G_{\phi\phi} = \sin^2\theta \, G_{\theta\theta}. \tag{10.17}$$

where $\Phi' := d\Phi/dr$, etc. All other components vanish.

10.3 Static perfect-fluid Einstein equations

Stress–energy tensor

We are interested in static stars, in which the fluid has no motion. The only nonzero component of $\vec{U}$ is therefore U^0. What is more, the normalization condition

$$\vec{U} \cdot \vec{U} = -1 \tag{10.18}$$

implies, as we have seen before,

$$U^0 = e^{-\Phi}, \qquad U_0 = -e^{\Phi}. \tag{10.19}$$

Then **T** has components given by Eq. 4.38:

$$T_{00} = \rho e^{2\Phi}, \tag{10.20}$$

$$T_{rr} = p e^{2\Lambda}, \tag{10.21}$$

$$T_{\theta\theta} = r^2 p, \tag{10.22}$$

$$T_{\phi\phi} = \sin^2\theta \, T_{\theta\theta}. \tag{10.23}$$

All other components vanish.

Equation of state

The stress–energy tensor involves both p and ρ, but these may be related by an equation of state. For a simple fluid in local thermodynamic equilibrium there always exists a relation of the form

$$p = p(\rho, S), \tag{10.24}$$

which gives the pressure p in terms of the energy density ρ and specific entropy S. One often deals with situations in which the entropy can be considered to be a constant (in particular, negligibly small), so that one has the relation

$$p = p(\rho). \tag{10.25}$$

These relations will of course have different functional forms for different fluids. We will suppose that *some* such relation always exists.

Equations of motion

The conservation laws are (Eq. 7.6)

$$T^{\alpha\beta}{}_{;\beta} = 0. \tag{10.26}$$

These are four equations, one for each value of the free index α. Because of the symmetries, only one of these equations does not vanish identically: the one for which $\alpha = r$. It implies that

$$(\rho + p)\frac{d\Phi}{dr} = -\frac{dp}{dr}. \tag{10.27}$$

This equation tells us that the pressure gradient needed to keep the fluid static in the gravitational field depends on $d\Phi/dr$.

Einstein equations

The $(0, 0)$ component of Einstein's equations can be found from Eqs. 10.14 and 10.20. It is convenient at this point to replace $\Lambda(r)$ with a different unknown function $m(r)$ defined as

$$m(r) := \tfrac{1}{2}r(1 - e^{-2\Lambda}), \tag{10.28}$$

or

$$g_{rr} = e^{2\Lambda} = \left(1 - \frac{2m(r)}{r}\right)^{-1}. \tag{10.29}$$

Then the $(0, 0)$ equation implies that

$$\frac{dm(r)}{dr} = 4\pi r^2 \rho. \tag{10.30}$$

It has the same form as the Newtonian equation, which calls $m(r)$ the mass inside the sphere of radius r. Therefore in relativity we call $m(r)$ the mass function, but it cannot be interpreted as the mass energy inside r since the total energy is not localizable in GR. We shall explore the Newtonian analogy in § 10.5 below.

The (r, r) equation, from Eqs. 10.15 and 10.21, can be cast in the form

$$\frac{d\Phi}{dr} = \frac{m(r) + 4\pi r^3 p}{r[r - 2m(r)]}.$$

(10.31)

If one has an equation of state of the form Eq. 10.25 then Eqs. 10.25, 10.27, 10.30, and 10.31 are four equations for the four unknowns Φ, m, ρ, p. If the more general equation of state Eq. 10.24 is needed then S is a completely arbitrary function. There is *no* additional information contributed by the (θ, θ) and (ϕ, ϕ) Einstein equation, because (i) it is clear from Eqs. 10.16, 10.17, 10.22, and 10.23 that the two equations are essentially the same, and (ii) the Bianchi identities ensure that this equation is a consequence of Eqs. 10.26, 10.30, and 10.31.

10.4 The exterior geometry

Schwarzschild metric

In the region outside the star we have $\rho = p = 0$, and we get the two equations

$$\frac{dm}{dr} = 0,$$

(10.32)

$$\frac{d\Phi}{dr} = \frac{m}{r(r - 2m)}.$$

(10.33)

These have the solutions

$$m(r) = M = \text{const.},$$

(10.34)

$$e^{2\Phi} = 1 - \frac{2M}{r},$$

(10.35)

where the requirement that $\Phi \to 0$ as $r \to \infty$ has been applied. We therefore see that the exterior metric has the following form, called the *Schwarzschild metric:*

$$ds^2 = -\left(1 - \frac{2M}{r}\right) dt^2 + \left(1 - \frac{2M}{r}\right)^{-1} dr^2 + r^2 d\Omega^2.$$

(10.36)

For large r this becomes

$$ds^2 \approx -\left(1 - \frac{2M}{r}\right) dt^2 + \left(1 + \frac{2M}{r}\right) dr^2 + r^2 d\Omega^2.$$

(10.37)

One can find coordinates (x, y, z) such that this becomes

$$ds^2 \approx -\left(1 - \frac{2M}{R}\right) dt^2 + \left(1 + \frac{2M}{R}\right) (dx^2 + dy^2 + dz^2),$$

(10.38)

where $R := (x^2 + y^2 + z^2)^{1/2}$. We see that this is the far-field metric of a star of total mass M (see Eq. 8.60). This justifies the definition in Eq. 10.28 and the choice of the symbol M.

Generality of the metric

A more general treatment, as in Misner *et al.* (1973), establishes *Birkhoff's theorem*, that the Schwarzschild solution, Eq. 10.36, is the only spherically symmetric asymptotically flat solution to Einstein's vacuum field equations, even if we drop our initial assumptions that the metric is static, i.e. if we start with Eq. 10.5. This means that even a radially pulsating or collapsing star will have a static exterior metric of constant mass M. One conclusion one can draw from this is that there are no gravitational waves from pulsating spherical systems. (This has an analogy in electromagnetism: there is no 'monopole' electromagnetic radiation either.) We found this result from linearized theory in Exercise 9.47, § 9.7.

10.5 The interior structure of the star

Inside the star we have $\rho \neq 0, p \neq 0$, and so we can divide Eq. 10.27 by $(\rho + p)$ and use it to eliminate Φ from Eq. 10.31. The result is called the Tolman–Oppenheimer–Volkov (T–O–V) equation:

$$\frac{dp}{dr} = -\frac{(\rho + p)(m + 4\pi r^3 p)}{r(r - 2m)}. \tag{10.39}$$

Combined with Eq. 10.30 for dm/dr and an equation of state of the form Eq. 10.25, this gives three equations for m, ρ, and p. We have reduced Φ to a subsidiary position; it can be found from Eq. 10.27 once the others equations have been solved.

General rules for integrating the equations

Since there are two first-order differential equations, Eqs. 10.30 and 10.39, they require two constants of integration, one being $m(r = 0)$ and the other $p(r = 0)$. We now argue that $m(r = 0) = 0$. A tiny sphere of radius $r = \varepsilon$ has circumference $2\pi\varepsilon$, and proper radius $|g_{rr}|^{1/2}\varepsilon$ (from the line element). Thus a small circle about $r = 0$ has ratio of circumference to radius equal to $2\pi|g_{rr}|^{-1/2}$. But if spacetime is locally flat at $r = 0$, as it must be at *any* point of the manifold, then a *small* circle about $r = 0$ must have ratio of circumference to radius equal to 2π. Therefore $g_{rr}(r = 0) = 1$, and so as r goes to zero, $m(r)$ must also go to zero, in fact *faster* than r. The other constant of integration, $p(r = 0) := p_c$ or, equivalently, ρ_c, from the equation of state, simply defines the stellar model. *For a given equation of state $p = p(\rho)$, the set of all spherically symmetric static stellar models forms a one-parameter sequence, the parameter being the central density.* This result follows just from the standard uniqueness theorems for first-order ordinary differential equations.

Once $m(r), p(r)$ and $\rho(r)$ are known, the *surface* of the star is defined as the place where $p = 0$. (Notice that, by Eq. 10.39, the pressure decreases monotonically outwards from the center.) The reason why $p = 0$ marks the surface is that p must be continuous everywhere, for otherwise there would be an infinite pressure gradient and infinite forces on fluid elements. Since $p = 0$ in the vacuum outside the star, the surface must also have $p = 0$. Therefore one stops integrating the interior solution there and requires that the exterior metric should be the Schwarzschild metric. Let the radius of the surface be R. Then, in order to have a smooth geometry, the metric functions must be continuous at $r = R$. Inside the star we have

$$g_{rr} = \left(1 - \frac{2m(r)}{r}\right)^{-1}$$

and outside we have

$$g_{rr} = \left(1 - \frac{2M}{r}\right)^{-1}.$$

Continuity clearly defines the constant M as follows:

$$M := m(R). \tag{10.40}$$

Thus the total mass of the star as determined by distant orbits is found to be the integral

$$M = \int_0^R 4\pi r^2 \rho \, dr, \tag{10.41}$$

just as in Newtonian theory. This analogy is rather deceptive, however, since the integral is over the volume element $4\pi r^2 dr$, which is *not* the element of *proper* volume. Proper volume in the hypersurface $t = $ const. is given by

$$|g|^{1/2} \, d^3x = e^\Lambda r^2 \sin\theta \, dr \, d\theta \, d\phi, \tag{10.42}$$

which, after doing the (θ, ϕ) integration, is just $4\pi r^2 e^{\Phi+\Lambda} \, dr$. Thus M is not in any sense just the sum of all the proper energies of the fluid elements. The difference between the proper and coordinate volume elements is where the 'gravitational potential energy' contribution to the total mass is to be found in these coordinates. We need not look in more detail at this; it only illustrates the care that one must take in applying Newtonian interpretations to relativistic equations.

Having obtained M, this determines g_{00} outside the star and hence g_{00} at its surface:

$$g_{00}(r = R) = -\left(1 - \frac{2M}{R}\right). \tag{10.43}$$

This serves as the integration constant for the final differential equation, the one which determines Φ inside the star: Eq. 10.27. We thereby obtain the complete solution.

Notice that solving for the structure of the star is the first time that we actually needed to assume that the point $r = 0$ is contained in the spacetime. Earlier, we argued that it need not be thus contained, and the discussion *before* we embarked on the interior solution made no such assumption. We make the assumption here because we want to talk about 'ordinary' stars, which we feel must have the same global topology as Euclidean space, differing from it only by being curved here and there. However, the exterior Schwarzschild solution is independent of assumptions about the point $r = 0$, and when we discuss black holes we shall see how different the point $r = 0$ can be.

Notice also that for our ordinary stars we always have $2m(r) < r$. This is certainly true near $r = 0$, since we have seen that we need $m(r)/r \to 0$ at $r = 0$. If it ever happened that near some radius r_1 we had $r - 2m(r) = \varepsilon$, with ε small and decreasing with r, then by the T–O–V equation, Eq. 10.39, the pressure gradient would be of order $1/\varepsilon$ and negative. This would cause the pressure to drop so rapidly from any finite value that it would pass through zero *before* ε reached zero. But as soon as p vanishes we have reached the surface of the star. Outside that point, m is constant and r increases. So nowhere in the spacetime of an ordinary star can $m(r)$ reach $\frac{1}{2}r$.

The structure of Newtonian stars

Before looking for solutions, we shall briefly look at the Newtonian limit of these equations. In Newtonian situations we have $p \ll \rho$, so we also have $4\pi r^3 p \ll m$. Moreover, the metric must be nearly flat, so in Eq. 10.29 we require $m \ll r$. These inequalities simplify Eq. 10.39 to

$$\frac{\mathrm{d}p}{\mathrm{d}r} = -\frac{\rho m}{r^2}. \tag{10.44}$$

This is exactly the same as the equation of hydrostatic equilibrium for Newtonian stars (see Chandrasekhar 1939), a fact which should not surprise us in view of our earlier interpretation of m and of the trivial fact that the Newtonian limit of ρ is just the mass density. Comparing Eq. 10.44 with its progenitor, Eq. 10.39, shows that all relativistic corrections tend to steepen the pressure gradient relative to the Newtonian value. In other words, in GR, for a fluid to remain static it must have stronger internal forces than in Newtonian gravity. This can be interpreted loosely as indicating a stronger field. An extreme instance of this is gravitational collapse: a field so strong that the fluid's pressure cannot resist it. We shall discuss this more fully in § 10.7 below.

10.6 Exact interior solutions

In Newtonian theory, Eqs. 10.30 and 10.44 are very hard to solve analytically for a given equation of state. Their relativistic counterparts are worse.[1] We shall discuss two

---

[1] If one does not restrict the equation of state, then Eqs. 10.44 and 10.39 are easier to solve. For example, one can arbitrarily assume a function $m(r)$, deduce $\rho(r)$ from it via Eq. 10.30, and hope to be able to solve Eq. 10.44 or 10.39 for p. The result, two functions $p(r)$ and $\rho(r)$, implies an 'equation of state' $p = p(\rho)$ by eliminating r. This is unlikely to be physically realistic, so most exact solutions obtained in this way are not of particular interest to the astrophysicist.

interesting exact solutions to the relativistic equations, one due to Schwarzschild and one by Buchdahl (1981).

The uniform-density interior solution

To simplify the task of solving Eqs. 10.30 and 10.39, we make the assumption

$$\rho = \text{const.} \tag{10.45}$$

This replaces the question of state and leads to what is called the *Schwarzschild interior solution*. There is no physical justification for it, of course. In fact, the speed of sound, which is proportional to $(dp/d\rho)^{1/2}$, is infinite! Nevertheless, the interiors of dense neutron stars are of *nearly* uniform density, so this solution has some interest for us in addition to its pedagogic value as an example of the method one uses to solve the system.

We can integrate Eq. 10.30 immediately:

$$m(r) = 4\pi\rho r^3/3, \quad r \leqslant R, \tag{10.46}$$

where R is the star's as yet undetermined radius. Outside R the density vanishes, so $m(r)$ is constant. By demanding the continuity of g_{rr} we find that $m(r)$ must be continuous at R. This implies

$$m(r) = 4\pi\rho R^3/3 := M, \quad r \geqslant R, \tag{10.47}$$

where we denote the constant by M, the Schwarzschild mass.

We can now solve the T–O–V equation, Eq. 10.39:

$$\frac{dp}{dr} = -\frac{4}{3}\pi r \frac{(\rho + p)(\rho + 3p)}{1 - 8\pi r^2\rho/3}. \tag{10.48}$$

This is easily integrated from an arbitrary central pressure p_c to give

$$\frac{\rho + 3p}{\rho + p} = \frac{\rho + 3p_c}{\rho + p_c}\left(1 - 2\frac{m}{r}\right)^{1/2}. \tag{10.49}$$

From this it follows that

$$R^2 = \frac{3}{8\pi\rho}[1 - (\rho + p_c)^2/(\rho + 3p_c)^2] \tag{10.50}$$

or

$$p_c = \rho\frac{[1 - (1 - 2M/R)^{1/2}]}{[3(1 - 2M/R)^{1/2} - 1]}. \tag{10.51}$$

Replacing p_c in Eq. 10.49 by this gives

$$p = \rho\frac{(1 - 2Mr^2/R^3)^{1/2} - (1 - 2M/R)^{1/2}}{3(1 - 2M/R)^{1/2} - (1 - 2Mr^2/R^3)^{1/2}}. \tag{10.52}$$

Notice that Eq. 10.51 implies $p_c \to \infty$ as $M/R \to 4/9$. We will see later that this is a very general limit on M/R, even for more realistic stars.

We complete the uniform-density case by solving for Φ from Eq. 10.27. Here we know the value of Φ at R, since it is implied by the continuity of g_{00}:

$$g_{00}(R) = -(1 - 2M/R).\tag{10.53}$$

Therefore we find

$$\exp(\Phi) = \tfrac{3}{2}(1 - 2M/R)^{1/2} - \tfrac{1}{2}(1 - 2Mr^2/R^3)^{1/2}, \quad r \leqslant R.\tag{10.54}$$

Note that Φ and m are monotonically increasing functions of r, while p decreases monotonically.

Buchdahl's interior solution

Buchdahl (1981) found a solution for the equation of state

$$\rho = 12(p_* p)^{1/2} - 5p,\tag{10.55}$$

where p_* is an arbitrary constant. While this equation has no particular physical basis, it does have two nice properties: (i) it can be made causal everywhere in the star by demanding that the local sound speed $(dp/d\rho)^{1/2}$ be less than 1; and (ii) for small p it reduces to

$$\rho = 12(p_* p)^{1/2},\tag{10.56}$$

which, in the Newtonian theory of stellar structure, is called an $n = 1$ polytrope. The $n = 1$ polytrope is one of the few exactly solvable Newtonian systems (see Exercise 10.14, § 10.8), so Buchdahl's solution may be regarded as its relativistic generalization. The causality requirement demands

$$p < p_*, \quad \rho < 7p_*.\tag{10.57}$$

Like most exact solutions[2] this one is difficult to deduce from the standard form of the equations. In this case, one requires a different radial coordinate r'. This is defined, in terms of the usual r, implicitly by Eq. 10.59 below, which involves a second arbitrary constant β and the function[3]

$$u(r') := \beta \frac{\sin Ar'}{Ar'}, \quad A^2 := \frac{288\pi p_*}{1 - 2\beta}.\tag{10.58}$$

Then we write

$$r(r') = r' \frac{1 - \beta + u(r')}{1 - 2\beta}.\tag{10.59}$$

---

[2] An exact solution is one which can be written in terms of simple functions of the coordinates, such as polynomials and trigonometric functions. Finding such solutions is an art which requires the successful combination of useful coordinates, simple geometry, good intuition, and in most cases luck. See Stephani et al. (2003) for a review of the subject.

[3] Buchdahl uses different notation for his parameters.

Rather than demonstrate how to obtain the solution (see Buchdahl 1981), we shall be content simply to write it down. In terms of the metric functions defined in Eq. 10.7 we have, for $Ar' \leqslant \pi$,

$$\exp(2\Phi) = (1 - 2\beta)(1 - \beta - u)(1 - \beta + u)^{-1}, \tag{10.60}$$

$$\exp(2\Lambda) = (1 - 2\beta)(1 - \beta + u)(1 - \beta - u)^{-1}(1 - \beta + \beta \cos Ar')^{-2}, \tag{10.61}$$

$$p(r) = A^2(1 - 2\beta)u^2[8\pi(1 - \beta + u)^2]^{-1}, \tag{10.62}$$

$$\rho(r) = 2A^2(1 - 2\beta)u(1 - \beta - \tfrac{3}{2}u)[8\pi(1 - \beta + u)^2]^{-1}, \tag{10.63}$$

where $u = u(r')$. The surface $p = 0$ is where $u = 0$, i.e. at $r' = \pi/A \equiv R'$. At this point we have

$$\exp(2\Phi) = \exp(-2\Lambda) = 1 - 2\beta, \tag{10.64}$$

$$R \equiv r(R') = \pi(1 - \beta)(1 - 2\beta)^{-1}A^{-1}. \tag{10.65}$$

Therefore β is the value of M/R on the surface, which in the light of Eq. 10.13 is related to the surface redshift of the star by

$$z_s = (1 - 2\beta)^{-1/2} - 1. \tag{10.66}$$

Clearly the nonrelativistic limit of this sequence of models is the limit $\beta \to 0$. The mass of the star is given by

$$M = \frac{\pi\beta(1 - \beta)}{(1 - 2\beta)A} = \left[\frac{\pi}{288p_*(1 - 2\beta)}\right]^{1/2}\beta(1 - \beta). \tag{10.67}$$

Since β alone determines how relativistic the star is, the constant p_* (or A) simply gives an overall dimensional scaling to the problem; it can be given any desired value by an appropriate choice of the unit for distance. It is β, therefore, whose variation produces nontrivial changes in the structure of the model. The lower limit on β is, as we remarked above, zero. The upper limit comes from the causality requirement, Eq. 10.57, and the observation that Eqs. 10.62 and 10.63 imply

$$p/\rho = \tfrac{1}{2}u(1 - \beta - \tfrac{3}{2}u)^{-1}, \tag{10.68}$$

whose maximum value is at the center, $r = 0$:

$$p_c/\rho_c = \beta(2 - 5\beta)^{-1}. \tag{10.69}$$

Demanding that this be less than $\tfrac{1}{7}$ gives

$$0 < \beta < \tfrac{1}{6}. \tag{10.70}$$

This range spans a spectrum of physically reasonable models, from the Newtonian ($\beta \approx 0$) to the very relativistic (surface redshift 0.22).

10.7 Realistic stars and gravitational collapse

Buchdahl's theorem

We saw in the previous section that there are no uniform-density stars with radii smaller than $9M/4$, because to support them in a static configuration would require pressures larger than infinite! This is in fact true of *any* stellar model, and is known as Buchdahl's theorem (Buchdahl 1959). Suppose that one manages to construct a star in equilibrium with radius $R = 9M/4$, and then gives it a (spherically symmetric) inward push. It has no choice but to collapse inwards: it cannot reach a static state again. But, during its collapse, the metric outside it is just the Schwarzschild metric. What is left, then, is the vacuum Schwarzschild geometry outside. This is the metric of a black hole, and we will study it in detail in the next chapter. First we look at some causes of gravitational collapse.

Evolution of normal stars into giants

Any realistic appraisal of the chances of the formation of a neutron star or a black hole must begin with an understanding of the way stars evolve. We give a brief summary here. See the bibliography in § 10.8 for books that cover the subject in detail.

An ordinary star like our Sun derives its luminosity from nuclear reactions taking place deep in its core, mainly the conversion of hydrogen to helium. Because a star is always radiating energy, it needs nuclear reactions to replace that energy in order to remain static. The Sun will burn hydrogen in a fairly steady way for some 10^{10} years. A more massive star, whose core has to be denser and hotter to support the greater mass, may remain steady only for a million years, because the nuclear reaction rates are very sensitive to temperature and density. Astronomers have a name for steady stars: they call them "main sequence stars" because they all fall in a fairly narrow band when one plots their surface temperature against luminosity: the luminosity and temperature of a normal star are determined mainly by its mass. Not surprisingly, more massive stars tend to be both hotter and brighter. Astronomers call this plot the Hertzsprung–Russell diagram.

The radiation of energy from the star's surface causes a very slow leakage of the thermal energy stored in its hot interior. One way of seeing this is that it takes about a million years for a photon produced in a nuclear reaction in our Sun's center to scatter its way out to the surface. All these scatterings reduce its energy from the gamma-ray part of the spectrum to the visible.

However, eventually every star uses up its original supply of hydrogen in the core. When this energy source turns off, the core of the star begins to shrink as the star gradually radiates its stored energy away. This shrinking compresses and therefore heats the core. It is worth noting here that this reflects a fundamental property of self-gravitating systems, which is that they have *negative* specific heat: as they lose energy they get hotter! Such systems are thermodynamically unstable; we speak of the *gravito-thermal instability*. Stars manage to remain quasi-stable for millions or billions of years, but eventually they fall prey to this instability.

When the temperature in the shrinking core gets high enough, it ignites another set of nuclear reactions, which fuse helium nuclei into carbon and oxygen, again releasing energy and providing thermal support against gravity. Because these nuclear reactions are extremely temperature-sensitive, the luminosity of the star increases dramatically. This higher energy flux forces the outer layers of the star to expand, and the star acquires a kind of 'core–halo' structure. Its surface area is typically so large that it cools below the surface temperature of the Sun, despite the immense temperatures inside and the high luminosity with which it must radiate. Such a star is called a *red giant*, because its lower surface temperature makes it radiate more energy in the red part of the spectrum. Giants are not on the main sequence: they get cooler as they get brighter.

The high luminosity of red giants leads to a steady blowing away of the outermost layers of their atmospheres, gradually reducing their masses. Stars showing such strong stellar winds form spectacular 'planetary nebulae', a favorite subject of astronomical photographs. This mass loss is crucial to what happens next, and that depends on the original mass of the star.

A star that starts with a mass up to about $8\,M_\odot$ blows away most of its envelope – the part that is not participating in the nuclear reactions – before the helium fusion in its core stops. The loss of all that weight gradually allows the core to expand and cool off, typically quenching the nuclear reactions before they use up all the helium. The result is a dense ball of helium, carbon, and oxygen, which we call a white dwarf. Its exact composition depends on the original mass of the star.

A white dwarf involves fascinating physics. It is supported by the quantum mechanical pressure that comes from electrons being packed closely together, which we describe in the next subsection. A white dwarf typically has a mass up to around $1\,M_\odot$ and is about the size of the Earth. Gravity is therefore much stronger on its surface than on the surface of the Sun, although not yet highly relativistic. Crucially, as we will see below, this pressure cannot support much more than about $1.3\,M_\odot$ against gravity: this is called the *Chandrasekhar limit*. It was first discovered in 1930 by the then-young Indian scientist, S. Chandrasekhar, whom we quoted in Chapter 8. This was only five years after the publication of the work on the quantum pressure of electrons by E. Fermi and P. Dirac. Chandrasekhar received the Nobel Prize in Physics for his application of this new physics to astronomy, but he had to wait for it until 1983.

A star more massive than $8\,M_\odot$ will lose some fraction of its total mass during helium burning, but its remaining weight keeps the helium fusion going. The helium is exhausted before the whole envelope is blown away. Because the star's mass is larger than the Chandrasekhar limit, the core contracts further, reaching the temperature at which carbon will fuse into silicon, and after that silicon into iron. These reactions provide thermal support for a short time, but eventually this has to stop, since ^{56}Fe is the most stable of all nuclei – any reaction converting iron into something else absorbs energy rather than releasing it. The core can then no longer support the weight of the envelope.

The result is a catastrophic core collapse, with the core going essentially into free fall from the size of the Earth to a ball about 10 km in radius. At this density there is no room for atoms. The electrons and protons are squeezed together, and it becomes energetically favorable for an electron and a proton to form a neutron, radiating a neutrino (as explained below). The imploding ball effectively becomes one giant nucleus with a mass close to that of the Sun, consisting almost entirely of neutrons: a proto-neutron star. At this density, the quantum-mechanical pressure (due this time to neutrons being packed closely together – see the next subsection) suddenly becomes strong enough again to support this ball against its own weight. It is the last chance for the core to avoid being trapped in a black hole of its own making.

But the core is not isolated: the star's envelope, which initially gets left behind by the collapse, starts crashing down onto it. There are now just three possibilities for the final outcome: an explosion that blows off the envelope and leaves a neutron star behind; a failed explosion that leads to a black hole into which the neutron core and the envelope both disappear; or a huge explosion that leads to the complete dispersal of the core and the envelope.

In order to understand which of these fates any particular star may undergo, we need first to understand the quantum-mechanical pressure forces that play such a crucial role in this story, supporting a white dwarf or a neutron star against the strong pull of gravity. These forces are the arbiters that decide whether a star becomes a white dwarf, or a neutron star, or a black hole. Furthermore, they determine the maximum masses of white dwarfs and neutron stars. And, as we shall see, the radically different densities of these two kinds of remnant stars can be traced to the huge difference between the mass of the electron and that of the neutron. It is one of the wonders of the natural world that physics at the particle level translates so directly into the nature of compact stellar remnants.

Quantum-mechanical pressure

We shall now give an elementary discussion of the forces that support white dwarfs and neutron stars. Consider an electron in a box of volume V. Because of the Heisenberg uncertainty principle, its momentum p is uncertain by an amount of order

$$\Delta p = hV^{-1/3}, \tag{10.71}$$

where h is Planck's constant. If its momentum has magnitude between p and $p + \mathrm{d}p$, it is in a region of momentum space of volume $4\pi p^2\,\mathrm{d}p$. The number of 'cells' in this region of volume Δp is

$$\mathrm{d}N = 4\pi p^2\,\mathrm{d}p/(\Delta p)^3 = \frac{4\pi p^2\,\mathrm{d}p}{h^3}V. \tag{10.72}$$

Since it is impossible to define the momentum of the electron more precisely than Δp, this is the number of possible momentum states with momentum between p and $p+\mathrm{d}p$ in a box of volume V. Now, electrons are Fermi particles, which means that they have the remarkable property that no two of them can occupy exactly the same state. (This is the basic reason for the variety of the periodic table and the solidity and relative impermeability of matter.) Electrons have spin $\frac{1}{2}$, which means that for each momentum state there are two spin states ('spin-up' and 'spin-down'), so there are a total of

$$V\frac{8\pi p^2\,\mathrm{d}p}{h^3} \tag{10.73}$$

states, which is then the *maximum* number of electrons that can have momenta between p and $p + dp$ in a box of volume V.

Now suppose we cool off a gas of electrons as far as possible, which means reducing each electron's momentum as far as possible. If there is a total of N electrons, then they are as cold as possible when they fill all the momentum states from $p = 0$ to some upper limit p_f, determined by the equation

$$\frac{N}{V} = \int_0^{p_f} \frac{8\pi p^2 \, dp}{h^3} = \frac{8\pi p_f^3}{3h^3}. \tag{10.74}$$

Since N/V is the number density, we see that a cold electron gas obeys the relation

$$n = \frac{8\pi p_f^3}{3h^3}, \quad p_f = \left(\frac{3h^3}{8\pi}\right)^{1/3} n^{1/3}. \tag{10.75}$$

The number p_f is called the Fermi momentum. Notice that it depends only on the number of particles per unit volume, not on their masses.

Each electron has mass m and energy $E = (p^2 + m^2)^{1/2}$. Therefore the total energy density in such a gas is

$$\rho = \frac{E_{total}}{V} = \int_0^{p_f} \frac{8\pi p^2}{h^3} (m^2 + p^2)^{1/2} \, dp. \tag{10.76}$$

The pressure can be found from Eq. 4.22 with ΔQ set to zero, since we are dealing with a closed system:

$$p = -\frac{d}{dV}(E_{total}) = -V \frac{8\pi p_f^2}{h^3}(m^2 + p_f^2)^{1/2} \frac{dp_f}{dV} - \rho.$$

For a constant number of particles N, we have

$$V\frac{dp_f}{dV} = -n\frac{dp_f}{dn} = \frac{1}{3}\left(\frac{3h^3}{8\pi}\right)^{1/3} n^{1/3} = \tfrac{1}{3}p_f,$$

and we get

$$p = \left(\frac{8\pi}{3h^3}\right) p_f^3 (m^2 + p_f^2)^{1/2} - \rho. \tag{10.77}$$

For a very relativistic gas where $p_f \gg m$ (which will be the case if the gas is compressed to small V) we have

$$\rho \approx \frac{2\pi p_f^4}{h^3} \tag{10.78}$$

$$p \approx \frac{1}{3}\rho. \tag{10.79}$$

This is the equation of state for a 'cold' electron gas. So the gas has a pressure comparable to its density even when it is as cold as possible. In Exercise 4.22, § 4.10, we saw that Eq. 10.79 is also the equation of state for a photon gas. The reason why this relativistic

Fermi gas behaves like a photon gas is essentially that the energy of each electron far exceeds its rest mass; the rest mass is unimportant, so setting it to zero (turning it into a photon) changes little.

The collection of electrons packed into the lowest-energy quantum states is called a *degenerate electron gas*, and their pressure is called the degeneracy pressure.

White dwarfs

When an ordinary star is compressed, it reaches a stage where the electrons have become free of the nuclei, and so there are two gases, one of electrons and one of nuclei. Since they have the same temperatures, and hence the same energies per particle, the much less massive electrons have far less momentum per particle. However, upon compression the Fermi momentum level rises (Eq. 10.75) until it becomes comparable with the momentum of the electrons. They are then effectively a cold electron gas, and supply the pressure for the star. We call this kind of star a *white dwarf*.

The nuclei have momenta well in excess of p_f, so they are a classical gas, but they supply little pressure. On the other hand, the nuclei supply most of the gravity, since there are more neutrons and protons than electrons, and they are much more massive. So the mass density for Newtonian gravity (which is adequate here) is

$$\rho = \mu m_p n_e, \tag{10.80}$$

where μ is the ratio of number of nucleons to the number of electrons (of order 1 or 2, depending on the composition of the core before it became so dense), m_p is the proton mass, and n_e the number density of electrons. The relation between pressure and density for the whole gas when the electrons are relativistic is therefore

$$p = k\rho^{4/3}, \quad k = \frac{2\pi}{3h^3}\left(\frac{3h^3}{8\pi\mu m_p}\right)^{4/3}. \tag{10.81}$$

The Newtonian structure equations for the star are

$$\left.\begin{array}{l} dm/dr = 4\pi r^2 \rho, \\ dp/dr = -\rho m/r^2. \end{array}\right\} \tag{10.82}$$

In order of magnitude, these are (for a star of mass M, radius R, typical density $\bar\rho$ and typical pressure $\bar p$)

$$\left.\begin{array}{l} M = R^3 \bar\rho, \\ \bar p/R = \bar\rho M/R^2. \end{array}\right\} \tag{10.83}$$

Setting $\bar p = k\bar\rho^{4/3}$, from Eq. 10.81, gives

$$k\bar\rho^{1/3} = \frac{M}{R}. \tag{10.84}$$

Using Eq. 10.83 in this, we see that R cancels out and we get an equation for M:

$$M = \left(\frac{3k^3}{4\pi}\right)^{1/2} = \frac{1}{32\mu^2 m_p^2}\left(\frac{6h^3}{\pi}\right)^{1/2}. \tag{10.85}$$

Using geometrized units, one finds (with $\mu = 2$)

$$M = 0.47 \times 10^5 \text{ cm} = 0.32 \, M_\odot. \tag{10.86}$$

From our derivation we should expect this to be the order of magnitude of the maximum mass supportable by a relativistic electron gas, when most of the gravity comes from a cold nonrelativistic gas of nuclei. This is called the *Chandrasekhar limit*, and a more precise calculation, based on integrating Eq. 10.82 more carefully, puts it at $M \approx 1.3 \, M_\odot$. Any star more massive than this cannot be supported by electron pressure, and so cannot be a white dwarf. In fact, the actual upper mass limit is marginally smaller, occurring at central densities of about $10^{10} \text{ kg m}^{-3}$, due to an instability that we describe in the next subsection.

Because white dwarfs are the endpoint of evolution for stars like our Sun and for all less-massive stars, there are a lot of them in the Galaxy: possibly about 10^{11}, which would constitute about half of the whole stellar population of the Milky Way. One does not see these stars when one looks up in the night sky because their luminosity is very low compared to normal stars, where nuclear reactions are resupplying the energy that the star radiates. White dwarfs have no nuclear energy generation source, so most just gradually cool off and disappear from view.

Neutron stars

If the material is compressed to a higher density than that characteristic of a white dwarf (which has $\rho \lesssim 10^{10} \text{ kg m}^{-3}$), the kinetic energy of the electrons increases to the extent that electrons combine with protons to forms neutrons, releasing energy in the form of neutrinos. So compression results in the loss of electrons from the electron gas which is providing the principal pressure: this pressure does not build up rapidly enough as the density increases, and the star is unstable to collapse. It does not have the possibility of being stable again until its central density reaches around $10^{17} \text{ kg m}^{-3}$. By the time this density is reached, essentially all the electrons are united with protons to form a gas of almost pure neutrons. These are also Fermi particles, so they obey exactly the same quantum equation of state as we derived for electrons, Eqs. 10.78 and 10.79. This is a *neutron star.*

The main differences between a neutron star and a white dwarf are two: first, a much higher density is required to push p_f up to the typical momentum of a neutron, which is more massive than an electron; and second, the total energy density is now provided by the neutrons themselves: there is no extra gas of ions providing most of the self-gravitation. So here, the total equation of state at high compression is Eq. 10.79:

$$p = \frac{1}{3}\rho. \tag{10.87}$$

Unfortunately, there exist no simple arguments giving an upper mass in this case, since the fully relativistic structure equations Eqs. 10.30 and 10.39 must be used.

Since a core will not collapse until its mass exceeds the Chandrasekhar limit, neutron stars will have masses above $1\,M_\odot$. A spherical star with this mass and a mean density of $10^{17}\,\mathrm{kg\,m^{-3}}$ will have a radius of only 17 km. We will see below that even this is a bit high: the physics of these stars, along with data from radio and gravitational wave observations, all suggest radii around 10–12 km. At this size, $M/R \sim 0.15$. So a neutron star packs the mass of the Sun into a ball the size of a big city. It is only three times larger than the smallest size allowed by Buchdahl's theorem. This is the 'last stop' that a star can make if it is to avoid becoming a black hole.

Notice also that a neutron star can rotate exceptionally rapidly. A star will begin to break apart if its equator rotates as fast as an orbiting particle would. In Newtonian terms this 'break-up velocity' is $v_{\max} = \sqrt{(M/R)} = 0.39$. With a size of 10 km, this would give a rotational period of 0.54 ms, or a spin rate of 1.8 kHz! Compare this to the rotational speed of an internal combustion engine in a car, which is limited to perhaps 6,000 r.p.m., or 100 Hz. Here we can have an entire star spinning 20 times faster.

We will see in the next chapter that a particle of rest mass m that is at rest on the surface of a star of mass M and radius R has an energy, as measured by an observer far away, equal to $m\sqrt{(1-M/R)}$. This means that the neutrons in a neutron star have given up about 8% of their rest mass energy in order to come to rest and form the star. This is an enormous fraction, much larger than the 0.1–1% that is released in nuclear reactions. Gravitational core collapse is a hugely efficient way to convert rest mass into energy.

The existence of neutron stars was first proposed in a remarkable paper by Baade & Zwicky (1934), just two years after the discovery of the neutron.[4] They suggested that the stars could form in supernova explosions, since the enormous energy that was available from forming them would be sufficient to power the explosions that astronomers were observing. However logical, the suggestion was not widely accepted by astronomers at the time. One reason was that nobody had observed a neutron star. They are indeed very hard to observe directly. The collapse and compression has made them very hot, 10^6 K or higher, but their radii are so small that their luminosity is small, and most of that energy comes out in the X-ray part of the spectrum, which was not accessible in 1934.

Then came the discovery of pulsars in 1967 by Jocelyn Bell and Antony Hewish (Hewish *et al.* 1968). As mentioned in the previous chapter, pulsars are spinning neutron stars, with observed spin frequencies from 0.04 Hz (PSR J0250+5854) up to 716 Hz (PSR J1748-2446ad), which is only 40% of the maximum we computed above. They also have very strong polar magnetic fields (10^8–10^{15} G). For reasons that are not well understood, the magnetic axis is not aligned with the spin axis, so the magnetic poles sweep through the pulsar's sky. The pulsar's ionized plasma atmosphere emits radiation

---

[4] The speed with which Chandrasekhar in 1930 and Baade & Zwicky in 1934 used the latest developments in quantum theory and nuclear theory to make major steps in astronomy is astonishing. Baade and Zwicky did not live long enough to receive Nobel Prizes for their suggestion, although Zwicky did live to see the discovery of pulsars and the validation of his prediction that neutron stars should come from supernovae.

preferentially in the polar directions: as the pole sweeps across the observer's position, the observer will register a pulse of radiation, which gives the objects their name. Radio and optical waves, X-rays and gamma-rays have been observed from different pulsars. They are detectable across a large fraction of our Galaxy. With this wealth of evidence, the collapse of the core of a massive star to a neutron star or black hole quickly became the standard picture for how a supernovae of Type II is initiated. It has been estimated that there are 10^8 neutron stars in our Milky Way galaxy, which would represent about 0.05% of all stars. For the discovery of pulsars, Hewish was awarded the Nobel Prize in Physics in 1974, but Bell was not.[5]

While the physics of how a pulsar's pulses are produced is outside the scope of this textbook, it is worth noting that it has been a challenging problem ever since the discovery of pulsars. Recently there has been significant progress using a model in which plasma instabilities in the outer magnetosphere launch large numbers of microsecond-long bursts of radio waves, the superposition of which becomes a single pulse in our radio observations (Philippov *et al.* 2019). This paper contains the following noteworthy statement:

> It is remarkable that a nanosecond-long radio spike produced from a region just tens of meters across in a single plasmoid merger in the magnetosphere of Crab pulsar 2 kpc away is so powerful that it can be detected at Earth. This is perhaps the smallest-size source of observable emission in all of astrophysics.

If the physics of the magnetospheres of neutron stars is extraordinary, the physics of their interiors is even more so. Neutron star matter is by any measure the most complex and fascinating state of matter that physicists have yet encountered. The dense degenerate gas of neutrons appears to be superfluid, despite temperatures of millions of kelvins. The neutrons are in chemical equilibrium with a much lower-density gas of protons and electrons, and the protons may also exhibit superconductivity at these temperatures! These properties are probably important for understanding why neutron stars develop such strong magnetic fields that do not align with their rotation axes, but the connection is not yet understood.

At the very center of a neutron star the density may be so large that the neutrons dissolve into essentially free quarks. It has proved extremely difficult for nuclear theorists to compute an equation of state for nuclear matter under these conditions. The physical conditions are out of the reach of laboratory experiment, and the short range and complexity of the nuclear force (the strong interaction) requires physicists to make one or another approximation and assumption in order to arrive at an equation of state. There

---

[5] This was one of the more controversial episodes in the history of the Nobel Prize. Bell was a graduate student at Cambridge University at the time of the discovery, and she clearly played a key role in the discovery. The Nobel committee decided that her PhD supervisor Hewish, who had built the radio antenna array that Bell used to acquire the data, deserved most of the credit. The later discovery of the first binary pulsar in 1975 had a different outcome, with (as we mentioned in § 9.4) a Nobel Prize in 1993 for both Hulse, the graduate student, and Taylor, his supervisor. It would seem that the Nobel committee in 1993 had perhaps taken note of the criticisms voiced in 1974.

are thus dozens of proposed equations of state, all leading to stars with properties that are very different from one another (Lattimer & Prakash 2000, Lattimer 2010). Different equations of state predict different relations between the mass and radius of a neutron star, and also vastly different maximum masses for neutron stars, ranging up to about $2.5\,M_\odot$. It is generally accepted that the maximum is less than $3\,M_\odot$, which means that any object that tries to squeeze more than that much mass into the size of a neutron star must collapse to a black hole.

From the point of view of understanding neutron star physics, the most interesting pulsars are those in binary systems. Radio astronomers can sometimes measure the mass of at least one of the neutron stars in a binary pulsar system with good accuracy. Remarkably, most such measurements cluster around $1.4\,M_\odot$ (Lorimer 2008, Stairs 2003). However, some pulsars have masses above $2\,M_\odot$; the largest pulsar mass measured so far is that of PSR J0740+6620, a millisecond pulsar in a binary, whose mass is $2.14^{+0.10}_{-0.09}\,M_\odot$ (Cromartie *et al.* 2020).

Gravitational wave observations of merging neutron stars, to be described in Chapter 12, are beginning to measure the masses and radii of both stars in the binary, although at present the uncertainties are rather large (Abbott *et al.* 2018c). One intriguing system (GW190814), which we will meet in § 12.4, seems to have contained a compact component with a mass of at least $2.5\,M_\odot$. Was it a neutron star of exceptionally large mass, or a black hole of exceptionally small mass? As gravitational wave detector sensitivities improve and more data comes in, observations are likely to resolve much of the uncertainty in the equation of state, answering this question and providing important guidance to nuclear physicists trying to understand the strong interaction.

Although neutron stars are observed primarily as radio pulsars, some X-ray and gamma-ray pulsars have also been discovered. Furthermore, some nonpulsating neutron stars are also known through their X-ray and gamma-ray emission. Neutron stars in some binary systems become X-ray sources when gas falls onto them from their companions, is heated by falling into the deep gravitational well, and emits X-rays. This also leads to the stars being spun up to very high rotation rates. The fastest known pulsar is PSR J1748-2446ad, which spins at 716 Hz, as we noted above, and was probably spun up this way.

Rotation can, in principle, considerably increase the upper limit on stellar masses (Stergioulas 2003), at least until rotation-induced relativistic instabilities set in (Friedman & Schutz 1978, Andersson *et al.* 1999, Kokkotas & Schmidt 1999). Realistically, however, this probably doesn't allow more than a factor of 1.5 in mass.

The association with supernova explosions has been confirmed for a number of pulsars, including one of the youngest known pulsars, PSR B0531+21, in the center of the Crab nebula. Studies of the motion of pulsars suggest that they get strong 'kicks' when they are born, with typical velocities of 400 to 1000 $\mathrm{km\,s^{-1}}$. This should be compared with the orbital speed of the Sun around the center of the Galaxy, which is about 200 $\mathrm{km\,s^{-1}}$, and the typical random speeds of stars relative to one another, which is some tens of $\mathrm{km\,s^{-1}}$. The kick must result from some kind of asymmetry in the gravitational collapse and subsequent initial explosion. Although most pulsars are probably formed in binary systems (since most stars are in binaries), a supernova explosion usually disrupts the binary system: the loss of mass reduces the gravitational attraction between the two stars suddenly, and typically leaves the velocity of the remnant neutron star or black hole higher than the new escape velocity. This can be compensated by a kick of the right size in an appropriate direction. To get a double-neutron-star binary requires the system to survive two such explosions. Pulsars in binaries are rare, and binaries with two neutron stars are even more rare.

Supernovae: forming compact objects

The ultimate fate of a star undergoing core collapse mainly depends on four things: its mass, rotation, magnetic field, and chemical composition.

Stars above about $8\,M_\odot$ and below about $25\,M_\odot$ probably form neutron stars, successfully blowing away their envelopes in a conventional Type II supernova explosion. The physical mechanism for blowing the gas away involves subtle details of neutrino physics, which allow the neutrinos emitted when neutrons are formed in the core to exert enough pressure on the initially collapsing envelope to turn it around and blow it away. The high temperatures and compression experienced initially by the envelope as it falls inwards lead to nuclear fusion, turning much of the hydrogen and helium in the envelope into heavier elements. Astronomers believe that the elements of the periodic table up to iron are mostly formed in supernova explosions. We shall see in Chapter 12 that the elements heavier than iron are probably formed mainly in the mergers of neutron stars.

If the original star had a mass greater than $25\,M_\odot$ (stars like this are quite rare), then its fate depends on what astronomers call its 'metallicity', which means the relative abundance of elements heavier than helium. These elements can inhibit radiation from leaving the hot core (this is known as the opacity), so that the higher the metallicity, the more of the envelope is expelled by the bounce. Very-high-metallicity massive stars can still form neutron stars. But lower-metallicity stars above $25\,M_\odot$ fail to expel their entire envelopes, and the material that falls back onto the neutron-star core forces the core to collapse to a black hole. These so-called 'stellar-mass' black holes may have masses anywhere from about 5 to $60\,M_\odot$, depending on the progenitor star. For initial masses well above about $40\,M_\odot$, the explosion may be accompanied by a long gamma-ray burst, the associated supernova being called a *hypernova*.

The facts that neutron stars have an upper mass limit well below $3\,M_\odot$ and supernovae do not seem to form black holes below about $5\,M_\odot$ give rise to the notion that there is a *mass gap* in the mass spectrum of compact objects. As we shall see, there is another mass gap at much larger masses, so this one is the 'lower mass gap'. Since black holes in the mass range 2.5–$5\,M_\odot$ can be formed by the merger of two neutron stars, this gap must have at least a sparse population. As we mentioned above, the gravitational wave event GW190814 has identified a merger component in this range.

Although stars with masses larger than $10\,M_\odot$ are rare, the explosions that lead to neutron stars and black holes are visible throughout the Universe. The progenitors' metallicity depends on the circumstances of their formation. Most stars formed today have relative element abundances similar to that of the Sun: they form from gas clouds that have been mixed with matter containing a whole spectrum of elements that were created by previous generations of stars and in previous supernova explosions and neutron-star mergers. However, the distribution of elements is not totally homogeneous, and there are regions of star formation even today in the Milky Way where the gas clouds are nearly pristine hydrogen and helium, the two principal elements formed in the Big Bang (see Chapter 13).

The higher opacity of the heavier elements affects star formation just as it affects the eventual death of the star. Higher-metallicity star-forming regions tend to produce less-massive stars, because when the inner part of the collapsing gas cloud heats up, its radiation will be trapped in the outer regions by their opacity, which tends to blow the outer regions away. Low-metallicity regions can form stars well above $60\,M_\odot$, collapsing eventually to black holes with masses upwards of perhaps $30\,M_\odot$. But it is thought that black holes above about $60\,M_\odot$ do not form by the collapse of massive stars, because the physics at these high densities leads to an instability that causes the stars to disintegrate completely in a very bright supernova explosion.

The very first generation of stars formed after the Big Bang had zero metallicity. They might have had masses of several thousand $M_\odot$, and they may have left behind so-called *intermediate-mass black holes* of hundreds or thousands of $M_\odot$. This and the instability mentioned in the previous paragraph suggest that there is an 'upper mass gap' in the black hole mass distribution in the range 60–$150\,M_\odot$. As with the lower mass gap, gravitational wave observations have already identified mergers with component masses in this range. We will discuss GW190521 in § 12.4.

Intermediate-mass black holes are given this name because their masses lie between those of stellar-mass black holes and those of the supermassive black holes that astronomers have discovered in the centers of most ordinary galaxies. We will discuss supermassive black holes in more detail in the next chapter.

This rather simple picture of stellar evolution can be altered by rotation and magnetic fields, and this is the subject of much current research. Rotation may induce currents that change the main-sequence evolution by mixing the inner and outer layers of a star. In the collapse phase, rotation becomes extremely important if significant angular momentum is preserved in the collapsing core. But substantial magnetic fields may allow the transfer of angular momentum from the core to the rest of the star, permitting a more spherical collapse.

Our discussion here has assumed that the evolving star was isolated. However, most stars are formed in binary or even triple systems, and when one of the stars enters the giant phase, it can have become so large that it either begins to shed mass onto a companion or even totally engulfs the companion. Stellar evolution therefore becomes a more complicated story in binaries, but binaries of neutron stars and black holes are key sources for today's gravitational wave detectors. We will therefore take up this story again in Chapter 12.

10.8 Bibliography

Our construction of the spherical coordinate system is similar to that in most other texts, but it is not particularly systematic nor is it clear how to generalize the method to other symmetries. Group theory affords a more systematic approach. See Stephani *et al.* (2003). This monograph also contains more exact solutions for compressible stars.

It is clear that the computation of Riemann and Einstein tensors of metrics more complicated than the simple static spherical metric will be time consuming. Modern computer-algebra systems, such as Maple or Mathematica, contain packages that can do this automatically. The most elegant way to do these computations by hand is the Cartan approach, described in Misner *et al.* (1973).

A full discussion of spherical stellar structure may be found in Shapiro & Teukolsky (1983). In deriving stellar solutions we demanded the continuity of g_{00} and g_{rr} across the surface of the star. A full discussion of the correct 'junction conditions' across a surface of discontinuity is given in Misner *et al.* (1973) or Wald (1984).

The evolution of stars through the different stages of nuclear burning is a huge subject, on which astrophysicists have made great progress with the help of computer simulations. But there are still many open questions, particularly concerning the evolution of stars in binary systems, where interaction with the companion becomes an issue. Good references include Hansen *et al.* (2004) and Tayler (1994). A free online textbook is G. W. Collins, *The Fundamentals of Stellar Astrophysics*: http://ads.harvard.edu/books/1989fsa..book/.

A more rigorous derivation of the equation of state of a Fermi gas may be found in quantum mechanics texts. See Chandrasekhar (1957) for a full derivation of his limit on white-dwarf masses. See also Shapiro & Teukolsky (1983). The instability which leads to the absence of stars intermediate in central density between white dwarfs and neutron stars is discussed in Harrison *et al.* (1965), Shapiro & Teukolsky (1983), and Zel'dovich & Novikov (1971).

It is believed that the main route to the formation of neutron stars and black holes is the collapse of the core of a massive star, which leads to a supernova of Type II. Such supernovae may be weak sources of gravitational radiation, as we discuss in Chapter 12. The more massive collapse events may lead to so-called hypernovae and gamma-ray bursts. A good introduction to contemporary research is in the collection by Höflich *et al.* (2004). Another good review is Woosley & Janka (2005). For an up-to-date view, the reader should consult the recent Proceedings of the Texas Symposium in Relativistic Astrophysics, a series of conferences that takes place every two years. Other types of supernovae arise from white-dwarf collapse and apparently do not lead to neutron stars. We will discuss the mechanism underlying supernovae of Type Ia in Chapter 13 on cosmology.

For a review of the fascinating story of pulsars, see Lyne & Graham-Smith (1998) or Lorimer & Kramer (2004). Pulsars are important places for testing our understanding of fundamental physics. See Kramer & Wex (2009), Kaspi & Kramer (2016), and Kramer (2016). Recent reviews of the observed properties of neutron stars and of what is known about their equation of state include Lattimer (2012), Özel & Freire (2016), and Lattimer & Prakash (2016).

There are many resources for neutron stars and black holes on the web, including websites about X-ray observations. Readers can find popular-style articles on the Einstein Online website: www.einstein-online.info/.

Exercises

10.1 Starting with $ds^2 = \eta_{\alpha\beta} \, dx^\alpha dx^\beta$, show that the coordinate transformation $r = (x^2 + y^2 + z^2)^{1/2}$, $\theta = \arccos(z/r)$, $\phi = \arctan(y/x)$ leads to Eq. 10.1, $ds^2 = -dt^2 + dr^2 + r^2(d\theta^2 + \sin^2\theta \, d\phi^2)$.

10.2 In deriving Eq. 10.5 we argued that if $\vec{e}_t$ were not orthogonal to $\vec{e}_\theta$ and $\vec{e}_\phi$, the metric would pick out a preferred direction. To see this, show that under rotations that hold t and r fixed, the pair $(g_{\theta t}, g_{\phi t})$ transforms as a vector field. If these do not vanish, they thus define a vector field on every sphere. Such a vector field cannot be spherically symmetric unless it vanishes: construct an argument to this effect, perhaps by considering the discussion of parallel transport on the sphere at the beginning of § 6.4.

10.3 The locally measured energy of a particle, given by Eq. 10.11, is the energy the same particle would have in SR if it passed the observer with the same speed. It therefore contains no information about gravity, about the curvature of spacetime. By referring to Eq. 7.34 show that the difference between E_* and E in the weak-field limit is, for particles with small velocities, just the gravitational potential energy.

10.4 Use the result of Exercise 6.35, § 6.8, to calculate the components of $G_{\mu\nu}$ in Eqs. 10.14–10.17.

10.5 Show that a static star must have $U^r = U^\theta = U^\phi = 0$ in our coordinates, by examining the result of the transformation $t \rightarrow -t$.

10.6 (a) Derive Eq. 10.19 from Eq. 10.18.
 (b) Derive Eqs. 10.20–10.23 from Eq. 4.37.

10.7 Describe how to construct a static stellar model in the case for which the equation of state has the form $p = p(\rho, S)$. Show that one must give an additional arbitrary function, such as $S(r)$ or $S(m(r))$.

10.8 (a) Prove that the expressions $T^{\alpha\beta}{}_{;\beta}$ for $\alpha = t, \theta$, or ϕ must vanish by virtue of the assumptions of a static geometric and spherical symmetry. (Do *not* calculate the expressions from Eqs. 10.20–10.23. Devise a much shorter argument.)
 (b) Derive Eq. 10.27 from Eqs. 10.20–10.23.
 (c) Derive Eq. 10.30 from Eqs. 10.14, 10.20, and 10.29.
 (d) Prove Eq. 10.31.
 (e) Derive Eq. 10.39.

10.9 (a) Define a new radial coordinate in terms of the Schwarzschild r by

$$r = \bar{r}(1 + M/2\bar{r})^2. \tag{10.88}$$

Notice that as $r \to \infty, \bar{r} \to r$, while at the horizon $r = 2M$ we have $\bar{r} = \frac{1}{2}M$. Show that the metric for spherical symmetry takes the form

$$ds^2 = -\left(\frac{1 - M/2\bar{r}}{1 + M/2\bar{r}}\right)^2 dt^2 + \left(1 + \frac{M}{2\bar{r}}\right)^4 (d\bar{r}^2 + \bar{r}^2 \, d\Omega^2). \qquad (10.89)$$

(b) Define quasi-Cartesian coordinates by the usual equations $x = \bar{r} \cos \phi \sin \theta$, $y = \bar{r} \sin \phi \sin \theta$, and $z = \bar{r} \cos \theta$ so that (as in Exercise 10.1), $d\bar{r}^2 + \bar{r}^2 \, d\Omega^2 = dx^2 + dy^2 + dz^2$.

Thus, the metric has been converted into coordinates (x, y, z), which are called *isotropic* coordinates. Now take the limit as $\bar{r} \to \infty$ and show that

$$ds^2 = -\left[1 - \frac{2M}{\bar{r}} + 0\left(\frac{1}{\bar{r}^2}\right)\right] dt^2$$

$$+ \left[1 + \frac{2M}{\bar{r}} + 0\left(\frac{1}{\bar{r}^2}\right)\right] \times (dx^2 + dy^2 + dz^2).$$

This proves Eq. 10.38.

10.10 Complete the calculation for a uniform-density star.

(a) Integrate Eq. 10.48 to get Eq. 10.49 and fill in the steps leading to Eqs. 10.50–10.52 and Eq. 10.54.

(b) Calculate e^Φ and the redshift experienced by light traveling to infinity from the center of the star if $M = 1 \, M_\odot = 1.47 \, \mathrm{km}$ and $R = 1 \, R_\odot = 7 \times 10^5 \, \mathrm{km}$ (for a star like the Sun), and again if $M = 1 \, M_\odot$ and $R = 10 \, \mathrm{km}$ (typical of a neutron star).

(c) Take $\rho = 10^{-11} \, \mathrm{m}^{-2}$ and $M = 0.5 \, M_\odot$, and compute R and e^Φ at the surface and center, and the redshift from the surface to the center. What is the density $10^{-11} \, \mathrm{m}^{-2}$ in $\mathrm{kg} \, \mathrm{m}^{-3}$?

10.11 Derive the restrictions in Eq. 10.57.

10.12 Prove that Eqs. 10.60–10.63 do indeed solve Einstein's equations, given by Eqs. 10.14–10.17 and Eqs. 10.20–10.23 or Eqs. 10.27, 10.30, and 10.39.

10.13 Derive Eqs. 10.66 and 10.67.

10.14 A Newtonian polytrope of index n satisfies Eqs. 10.30 and 10.44, with equation of state $p = K\rho^{(1+1/n)}$ for some constant K. Polytropes were discussed in detail by Chandrasekhar (1957). Consider the case $n = 1$, to which Buchdahl's equation of state reduces as $\rho \to 0$.

(a) Show that ρ satisfies the equation

$$\frac{1}{r^2} \frac{d}{dr}\left(r^2 \frac{d\rho}{dr}\right) + \frac{2\pi}{K} \rho = 0, \qquad (10.90)$$

and show that its solution is

$$\rho = \alpha u(r), \quad u(r) = \frac{\sin Ar}{Ar}, \quad A^2 = \frac{2\pi}{K},$$

where α is an arbitrary constant.

(b) Find the relation of the Newtonian constants α and K to the Buchdahl constants β and p_* by examining the Newtonian limit ($\beta \to 0$) of Buchdahl's solution.

(c) From the Newtonian equations find $p(r)$, the total mass M, and the radius R, and show them to be identical to the Newtonian limits of Eqs. 10.62, 10.67, and 10.65.

10.15 Calculations of stellar structure more realistic than Buchdahl's solution must be done numerically. But Eq. 10.39 has a zero denominator at $r = 0$, so the numerical calculation must avoid this point. One approach is to find a power-series solution to Eqs. 10.30 and 10.39, valid near $r = 0$, of the form

$$m(r) = \sum_j m_j r^j,$$

$$p(r) = \sum_j p_j r^j, \tag{10.91}$$

$$\rho(r) = \sum_j \rho_j r^j.$$

Assume that the equation of state $p = p(\rho)$ has the following expansion near the central density ρ_c:

$$p = p(\rho_c) + (p_c \Gamma_c / \rho_c)(\rho - \rho_c) + \cdots , \tag{10.92}$$

where Γ_c is the adiabatic index $d(\ln p)/d(p)$ evaluated at ρ_c. Find the first two nonvanishing terms in each power series in Eq. 10.91, and estimate the largest radius r at which these terms give an error no larger than 0.1% in any power series. Numerical integrations may be started at such a radius using the power series to provide the initial values.

10.16 (a) The two simple equations of state derived in § 10.7, $p = k\rho^{4/3}$ (Eq. 10.81) and $p = \rho/3$ (Eq. 10.87), differ in a fundamental way: the first has an arbitrary dimensional constant k, the second does not. Use this fact to argue that a stellar model constructed using only the second equation of state can only have solutions in which $\rho = \mu/r^2$ and $m = \nu r$, for some constants μ and ν. The key to the argument is that $\rho(r)$ may be given any value by a simple change of the unit of length, but there are no other constants in the equations whose values are affected by such a change.

(b) Show from this that the only nontrivial solution of this type is for $\mu = 3/(56\pi)$, $\nu = 3/14$. This is physically unacceptable, since it is singular at $r = 0$ and has no surface.

(c) Do there exist solutions which are nonsingular at $r = 0$ or which have finite surfaces?

10.17 (This problem requires programming a computer, or at least using a spreadsheet.) Numerically construct a sequence of stellar models using the equation of state

$$p = \begin{cases} k\rho^{4/3}, & \rho \leqslant (27k^3)^{-1}, \\ \frac{1}{3}\rho, & \rho \geqslant (27k^3)^{-1}, \end{cases} \tag{10.93}$$

where k is given by Eq. 10.81. This is a crude approximation to a realistic 'stiff' neutron-star equation of state. Construct the sequence by using the following values for ρ_c: $\rho_c/\rho_* = 0.1, 0.8, 1.2, 2, 5, 10$, where $\rho_* = (27k^3)^{-1}$. Use the power series developed in Exercise 10.15 to start the integration. Does the sequence seem to approach a limiting mass, a limiting value of M/R, or a limiting value of the central redshift?

10.18 Show that the remark made before Eq. 10.80, that the nuclei supply little pressure, is true for the regime under consideration, i.e. where $m_e < p_f^2/3k_BT < m_p$, in which k_B is Boltzmann's constant (not the same k as in Eq. 10.81). What temperature range is this for white dwarfs, where $n \approx 10^{37}\,\mathrm{m}^{-3}$?

10.19 Our Sun has an equatorial rotation velocity of about $2\,\mathrm{km\,s^{-1}}$.

(a) Estimate its angular momentum, on the assumption that the rotation is rigid (uniform angular velocity) and the Sun is of uniform density. As the true angular velocity is likely to increase inwards, this is a lower limit on the Sun's angular momentum.

(b) If the Sun were to collapse to neutron-star size (say 10 km radius), conserving both mass and total angular momentum, what would its angular velocity of rigid rotation be? In nonrelativistic language, would the corresponding centrifugal force exceed the Newtonian gravitational force on the equator?

(c) A neutron star of $1\,M_\odot$ and radius 10 km rotates 30 times per second (typical of young pulsars). Again in Newtonian language, what is the ratio of centrifugal to gravitational force on the equator? In this sense the star is slowly rotating.

(d) Suppose that a main-sequence star of $1\,M_\odot$ has a dipole magnetic field with typical strength 1 gauss in the equatorial plane. Assuming flux conservation in this plane, what field strength should we expect if the star collapses to radius of 10 km? (The Crab Pulsar's field is of the order of 10^{11} gauss.)

11   Schwarzschild geometry and black holes

11.1   Trajectories in the Schwarzschild spacetime

The Schwarzschild geometry is the geometry of the vacuum spacetime outside a spherical star. It is determined by one parameter, the mass M, and has the line element

$$ds^2 = -\left(1 - \frac{2M}{r}\right)dt^2 + \left(1 - \frac{2M}{r}\right)^{-1} dr^2 + r^2\, d\Omega^2 \tag{11.1}$$

in the coordinate system developed in the previous chapter. Its importance is not just that it describes the gravitational field outside a star: we shall see that it also gives the geometry of a spherical black hole. A careful study of its timelike and null geodesics – the paths of freely moving particles and photons – will be our route to understanding the physical meaning of this metric and the important role that it now plays in astronomy.

Black holes in Newtonian gravity

Before we embark on the study of fully relativistic black holes, it is as well to understand that the physics is not really exotic, and that speculations on analogous objects go back two centuries. The idea of a black hole follows inevitably from the weak equivalence principle, which was part of Newtonian gravity: that trajectories in the gravitational field of any body depend only on the position and velocity of the particle, not its internal composition. The question of whether a particle can escape from the gravitational field of a body is, then, only an issue of velocity: does it have a speed larger than the escape velocity from the location where it starts? For a spherical body like a star, the escape velocity depends only on how far one is from the center of the body.

Now, a star is visible to us because light escapes from its surface. As long ago as the late 1700s, the British physicist John Michell and the French mathematician and physicist Pierre Laplace speculated (independently) on the possibility that stars might exist whose escape velocity was larger than the speed of light. At that period in history, it was popular to regard light as a particle traveling at a finite speed. Michell and Laplace both understood that if one were able to make a star more compact than the Sun, but with the same mass, then it would have a larger escape velocity. One could therefore imagine a star so compact that the escape velocity from its surface would exceed the velocity of light. The star would then be dark, invisible.

For a spherical star, this is a simple computation. By the conservation of energy, a particle launched from the surface of a star with mass M and radius R will just barely escape if its gravitational potential energy balances its kinetic energy (using Newtonian language):

$$\frac{1}{2}v^2 = \frac{GM}{R}. \tag{11.2}$$

Setting $v = c$ in this relation gives the criterion for the size of a star that would be invisible:

$$R = \frac{2GM}{c^2}. \tag{11.3}$$

Remarkably, as we shall see in § 11.2, this is exactly the modern formula for the radius of a black hole in general relativity. Now, both Michell and Laplace knew the mass of the Sun and the speed of light to sufficient accuracy to realize that this formula gives an absurdly small size, of order a few kilometers, so that to them the calculation was in the end nothing more than an amusing speculation.

Today this is far more than an amusing speculation: objects of this compactness that trap light exist all over the Universe, with masses ranging from a few solar masses up to 10^9 or $10^{10}\,M_\odot$. (We will discuss this in § 11.4 below.) The small size of a few kilometers is not as absurd as it once seemed. For a $1\,M_\odot$ star, using modern values for c and G, the radius is about 3 km. We saw in the last chapter that neutron stars have radii perhaps three or four times as large, with comparable masses, and that they cannot support more than three solar masses, perhaps less. So, when neutron stars accrete large amounts of material from a companion star, or when neutron stars merge together (as LIGO and Virgo observed for the first time in 2017 – see the next chapter), the formation of something even more compact is possible, and indeed may be inevitable.

What is more, it takes even less exotic physics to form a more massive black hole. Consider the mean density of an object (again in Newtonian terms) with size given by Eq. 11.3:

$$\bar{\rho} = \frac{M}{\frac{4}{3}\pi R^3} = \frac{3c^6}{32\pi G^3 M^2}. \tag{11.4}$$

This scales as M^{-2}, so that the density needed to form such an object goes down as its mass goes up. It is not hard to show (see Exercise 11.1) that a contracting spherical object with a mass of $10^9\,M_\odot$ would become a Newtonian 'dark star' even before its density had risen to the density of water! Astronomers believe that this is a typical mass for the black holes that are thought to power quasars (see § 11.4 below), so these objects might not have required any exotic physics to form.

Although there is a basic similarity between the old concept of a Newtonian dark star and the modern black hole, which we will explore in this chapter, there are big differences too. Most fundamentally, for Michell and Laplace the star was dark because light could not escape to infinity. The star was still there, emitting light. The light would still leave the surface, but gravity would eventually pull it back, the way a ball thrown upwards on Earth turns around and falls back to the surface. But in our modern view, this would violate the principles of relativity. Consider what the turning around of this photon would look like to an inertial observer who happens to be momentarily at rest right next to the turning point. The photon would simply come to rest relative to that observer. But we know that photons do not come to rest with respect to any inertial observer: they always have speed c relative to *any* local inertial oberver.

So if such a black hole really exists in general relativity, then the trapped light never even leaves its surface. We shall see that this is exactly the case, and moreover that this 'surface' is itself not the edge of a massive body but is just a location in empty space, a location vacated by the inexorable collapse of the material that formed the hole, a mere separatrix between what can get out and what cannot. Light that starts out just marginally outside this location does indeed escape if it is directed outwards. Light that starts out inside never progresses across this surface: it immediately starts falling further toward the center.

Conserved quantities along geodesics

To verify the description in the previous paragraph of how light gets trapped, we need to begin by examining the trajectories of freely falling particles. This will allow us to calculate whether light rays are trapped or can escape from any given location.

We have seen (in Eq. 7.29 and the associated discussion) that, when a spacetime has a certain symmetry, there is an associated conserved momentum component for trajectories. Because our space has so many symmetries – time independence and spherical symmetry – the values of the conserved quantities turn out to determine the trajectory completely. We shall treat 'particles' with mass and 'photons' without mass in parallel.

Time independence of the metric means that the energy $-p_0$ is constant on the trajectory. For massive particles with rest mass $m \neq 0$, we work with the energy per unit mass (the specific energy) $\tilde{E}$, while for photons we just use the energy E:

$$\text{particle: } \tilde{E} := -p_0/m; \quad \text{photon: } E = -p_0. \tag{11.5}$$

Because the angle ϕ about the axis is independent of the metric, this implies that the angular momentum p_ϕ is constant. We again define the specific angular momentum $\tilde{J}$ for massive particles and the ordinary angular momentum J for photons:

$$\text{particle: } \tilde{J} := p_\phi/m; \quad \text{photon: } J = p_\phi. \tag{11.6}$$

Because of spherical symmetry, the motion is always confined to a single plane, and we can choose that plane to be the equatorial plane. Then θ is constant ($\theta = \pi/2$) for the orbit, so $d\theta/d\lambda = 0$, where λ is any parameter on the orbit. But p^θ is proportional to this, so it also vanishes. The other components of momentum are:

$$\text{particle: } p^0 = g^{00}p_0 = m\left(1 - \frac{2M}{r}\right)^{-1}\tilde{E},$$

$$p^r = m\,dr/d\tau,$$

$$p^\phi = g^{\phi\phi}p_\phi = m\frac{1}{r^2}\tilde{J}; \tag{11.7}$$

$$\text{photon: } p^0 = \left(1 - \frac{2M}{r}\right)^{-1} E,$$

$$p^r = dr/d\lambda,$$

$$p^\phi = d\phi/d\lambda = J/r^2. \tag{11.8}$$

The equation for a photon's p^r should be regarded as *defining* the affine parameter λ. The equation $\vec{p} \cdot \vec{p} = -m^2$ implies that

particle:

$$-m^2 \tilde{E}^2 \left(1 - \frac{2M}{r}\right)^{-1} + m^2 \left(1 - \frac{2M}{r}\right)^{-1} \left(\frac{dr}{d\tau}\right)^2 + \frac{m^2 \tilde{J}^2}{r^2} = -m^2; \tag{11.9}$$

photon:

$$-E^2 \left(1 - \frac{2M}{r}\right)^{-1} + \left(1 - \frac{2M}{r}\right)^{-1} \left(\frac{dr}{d\lambda}\right)^2 + \frac{J^2}{r^2} = 0. \tag{11.10}$$

These can be solved to give the basic equations for orbits,

$$\text{particle: } \left(\frac{dr}{d\tau}\right)^2 = \tilde{E}^2 - \left(1 - \frac{2M}{r}\right)\left(1 + \frac{\tilde{J}^2}{r^2}\right); \tag{11.11}$$

$$\text{photon: } \left(\frac{dr}{d\lambda}\right)^2 = E^2 - \left(1 - \frac{2M}{r}\right)\frac{J^2}{r^2}. \tag{11.12}$$

Types of orbits

Both of Eqs. 11.11 and 11.12 have the same general form, and they allow us to define *effective potentials*:

$$\text{particle: } \tilde{V}^2(r) = \left(1 - \frac{2M}{r}\right)\left(1 + \frac{\tilde{J}^2}{r^2}\right); \tag{11.13}$$

$$\text{photon: } V^2(r) = \left(1 - \frac{2M}{r}\right)\frac{J^2}{r^2}. \tag{11.14}$$

Their typical forms are plotted in Figures 11.1 and 11.2, in which various points have been labeled and possible trajectories drawn (dashed lines).

Both Eqs. 11.11 and 11.12 imply that, since the left side must be positive or zero, the energy of a trajectory must not be less than the potential V. (Here and until Eq. 11.17 we will take E and V to refer to $\tilde{E}$ and $\tilde{V}$ as well, since the remarks for the two cases are identical.) So, for an orbit of given E, the radial range is restricted to those radii for which V is smaller than E. For instance, consider a trajectory which has the value of E indicated by point G (in either diagram). If the particle comes in from $r = \infty$, then it cannot reach a smaller value of r than where the dashed line hits the V^2 curve, at point G.

Point G is called a *turning* point, since $E^2 = V^2$ there, and Eq. 11.12 then implies $(dr/d\lambda)^2 = 0$. Do not forget that Eq. 11.12 governs only the radial part of the motion.

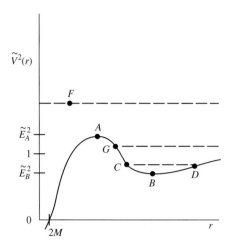

Figure 11.1   Typical effective potential for a massive particle of fixed specific angular momentum in the Schwarzschild metric.

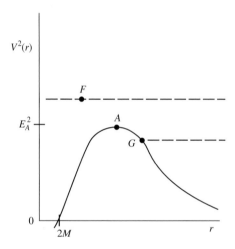

Figure 11.2   The same as Figure 11.1 for a massless particle.

The angular motion of the particle will continue because J is conserved, so the particle will be on a trajectory that reaches a minimum radius at G and then moves out again. Similar conclusions apply to Eq. 11.11.

To see how this happens, we differentiate Eqs. 11.11 and 11.12. For particles, differentiating the equation

$$\left(\frac{dr}{d\tau}\right)^2 = \tilde{E}^2 - \tilde{V}^2(r)$$

with respect to τ gives

$$2\left(\frac{dr}{d\tau}\right)\left(\frac{d^2r}{d\tau^2}\right) = -\frac{d\tilde{V}^2(r)}{dr}\frac{dr}{d\tau},$$

or

$$\text{particles:} \quad \frac{d^2r}{d\tau^2} = -\frac{1}{2}\frac{d}{dr}\tilde{V}^2(r). \tag{11.15}$$

Similarly, the photon equation gives

$$\text{photons:} \quad \frac{d^2r}{d\lambda^2} = -\frac{1}{2}\frac{d}{dr}V^2(r). \tag{11.16}$$

These are the analogs in relativity of the equation

$$m\mathbf{a} = -\nabla\phi,$$

where ϕ is the potential for some force. If we now look again at point G, we see that the radial acceleration of the trajectory is outwards, so that the particle (or photon) comes in to the minimum radius, but is accelerated outward as it turns around, and so it returns to $r = \infty$. This is a 'hyperbolic' orbit – the analog of orbits which are true hyperbolae in Newtonian gravity.

It is clear from Eq. 11.15 or 11.16 that a circular orbit ($r = $ const.) is possible only at a minimum or maximum of V^2. These occur at points A and B in the diagrams (there is no point B for photons). A maximum is, however, unstable, since any small change in r results in an acceleration away from the maximum, by Eqs. 11.15 and 11.16. So, for particles, there is one stable (B) and one unstable circular orbit (A) for this value of $\tilde{J}$. For photons, there is only one circular orbit for this J, and it is unstable.

We can be quantitative by evaluating

$$0 = \frac{d}{dr}\left[\left(1 - \frac{2M}{r}\right)\left(1 + \frac{\tilde{J}^2}{r^2}\right)\right]$$

and

$$0 = \frac{d}{dr}\left[\left(1 - \frac{2M}{r}\right)\frac{J^2}{r^2}\right].$$

These give, respectively

$$\text{particles:} \quad r = \frac{\tilde{J}^2}{2M}\left[1 \pm \left(1 - \frac{12M^2}{\tilde{J}^2}\right)^{1/2}\right]; \tag{11.17}$$

$$\text{photons:} \quad r = 3M. \tag{11.18}$$

For particles, there are two radii, as we expect, but only if $\tilde{J}^2 > 12M^2$. The two radii are identical for $\tilde{J}^2 = 12M^2$ and do not exist at all for $\tilde{J}^2 < 12M^2$. This indicates a qualitative change in the shape of the curve for $\tilde{V}^2(r)$ for small $\tilde{J}$. The squared potentials for the two cases $\tilde{J}^2 = 12M^2$ and $\tilde{J}^2 < 12M^2$ are illustrated in Figure 11.3. Since there is a minimum $\tilde{J}^2$ for a circular particle orbit, there is also a minimum r for a stable circular orbit, obtained by taking $\tilde{J}^2 = 12M^2$ in Eq. 11.17:

$$\text{particle:} \quad r_{\min} = 6M. \tag{11.19}$$

Notice that this location does not depend on any property of the particle, apart from the fact that it has nonzero rest mass: it is a property of the Schwarzschild geometry itself. This orbit

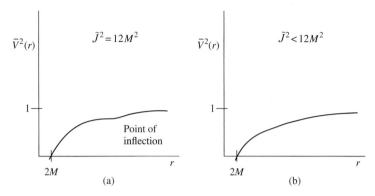

As Figure 11.1 for the indicated values of specific angular momentum.

is referred to as the *innermost stable circular orbit*, or ISCO. No massive particle can stay on a circular orbit in the Schwarzschild geometry inside $r = 6M$.

For photons, the unstable circular orbit is always at the same radius, $r = 3M$, regardless of J. The angular momentum and energy of a photon affect its wavelength but not its speed, so the dynamics of photons do not depend on E or J.

The last kind of orbit we need to consider is the orbit whose energy is given by the line which passes through the point F in Figures 11.1 and 11.2. Since this nowhere intersects the potential curve, this orbit plunges right through $r = 2M$ and never returns. From Exercise 11.2, § 11.6, we see that for such an orbit the impact parameter b is small: it is aimed more directly at the hole than are orbits of smaller $\tilde{E}$ for the same $\tilde{J}$.

Of course, if the geometry under consideration is a star, its radius R will exceed $2M$, and the potential diagrams, Figures 11.1–11.3, will be valid only outside R. If a particle reaches R it will hit the star. Depending on R/M, then, only certain kinds of orbits will be possible.

Perihelion shift: noncircular orbits

A particle (or planet) in a (stable) *circular* orbit around a star will make one complete orbit and come back to the same point (i.e. the same value of ϕ) in a fixed amount of coordinate time, which is called its period P. This period can be determined as follows. From Eq. 11.17 it follows that a stable circular orbit at radius r has angular momentum

$$\tilde{J}^2 = \frac{Mr}{1 - 3M/r},\tag{11.20}$$

and since $\tilde{E}^2 = \tilde{V}^2$ for a circular orbit, it also has energy

$$\tilde{E}^2 = \left(1 - \frac{2M}{r}\right)^2 \bigg/ \left(1 - \frac{3M}{r}\right).\tag{11.21}$$

Now, we have

$$\frac{\mathrm{d}\phi}{\mathrm{d}\tau} := U^\phi = \frac{p^\phi}{m} = g^{\phi\phi}\frac{p_\phi}{m} = g^{\phi\phi}\tilde{J} = \frac{1}{r^2}\tilde{J}\tag{11.22}$$

and

$$\frac{dt}{d\tau} := U^0 = \frac{p^0}{m} = g^{00}\frac{p_0}{m} = g^{00}(-\tilde{E}) = \frac{\tilde{E}}{1 - 2M/r}. \tag{11.23}$$

We obtain the angular velocity by dividing these:

$$\frac{dt}{d\phi} = \frac{dt/d\tau}{d\phi/d\tau} = \left(\frac{r^3}{M}\right)^{1/2}. \tag{11.24}$$

The period, which is the time taken for ϕ to change by 2π, is

$$P = 2\pi \left(\frac{r^3}{M}\right)^{1/2}. \tag{11.25}$$

This is the coordinate time, of course, not the particle's proper time. But, as discussed in Exercise 11.8, § 11.6, coordinate time is the same as proper time far away, so this is the orbital period that a distant observer will measure. It happens, coincidentally, that this is identical to the Newtonian expression.

Now, a slightly noncircular orbit of a planet around the Sun will oscillate in and out about a central radius r. In Newtonian gravity the orbit is a perfect ellipse (if we ignore perturbations by other planets), which means, among other things, that it is *closed*: after a fixed amount of time it returns to the same point (the same r and ϕ). In GR, this does not happen; a typical orbit looks like those in Figure 11.4(b), (c). However, when the effects of relativity are small and the orbit is nearly circular, the relativistic orbit must be almost closed: it must look like an ellipse which slowly rotates about the center. One way to describe this is to look at the *perihelion* of the orbit, the point of closest approach to the Sun.

It is worth taking a diversion for a moment to talk about nomenclature. *Peri* means closest and *helion* refers to the Sun; for orbits about any star the name *periastron* is more appropriate. For orbits around Earth – *geo* – one speaks of the *perigee*. An orbit around a black hole has a *peribothon*. The opposite of *peri* is *ap*: the furthest distance. Thus, an orbit may also have an aphelion, apastron, apogee, or apbothon, depending on what it is orbiting around. The general terms, not specific to any particular central object, are *perapsis* and *apapsis*.

The location of the perihelion will move steadily around the Sun, so even if the shift after one orbit is tiny, it is an effect that accumulates with time; observers can therefore hope to measure it. It will be largest for Mercury because all the other planets are further from the Sun and are therefore under the influence of significantly smaller relativistic corrections to Newtonian gravity. The measurement of Mercury's precession was a herculean task, first accomplished in the 1800s. Owing to various other effects, such as the perturbations of Mercury's orbit due to the other planets, the observed precession is about $5600''$/century. Astronomers calculated, from Newton's laws, the contributions of all the other planets, and they were left with a mere $43''$/century that was unaccounted for. Einstein used his new theory of GR to calculate the relativistic shift, and found exactly the same amount. This was the first observational evidence supporting general relativity. Einstein was so excited by this striking confirmation of his theory that, as he later remarked, he had palpitations of the heart for some days afterwards.

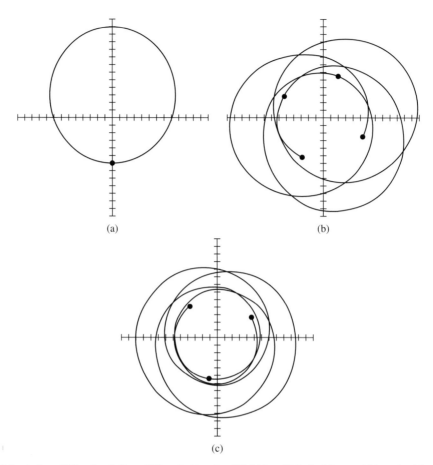

(a)　　　　　　　　　　(b)

(c)

Figure 11.4 (a) A Newtonian orbit is a closed ellipse. Grid marked in units of M. (b) An orbit in the Schwarzschild metric with pericentric and apcentric distances similar to those in (a). Pericenters (heavy dots) advance by about 97° per orbit. (c) A moderately more compact orbit than in (b) has a considerably larger pericenter shift, about 130°.

Modern astronomy has access to systems that are much more highly relativistic than the Solar System. The Hulse–Taylor binary pulsar, which we met in § 9.5, is in a highly elliptical orbit where the stars at periastron are as close to one another as 1.1 solar radii. The precession is therefore much larger than Mercury's: 4.2° per year. The black hole in the center of our Galaxy, called Sagittarius A*, and which we will discuss later in this chapter, is orbited by many resolvable stars. In one of the highest-precision measurements ever made in astronomy, the GALAXY team (Abuter *et al.* 2020) measured the peribothon shift of one of them, called S2, to be 13 arcminutes in a single orbit, with an accuracy of about 20%.

We wish to follow Einstein and calculate the precession from our orbital equations. Let us begin by getting an equation for the particle's path in the r–ϕ plane. We have $dr/d\tau$ from Eq. 11.11. We then find $d\phi/d\tau$ from Eq. 11.22 and divide to get

$$\left(\frac{dr}{d\phi}\right)^2 = \frac{\tilde{E}^2 - (1 - 2M/r)\left(1 + \tilde{J}^2/r^2\right)}{\tilde{J}^2/r^4}. \tag{11.26}$$

It is convenient to define

$$u := 1/r, \tag{11.27}$$

in terms of which we obtain

$$\left(\frac{du}{d\phi}\right)^2 = \frac{\tilde{E}^2}{\tilde{J}^2} - (1 - 2Mu)\left(\frac{1}{\tilde{J}^2} + u^2\right). \tag{11.28}$$

The *Newtonian* orbit is found by neglecting u^3 terms (see Exercise 11.12, § 11.6):

$$\text{Newtonian:} \quad \left(\frac{du}{d\phi}\right)^2 = \frac{\tilde{E}^2}{\tilde{J}^2} - \frac{1}{\tilde{J}^2}(1 - 2Mu) - u^2. \tag{11.29}$$

A circular orbit in Newtonian theory has $u = M/\tilde{J}^2$ (take the square root as equal to 1 in Eq. 11.17), so we define

$$y = u - \frac{M}{\tilde{J}^2}, \tag{11.30}$$

the variable that represents the deviation from circularity. We then get

$$\left(\frac{dy}{d\phi}\right)^2 = \frac{\tilde{E}^2 - 1}{\tilde{J}^2} + \frac{M^2}{\tilde{J}^4} - y^2. \tag{11.31}$$

It is easy to see that this is satisfied by

$$\text{Newtonian:} \quad y = \left[\frac{\tilde{E}^2 + M^2/\tilde{J}^2 - 1}{\tilde{J}^2}\right]^{1/2} \cos(\phi + B), \tag{11.32}$$

where B is arbitrary. This is clearly periodic: as ϕ advances by 2π, y returns to its value and, therefore, so does r. The constant B just determines the initial orientation of the orbit. It is interesting, but unimportant for our purposes, that by expressing this in terms of r we get

$$\text{Newtonian:} \quad \frac{1}{r} = \frac{M}{\tilde{J}^2} + \left[\frac{\tilde{E}^2 + M^2/\tilde{J}^2 - 1}{\tilde{J}^2}\right]^{1/2} \cos(\phi + B), \tag{11.33}$$

which is the equation of an *ellipse*.

We now consider the relativistic case and make the same definition of y, but instead of throwing away the u^3 term in Eq. 11.28 we assume that the orbit is nearly circular, so that y is small, and we neglect only the terms in y^3. Then we get

$$\left(\frac{dy}{d\phi}\right)^2 = \frac{\tilde{E}^2 + M^2/\tilde{J}^2 - 1}{\tilde{J}^2} + \frac{2M^4}{\tilde{J}^6} + \frac{6M^3}{\tilde{J}^2}y - \left(1 - \frac{6M^2}{\tilde{J}^2}\right)y^2. \tag{11.34}$$

This can be made analogous to Eq. 11.31 by completing the square on the right-hand side. The result is the solution

$$y = y_0 + A\cos(k\phi + B), \tag{11.35}$$

where B is arbitrary and the other constants are

$$k = \left(1 - \frac{6M^2}{\tilde{J}^2}\right)^{1/2},$$

$$y_0 = \frac{3M^3}{k^2 \tilde{J}^2},$$

$$A = \frac{1}{k}\left(\frac{\tilde{E}^2 + M^2/\tilde{J}^2 - 1}{\tilde{J}^2} + \frac{2M^4}{\tilde{J}^6} - y_0^2\right)^{1/2}. \tag{11.36}$$

The appearance of the constant y_0 just means that the orbit oscillates not about $y = 0$ ($u = \tilde{M}/J^2$) but about $y = y_0$: Eq. 11.30 doesn't use the correct radius for a circular orbit in GR. The amplitude A is also somewhat different, but what is most interesting here is the fact that k does not equal 1. The orbit returns to the same r when $k\phi$ goes through 2π, from Eq. 11.35. Therefore the change in ϕ from one perihelion to the next is

$$\Delta\phi = \frac{2\pi}{k} = 2\pi\left(1 - \frac{6M^2}{\tilde{J}^2}\right)^{-1/2}, \tag{11.37}$$

which, for nearly Newtonian orbits, is

$$\Delta\phi \simeq 2\pi\left(1 + \frac{3M^2}{\tilde{J}^2}\right). \tag{11.38}$$

The perihelion *advance*, then, from one orbit to the next, is

$$\Delta\phi = 6\pi M^2/\tilde{J}^2 \text{ radians per orbit.} \tag{11.39}$$

We can use Eq. 11.20 to obtain $\tilde{J}$ in terms of r, since the corrections for noncircularity will cause changes in Eq. 11.39 of the same order as terms we have already neglected. Moreover, if we consider orbits about a nonrelativistic star, we can approximate Eq. 11.20 as follows:

$$\tilde{J}^2 = \frac{Mr}{1 - 3M/r} \approx Mr,$$

so that we get

$$\Delta\phi \approx 6\pi\frac{M}{r}. \tag{11.40}$$

For Mercury's orbit, $r = 5.55 \times 10^7$ km and $M = 1\,M_\odot = 1.47$ km, so that

$$(\Delta\phi)_{\text{Mercury}} = 4.99 \times 10^{-7} \text{ radians per orbit.} \tag{11.41}$$

This single-orbit shift was impossibly small to measure in the nineteenth century, but – as noted above – it accumulates with time. Each orbit takes 0.24 yr, so over a century the shift is what Einstein calculated in 1915:

$$(\Delta\phi)_{\text{Mercury}} = 0.43''/\text{yr} = 43''/\text{century.} \tag{11.42}$$

Binary pulsars

Another system in which the pericenter shift is observable is the Hulse–Taylor binary pulsar system PSR B1913+16, which we introduced in § 9.5. While this is not a 'test particle' orbiting a spherical star, but is rather two roughly equal-mass stars orbiting their common center of mass, the pericenter shift (in this case, the periastron shift) still happens in the same way. The two neutron stars of the Hulse–Taylor system have mean separation 1.2×10^9 m, so using Eq. 11.40 with $M = 1.4 \, M_\odot = 2.07$ km gives a crude estimate of $\Delta\phi = 3.3 \times 10^{-5}$ radians per orbit $= 2.1°$ per year. This is much easier to measure than Mercury's shift! In fact, a more careful calculation, taking into account the high eccentricity of the orbit and the fact that the two stars are of comparable mass, predicts $4.2°$ per year.

For our purposes here we have calculated the periastron shift from the known masses of the star. But in fact the observed shift of $4.2261° \pm 0.0007$ per year is one of the measurements which enable us to calculate the masses of the neutron stars in the PSR B1913+16 system. The other measurement is another relativistic effect: the redshift of the pulsar's apparent pulse rate, which results from two effects. One is the special-relativistic 'transverse-Doppler' term: the $O(v)^2$ term in Eq. 2.39. The other is the changing gravitational redshift as the pulsar's eccentric orbit brings it in and out of its companion's gravitational potential. These two redshift effects are observationally indistinguishable from one another, but their combined resultant redshift gives one more number that depends on the masses of the stars. Using it and the periastron shift and the Newtonian mass function for the orbit allows one to determine the stars' masses and the orbit's inclination (see Stairs 2003).

While PSR B1913+16 was the first binary pulsar to be discovered, astronomers now know of many more (Lorimer 2008, Stairs 2003). The most dramatic system is the so-called double pulsar system PSR J0737-3039, the first binary discovered in which *both* members are seen as radio pulsars (Kramer *et al.* 2006). In this system the pericenter shift is around $17°$ per year! Because both pulsars can be tracked, this system has become the best testing ground so far for the deviations of general relativity from simple Newtonian behavior (Wex 2014). We discuss some of these deviations now.

Post-Newtonian gravity

The pericenter shift is an example of corrections that general relativity makes to Newtonian orbital dynamics. In going from Eq. 11.37 to Eq. 11.38, we made an approximation that the orbit was 'nearly Newtonian', and we did a Taylor expansion in the small quantity $M^2/\tilde{J}^2$, which by Eq. 11.20 is $(M/r)(1 - 3M/r)$. This is indeed small if the particle's orbit is far from the black hole, $r \gg M$. So, orbits far from a black hole are very nearly like Newtonian orbits, with small corrections such as the pericenter shift. These are orbits where, in Newtonian language, the gravitational potential M/r is small, as is the particle's orbital velocity, which for a circular orbit is related to the potential by $v^2 = M/r$. Far from even an extremely relativistic source, gravity is close to being Newtonian. We say that the pericenter shift is a post-Newtonian effect, a correction to Newtonian motion in the limit of weak fields and slow motion.

We showed in § 8.4 that Newton's field equations emerge from the full equations of general relativity in this same limit, where the field (we called it h in that calculation) is weak and the velocities small. It is possible to follow that approximation to higher order and to keep the first post-Newtonian corrections to the Einstein field equations in this same limit. If one did this, one would be able to show the result we have asserted, that the pericenter shift is a general feature of orbital motion. We showed in that section that the metric of spacetime that describes a Newtonian system with gravitational potential ϕ is, to first order in ϕ, given by the metric

$$ds^2 = -(1 + 2\phi)dt^2 + (1 - 2\phi)(dx^2 + dy^2 + dz^2). \tag{11.43}$$

For comparison, let us take the limit for weak fields of the Schwarzschild metric given in Eq. 11.1:

$$ds^2 \approx -\left(1 - \frac{2M}{r}\right)dt^2 + \left(1 + \frac{2M}{r}\right)dr^2 + r^2\,d\Omega^2, \tag{11.44}$$

where we have expanded g_{rr} and kept only the lowest order term in M/r. These two metrics look similar – their g_{00} terms are identical – but they appear not to be identical because if we transform the spatial line element of Eq. 11.43 to polar coordinates in the usual way then we would expect to see in Eq. 11.44 the spatial line element

$$d\ell^2 = \left(1 + \frac{2M}{r}\right)\left(dr^2 + r^2\,d\Omega^2\right). \tag{11.45}$$

(Readers who want to be reminded about the transformation from Cartesian to spherical coordinates in flat space should see Eq. 6.19 and Exercise 6.28.) Is this difference from the spatial part of Eq. 11.44 an indication that the Schwarzschild metric does not obey the Newtonian limit? The answer is no – we are dealing with a coordinate effect.

Coordinates in which the line element has the form of Eq. 11.45 are called *isotropic coordinates* because there is no distinction in the metric among the directions x, y, and z. Isotropic coordinates are related to Schwarzschild coordinates by a change in the definition of the radial coordinate. If we call the isotropic coordinates $(t, \bar{r}, \theta, \phi)$, where (t, θ, ϕ) are the same as in the Schwarzschild case, then we can make a simple transformation for large r given by

$$\bar{r} = r - M. \tag{11.46}$$

Then to lowest order in M/r or $M/\bar{r}$, the expression $1 + 2M/r$ in Eq. 11.44 equals $1 - 2M/\bar{r}$. But the factor in front of $d\Omega^2$ is r^2, which becomes (again to first order) $\bar{r}^2(1 + 2M/r)$. It follows that this simple transformation changes the spatial part of the line element of Eq. 11.44 to Eq. 11.45 in terms of the radial coordinate $\bar{r}$. This demonstrates that the far field of the Schwarzschild solution does indeed conform to the form in Eq. 8.50.

There are of course many other post-Newtonian effects in general relativity besides the pericenter shift. In the next paragraph we will explore the deflection of light. Later in this chapter we will meet the dragging of inertial frames due to the rotation of the source of gravity (also called gravitomagnetism). It is possible to expand the Einstein equations beyond the first post-Newtonian equations and find second and higher post-Newtonian effects. Physicists have put considerable work into doing such expansions, because they provide highly accurate predictions of the orbits of inspiralling neutron stars and black holes, which are needed for the detection of the gravitational waves they emit. (See Blanchet (2014) and Chapter 12.)

The post-Newtonian expansion assumes not just weak gravitational fields but also slow velocities. The effect of this is that the Newtonian (Keplerian) motion of a planet depends only on the g_{00} part of the metric. The reason is that, in computing the total elapsed proper time along the world line of the particle, the spatial distance increments that the particle makes (for example $\mathrm{d}r$) are much smaller than the time increments $\mathrm{d}t$, since $\mathrm{d}r/\mathrm{d}t \ll 1$. Recall that g_{00} is also responsible for the gravitational redshift. It follows, therefore, that Newtonian gravity can be identified with the gravitational redshift: knowing one determines the other fully. *Newtonian gravity is produced exclusively by the curvature of time in spacetime.* Spatial curvature comes in only at the level of post-Newtonian corrections.

Gravitational deflection of light

In our discussion of orbits we treated only particles, not photons, because photons do not have bound orbits in Newtonian gravity. In this section we treat the analogous effect for photons, their deflection from straight-line motion as they pass through a gravitational field. Historically, this was the first general relativistic effect to have been predicted before it was observed, and its confirmation in the eclipse of 1919 (see McCrea 1979) made Einstein an international celebrity. The fact that it was a British team (led by Eddington) who made the observations to confirm the theories of a German physicist incidentally helped to alleviate post-war tension between the scientific communities of the two countries. In modern times, the light-deflection phenomenon has become a key tool of astronomy, as we describe in a separate subsection on gravitational lensing below. But first we need to understand how and why gravity deflects light.

We begin by calculating the trajectory of a photon in the Schwarzschild metric under the assumption that M/r is everywhere small along the trajectory. The equation of the orbit is the ratio of Eq. 11.8 to the square root of Eq. 11.12:

$$\frac{\mathrm{d}\phi}{\mathrm{d}r} = \pm \frac{1}{r^2} \left[\frac{1}{b^2} - \frac{1}{r^2} \left(1 - \frac{2M}{r} \right) \right]^{-1/2}, \tag{11.47}$$

where we have defined the *impact parameter*

$$b := J/E. \tag{11.48}$$

In Exercise 11.2, § 11.6, it is shown that b would be the minimum value of r in Newtonian theory, where there is no deflection. It therefore represents the 'offset' of the photon's initial trajectory from a parallel one moving purely radially. An incoming photon with $J > 0$ obeys the equation

$$\frac{\mathrm{d}\phi}{\mathrm{d}u} = \left(\frac{1}{b^2} - u^2 + 2Mu^3 \right)^{-1/2}, \tag{11.49}$$

with the same definition as before,

$$u = 1/r. \tag{11.50}$$

If we neglect the u^3 term in Eq. 11.49, all effects of M disappear, and the solution is

$$r \sin(\phi - \phi_0) = b, \tag{11.51}$$

a straight line. This is, of course, the Newtonian result.

Suppose now we assume $Mu \ll 1$ but is not entirely negligible. Then if we define

$$y := u(1 - Mu), \qquad u = y(1 + My) + O(M^2 u^2), \tag{11.52}$$

Eq. 11.49 becomes

$$\frac{d\phi}{dy} = \frac{(1 + 2My)}{(b^{-2} - y^2)^{1/2}} + O(M^2 u^2). \tag{11.53}$$

This can be integrated to give

$$\phi = \phi_0 + \frac{2M}{b} + \arcsin(by) - 2M \left(\frac{1}{b^2} - y^2 \right)^{1/2}. \tag{11.54}$$

The initial trajectory has $y \to 0$, so $\phi \to \phi_0$: ϕ_0 is the incoming direction. The photon reaches its smallest r when $y = 1/b$, as one can see from setting $dr/d\lambda = 0$ in Eq. 11.22 and using our approximation $Mu \ll 1$. This occurs at the angle $\phi = \phi_0 + 2M/b + \pi/2$. It has thus passed through an angle $\pi/2 + 2M/b$ as it travels to its point of closest approach. By symmetry, it passes through a further angle of the same size as it moves outwards from its point of closest approach (see Figure 11.5). It thus passes through a total angle of $\pi + 4M/b$. If it were going along a straight line, this angle would be π, so the net deflection is

$$\Delta\phi = 4M/b. \tag{11.55}$$

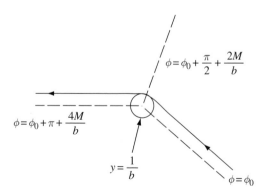

Figure 11.5   Deflection of a photon.

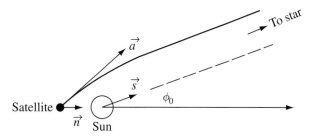

An observation from Earth of a star not at the limb of the Sun does not need to correct for the full deflection of Figure 11.5.

To the accuracy of our approximations, we may use for b the radius of closest approach rather than the impact parameter J/E. For the Sun, the maximum effect is for trajectories for which $b = R_\odot$, the radius of the Sun. Given $M = 1\,M_\odot = 1.47\,\text{km}$ and $R_\odot = 6.96 \times 10^5$ km, we find

$$(\Delta\phi)_{\odot,\text{max}} = 8.45 \times 10^{-6} \text{ rad} = 1''.74. \tag{11.56}$$

For Jupiter, with $M = 1.12 \times 10^{-3}$ km and $R = 6.98 \times 10^4$ km we have

$$(\Delta\phi)_{\Psi,\text{max}} = 6.42 \times 10^{-8} \text{ rad} = 0''.013. \tag{11.57}$$

This deflection was measured by the Hipparcos astrometry satellite and is a standard effect that must be corrected for in the measurements of star positions by Hipparcos' successor, the much higher-precision Gaia astrometry satellite (Brown *et al.* 2018).

Of course, satellite observations of stellar positions are made from a position near Earth, and for stars that are not near the Sun in the sky the satellite will receive their light before the total deflection, given by Eq. 11.55, has taken place. This situation is illustrated in Figure 11.6. An observer *at rest* at the position of the satellite observes an apparent position in the direction of the vector $\vec{a}$, tangent to the path of the light ray, and if they know their distance r from the Sun then they can calculate the true direction to the star, $\vec{s}$. Exercise 11.17, § 11.6, derives the general result.

Although the deflection of light was a key to establishing the correctness of general relativity, it is interesting that there is a purely Newtonian argument for light deflection. This was first predicted by Cavendish in 1784 and independently by the German astronomer J. G. von Söldner (1776–1833) in 1801. The argument relies on the same view of light as we mentioned in § 11.1, where we discussed the Newtonian version of a black hole: light behaving like a particle moving at speed c. Since the motion of a particle in a gravitational field depends only on its velocity, Cavendish and von Söldner were able to compute the simple result that the deflection would be $2M/b$, which is exactly *half* the prediction of general relativity in Eq. 11.55. Einstein himself, unaware of this previous work, derived the same result as Cavendish and von Söldner in 1908, just at the beginning of his quest for a relativistic theory of gravity. So the triumph of general relativity was not that it predicted a deflection, but that, once Einstein had fully formulated the theory, it predicted the *right amount* of deflection.

The reason that this Newtonian result is half the fully relativistic result is not hard to understand. Recall the remark at the end of the paragraph on post-Newtonian effects, that the orbits of planets depend only on g_{00} because their velocities are small. But light is not slow: the spatial increments dr are comparable to the time increments dt. This means that the spatial part of the metric, say g_{rr}, is of equal weight with g_{00} in determining the motion of a photon. Looking at Eq. 11.44, we see that the deviations from flatness in both g_{00} and g_{rr} are the same. Whatever deflection is produced by g_{00} (the only part of the metric that the Newtonian computations were sensitive to) is doubled by including the contribution of g_{rr}. *The extra-large deflection of light in general relativity compared with Newtonian gravity is direct evidence for the curvature of space as well as time.*

Gravitational lensing

It may of course happen that photons from the same star will travel on trajectories that pass on opposite sides of the deflecting star and intersect each other after deflection, as illustrated in Figure 11.7. Rays 1 and 2 are essentially parallel if the star ($*$) is far from the deflecting object (S). An observer at position B would then see *two* images of the star, coming from apparently different directions.

This is a very simple and special arrangement of the objects, but it illustrates the principle that gravitating bodies act as lenses. Lensing is essentially universal: no matter how weak the deflection, it would always be possible to place the observer B far enough away from S to see two rays from the same point on the source. We don't get such 'double vision' when we look at the heavens because the probability of being in exactly the right spot B for any given star and lens is small, and because many sources are not pointlike like our star: if the angular separation of the images at B (which is of order the deflection angle) is small compared with the angular size of the object on the sky, then we are not likely to be able to tell the difference between the two rays.

But as astronomers have built larger and more powerful telescopes that are able to see much greater distances into the Universe, they have revealed a sky filled with lensed images. Of particular importance is lensing by clusters of galaxies. There are so many galaxies in the Universe that, beyond any given relatively distant cluster, there is a high probability that there will be another group of galaxies located in just the right position to be lensed into multiple images in our telescope: the probability of 'being in exactly the right spot' has become reasonably large. What is more, the masses of galaxy clusters are

Figure 11.7 Deflection can produce multiple images.

Figure 11.8   A picture taken with the Hubble Space Telescope of a cluster called 0024+1654 (the fuzzy round galaxies) showing many images of a much more distant galaxy, owing to lensing by the gravity of the cluster. The images are distorted, all of them stretched in the direction transverse to the line joining the image and the center of the cluster, and compressed along this line. Picture courtesy of W. N. Colley and E. Turner (Princeton University), J. A. Tyson (Bell Labs, Lucent Technologies), and NASA.

huge, so the deflections are much bigger than the sizes of the more distant galaxy images; therefore separating them is not a problem. In fact, one often sees multiple images of the same object, created by the irregularities of the lensing mass distribution. A spectacular example is shown in Figure 11.8.

This is an aspect of what astronomers call *weak lensing*, because the deflection angles are quite small: the effect is measurable because the distances are so large. Sometimes weak lensing does not produce distinct images, but rather magnifies the brightness of distant objects: the lens focuses more light on the telescope than would reach it if the lens were not there. Very bright objects seen over large cosmological distances are typically affected by weak lensing.

Thus, even more important than the creation of separate images can be the brightening of single images by the focusing of light from them. This is called magnification.

The magnification of galaxy images by lensing makes it possible for astronomers to see galaxies at greater distances, and has helped studies of the very early Universe. Lensing also helps us map the mass distribution of the lensing cluster, and this has shown that clusters have much more mass than can be associated with their luminous stars. Astronomers call this *dark matter*, and it will be an important subject in Chapter 13, when we discuss cosmology.

Detailed modeling of observed lenses has shown that the dark matter is distributed more smoothly inside clusters than the stars, which clump into the individual galaxies. The mass of a cluster may be ten or more times the mass of its stars. The dark matter is presumably made of something that carries no electric charge, because light from the distant galaxies passes through it without absorption or scattering, and it seems to emit no electromagnetic radiation. Physicists do not know of any elementary particle that could serve as dark matter in this way: apart from neutrinos, all electrically neutral particles known at present (mid-2021) are unstable and decay quickly. And neutrinos are so light that their velocities in the early Universe, where they were in thermal equilibrium with other particles, would have been far too large for them to have been trapped in the gravitational fields of nascent clusters. Laboratory searches for dark matter particles that randomly pass through a laboratory have so far not found convincing detections, despite considerable effort. Dark matter is one of the biggest mysteries at the intersection of astronomy and particle physics today.

Studies of gravitational lensing have the potential to reveal any matter that clumps in some way. Lensing observations of individual stars in our galaxy, called *microlensing,* have raised the possibility of a population of objects with masses of the order of $0.5\,M_\odot$ that are also dark: they are observed only because they pass across more distant stars in our Galaxy and produce lensing. In this case, observers do not see two separate images. Instead, they see the brightening of the distant star. This brightening is temporary because the lensing object itself moves, and so the lensing event is transitory. The name *massive compact halo objects*, or MACHOs, has been coined for this hypothetical population of lensing objects. MACHOs might be small black holes, formed by some unknown process in the early Universe, or brown dwarfs, which are failed stars, too small to reach the nuclear ignition temperature. Observations so far are still inconclusive about the existence and nature of this population.

Gravitational lensing has even given us our first 'picture' of a black hole, shown in Figure 11.9. This was made by the Event Horizon Telescope (EHT: Akiyama *et al.* 2019). The black hole is the supermassive one ($2 \times 10^9\,M_\odot$) at the center of the galaxy M87, an elliptical galaxy in the Virgo Cluster of galaxies, about 16 Mpc distant. The image represents the intensity of mm-wave radio emission from hot gas near the black hole, forming a ring in orbit around it. We are viewing it from an oblique angle so that some of the ring is actually behind the hole, but lensing brings its radio emission over the hole and toward us, giving us a complete image of the ring. The dark center is, of course, where the hole itself is. The inner edge of the ring is at a radius inside $6M$, which for the given mass and distance subtends an angle less than 10^{-10} radians. This extraordinary resolution was achieved by coherently combining radio observations from telescopes at locations all around the Earth.

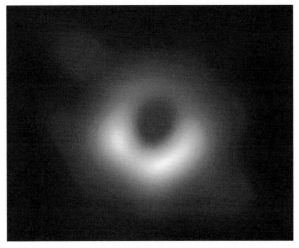

Figure 11.9   The first black-hole image made by the Event Horizon Telescope, a synthesis of mm-wave radio observations by telescopes all over the world. The radio emission comes from gas very near the black hole, some of it from behind the hole but gravitationally lensed toward us. The dark center is where the hole itself resides. Credit: EHT Collaboration.

11.2 Nature of the surface $r = 2M$

Having explored the geometry outside the surface $r = 2M$, we now turn to studying the geometry at that surface. It is clear that something goes wrong with the line element, Eq. 11.1, at $r = 2M$, but what is not immediately clear is whether the problem is with the geometry or just with the coordinates. We shall see that the bad behavior of the line element is an example of a coordinate singularity. To prove this, we shall conduct thought experiments, dropping particles toward the surface and seeing that they reach it in a finite proper time and that nothing disastrous happens to them when they get there. We will then explore other coordinates that do not go singular there, and which explicitly demonstrate not only the regularity of spacetime at this surface but also its causal nature, that anything can fall across it but nothing can emerge from inside.

Coordinate singularities

Coordinate singularities – places where the coordinates do not describe the geometry properly – are not unknown in ordinary calculus. Consider spherical coordinates at the poles. The north pole on a sphere has coordinates $\theta = 0, 0 \leqslant \phi < 2\pi$. That is, although ϕ can have any value for $\theta = 0$, all values really correspond to a single point.

We might draw a coordinate diagram of the sphere as follows (Figure 11.10 – maps of the globe are sometimes drawn this way), in which it would not be at all obvious that all points at $\theta = 0$ are really the same point. We *could*, however, convince ourselves of this by calculating the circumference of every circle of constant θ and verifying that these

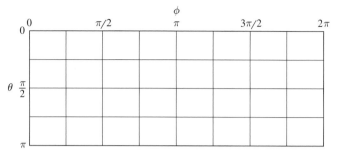

Figure 11.10 One way of drawing a sphere on a flat piece of paper. Not only are $\phi = 0$ and $\phi = 2\pi$ really the same lines, but the lines $\theta = 0$ and $\theta = \pi$ are each really just one point. Spherical coordinates are therefore not faithful representations of the sphere everywhere.

approached zero as $\theta \to 0$ and $\theta \to \pi$. That is, *by asking questions that have an invariant geometrical meaning, one can tell whether the coordinates are bad.* For the sphere, the metric is positive-definite, so if two points have zero distance between them then they are the same point (e.g. $\theta = 0, \phi = \pi$ and $\theta = 0, \phi = 2\pi$: see Exercise 11.19, § 11.6).

In relativity, the situation is more subtle, since there are curves (null curves) where distinct points have zero invariant distance between them. In fact, the whole question of the nature of the surface $r = 2M$ is so subtle that it was not answered satisfactorily until 1960. (This was just in time, too, since black holes began to be of importance in astronomy within the following decade, as new technology made observations of quasars, pulsars, and X-ray sources possible.) We shall explore the problem by asking a few geometrical questions about the metric and then demonstrating a coordinate system which has no singularity at this surface.

Infalling particles

Let a particle fall to the surface $r = 2M$ from any finite radius R. How much proper time does that take? That is, how much time is elapsed on the particle's clock? The simplest case to discuss is the one where a particle falls in radially. Since $d\phi = 0$ we have $\tilde{L} = 0$ and, from Eq. 11.11,

$$\left(\frac{dr}{d\tau}\right)^2 = \tilde{E}^2 - 1 + \frac{2M}{r}, \tag{11.58}$$

or

$$d\tau = -\frac{dr}{\left(\tilde{E}^2 - 1 + 2M/r\right)^{1/2}} \tag{11.59}$$

(the minus sign is there because the particle falls inward). It is clear that if $\tilde{E}^2 > 1$ (an unbound particle) the integral of the right-hand side from R to $2M$ is finite. If $\tilde{E} = 1$ (a particle falling from rest at ∞) the integral is simply

$$\Delta\tau = \frac{4M}{3}\left[\left(\frac{r}{2M}\right)^{3/2}\right]_{2M}^{R}, \tag{11.60}$$

which is again finite. And if $\tilde{E} < 1$, there is again no problem since the particle cannot be at larger r than where $1 - \tilde{E}^2 = 2M/r$ (see Eq. 11.58). So the answer is that any particle can reach the horizon in a finite amount of proper time. In fact there is nothing about the integral that prevents us placing its lower limit smaller than $2M$, i.e. on the other side of the surface $r = 2M$. The particle can go *inside* $r = 2M$ in a finite proper time.

We now ask how much coordinate time elapses as the particle falls in. For this we use

$$U^0 = \frac{dt}{d\tau} = g^{00}U_0 = g^{00}\frac{p_0}{m} = -g^{00}\tilde{E} = \left(1 - \frac{2M}{r}\right)^{-1}\tilde{E}.$$

Therefore we have

$$dt = \frac{\tilde{E}\,d\tau}{1 - 2M/r} = -\frac{\tilde{E}\,dr}{(1 - 2M/r)(\tilde{E}^2 - 1 + 2M/r)^{1/2}}. \tag{11.61}$$

For simplicity, we again consider the case $\tilde{E} = 1$ and examine this near $r = 2M$ by defining the new variable

$$\varepsilon := r - 2M.$$

Then we get

$$dt = \frac{-(\varepsilon + 2M)^{3/2}d\varepsilon}{(2M)^{1/2}\varepsilon}. \tag{11.62}$$

It is clear that as $\varepsilon \to 0$ the integral of this goes as $\log \varepsilon$, which diverges. One would also find this for $\tilde{E} \neq 1$, because the divergence comes from the $[1 - (2M/r)]^{-1}$ factor, which does not contain $\tilde{E}$. Therefore a particle reaches the surface $r = 2M$ only after an infinite coordinate time has elapsed. Since the proper time is finite, the coordinate time must be behaving badly.

Inside $r = 2M$

To see just how badly it behaves, let us ask what happens to a particle after it reaches $r = 2M$. It must clearly pass to smaller r unless it is destroyed. This might happen if at $r = 2M$ there were a *curvature singularity*, where the gravitational forces were strong enough to tear anything apart. But calculation of the components $R^\alpha{}_{\beta\mu\nu}$ of Riemannian tensor in the local inertial frame of the infalling particle shows them to be perfectly finite: Exercise 11.21, § 11.6. So we must conclude that the particle will just keep going.

If we look at the geometry inside but near $r = 2M$, by introducing $\varepsilon := 2M - r$, then the line element is

$$ds^2 = \frac{\varepsilon}{2M - \varepsilon}dt^2 - \frac{2M - \varepsilon}{\varepsilon}d\varepsilon^2 + (2M - \varepsilon)^2\,d\Omega^2. \tag{11.63}$$

Since $\varepsilon > 0$ inside $r = 2M$ we see that a line on which t, θ, ϕ are constant has $ds^2 < 0$: it is timelike. Therefore ε (and hence r) is a *timelike* coordinate. Similarly, it is evident from the line element that t has become spacelike, because $g_{00} > 0$. This is even more evidence for the strangeness of t and r!

Since the infalling particle *must* follow a timelike world line, it must constantly change r, and of course this means *decrease r*. So a particle inside $r = 2M$ will inevitably reach $r = 0$, where a *true* curvature singularity awaits it: sure destruction by infinite forces (Exercise 11.21, § 11.6).

But what happens if the particle inside $r = 2M$ tries to send out a photon to someone outside $r = 2M$ in order to describe its impending doom? This photon, no matter how directed, must also go forward in 'time' as seen locally by the particle, and this means to decreasing r. So the photon will not get out either. Everything inside $r = 2M$ is trapped and, moreover, doomed to encounter the singularity at $r = 0$, since $r = 0$ is in the future of every timelike and null world line inside $r = 2M$.

Once a particle crosses the surface $r = 2M$, it cannot be seen by *any* external observer, even one hovering in a spaceship just outside $r = 2M$, since to be seen means to send out a photon which reaches the external observer. This shows the difference between the black hole in general relativity and the 'dark star' of Newtonian gravity. In Newtonian gravity light could start inside $r = 2M$, move outwards, and be seen by an observer near it, even though it was destined to turn around and fall back if it got further away. In relativity nothing crosses from inside $r = 2M$ to outside.

In relativity, this surface is therefore called a *horizon*, since a horizon on Earth has the same effect (for different reasons!). We shall henceforth refer to $r = 2M$ as the *Schwarzschild horizon.*

Notice that there is nothing locally remarkable about a horizon on Earth or in spacetime. A ship may be sailing just on the horizon, as seen by yourself standing on the shore, but people on the ship do not see anything different about the ocean near them that would tell them it is a horizon; there is no marker in the sea saying "Beware! You are about to cross the horizon!" It is the same in spacetime. We saw above that the local curvature, as expressed by the components of the Riemann tensor in a local inertial frame, are all finite at $r = 2M$. An astronaut unlucky enough to be falling in will just keep falling, and there is nothing local there to warn the astronaut that they are crossing a horizon. Unlike on Earth, however, the Schwarzschild horizon is a causal horizon: the astronaut will not be able to turn around and come back.

Coordinate systems

So far, our approach has been purely algebraic – we have no 'picture' of the geometry yet. To develop a picture we will first draw a coordinate diagram in Schwarzschild coordinates, and on it we will draw the light cones, or at least the paths of the radially ingoing and outgoing null lines emanating from certain events (Figure 11.11). These light cones may be calculated by solving $ds^2 = 0$ for θ and ϕ constant:

$$\frac{dt}{dr} = \pm \left(1 - \frac{2M}{r}\right). \tag{11.64}$$

In a t–r diagram, these lines have slope ± 1 far from the star (according to the standard SR light cone) but their slope approaches $\pm \infty$ as $r \to 2M$. This means that they become more vertical: the cone 'closes up'.

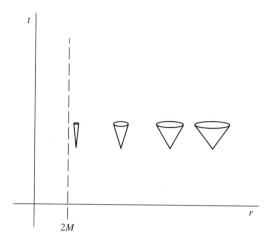

Figure 11.11   Light cones drawn in Schwarzschild coordinates close up near the surface $r = 2M$.

Since particle world lines are confined within the local light cone (a particle must move more slowly than light) this closing up of the cones forces the world lines of particles to become more vertical: if they reach $r = 2M$, they reach it at $t = \infty$. This is the 'picture' behind the algebraic result that a particle takes infinite coordinate time to reach the horizon.

Notice that *no* particle world line reaches the line $r = 2M$ for any finite value of t. This might suggest that the line $r = 2M, -\infty < t < \infty$, is really not a line at all but a single point in spacetime. That is, our coordinates may go bad by expanding a single event into the whole line $r = 2M$, which would have the effect that if any particle reached the horizon after that event then it would have to cross $r = 2M$ 'after' $t = +\infty$. This singularity would then be very like the one in Figure 11.10 for spherical coordinates at the pole: a whole line in the bad coordinates representing a point in the real space.

The fact is that the coordinate diagram in Figure 11.11 makes no attempt to represent the *geometry* properly, only the coordinates. It clearly does a poor job on the geometry because the light cones close up. Since we have already decided that they *don't* really close up (particles reach the horizon at finite *proper* time and encounter a perfectly well-behaved geometry there), the remedy is to find coordinates which do not close up the light cones.

Kruskal–Szekeres coordinates

The search for these coordinates was a long and difficult one, and ended in 1960. The good coordinates are known as Kruskal–Szekeres coordinates; they are called u and v, and are defined by

$$\left.\begin{aligned} u &= (r/2M - 1)^{1/2}e^{r/4M}\cosh(t/4M), \\ v &= (r/2M - 1)^{1/2}e^{r/4M}\sinh(t/4M), \end{aligned}\right\} \tag{11.65}$$

for $r > 2M$ and

$$\left.\begin{aligned} u &= (1 - r/2M)^{1/2}e^{r/4M}\sinh(t/4M), \\ v &= (1 - r/2M)^{1/2}e^{r/4M}\cosh(t/4M), \end{aligned}\right\} \tag{11.66}$$

for $r < 2M$. (This transformation is singular at $r = 2M$, but that is necessary in order to eliminate the coordinate singularity there.) The Schwarzschild metric in these coordinates is found to be

$$ds^2 = -\frac{32M^3}{r} e^{-r/2M} (dv^2 - du^2) + r^2 d\Omega^2, \qquad (11.67)$$

where, now, r is not to be regarded as a coordinate but as a function of u and v, given implicitly by the inverse of Eqs. 11.65 and 11.66:

$$\left(\frac{r}{2M} - 1\right) e^{r/2M} = u^2 - v^2. \qquad (11.68)$$

Although r is no longer a coordinate, it does have a geometrical meaning, since the geometry is spherically symmetric. At any event in spacetime, there is a unique sphere of symmetry containing that event. The value of r for that event can be inferred from a measurement of the circumference $(2\pi r)$ or the area $(4\pi r^2)$ of that sphere.

Notice several things about Eq. 11.67. There is nothing singular about any metric term at $r = 2M$. There *is*, however, a singularity at $r = 0$, where we expect it. A radial null line $(d\theta = d\phi = ds = 0)$ is a line

$$dv = \pm du. \qquad (11.69)$$

This last result is very important. It means that in a (u, v) diagram, the light cones are all just as open as in SR. This result makes these coordinates particularly useful for visualizing the geometry in a coordinate diagram. The (u, v) diagram is, then, given in Figure 11.12. Compare this with the result of Exercise 5.21, § 5.7.

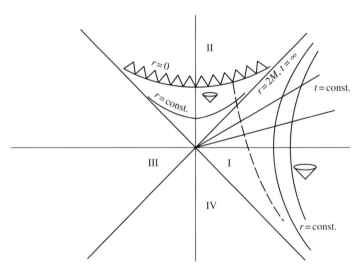

Figure 11.12   Kruskal-Szekeres coordinates keep the light cones at $45°$ everywhere. The singularity at $r = 0$ (toothed line) bounds the future of all events inside (above) the line $r = 2M, t = +\infty$. Events outside this horizon have part of their future free of singularities.

Much needs to be said about this figure. First, two light cones are drawn for illustration. *Any* 45° line is a radial null line. Second, only u and v are plotted: θ and ϕ are suppressed; therefore each point is really a two-sphere of events. Third, lines of constant r are hyperbolae, as is clear from Eq. 11.68. For $r > 2M$ these hyperbolae run roughly vertically, being asymptotic to the 45° line from the origin $u = v = 0$. For $r < 2M$ the hyperbolae run roughly horizontally, with the same asymptotes. This means that, for $r < 2M$, a timelike line (confined within the light cone) cannot remain at constant r. This is the result we had before. The hyperbola $r = 0$ is the end of the spacetime, since a true singularity is there. Note that although $r = 0$ is a 'point' in ordinary space, it is a whole hyperbola here. However, not too much can be made of this, since it is a singularity of the geometry: we should not glibly speak of it as a part of spacetime with a well-defined dimensionality.

Our fourth remark is that lines of constant t, being orthogonal to lines of constant r, are straight lines in this diagram, radiating outwards from the origin $u = v = 0$. (They are orthogonal to the hyperbolae $r = $ const. in the *spacetime* sense of orthogonality; recall our diagrams in § 1.7 of invariant hyperbolae in SR, which had the same property of being orthogonal to lines radiating out from the origin.) In the limit as $t \to \infty$, these lines approach the 45° line from the origin. Since all the lines $t = $ const. pass through the origin, the origin would be expanded into a whole line in a (t, r) coordinate diagram such as Figure 11.11, which is what we guessed after discussing that diagram. A world line crossing this $t = \infty$ line in Figure 11.12 enters the region in which r is a time coordinate, and so cannot get out again. The true horizon, then, is this line $r = 2M, t = +\infty$.

Fifth, since for a distant observer t really does measure proper time, and an object that falls to the horizon crosses all the lines $t = $ const. up to $t = \infty$, a distant observer would conclude that it takes an infinite time for the infalling object to reach the horizon. We have already drawn this conclusion, but here we see it displayed clearly in the diagram. There is nothing 'wrong' in this statement: the distant observer does wait an infinite time to get the information that the object has crossed the horizon. But the object reaches the horizon in a finite time on its own clock. If the infalling object sends out radio pulses each time its clock ticks, then it will emit only a finite number before reaching the horizon, so the distant observer can receive only a finite number of pulses. Since these are stretched out over a very large amount of the distant observer's time, the observer concludes that time on the infalling clock is slowing down and eventually stopping. If the infalling 'clock' is a photon, the observer will conclude that the photon experiences an infinite gravitational redshift. This will also happen if the infalling 'object' is a gravitational wave of short wavelength compared with the size of the horizon.

Sixth, this horizon is itself a null line. This *must* be the case, since the horizon is the boundary between null rays that cannot get out and those that can. It is therefore the path of the *marginal* null ray.

Seventh, the 45° lines from the origin divide spacetime up into four regions, labeled I, II, III, IV. Region I is clearly the *exterior*, $r > 2M$, and region II is the interior of the horizon. But what about III and IV? To discuss them is beyond our scope in this text (see Misner *et al.* 1973, Box 33.2G and Chapter 34, and Hawking & Ellis 1973), but one remark must be made. Consider the dashed line in Figure 11.12, which could be the path of an infalling particle. If this black hole were formed by the collapse of a star, then we know that outside the star the geometry is the Schwarzschild geometry, but inside the star it may be quite different. The dashed line may be taken to be the path of the *surface* of the collapsing star, in which case the region of the diagram to the right of it is *outside* the star and so correctly describes the spacetime geometry, but everything to the left would be inside the star (smaller r) and hence has possibly no relation to the true geometry of the spacetime. This includes all of regions III and IV, so they are to be ignored by the astrophysicist (though they can be interesting to the mathematician!). Note that parts of I and II are also to be ignored, but there is still a singularity and horizon outside the star.

The eighth and last remark we will make is that the coordinates u and v are *not* particularly good for describing the geometry far from the star, where g_{uu} and g_{vv} fall off exponentially in r. The coordinates t and r are best there; indeed, they were constructed in order to be well behaved there. But if one is interested in the horizon, then one uses u and v.

11.3 General black holes

Formation of black holes

The formation of a horizon has to do with the collapse of matter to such small dimensions that the gravitational field traps everything within a certain region, which is called the interior of the horizon. We have explored the structure of the black hole in one particular case – the static, spherically symmetric situation – but the formation of a horizon is a much more general phenomenon. When we discuss astrophysical black holes in § 11.4 we will address the question of how Nature might arrange to get so much mass into such a small region. But it should be clear that, in the real world, black holes are not formed from perfectly spherical collapse. Black holes form in complicated dynamical circumstances, and after they form they continue to participate in dynamics: as members of binary systems, as centers of gas accretion, as sources of gravitational radiation. In this section we learn how to define black holes and what we know about them in general.

The central astonishing property of the Schwarzschild horizon is that anything that crosses it cannot get back outside it. The definition of a general horizon (called an event horizon) focuses on this property. *An event horizon is the boundary in spacetime between events that can communicate with distant observers and events that cannot.* This definition assumes that distant observers exist and that the spacetime is asymptotically flat, as defined in § 8.4. And it permits the communication to take an arbitrarily long time: an event is considered to be outside the horizon provided that it can emit a photon in even just one special direction that eventually escapes out to a distant observer. The most important part of the definition to think about is that the horizon is a boundary in *spacetime*, not just in the space defined by one moment of time. It is a three-dimensional surface that separates the events of spacetime into two regions: trapped events inside the horizon and untrapped events outside.

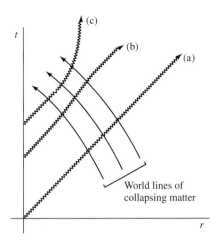

Figure 11.13   Schematic spacetime diagram of spherical collapse. Light ray (a) hardly feels anything, (b) is delayed, and (c) is marginally trapped. The horizon is defined as the ray (c), so it grows continuously from zero radius as the collapse proceeds.

Since no form of communication can go faster than light, the test of whether events can communicate with distant observers is whether they can send light rays, i.e. whether there are null rays that can get arbitrarily far away. As the boundary between null rays that can escape and null rays that are trapped, the horizon itself is composed of null world lines. These are the marginally trapped null rays, the null rays that neither move away to infinity nor fall inwards. By definition these marginal null rays stay on the horizon forever, because if a ray were to leave it toward the exterior or interior then it would not mark the horizon. It is not hard to see that this definition fits the Schwarzschild horizon, which is static and unchanging, but when we consider dynamical situations, there are some surprises.

The formation of a horizon from a situation where there is initially no black hole illustrates well the dynamical nature of the horizon. Consider the collapse of a spherical star to form a black hole. In the end there is a static Schwarzschild horizon, but before that there is an intermediate period of time in which the horizon is growing from zero radius to its full size. This is easy to see by considering Figure 11.13, which illustrates (*very* schematically) the collapsing situation. (The time coordinate is a kind of Schwarzschild time, but this isn't be taken too literally.) As matter falls in, the trajectories of photons that start from the center of the collapsing star (the wavy lines) are more and more affected. Photon (a) gets out with little trouble, photon (b) has some delay, and photon (c) is the *marginal* one, which just gets trapped and remains on the Schwarzschild horizon. Anything later than (c) is permanently trapped, anything earlier gets out. So photon (c) does in fact represent the horizon at all times, by definition, since it is the boundary between trapped and untrapped. Thus, one sees the horizon grow from zero radius to $2M$ by watching photon (c)'s progress outwards. For this spherically symmetric situation, if one knew the details of the collapse, one could easily determine the position of the horizon. But if there were no symmetry – particularly if the collapse produced a large amount of gravitational radiation – then the calculation would be far more difficult, although the principle would be the same.

As a more subtle example of a dynamical horizon, consider what happens to the Schwarzschild horizon of the 'final' black hole in Figure 11.13 if at a much later time some more gas falls in (not illustrated in the figure). For simplicity, we again assume the infalling gas is perfectly spherical. Let the mass of the hole before the gas falls in be called M_0. The surface $r = 2M_0$ is static and appears to be the event horizon: photons inside it fall towards the singularity at the center, and photons infinitesimally outside it gradually move further and further away from it. But then the new gas falls across this surface and increases the mass of the hole to M_1. Clearly the new final state will be a Schwarzschild solution of mass M_1, where the larger surface $r = 2M_1$ looks like an event horizon. It consists of null rays neither falling in nor diverging outwards. Now, what is the history of these rays? What happens if we trace backwards in time along a ray that just stays at $r = 2M_1$? We would find that, before the new gas arrived, it was one of those null world lines just *outside* the surface $r = 2M_0$, one of the rays that were very gradually diverging from it. The extra mass has added more gravitational attraction (more curvature), which stopped the ray from moving away and now holds it exactly at $r = 2M_1$. Moreover, the null rays that formed the static surface $r = 2M_0$ before the gas fell in are now inside $r = 2M_1$ and are therefore falling toward the singularity, again pulled in by the gravitational attraction of the extra mass.

The boundary in *spacetime* between what is trapped and what is not consists therefore of the null rays that in the end sit at $r = 2M_1$, including their continuation backwards in time. The null rays on $r = 2M_0$ were not actually part of the true horizon of spacetime even at the earlier time: they are just trapped null rays that took a long time to find out that they were trapped! Before the gas fell in, what looked like a static event horizon ($r = 2M_0$) was not an event horizon at all, even though it was temporarily a static collection of null rays. Instead, the true boundary between trapped and untrapped was even at that early time gradually expanding outwards, traced by the null world lines that eventually became the surface $r = 2M_1$.

This example illustrates the fact that the horizon is not a location in space but a boundary in spacetime. It is not possible to determine the location of the horizon by looking at a system at one particular time; instead one must look at its entire evolution in time, find out which null rays eventually really do escape and which ones are trapped, and then trace out the boundary between them. The horizon is a property of the spacetime as a whole, not of space at any one moment of time. In our example, if many years later some further gas falls into the hole, then we would find that even those null rays at $r = 2M_1$ would actually be trapped and would not have been part of the true event horizon after all. The only way to find the true horizon is to know the *entire* future evolution of the spacetime and then trace out the boundary between trapped and untrapped regions.

This definition of the event horizon is mathematically consistent and logical, but it is difficult to work with in practical situations. If one knows only a limited amount about a spacetime, it can in principle be impossible to locate the horizon. For example, computer simulations of solutions of Einstein's equations containing black holes cannot run forever, so they don't have all the information needed even to locate the horizons of their black holes. So physicists also work with another kind of surface that can be defined at any one moment of time. It uses the other property of the Schwarzschild horizon, that it is a static collection of null rays. Physicists define a *locally trapped surface* to be a two-dimensional surface at any particular time whose outwardly directed null rays are neither expanding nor contracting at that particular moment. Roger Penrose (1965) proved that locally trapped surfaces are always *inside* true horizons in dynamical situations, just as $r = 2M_0$ turned out to be inside the true horizon. Of course, very often they are so close to the true horizon that they offer an excellent approximation to it. This proof established for the first time that no particular symmetry was needed in order to form a black hole: if there was strong enough gravity, a black hole was inevitable. For this proof, Penrose shared the 2020 Nobel Prize in Physics.

General properties of black holes

While the detailed structure of an event horizon is not easy to compute, some important general properties of horizons are understood, and they underpin the confidence with which astronomers now employ black holes in models of complex astrophysical phenomena. Here are some important theorems and conjectures.

(1) It is believed that any horizon in an asymptotically flat spacetime will eventually become stationary, provided that it is not constantly disturbed by outside effects such as gas accretion. So an isolated black hole should become stationary. (The calculations in items (3) and (4) below support this idea.) The *stationary* horizons, the endpoints of collapse, are actually completely known in GR. The principal result is that a stationary vacuum black hole is characterized by just two numbers: its total mass M and total angular momentum J. These parameters are defined not by any integrals over the 'interior' of the horizon, but by the gravitational field far from the hole. We defined the mass M of any metric in this fashion in § 8.4, and in Exercise 8.19, § 8.5, we saw how J can be similarly defined. The *unique stationary vacuum black hole is the Kerr solution,* Eq. 11.71, which we study in detail below. If the angular momentum is zero, the Kerr solution becomes the Schwarzschild metric. This uniqueness theorem results from work done by Hawking (1972), Carter (1973), and Robinson (1975). See Chruściel (1996) and Heusler (1998) for reviews.

(2) If the black hole is not in vacuum, its structure may be more general. It may carry an electric charge Q and, in principle, a magnetic monopole moment F, although magnetic monopoles have not been found in Nature. Both of these charges can be measured by Gauss' law integrals over surfaces surrounding the hole and far from it. It is also felt that collapse is unlikely to lead to a significant residual charge Q, so astrophysicists normally take only M and J to be nonzero. But other kinds of fields, such as those encountered in particle physics theories, can add other complications: self-gravitating interacting scalar fields, Yang–Mills fields, and so on. See Heusler (1998) for a review. Again, these complications are not usually thought to be relevant in astrophysics. What may be relevant, however, is the distortion of the horizon produced by the tidal effect of matter surrounding the black hole or by the presence of another nearby black hole (say, in a binary system). If a massive stationary disk of gas surrounds the hole, then the metric will not be exactly Kerr (e.g. Will 1974, 1975, Ansorg 2005). Binary black holes, when they approach closely on their way to merger, can become highly distorted:

Snapshots of the shapes of the horizons of two spinning black holes colliding head-on. The left image is just before merger, showing the tidal distortion of the horizons. The right image is just after merging, before the hole has had time to settle down into a Kerr black hole. Taken from Figure 14 of Lovelace *et al.* (2010). Copyright 2010 by the American Physical society. (Reproduced with permission.)

see Figure 11.14. This figure also shows that horizons, when they are initially formed, can be very far from the smooth axisymmetric Kerr horizon.

(3) If gravitational collapse is *nearly* spherical, then it is believed that all nonspherical parts of the mass distribution – the quadrupole moment, octupole moment – except possibly for angular momentum, are radiated away in gravitational waves, and a stationary black hole of the Kerr type is left behind. If there is no angular momentum, a Schwarzschild hole is left behind (Price 1972a,1972b). A full proof of this linear stability in the case of Schwarzschild was finally published in 2016 (Dafermos *et al.* 2016), employing clever coordinate conditions to control solutions arising from arbitrary (linear) initial perturbations of Schwarzschild. It is known that the Kerr metric is linearly stable at least for sufficiently small values of its angular momentum parameter a (Häfner *et al.* 2020).

(4) Numerical simulations of a black-hole merger, in full nonlinear GR, always show that the initially distorted horizon, as in Figure 11.14, quickly settles down to Kerr. There is no analytic proof that this has to happen, but since Kerr is the unique uncharged black hole, any non-Kerr outcome of a merger would need to be something other than a single black hole. As we will see in the next chapter, gravitational wave observations of mergers have so far been consistent with the formation of a single Kerr black hole.

(5) An important general result concerning nonstationary horizons is the area theorem of Hawking (Hawking & Ellis 1973): in any dynamical process involving black holes, the total area of all the horizons cannot decrease in time. We saw this in a qualitative way in our earlier discussion of how to define a horizon when matter is falling into a black hole: the area is actually increasing during the period before the infalling gas reaches the hole. We shall see below how to quantify this theorem by calculating the area of the Kerr horizon. The area theorem implies that, while two black holes can collide and coalesce, a single black hole can never bifurcate spontaneously into two smaller ones. (A restricted proof of this is to be found in Exercise 11.29, § 11.6, using the Kerr area formula below; a full proof is outlined in Misner *et al.* 1973, Exercise 34.4, and requires techniques beyond the scope of this book.) The theorem assumes that the local energy density of matter in spacetime (ρ) is positive.

(6) The analogy between an ever-increasing area and the ever-increasing entropy of thermodynamics has led to the development of *black-hole thermodynamics*, and to

the understanding that black holes fit into thermal physics in a very natural way. The connection comes from the Hawking radiation. The Hawking radiation has a thermal spectrum, which allows one to define a temperature for the black hole, and hence an entropy for the horizon. An isolated black hole will actually radiate away mass, and hence area, converting its entropy into the even larger entropy of the escaping radiation. This violation of the area theorem by quantum effects happens because in quantum mechanics, energies are not always required to be positive. We study this Hawking radiation in § 11.5 below. The entropy associated with the area of the horizon is given in Eq. 11.114.

(7) Inside the Schwarzschild and Kerr horizons there are curvature singularities where the curvature, and hence the tidal gravitational force, becomes infinite even when measured by a freely falling inertial observer. General theorems, mostly due to Hawking and Penrose, imply that any horizon will contain a singularity within it of some kind (Hawking & Ellis 1973). But it is not known whether such singularities will always have infinite curvature; all that is known is that infalling geodesics are incomplete and cannot be continued for an infinite amount of proper time or affine distance. The existence of these singularities is generally regarded as a serious shortcoming of general relativity: that its predictions have limited validity in time inside horizons. Many physicists expect this shortcoming to be remedied by a quantum theory of gravity. The uncertainty principle of quantum mechanics, so the expectation goes, will make the singularity a little 'fuzzy', the tidal forces will not quite reach infinite strength, and the waveform will continue further into the future. In the absence of an acceptable theory of quantum gravity this remains only a hope, but there are some computations in restricted quantum models that suggest that this might in fact happen (for example: Bojowald 2008; Ashtekar & Bojowald 2006; Bramberger et al. 2017).

(8) The generic existence of singularities inside horizons, hidden from the view of distant observers, prompts the question of whether there can be so-called *naked singularities*, i.e. singularities outside horizons. These would be far more problematic for general relativity, for it would mean that situations could arise in which general relativity could make no predictions beyond a certain time even for normal regions of spacetime. Having singularities in unobservable regions inside horizons is bad enough, but if singularities arose outside horizons then general relativity would be even more flawed. In response to this concern, Penrose (1979) formulated the *cosmic censorship conjecture*, according to which no naked singularities can arise out of nonsingular initial conditions in asymptotically flat spacetimes. Penrose had no proof of this, and offered it as a challenge to relativity theorists. It has turned out to be very difficult to prove this conjecture, and even to formulate it precisely (Berger 2002, Rendall 2005). A disproof of one important form of the conjecture, which however suggests a new way to formulate a potentially correct version, is given in Dafermos & Luk (2017), where it is shown that a naked but weak null singularity can evolve. One naked singularity seems inescapable in general relativity: the Big Bang of standard cosmology is naked to our view. If the Universe re-collapses there will similarly be a Big Crunch in the future of all our world lines. These issues are discussed in Chapter 13.

The first item in the list above is truly remarkable: a massive black hole, possibly formed from 10^{60} individual atoms and molecules – whose history as a gas may have included complex gas motions, shock waves, magnetic fields, nucleosynthesis, and all kinds of other complications – is described fully and exactly by just two numbers, its mass and spin. The horizon and the entire spacetime geometry outside it depend on just these two numbers. All the complication of the formation process is effaced, forgotten, reduced to two simple numbers.

No other macroscopic body is even remotely so simple to describe. We might characterize a star by its mass, luminosity, and color, but these are just a start, just categories that contain an infinite potential variety within them. Stars can have magnetic fields, spots, winds, differential rotation, and many other large-scale features, to say nothing of the different motions of individual atoms and ions.

While this variety may not be relevant in most circumstances, it is there. For a black hole, it is simply not there. There is nothing but mass and spin, no individual structure or variety revealed by microscopic examination of the horizon. The reaction of Chandrasekhar to this fact, quoted in § 11.4 below, sums up how extraordinary this situation is.

In fact, as we noted earlier, the horizon is not even a real surface; it is just a boundary in empty space between trapped and untrapped regions. The fact that no information can escape from inside the hole means that no information about what has fallen in is visible from the outside. The only quantities that remain are those that are conserved by the fundamental laws of physics: mass and angular momentum. This loss of information accounts for the extremely large value of a black hole's entropy, Eq. 11.114.

Kerr black hole

The Kerr black hole (Kerr 1963) is axially symmetric but not spherically symmetric (i.e. rotationally symmetric about one axis only, which is the angular momentum axis), and is characterized by two parameters, M and J. Since J has dimension m^2, one conventionally defines

$$a := J/M, \tag{11.70}$$

which then has the same dimensions as M. The line element is

$$ds^2 = -\frac{\Delta - a^2 \sin^2\theta}{\rho^2}\, dt^2 - 2a\frac{2Mr\sin^2\theta}{\rho^2}\, dt\, d\phi$$
$$+ \frac{(r^2+a^2)^2 - a^2\Delta\sin^2\theta}{\rho^2}\sin^2\theta\, d\phi^2 + \frac{\rho^2}{\Delta}dr^2 + \rho^2\, d\theta^2, \tag{11.71}$$

where

$$\Delta := r^2 - 2Mr + a^2,$$
$$\rho^2 := r^2 + a^2\cos^2\theta. \tag{11.72}$$

The coordinates are called Boyer–Lindquist coordinates; ϕ is the angle around the axis of symmetry, t is the time coordinate in which everything is stationary, and r and θ are similar to the spherically symmetric r and θ but are not so readily associated with any geometrical definition. In particular, since there are no metric two-spheres, the coordinate r cannot be defined as an 'area' coordinate as we did before. The following points are important:

(1) Surfaces $t = \text{const.}, r = \text{const.}$ do not have the metric of the two-sphere, Eq. 10.2.
(2) The metric for $a = 0$ is identically equal to the Schwarzschild metric.
(3) There is an off-diagonal term in the metric, in contrast to Schwarzschild:

$$g_{t\phi} = -a \frac{2Mr \sin^2 \theta}{\rho^2};\tag{11.73}$$

It is just half of the coefficient of $\mathrm{d}t \, \mathrm{d}\phi$ in Eq. 11.71 because the line element contains two terms,

$$g_{t\phi} \, \mathrm{d}t \, \mathrm{d}\phi + g_{\phi t} \, \mathrm{d}\phi \, \mathrm{d}t = 2g_{t\phi} \, \mathrm{d}\phi \, \mathrm{d}t,$$

by the symmetry of the metric. *Any* axially symmetric stationary metric has preferred coordinates t and ϕ, namely those which have the property $g_{\alpha\beta,t} = 0 = g_{\alpha\beta,\phi}$. But the coordinates r and θ are more or less arbitrary, except that they may be chosen to be (i) orthogonal to t and ϕ ($g_{rt} = g_{r\phi} = g_{\theta t} = g_{\theta\phi} = 0$) and (ii) orthogonal to each other ($g_{\theta r} = 0$). In general, one cannot choose t and ϕ orthogonal to each other ($g_{t\phi} \neq 0$). Thus Eq. 11.71 has the minimum number of nonzero $g_{\alpha\beta}$. (See Carter 1969.)

Dragging of inertial frames

The presence of $g_{t\phi} \neq 0$ in the metric introduces qualitatively new effects on particle trajectories. Because $g_{\alpha\beta}$ is independent of ϕ, a particle's trajectory still conserves p_ϕ. But now we have

$$p^\phi = g^{\phi\alpha} p_\alpha = g^{\phi\phi} p_\phi + g^{\phi t} p_t,\tag{11.74}$$

and similarly for the time components:

$$p^t = g^{t\alpha} p_\alpha = g^{tt} p_t + g^{t\phi} p_\phi.\tag{11.75}$$

Consider a zero-angular-momentum particle, $p_\phi = 0$. Then, using the definitions (for nonzero rest mass)

$$p^t = m \, \mathrm{d}t/\mathrm{d}\tau, \quad p^\phi = m \, \mathrm{d}\phi/\mathrm{d}\tau,\tag{11.76}$$

we find that the particle's trajectory has

$$\frac{\mathrm{d}\phi}{\mathrm{d}t} = \frac{p^\phi}{p^t} = \frac{g^{\phi t}}{g^{tt}} := \omega(r, \theta).\tag{11.77}$$

This equation defines what we mean by ω, the angular velocity of a zero-angular-momentum particle. We shall find ω explicitly for the Kerr metric when we obtain the contravariant components $g^{\phi t}$ and g^{tt} below. But it is clear that this effect will be present in any metric for which $g_{t\phi} \neq 0$, which in turn happens whenever the source is rotating (e.g.

a rotating star as in Exercise 8.19, § 8.5). So we have the remarkable result that a particle dropped 'straight in' ($p_\phi = 0$) from infinity is 'dragged' just by the influence of gravity, so that it acquires an angular velocity in the same sense as that of the source of the metric (we'll see below that, for the Kerr metric, ω has the same sign as a). This effect weakens with distance (roughly as $1/r^3$; see Eq. 11.90 below for the Kerr metric), but it makes the angular momentum of the source measurable in principle, although in most situations the effect is small, as we saw in Exercise 8.19, § 8.5. This effect is often called the *dragging of inertial frames.*

This effect has a close analogy with magnetism. Newtonian gravity is, of course, very similar to electrostatics, with the sign change that ensures that 'charges' of the same sign (*i.e.* masses) attract one another. In electromagnetism, a spinning charge creates additional effects which we call magnetism. Here we have a gravitational analog, the gravitational effects due to a spinning mass. For that reason these effects are called *gravitomagnetism.* The analogy between gravitomagnetism and standard magnetism is perhaps easier to see in the *Lense–Thirring effect:* a gyroscope placed in orbit around a rotating star will precess by a small amount that is proportional to the angular momentum of the star, just as a spinning electron precesses if it orbits through a magnetic field.

Small as it is, the Lense–Thirring effect created by the spin of the Earth has been measured. Detailed studies of the orbits of three geodesy satellites have so far verified the prediction of GR with the remarkable accuracy at the 0.7% level (Ciufolini *et al.* 2016). A satellite called Gravity Probe B (GP-B) was launched in 2004, and it measured the same effect by tracking the precession of on-board gyroscopes. The experiment was troubled, however, by unexpected problems with the gyroscopes; the measurement of frame-dragging could only be performed with 20% accuracy (Everitt *et al.* 2011). Gravitomagnetic effects are regularly taken into account in modeling the details of the emission of X-rays near black holes (e.g. Brenneman & Reynolds 2006), and they may also sometime be seen in the double pulsar system PSR J0737-3039 referred to earlier in this chapter (Kramer & Wex 2009).

Ergoregion

Consider photons emitted in the equatorial plane ($\theta = \pi/2$) at some given r. In particular, consider those initially going in the $\pm\phi$ direction, i.e. tangent to a circle of constant r. Then they generally have only dt and $d\phi$ nonzero on the path at first and, since $ds^2 = 0$, we have

$$0 = g_{tt}\, dt^2 + 2g_{t\phi}\, dt\, d\phi + g_{\phi\phi}\, d\phi^2$$

$$\Rightarrow \frac{d\phi}{dt} = -\frac{g_{t\phi}}{g_{\phi\phi}} \pm \left[\left(\frac{g_{t\phi}}{g_{\phi\phi}}\right)^2 - \frac{g_{tt}}{g_{\phi\phi}}\right]^{1/2}. \tag{11.78}$$

Now, a remarkable thing happens if $g_{tt} = 0$: the two solutions are

$$\frac{d\phi}{dt} = 0 \quad \text{and} \quad \frac{d\phi}{dt} = \frac{2g_{t\phi}}{g_{\phi\phi}}. \tag{11.79}$$

We will see below that for the Kerr metric the second solution gives dϕ/dt the *same* sign as the parameter a, and so represents the photon sent off in the same direction as that in which the hole is rotating. The other solution means that the other photon – the one sent 'backwards' – initially doesn't move at all. The dragging of orbits has become so strong that this photon *cannot* move in the direction opposite to the rotation. Clearly any particle, which must move slower than a photon, will therefore have to rotate with the hole, even if it has an angular momentum that is arbitrarily large in the opposite sense to that of the hole!

We shall see that the surface where $g_{tt} = 0$ lies outside the horizon; it is called the *ergosphere*. It is sometimes also called the *static limit*, since inside it no particle can remain at fixed r, θ, ϕ. From Eq. 11.71 we conclude that it occurs at

$$r_0 := r_{\text{ergosphere}} = M + \sqrt{(M^2 - a^2 \cos^2 \theta)}. \tag{11.80}$$

Inside this radius, since $g_{tt} > 0$, all particles and photons must rotate with the hole.

This effect can in principle occur in other situations. Models for certain rotating stars are known where there are toroidal regions of space in which $g_{tt} > 0$ (Butterworth & Ipser 1976). These will also have these super-strong frame-dragging effects. They are called ergoregions, and their boundaries are ergotoroids. They can exist in solutions which have no horizon at all. But it seems unlikely that real neutron stars would have ergoregions, because they would have to be even more compact (and therefore relativistic) than neutron stars, and very rapidly rotating.

The Kerr horizon

In the Schwarzschild solution the horizon was the place where $g_{tt} = 0$ and $g_{rr} = \infty$. In the Kerr solution the ergosphere occurs at $g_{tt} = 0$ and the horizon is at $g_{rr} = \infty$, i.e. where $\Delta = 0$:

$$r_+ := r_{\text{horizon}} = M + \sqrt{(M^2 - a^2)}. \tag{11.81}$$

It is clear that the Kerr ergosphere lies outside the horizon except at the poles, where it is tangent to it. The full proof that this is the horizon is beyond our scope here: one needs to verify that no null lines can escape from inside r_+. We shall simply take it as given. (See the next subsection for a partial justification.) Since the area of the horizon is important (by Hawking's area theorem), we shall calculate it.

The horizon is a surface of constant r and t, by Eq. 11.81 and the fact that the metric is stationary. Any surface of constant r and t has an intrinsic metric whose line element comes from Eq. 11.71 with dt = dr = 0:

$$dl^2 = \frac{(r^2 + a^2)^2 - a^2 \Delta}{\rho^2} \sin^2 \theta \, d\phi^2 + \rho^2 \, d\theta^2. \tag{11.82}$$

The proper area of this surface is given by integrating the square root of the determinant of this metric over all θ and ϕ:

$$A(r) = \int_0^{2\pi} d\phi \int_0^\pi d\theta \sqrt{[(r^2 + a^2)^2 - a^2 \Delta]} \sin \theta. \tag{11.83}$$

Since nothing in the square root depends on θ or ϕ, and since the area of a unit two-sphere is

$$4\pi = \int_0^{2\pi} d\phi \int_0^{\pi} d\theta \sin \theta,$$

we immediately conclude that

$$A(r) = 4\pi \sqrt{[(r^2 + a^2)^2 - a^2\Delta]}. \qquad (11.84)$$

Since the horizon is defined by $\Delta = 0$, we get

$$A(\text{horizon}) = 4\pi(r_+^2 + a^2). \qquad (11.85)$$

Equatorial photon motion in the Kerr metric

A detailed study of the motion of photons in the equatorial plane gives insight into the ways in which rotating metrics differ from nonrotating ones. First, we must obtain the inverse of the metric, Eq. 11.71, which we write in the general stationary, axially symmetric, form:

$$ds^2 = g_{tt} \, dt^2 + 2g_{t\phi} \, dt \, d\phi + g_{\phi\phi} \, d\phi^2 + g_{rr} \, dr^2 + g_{\theta\theta} \, d\theta^2.$$

The only off-diagonal element involves t and ϕ; therefore

$$g^{rr} = \frac{1}{g_{rr}} = \Delta\rho^{-2}, \qquad g^{\theta\theta} = \frac{1}{g_{\theta\theta}} = \rho^{-2}. \qquad (11.86)$$

We need to invert the matrix

$$\begin{pmatrix} g_{tt} & g_{t\phi} \\ g_{t\phi} & g_{\phi\phi} \end{pmatrix}.$$

Calling its determinant D, the inverse is

$$\frac{1}{D}\begin{pmatrix} g_{\phi\phi} & -g_{t\phi} \\ -g_{t\phi} & g_{tt} \end{pmatrix}, \qquad D = g_{tt}g_{\phi\phi} - (g_{t\phi})^2. \qquad (11.87)$$

Notice one important deduction from this: the angular velocity of the dragging of inertial frames is the expression we wrote down earlier, Eq. 11.77:

$$\omega = \frac{g^{\phi t}}{g^{tt}} = \frac{-g_{t\phi}/D}{g_{\phi\phi}/D} = -\frac{g_{t\phi}}{g_{\phi\phi}}. \qquad (11.88)$$

This makes Eqs. 11.78 and 11.79 more meaningful. For the metric Eq. 11.71, some algebra gives

$$D = -\Delta \sin^2 \theta, \qquad g^{tt} = -\frac{(r^2 + a^2)^2 - a^2\Delta \sin^2 \theta}{\rho^2\Delta},$$

$$g^{t\phi} = -a\frac{2Mr}{\rho^2\Delta}, \qquad g^{\phi\phi} = \frac{\Delta - a^2 \sin^2 \theta}{\rho^2\Delta \sin^2 \theta}. \qquad (11.89)$$

Then the frame dragging is

$$\omega = \frac{2Mra}{(r^2 + a^2)^2 - a^2\Delta \sin^2 \theta}. \tag{11.90}$$

The denominator is positive everywhere, by Eq. 11.72, so this has the same sign as a, and it falls off for large r as r^{-3}, as we noted earlier.

A photon whose trajectory is in the equatorial plane has $d\theta = 0$; but, unlike the Schwarzschild case, this is only a special kind of trajectory: photons not in the equatorial plane may have qualitatively different orbits. Nevertheless, a photon for which $p^\theta = 0$ initially in the equatorial plane always has $p^\theta = 0$, since the metric is reflection symmetric through the plane $\theta = \pi/2$. By stationarity and axial symmetry the quantities $E = -p_t$ and $J = p_\phi$ are constants of the motion. Then the equation $\vec{p} \cdot \vec{p} = 0$ determines the motion. Denoting p^r by $dr/d\lambda$ as before, we find, after some algebra,

$$\left(\frac{dr}{d\lambda}\right)^2 = g^{rr}[(-g^{tt})E^2 + 2g^{t\phi}EJ - g^{\phi\phi}J^2]$$

$$= g^{rr}(-g^{tt})\left[E^2 - 2\,\omega EJ + \frac{g^{\phi\phi}}{g^{tt}}J^2\right]. \tag{11.91}$$

Using Eqs. 11.72, 11.86, and 11.90 for $\theta = \pi/2$, we get

$$\left(\frac{dr}{d\lambda}\right)^2 = \frac{(r^2 + a^2)^2 - a^2\Delta}{r^4}\left[E^2 - \frac{4Mra}{(r^2 + a^2)^2 - a^2\Delta}EJ\right.$$

$$\left. - \frac{r^2 - 2Mr}{(r^2 + a^2)^2 - a^2\Delta}J^2\right]. \tag{11.92}$$

This is to be compared with Eq. 11.12, to which it reduces when $a = 0$. Apart from the complexity of the coefficients, Eq. 11.92 differs from Eq. 11.12 in a qualitative way in the presence of a term in EJ. So we cannot simply define an effective potential V^2 and write $(dr/d\lambda)^2 = E^2 - V^2$. What we can do is nearly as good. We can factor Eq. 11.91:

$$\left(\frac{dr}{d\lambda}\right)^2 = \frac{(r^2 + a^2)^2 - a^2\Delta}{r^4}(E - V_+)(E - V_-). \tag{11.93}$$

Then $V_\pm$, by Eqs. 11.91 and 11.92, are given by

$$V_\pm(r) = [\omega \pm (\omega^2 - g^{\phi\phi}/g^{tt})^{1/2}]J \tag{11.94}$$

$$= \frac{2Mra \pm r^2\Delta^{1/2}}{(r^2 + a^2)^2 - a^2\Delta}J. \tag{11.95}$$

This is the analog of the *square root* of Eq. 11.14, to which it reduces when $a = 0$. Now, the square root of Eq. 11.14 becomes imaginary inside the horizon; similarly, Eq. 11.95 is complex when $\Delta < 0$. In each case the meaning is that in such a region there are *no* solutions to $dr/d\lambda = 0$, *no* turning points regardless of the energy of the photon. Once a photon crosses the line $\Delta = 0$ it *cannot* turn around and get back outside that line. Clearly,

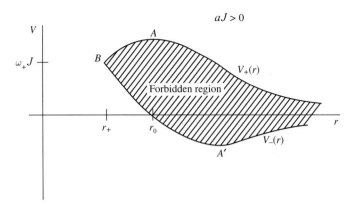

Factored potential diagram for equatorial photon orbits of positive angular momentum in the Kerr metric. As $a \to 0$, the upper and lower curves approach the two square roots of Figure 11.2 outside the horizon.

$\Delta = 0$ marks the *horizon* in the equatorial plane. What we haven't shown, but what is also true, is that $\Delta = 0$ marks the horizon for trajectories not in the equatorial plane.

We can discuss the qualitative features of photon trajectories by plotting $V_{\pm}(r)$. We choose first the case $aJ > 0$ (angular momentum in the same sense as the hole), and of course we confine attention to $r \geqslant r_+$ (outside the horizon). Notice that for large r the curves (in Figure 11.15) are asymptotic to zero, falling off as $1/r$. This is the regime in which the rotation of the hole makes almost no difference. For small r we see features not present without rotation: V_- goes through zero (easily shown to be at r_0, the location of the ergosphere) and meets V_+ at the horizon, both curves having the value $aJ/2Mr_+ := \omega_+ J$, where ω_+ is the value of ω on the horizon. From Eq. 11.93 it is clear that a photon can move only in regions where $E > V_+$ or $E < V_-$. We are used to photons with positive E: they may come in from infinity and either reach a minimum r or plunge in, depending on whether they encounter the hump in V_+. There is nothing qualitatively new here. But what of those for which $E \leqslant V_-$? Some of these have $E > 0$. Are they to be allowed?

To discuss negative-energy photons we must digress a moment and talk about moving along a geodesic backwards in time. We have associated our particles' paths with the mathematical notion of a geodesic. Now a geodesic is a curve, and the path of a curve can be traversed in either of two directions; for timelike curves one direction is forwards and the other backwards in time. The tangents to these two motions are simply opposite in sign, so one will have four-momentum $\vec{p}$ and the other $-\vec{p}$. The energies measured by observer $\vec{U}$ will be $-\vec{U} \cdot \vec{p}$ and $+\vec{U} \cdot \vec{p}$. So one particle will have positive energy and the other negative energy. In flat spacetime we conventionally take all particles to travel forwards in time; since all known particles have positive or zero rest mass, this causes them all to have positive energy relative to any Lorentz observer who also moves forwards in time. Conversely, if $\vec{p}$ has positive energy relative to some Lorentz observer, it has positive energy relative to *all* observers that go forwards in time.

In the Kerr metric, however, it will not do simply to demand positive E. The reason is that E is the energy relative to an observer at infinity; the particle near the horizon is far from infinity, so the direction of forward time isn't so clear. What we must do is set up some observer $\vec{U}$ near the horizon who will have a clock, and demand that $-\vec{p} \cdot \vec{U}$ be positive for particles that pass near the observer. A convenient observer (but any will do) is one who has zero angular momentum and resides at fixed r, circling the hole at the angular velocity ω.

This *zero-angular-momentum observer* (ZAMO) is not on a geodesic, so they must have a rocket engine to remain on this trajectory. (In this respect the observer is no different from us, who must use our legs to keep us at constant r in Earth's gravitational field.) It is easy to see that the observer has four-velocity $U^0 = A$, $U^\phi = \omega A$, $U^r = U^\theta = 0$, where A is found from the condition $\vec{U} \cdot \vec{U} = -1 : A^2 = g_{\phi\phi}/(-D)$. This is nonsingular for $r > r_+$. Then the observer measures the energy of a particle to be

$$E_{\text{ZAMO}} = -\vec{p} \cdot \vec{U} = -(p_0\, U^0 + p_\phi\, U^\phi)$$
$$= A(E - \omega J). \tag{11.96}$$

This is the energy we must demand be positive-definite. Since A is positive, we require

$$E > \omega J. \tag{11.97}$$

From Eq. 11.95 it is clear that any photon with $E > V_+$ also satisfies Eq. 11.97 and so is allowed, while any with $E < V_-$ violates Eq. 11.97 and is moving backwards in time. So in Figure 11.15 we consider only trajectories for which E lies above V_+; for these there is nothing qualitatively different from Schwarzschild.

The Penrose process

For negative-angular-momentum particles, however, new features do appear. If $aJ < 0$ it is clear from Eq. 11.95 that the shape of the $V_\pm$ curves is just turned over, so they look like Figure 11.16. Again, of course, condition Eq. 11.97 means that forward-going photons must lie above $V_+(r)$, but now some of these can have $E < 0$! This happens only for $r < r_0$, i.e. inside the ergosphere. Now we see the origin of the name ergoregion: it is from the Greek *ergo-*, meaning energy, a region in which energy has peculiar properties.

The existence of orbits with negative total energy leads to the following interesting thought experiment, originally suggested by Roger Penrose (1969). Imagine dropping an

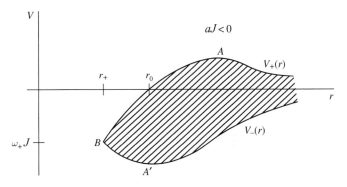

Figure 11.16 As Figure 11.15 for negative-angular-momentum photons.

unstable particle toward a Kerr black hole. When it is inside the ergosphere it decays into two photons, one of which finds itself on a null geodesic with negative energy with respect to infinity. It is trapped in the ergoregion and inevitably falls into the black hole. The other particle must have positive energy, since energy is conserved. If the positive-energy photon can be directed in such a way as to escape from the ergoregion, then the net effect is that it leaves the hole with less energy than it had to begin with: the infalling particle has 'pumped' the black hole and extracted energy from it! By examining Figures 11.15 and 11.16, one can convince oneself that this only works if the negative-energy particle also has a negative J, so that the process involves a decrease in the angular momentum of the hole. The energy has come at the expense of the spin of the black hole.

Blandford & Znajek (1977) suggested that the same process could operate in a practical way if a black hole is surrounded by an accretion disk (§ 11.4) containing a magnetic field. If the field penetrates inside the ergosphere, then it could facilitate the creation of pairs of electrons and positrons, and some of them could end up on negative-energy trapped orbits in the ergoregion. The escaping particles might form the energetic jets of charged particles that are known to be emitted from near black holes, especially in quasars. In this way, quasar jets might be powered by energy extracted from the rotation of black holes. As of this writing (mid-2021) this is probably the most favored models for powering quasar jets.

The Penrose process is not peculiar to the Kerr black hole; it happens whenever there is an ergoregion. If a rotating star has an ergoregion without a horizon (bounded by an ergotoroid rather than an ergosphere) the effective potentials look like Figure 11.17, drawn for $aJ < 0$ (Comins & Schutz 1978). (For $aJ > 0$ the curves just turn over, of course.)

The curve for V_+ dips below zero in the ergoregion. Outside the ergoregion it is positive, climbing to infinity as $r \to 0$. The curve for V_- never changes sign, and also goes to infinity as $r \to 0$. The Penrose process operates inside the ergoregion, where the negative-energy photons are trapped.

In stars, the existence of an ergoregion leads to an instability (Friedman 1978). Roughly speaking, here is how it works. Imagine a small 'seed' gravitational wave that has, perhaps by scattering from the star, begun to travel on a negative-energy null line in the ergoregion.

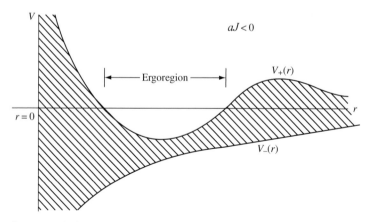

Figure 11.17   As Figure 11.16 for equatorial orbits in the spacetime of a star rotating rapidly enough to have an ergoregion.

It is trapped there, and must circle around the star forever within the ergotoroid. But, being a wave, it is not localized perfectly, and so part of the wave inevitably creates disturbances outside the ergoregion, i.e. with positive energy, which radiate to infinity. By conservation of energy, the wave in the ergoregion must get stronger because of this 'leakage' of energy, since strengthening it means creating more negative energy to balance the positive energy radiated away. But then the cycle continues: the stronger wave in the ergoregion leaks even more energy to infinity, and so its amplitude grows even more. This is the hallmark of an exponential process: the stronger it gets, the faster it grows. Of course, these waves carry angular momentum away as well, so in the long run the process results in the spin-down of the star and the complete disappearance of the ergoregion.

In principle, this might also happen to Kerr black holes, but here there is a crucial difference: the wave in the ergoregion does not have to stay there forever, but can instead travel across the horizon into the hole. There it can no longer create waves outside the ergosphere. Although a complete proof of the stability of Kerr black holes is not yet available, there is a proof (Dafermos *et al.* 2017) that solutions for linear gravitational waves on a Kerr background with small a/M (i.e. a slowly rotating black hole) are in fact stable, which would therefore mean that the ergoregion instability does not happen in Kerr black holes, or is at least not generic as it would be in a rotating star with an ergoregion.

11.4 Real black holes in astronomy

As we noted before, the unique stationary (time-independent) solution for a black hole is the Kerr metric. This extraordinary result makes studying black holes in the real world much simpler than one might have expected. Most real-world systems are so complicated that they can only be studied through idealizations; each star, for example, is individual, and astronomers work very hard to classify them, discover patterns in their appearance that give clues to their interiors, and generally build models that are complex enough to capture their important properties. But these models are still simplifications, since with 10^{57} particles a star has an enormous number of physical degrees of freedom.

Not so with black holes: provided they are stationary, they have just *two* intrinsic degrees of freedom: their mass and spin. In his 1975 Ryerson Lecture at the University of Chicago, long before the majority of astronomers had accepted that black holes were commonplace in the Universe, the astrophysicist S. Chandrasekhar expressed his reaction to the uniqueness theorem in this way (Chandrasekhar 1987):

> In my entire scientific life, extending over forty-five years, the most shattering experience has been the realization that an exact solution of Einstein's equations of general relativity, discovered by the New Zealand mathematician, Roy Kerr, provides the absolutely exact representation of untold numbers of massive black holes that populate the Universe. This shuddering before the beautiful, this incredible fact that a discovery motivated by a search after the beautiful in mathematics should find its exact replica in Nature, persuades me to say that beauty is that to which the human mind responds at its deepest and most profound.

The simplicity of the black hole model has made it possible to search for and identify with confidence the coalescing black holes which formed the bulk of LIGO and Virgo's gravitational wave detections: see the next chapter. This simplicity has also allowed indirect identifications of black holes from their effects on nearby gas and stars, and that is our subject in this section.

The small size of black holes means that in order to gain reasonable confidence in such an identification, one has normally to make an observation either with very high angular resolution to see matter near the horizon or by using photons of very high energy which originate from strongly compressed and heated matter near the horizon. Using orbiting X-ray observatories, astronomers since the 1970s have been able to identify a number of black-hole candidates, particularly in binary systems in our Galaxy. These have (often rather uncertain) masses in the range 6–20 $M_\odot$, which is smaller than the typical mass of the black holes in binaries detected by LIGO and Virgo. The X-rays emitted by X-ray binaries come from mass falling onto the compact object (either a neutron star or a black hole) from their companion star. These systems have arisen from the gravitational collapse of a massive star in a binary. We discussed in § 10.7 what happens when a supernova explodes in a binary system.

Since the 1990s, optical astronomers have been able to identify *supermassive black holes* in the centers of galaxies, with masses ranging from 10^6 to 10^{10} $M_\odot$. Much to their initial surprise, astronomers have found that almost all galaxies have a central black hole, and that the very bright emissions from active galaxies and quasars are powered by black holes. Our own Galaxy has a central black holes, whose mass of 4.15×10^6 $M_\odot$ is more modest by comparison with quasars, but is typical for normal galaxies. This black hole is called Sagittarius A* (Sgr A*), and we shall discuss it in more detail below. The ubiquity of these black holes in galaxies suggests that galaxy formation and the formation of supermassive black holes are linked. We will explore this in the context of our discussion in Chapter 13 of the formation of structure in the expanding Universe.

Black holes of stellar mass

An isolated black hole, formed by the collapse of a massive star, would be very difficult to identify. It might accrete a small amount of gas as it moves through the interstellar medium, but this gas would not emit much X-radiation before being swallowed. No such candidates have been identified. All known stellar-mass black holes are in binary systems whose companion star is so large or the orbit is so tight that the companion star begins dumping gas onto the black hole. Being in a binary system, the gas has angular momentum and so it forms a disk around the black hole. But within this disk there is friction, possibly caused by turbulence or by magnetic fields. Friction leads material to spiral inwards through the disk, giving up angular momentum and energy. Some of this energy goes into the internal energy of the gas, heating it up to temperatures in excess of 10^6 K, so that the peak of its emission spectrum is at X-ray wavelengths. The gas that spirals in eventually falls into the black hole, but the amount of mass is rather small and does not make a noticeable change in the hole's mass.

Many such X-ray binary systems are now known. Not all of them contain black holes: a neutron star is compact enough that gas accreting onto it will also reach X-ray temperatures. Astronomers distinguish black holes from neutron stars in these systems by two means: mass and pulsations. If the accreting object pulsates in X-rays at a very steady rate, then it is a pulsar and it cannot be a black hole: black holes cannot hold onto a nonaligned magnetic field and make it rotate with the hole's rotation. Most systems do not show such pulsations, however. In these cases, astronomers try to estimate the mass of the accreting object from observations of the velocity and orbital radius of the companion star (obtained by monitoring the Doppler shift of its spectral lines) and from an estimate of the companion's mass (again from its spectrum). These estimates are uncertain, particularly because it is usually hard to estimate the inclination of the plane of the orbit to the line of sight, but if the accreting object has an estimated mass much more than about $5\,M_\odot$ then it is likely to be a black hole. This is so because the maximum mass of a neutron star cannot be much more than $3\,M_\odot$ and is likely to be much smaller.

As Figure 11.18 shows, there is a gap between the known masses of neutron stars and of black holes, with neutron stars falling below about $2\,M_\odot$ and black holes starting around $5\,M_\odot$. The only objects securely known to be in this gap as of this writing (mid-2021) are objects that have been formed by the merger of two neutron stars, observed by the LIGO and Virgo detectors. When this figure was drawn there was only one, formed in the event GW170817. We will discuss this further in the next chapter, but it is important to remember that all the other objects in the figure in the mass range $1–10\,M_\odot$ are members of binary systems. The remnant of the merger that produced GW170817 is an isolated object, formed from binary stars. So there may well be many objects in the apparent mass gap which we do not know about because they are isolated and therefore do not emit X-rays.

The black holes that LIGO and Virgo have detected through their gravitational wave observations, described in the next chapter, are typically much more massive than those seen in X-ray binaries. This must have to do with their different formation scenarios, but there is much more work to be done before it is fully understood.

How do we know that these massive objects are black holes? The answer is that any other explanation is less plausible. They cannot be neutron stars: no equation of state that is causal (i.e. has a sound speed less than the speed of light) can support more than about $3\,M_\odot$. It would be possible to invent some kind of exotic matter (sometimes called bosonic matter: see the review by Liebling & Palenzuela 2017) that might just make a massive compact object that does not collapse, but we have no evidence for such matter. Eventually we expect to detect the unique signature of black holes in gravitational waves, i.e. their ringdown radiation (see below), and that will remove any lingering doubts.

Supermassive black holes

The best evidence for supermassive black holes is for the closest one to us: the black hole in the center of the Milky Way (see Figure 11.19). This is being studied by two independent teams, which have been making repeated high-resolution measurements of the positions

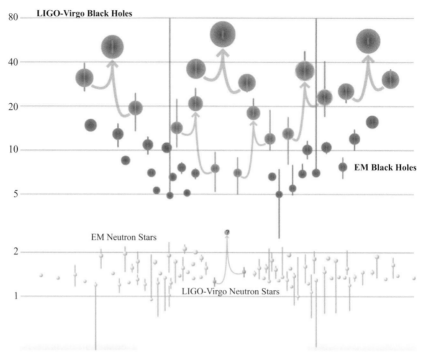

Figure 11.18 This diagram shows the masses of all the black holes and neutron stars known to astronomers as of the end of 2018. The vertical (logarithmically spaced) scale is the mass, in solar masses. The horizontal axis simply spreads out the objects so they can be seen more easily. The black holes detected by LIGO and Virgo are the most massive population. Below them are the black holes detected by X-ray astronomy. Black-hole merger events detected through their gravitational waves are rare, and therefore typically very far away; to be registered by LIGO and Virgo they have to have large masses. Black holes detected by X-ray satellites are not emitting X-rays; these come from an accretion disk around the black hole, fed by gas from a binary companion star. Our satellites are not sensitive enough to see them outside our own Galaxy and nearby galaxies, so these observations sample the more 'normal' black-hole population. The neutron stars at the bottom are no more massive than the (poorly known) upper mass limit on neutron stars. The single object in the 'mass gap' between neutron stars and black holes is the remnant of the merger GW170817, which will be discussed in the next chapter. It is not known whether it remains a very massive neutron star or whether it collapsed to a black hole. Credit: LIGO-Virgo and F. Elasky, Northwestern University.

of stars in the very center of the Galaxy (Eckart & Genzel 1997, Ghez *et al.* 1998, 2008, Schoedel *et al.* 2003). The observations have to be made using infrared light, because visible light from the center is obscured by interstellar dust.

Followed for over decades, these stars have been observed to follow clearly elliptical orbits, all of which have the same focus, which is the location of their common gravitating center. But that center is dark. The stars' spectra reveal their radial velocities, so with three-dimensional velocities and a measurement of the distance to the Galactic center, it is possible to measure the mass of the central object using these orbits: $4.15 \times 10^6 \, \mathrm{M_\odot}$.

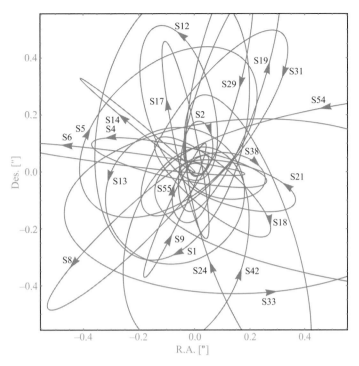

Figure 11.19 The cluster of stars orbiting very close to the massive black hole in the Galactic center, Sgr A*, which is located at the origin of this plot. The coordinate system is tied to the black hole. (It will therefore move slowly across the sky relative to our observing position, owing to the Sun's orbital motion about the center of the Galaxy.) Spectral information (the velocity Doppler shift) allows the orbits to be reconstructed in three dimensions, and they are then projected onto the plane of the sky to make this plot. The orbits all go around the location of the black hole at the origin, but the origin is not at an apparent focus of each ellipse; this is an effect of the projection. The ellipse of star S2 has a measured precession, as one expects of a relativistic orbit, but this effect is too small to be seen in these diagrams. Each orbit is therefore drawn as the best-fit Newtonian ellipse to the data for its star at the epoch of this diagram (2018). One mark on the axis scales is 0.1 arcsec, which corresponds to about 10^4 Schwarzschild horizon radii of the central black hole. (See Exercise 11.42.) A star coming within that distance therefore will be moving faster than $0.01c$. Credit: MPE Galactic Center Team.

The latest optical interferometric observations with the GRAVITY instrument of the European VLT array of telescopes confine the mass to within a sphere not much larger than the last stable circular orbit for a black hole of that mass (Abuter *et al.* 2018b). These remarkable observations earned the respective leaders of the two teams, Reinhard Genzel and Andrea Ghez, the 2020 Nobel Prize in Physics, shared with Roger Penrose (see earlier).

Many other black holes of similar masses are inferred in external galaxies, including in our nearest neighboring spiral galaxy, M31, known as the Andromeda Galaxy. And as we saw at the beginning of this chapter, thanks to the Event Horizon Telescope we even have an image of a nearby black hole in the quasar class: the $2 \times 10^9 \, M_\odot$ black hole in the center of the galaxy M87. While M87 is not a quasar, it has some characteristics, such as

The galaxy NGC6240 has two bright cores, which are thought to surround supermassive black holes that were brought together by the galaxy merger that formed the present galaxy. Notice how irregular NGC6240 is: it does not look like a standard spiral or elliptical galaxy. This is evidence of a recent merger. The image of the cores is an X-ray image by the Chandra satellite. It is superimposed on an optical image of the galaxy. Credit: NASA.

the expulsion of a jet of gas from the center, that might indicate a relationship to quasars or even that it was a quasar in the distant past that is quieter now because it has run out of fuel.

Although the observations of Sagittarius A* and M87 are indirect, in that we are observing the behavior of objects and light near the black hole, they probe the curved geometry so close to the horizon that it seems very unlikely that any other physical model could duplicate these observations. In Chapter 12 we will discuss the detections of gravitational waves from binary black-hole systems. These are direct detections of waves of gravity originating in the curvature of the black holes themselves, and they are removing any residual doubt that these ultra-compact objects truly are black holes.

Astronomers know that galaxies frequently merge, and this ought to bring their central black holes to merge as well. In fact, as we shall see in Chapter 13, current models for galaxy formation suggest that all galaxies are themselves the products of repeated mergers with smaller clusters of stars, and so it is possible that in the course of the formation and growth of galaxies the central black holes grew larger and larger by merging with incoming black holes. See Figure 11.20.

The space-based LISA gravitational wave detector should decide this issue, because mergers of black holes in the mass range 10^4–$10^7 \, M_\odot$ should produce detectable signals in the LISA frequency band, anywhere such mergers take place in the Universe. Even without LISA, there is a growing body of evidence for mergers of black holes. A number of galaxies with two distinct bright cores are known, and there is remarkable evidence for the ejection of a supermassive black hole at something like 1% of the speed of light from the center of a galaxy (Komossa *et al.* 2008). Speeds like this can only be achieved as a result of the 'kick' that the final black hole can get as a result of the merger (see below).

Intermediate-mass black holes

Since there are black holes of mass $10\,M_{\odot}$ and of mass $10^6\,M_{\odot}$ and more, are there black holes with masses between these values, around 100–$10^4\,M_{\odot}$? Astronomers call these intermediate-mass black holes, but not many are known. Some bright X-ray sources may be large black holes, and perhaps there are black holes in this mass range in globular clusters. The globular cluster Omega Centauri seems to have a central black hole of mass $4 \times 10^4\,M_{\odot}$ (Noyola *et al.* 2008). And LIGO and Virgo have detected the formation of a black hole with a mass of about $150\,M_{\odot}$ by the merger of two smaller (but still very massive) black holes (Abbott *et al.* 2020d, 2020e). We shall discuss this observation in more detail in the next chapter.

Astronomers are searching intensively for better evidence for these kinds of objects. Some might have formed in the first generation of star formation, when clouds made up of pure hydrogen and helium first collapsed. These first stars were probably a lot more massive than the stars, like our Sun, that formed later from gas clouds that had been polluted by the heavier elements made by the first generation of stars. Numerical simulations suggest that some of these first stars (called Population III stars) could have rapidly collapsed to black holes. But finding these holes will not be easy. Gravitational wave observations by LISA may be the best way of discovering holes in the upper part of this mass range.

Dynamical black holes

Although stationary black holes are simple, there are situations where black holes are expected to be highly dynamical, and these are more difficult to treat analytically. When a black hole is formed, any initial asymmetry (such as quadrupole moments) must be radiated away in gravitational waves, until finally only the mass and angular momentum are left behind. This happens extremely quickly: the linear oscillations typically damp out (ring down) exponentially after only a few cycles (Kokkotas & Schmidt 1999). The Kerr metric takes over very quickly.

These oscillations are called *quasi-normal modes* and they have a characteristic spectrum that depends on the two parameters M and a. The damping is much faster than modes of, say, neutron stars, which may take thousands of cycles to damp away. Gravitational wave observations of merging black holes should at some point be able to measure the ringdown of these oscillations immediately after merger. If they match the spectrum of Kerr, and in particular if they damp away as rapidly as expected, they will provide convincing evidence that the object formed in the merger is a Kerr black hole.

Even more dynamical, of course, are black holes in collision, either with other black holes or with stars. These have been studied numerically in large-scale supercomputer simulations. Numerical relativity has a long history, but interest and activity increased rapidly with the development of fast computers, triggering the first investigations aimed at full black-hole simulations in the 1970s (e.g. Eardley & Smarr 1979). Simulating spacetime is demanding, because each location in space, even outside the star or black hole, has a value of the metric tensor that needs to be treated as a dynamical entity that will evolve in time. The computational and visualization requirements of these problems helped to drive improvements in computer techniques and technology throughout the 1980s and 1990s.

Evolving spacetime presented challenges to numerical algorithms that had not been faced in other fields, such as: the causality issues of grids crossing horizons and of the merging of horizons; the avoidance of coordinate singularities; and the avoidance of a true infinite-curvature singularity inside the horizon. One by one these and other problems were solved by the worldwide community, until the first successful simulation of orbiting black holes that remained stable was performed by Pretorius (2005). Quickly thereafter the simulations were made robust and more efficient using techniques introduced independently by two other groups, Baker *et al.* (2006) and Campanelli *et al.* (2006).

Workers in the field immediately concentrated on consolidating its techniques and ensuring that the simulations by many different groups around the world agreed with one another (see Centrella *et al.* 2010). Computational relativists have since then made rapid progress in producing accurate predictions of waveforms for a range of black-hole merger systems, which have been integrated into the search techniques for gravitational waves from coalescing black-hole binaries, as we shall see in the next chapter.

Numerical simulations are also a laboratory for learning more about the physics of merging black holes and extreme nonlinear gravity. This includes also gravity in more than four spacetime dimensions, which is a theme of research toward quantum gravity (Cardoso *et al.* 2015). In four spacetime dimensions, one of the most interesting aspects of black-hole mergers that has emerged from simulations is the so-called 'kick'. In Chapter 10 we mentioned kicks from asymmetries in gravitational collapse. Similarly here, when there is no particular symmetry in the initial black-hole binary system, the emitted gravitational radiation will emerge asymmetrically, so that the waves will carry away a net linear momentum in some direction. The result will be that the final merged black hole recoils in the opposite direction. The velocity of the recoil, being dimensionless, does not depend on the overall mass scale of the system. It depends only on dimensionless initial data: the ratio of the masses of the initial holes, the dimensionless spin parameters of the holes (a/M), and the directions of the spins. Normally these recoil velocities are of order a few hundred km s^{-1}, which could be enough to expel the black hole from the center of a star cluster or even a spiral galaxy. Even more remarkably, recoil velocities exceeding 10% of the speed of light are inferred from simulations for some coalescences (Campanelli *et al.* 2007).

Numerical relativity is also addressing the next big objective in assisting gravitational wave astronomy: simulating mergers of neutron stars with one another (Faber & Rasio 2021) and of a neutron star with a black hole (Shibata & Taniguchi 2011). These problems add the complexity of hydrodynamics, nuclear physics, and magnetic fields to the already challenging issues of simulating dynamical spacetimes (Andersson & Comer 2007, Chamel & Haensel 2008, Font 2008). Simulations of observed binary-neutron-star coalescences should be very helpful in distinguishing different neutron-star equations of state from one another.

11.5 Hawking radiation

In 1975 Stephen Hawking startled the physics community by announcing that black holes aren't black: they radiate energy continuously! This doesn't come from any mistake in what we have already done; it arises in the application of quantum theory to the electromagnetic fields near a black hole.

General relativity is a classical theory and, although the search for a quantum version of it has been going on since the early twentieth century, we still have no convincing theory of quantum gravity. Standard quantum field theory (for example, quantum electrodynamics) is formulated on Minkowski spacetime, so it does not include effects of gravitation. Hawking's inspired idea was to formulate quantum field theory on the background spacetime of a black hole. This would not quantize gravity, but it would at least probe for new effects on quantum fields that gravity might bring.

Hawking did not simply put quantum fields into a spacetime that has some curvature. He chose to use the black hole, because it fundamentally changes causality. Light cones are central to formulations of quantum electrodynamics, and Hawking wanted to see what would happen if these light cones were distorted enough to include a horizon.

The result couldn't have been more counterintuitive for theorists who in 1975 had only recently begun to understand the classical horizon as a one-way street. And, as we shall see, it couldn't have been more satisfying for theorists who had noticed the analogy between the Hawking area theorem (see § 11.3) and thermodynamic entropy but who could not see how to make it more than an analogy. Hawking unified black holes and horizons with the nineteenth-century physics of thermodynamics.

Hawking's calculation (Hawking 1975) works within the full apparatus of quantum field theory, but we can derive its main prediction, that black holes radiate, rather simply from elementary considerations involving only one key aspect of quantum theory: the Heisenberg uncertainty principle. What follows, therefore, is a plausibility argument, not a rigorous discussion of the effect.

We have until now spoken of photons as particles following a geodesic trajectory in spacetime, but according to the uncertainty principle these 'particles' cannot be localized to arbitrary precision. Near the horizon this markedly changes the behavior of 'real' photons from what we have already described for idealized null particles.

One form of the Heisenberg uncertainty principle is $\Delta E \, \Delta t \geq \hbar/2$, where ΔE is the uncertainty in a particle's energy if it is measured within a time Δt. According to quantum field theory, ordinary space is filled with *vacuum fluctuations* in electromagnetic fields, which can be thought of as pairs of photons being produced at one event and recombining at another. Such pairs violate the conservation of energy, but if they recombine in a time less than $\Delta t = \hbar/2\Delta E$, where ΔE is the amount of energy in the pair, they violate no physical law. Thus, in the large, energy conservation holds rigorously, while, on a small scale, it is always being violated.

Now, as we have emphasized before, the spacetime near the horizon of a black hole is perfectly ordinary and, in particular, locally flat. Therefore these fluctuations will also be happening there. Consider a fluctuation which produces two photons, one of energy E and the other with energy $-E$, where E is the (conserved) energy of a photon as measured by a very distant observer. Although a local observer measures a different energy, the conversion to E does not change the sign of the energy outside the horizon, and we are here most interested in the sign.

Now, in flat spacetime the negative-energy photon would not be able to propagate freely, so it would necessarily recombine with the positive-energy one within a time $\hbar/2E$. But if produced just outside the horizon, it has a chance of crossing the horizon before the time $\hbar/2E$ elapses; once inside the horizon it *can* propagate freely, as we shall now show.

Consider the Schwarzschild metric for simplicity, and recall from our earlier discussion of orbits in the Kerr metric that negative energy is normally excluded because it corresponds to a particle that propagates backwards in time. Inside the event horizon, let us describe the physics from the point of view of an observer going forwards in local physical time, therefore going toward decreasing r. Recall that r is now the time coordinate, and t is a spatial coordinate. For simplicity, we choose a zero-angular-momentum observer, with four-velocity $\vec{U}$, on a trajectory for which, momentarily, $U^0 = 0$. Then U^r is the only nonzero component of $\vec{U}$ at that moment, so the normalization condition $\vec{U} \cdot \vec{U} = -1$ determines U^r:

$$U^r = -\left(\frac{2M}{r} - 1\right)^{1/2}, \qquad r < 2M, \tag{11.98}$$

which is negative because the observer's trajectory must go to smaller r in order to go forward in time.

Now consider a photon with four-momentum $\vec{p}$ that is observed by this observer. Any photon orbit is allowed for which $-\vec{p} \cdot \vec{U} > 0$. In fact we just take a zero-angular-momentum photon, moving radially inside the horizon. By Eq. 11.12 with $J = 0$, it clearly has a constant of the motion $E = \pm p^r$. Then its energy relative to our local observer is

$$-\vec{p} \cdot \vec{U} = -p^r U^r g_{rr} = -\left(\frac{2M}{r} - 1\right)^{-1/2} p^r. \tag{11.99}$$

This is positive if and only if the photon is also ingoing: $p^r < 0$. But it sets *no* restriction at all on E. Photons may travel on null geodesics inside the horizon, which have either sign of E, as long as $p^r < 0$. This should not be surprising: E is a spatial momentum component because t is a spatial coordinate inside the horizon. What we conclude is that there are photon trajectories inside the horizon with $E < 0$.

But E is a constant of the motion, so if a photon is created by a fluctuation outside the horizon with negative energy $E < 0$ (E is energy outside the horizon but momentum inside), then there do exist trajectories inside the horizon on which that fluctuation can become a 'real' photon, provided that it can reach them in the time Δt available. If this happens, then its positive-energy partner fluctuation is allowed to escape to infinity. This is the one we could in principle detect, so let us see what we can say about its energy.

We first look at the fluctuations in a freely falling inertial frame, which is the one for which spacetime is locally flat and in which the fluctuations should look normal. A frame that is momentarily at rest at coordinate $2M + \varepsilon$, with $\varepsilon \ll M$, will immediately begin falling inwards, following the trajectory of a particle with $\tilde{J} = 0$ and $\tilde{E} = [1 - 2M/(2M + \varepsilon)]^{1/2} \approx (\varepsilon/2M)^{1/2}$, from Eq. 11.11. A negative-energy virtual fluctuation that manages to avoid recombining until this frame has reached the horizon can then become real, because negative energies are now just negative spatial momenta.

The proper time that elapses for the infalling observer before it reaches the horizon, $\Delta\tau$, is obtained by integrating Eq. 11.59:

$$\Delta\tau = -\int_{2M+\varepsilon}^{2M} \left(\frac{2M}{r} - \frac{2M}{2M + \varepsilon} \right)^{-1/2} dr. \tag{11.100}$$

To first order in ε this is

$$\Delta\tau = 2(2M\varepsilon)^{1/2}. \tag{11.101}$$

We can find the energy $\mathcal{E}$ of the photon in this frame by setting this equal to the fluctuation time $\hbar/2\mathcal{E}$. The result is

$$\mathcal{E} = \tfrac{1}{4}\hbar(2M\varepsilon)^{-1/2}. \tag{11.102}$$

This is the energy of the outgoing photon, the one which reaches infinity, as calculated in the local inertial frame.

To find its energy when it gets to infinity we recall that

$$\mathcal{E} = -\vec{p} \cdot \vec{U},$$

with $-U_0 = \tilde{E} \approx (\varepsilon/2M)^{1/2}$. Therefore

$$\mathcal{E} = -g^{00}p_0U_0 = U_0 g^{00}E, \tag{11.103}$$

where E is the conserved energy on the photon's trajectory and is the energy that it is measured to have when it arrives at infinity. Evaluating g^{00} at $2M + \varepsilon$ gives, finally,

$$E = \mathcal{E}(\varepsilon/2M)^{1/2} = \hbar/8M. \tag{11.104}$$

Remarkably, it doesn't matter in this simple calculation at what value of $\varepsilon = r - 2M$ the photon originated: it always comes out with this characteristic energy! This is a sign that our simple model may be touching on something fundamental.

The rigorous calculation which Hawking performed showed that the photons which come out have the spectrum characteristic of a black body with temperature

$$T_H = \hbar/8\pi k_B M, \tag{11.105}$$

where (as before) k_B is Boltzmann's constant. The average energy of a photon in the black-body spectrum is

$$E = 2.701 k_B T = 0.860\hbar/8M, \tag{11.106}$$

rather close to our crude result, Eq. 11.104. This is quite good agreement, considering that we looked only at a subset of the outgoing photons, those which are radially moving and which have the maximum fluctuation energy allowed by the uncertainly principle for the given infall time from ε. Our argument does not show that the photons should have a black-body spectrum; but the fact that the spectrum originates in random fluctuations, plus the fact that the black hole is, classically, a perfect absorber, makes that result plausible as well.

Notice that for every radiated photon of energy E, its fluctuation partner is a photon of energy $-E$ inside the horizon. This means that the mass of the black hole decreases by the amount radiated away: energy is conserved. It is also interesting to compare the local wavelength λ of the photon with the distance ε. Since $\lambda = 2\pi\hbar/\mathcal{E}$, we get $\lambda = 8\pi(2M\varepsilon)^{1/2}$. This is much greater than ε, so the emitted photon cannot be associated with any local spot on the horizon of size ε: it is nonlocal as seen in the infalling local inertial frame. That might make one worry that the calculation doesn't work, because the frame is only locally inertial. But, near the horizon, the assumption of local flatness fails only over distances of order M, which is a measure of the radius of curvature of the local geometry. Since $\lambda \ll M$, it is perfectly valid to treat our photon as a flat-space fluctuation.

It is important to understand that the negative-energy photons in the Hawking effect are not the same as the negative-energy photons that we discussed in the Penrose process above. The Penrose process works only inside an ergoregion, and uses negative-energy orbits that are outside the horizon of the black hole. The Hawking result is more profound: it operates even for a non-spinning black hole and connects negative-energy photons *inside* the horizon with positive-energy counterparts outside. It operates in the Kerr metric as well, but again it happens across the horizon, not the ergosphere.

The Hawking effect does not lead to an unstable runaway in the way that the Penrose process does for a star with an ergoregion. The reason is that Hawking's negative-energy photon is already inside the horizon and does not create any further positive-energy photons outside. So the Hawking radiation is a steady thermal radiation, created by ever-present quantum fluctuations near the horizon.

Notice that the Hawking temperature of the hole is proportional to M^{-1}. The rate of radiation from a black body is proportional to AT^4, where A is the area of the body. In this case the radiating surface is the horizon, whose area is proportional to M^2 (see Eq. 11.85). So the luminosity of the hole is proportional to M^{-2}. This energy must come from the mass of the hole (every negative-energy photon falling into it reduces M), so we have

$$\left. \begin{array}{l} \mathrm{d}M/\mathrm{d}t \sim M^{-2}, \\ M^2\,\mathrm{d}M \sim \mathrm{d}t, \end{array} \right\} \tag{11.107}$$

from which we infer that the lifetime of the hole scales as

$$\tau \sim M^3. \tag{11.108}$$

The bigger the hole the longer it lives, and the cooler its temperature.

The numbers work out to show that a black hole of mass 10^{12} kg has a lifetime of 10^{10} yr, about the age of the Universe. Thus

$$\left(\frac{\tau}{10^{10}\ \mathrm{yr}}\right) = \left(\frac{M}{10^{12}\ \mathrm{kg}}\right)^3. \tag{11.109}$$

The smallest black holes formed from stars are surely larger than a solar mass. Since a solar mass is about 10^{30} kg, black holes formed from stellar collapse are essentially unaffected by this radiation, which has a temperature of about 10^{-7} K.

On the other hand, perhaps it was possible for holes of 10^{12} kg to form in the very early Universe. To see the observable effect of their *evaporation*, let us calculate the energy radiated in the last second of the evaporation of *any* black hole by setting $\tau = 1\,\text{s} = (3 \times 10^7)^{-1}$ yr in Eq. 11.109. We get

$$M \approx 10^6 \,\text{kg} \sim 10^{23} \,\text{J}. \tag{11.110}$$

So, for a brief second it would have a luminosity about 0.1% of the Sun's luminosity, but its spectrum would be very different. Its temperature would be 10^{11} K, emitting primarily in γ-rays! One might be tempted to try to explain the gamma-ray bursts mentioned earlier in this chapter as primordial black-hole evaporations, but the observed gamma bursts are in fact billions of times more luminous. A primordial black-hole evaporation would probably be visible only if it happened in our own Galaxy. No such events have been identified.

It must be pointed out that all derivations of Hawking's result are valid only if the typical photon has $E \ll M$, since they involve treating the spacetime of the black hole as a fixed background in which one solves the equations of quantum mechanics, unaffected to first order by the propagation of these photons. This approximation fails for $M \approx h/M$, or for black holes of mass

$$M_{\text{Pl}} = h^{1/2} = 1.6 \times 10^{-35} \,\text{m} = 2.2 \times 10^{-8} \,\text{kg}. \tag{11.111}$$

This is called the *Planck mass*, since it is a mass derived only from Planck's constant (and c and G). We first met it in Exercise 8.2. To treat quantum effects involving such black holes, one needs a consistent theory of quantum gravity, which is one of the most active areas of research in relativity today. What we can say here is that Hawking's calculation appears to have been one of the most fruitful steps in the direction of a full quantum theory of gravity.

As hinted above, the Hawking effect has provided a remarkable unification of gravity and thermodynamics. Consider Hawking's area theorem, which we may write as

$$\frac{\mathrm{d}A}{\mathrm{d}t} \geqslant 0. \tag{11.112}$$

For a Schwarzschild black hole,

$$A = 16\pi M^2,$$

$$\mathrm{d}A = 32\pi M \,\mathrm{d}M,$$

or (if we arrange factors in an appropriate way)

$$\mathrm{d}M = \frac{1}{32\pi M} \,\mathrm{d}A = \frac{\hbar}{8\pi k M} \,\mathrm{d}\left(\frac{kA}{4\hbar}\right). \tag{11.113}$$

Since $\mathrm{d}M$ is the change in the hole's total energy, and since $\hbar/8\pi k M$ is its Hawking temperature T_H, we may write Eq. 11.113 in the form

$$\mathrm{d}E = T_H \,\mathrm{d}S,$$

with

$$S = kA/4\hbar. \tag{11.114}$$

Since, by Eq. 11.112, this quantity S can never decrease, we have in Eqs. 11.113 and 11.112 the first and second laws of thermodynamics as they apply to black holes! That is, a black hole behaves in every respect as a thermodynamic black body with temperature $T_H = \hbar/8\pi kM$ and entropy $kA/4\hbar$. This analogy had been noticed as soon as the area theorem was discovered (see Bekenstein 1973, 1974, and Misner *et al.* 1973, Box 33.4), but at that time it was thought to be an incomplete analogy because black holes did not have a true temperature. The Hawking radiation fits the missing piece into the puzzle.

But the Hawking radiation has raised other questions. One of them concerns information. The emission of radiation from the black hole raises the possibility that the radiation could carry information that, in the classical picture, is effaced by the formation of the horizon. If the radiation is perfectly thermal, then it contains almost no information. But it is possible that the outgoing photons and gravitons have a thermal black-body spectrum but also have weak correlations that contain information. This information would not have come from inside the hole, but from the virtual photons and gravitons just outside the horizon, which are affected by the details of the collapsing matter that passes through them on its way to forming the black hole, and which then become real photons and gravitons by the process described above. Whether this picture is indeed correct, and what kind of information can in principle be recovered from the outgoing radiation, are still matters of considerable debate among physicists.

The Hawking radiation has also become a touchstone for the development of full theories of quantum gravity. Physicists test new quantization methods by showing that they can predict the Hawking radiation and associated physics, such as the entropy of the black hole. This is not sufficient to prove that a method will work in general, but it is regarded as necessary.

11.6 Bibliography

The story of Karl Schwarzschild's discovery of the solution named after him is an extraordinary one. See the online biography of Schwarzschild by J. J. O'Connor & E. F. Robertson in the MacTutor History of Mathematics archive, at the URL www-history.mcs.st-andrews.ac.uk/Biographies/Schwarzschild.html.

Black holes are the subject of a series of ten-minute video interviews with leading specialists in different aspects of black holes, from the history of their discovery to their role in modern astrophysics. The interviews are conducted by a teenaged interviewer and address high-school and university-level students. These were recorded by the Scienceface project and may be viewed on the project's YouTube channel: www.youtube.com/user/scienceface/featured/.

Other accessible articles about black holes may be found on the Einstein Online website www.einstein-online.info/.

The discovery of the Kerr metric, Kerr (1963), came more than 50 years after that by Schwarzschild, but still at a time when black holes were not fully understood. A history of the way in which scientists eventually came to understand black holes, with an afterword by Roy Kerr, is given in Melia (2009). A conference to commemorate the fiftieth anniversary of Kerr's discovery was recorded on video by the Scienceface project, including a talk by Kerr himself; the videos may be viewed on the project's YouTube channel, mentioned above.

The perihelion shift and deflection of light are the two classical tests of GR. Other theories predict different results: see Will (2014, 2018). A short, entertaining account of the observation of the deflection of light and its impact on Einstein's fame is to be found in McCrea (1979). To learn more about the theory of gravitational lensing, see Wambsganss (1998), Schneider *et al.* (1992), Schneider (2006), or Perlick (2004).

The Kerr metric has less symmetry than the Schwarzschild metric, so it might be expected that particle orbits would have fewer conserved quantities and therefore be harder to calculate. This is, quite remarkably, false. Even orbits out of the equator have three conserved quantities: energy, angular momentum, and a difficult-to-interpret quantity associated with the θ motion. The same remarkable property carries over to the wave equations that govern electromagnetic fields and gravitational waves in the Kerr metric: these equations separate completely in certain coordinate systems. See Teukolsky (1972) for the first general proof of this and Chandrasekhar (1983) for full discussions.

Black-hole thermodynamics is treated thoroughly in Carter (1979), while the related theory of quantum fields in curved spacetimes is reviewed by Wald (1984). This relates to work on quantum gravity. See the references cited in § 7.5. A much more rigorous derivation of the Hawking radiation from the notion of fluctuations near the horizon is given in Parikh & Wilczek (2000).

The field of numerical relativity is very active. See the books by Bona & Palenzuela-Luque (2005), Alcubierre (2008) and Shibata (2015) for surveys. Most numerical simulations are performed using the open-source community tools of the Einstein Toolkit: `https://einsteintoolkit.org/`. Students wishing to experiment with simple simulations will find examples and support on the website.

The history of numerical relativity is entangled with the history of supercomputing, because often the first users of new generations of supercomputers were the scientists working on a few 'grand challenge' problems, which included numerical relativity but also numerical astrophysics, climate science, the theory of the strong interaction, plasma fusion physics, and a handful more.

One of the most remarkable stories in this history is how numerical astrophysics and numerical relativity played a key role in the development in 1993 of the first graphical web browser, Mosaic, at the NCSA supercomputer center of the University of Illinois. NCSA's first director was Larry Smarr, a pioneer of numerical relativity, and Mosaic was developed to allow scientists at distant universities to share visualizations of numerical results on the World Wide Web, then only a few years old. Microsoft licenced Mosaic's code for its first version of the Internet Explorer browser, and the first Netscape browser was developed by some of the same team who produced Mosaic. This story is told by E. Seidel in one of the Scienceface interviews mentioned above: `http://tinyurl.com/y9t43xqk/`.

These early web browsers changed the Web from a tool for scientists into one that the general public wanted to use, and today their modern descendants support e-commerce, multimedia, social media, and the apps that run on mobile phones. It all started with research into black holes!

Exercises

11.1 Starting with Eq. 11.4, calculate the density at which a black hole of $10^9\,\mathrm{M}_\odot$ would form. Verify the statement, made in the text, that this is less than the density of water.

11.2 Consider a particle or photon in an orbit in the Schwarzschild metric with a certain E and J, at a radius $r \gg M$. Show that if spacetime were really *flat*, the particle would travel in a straight line which would pass a distance $b := J/(E^2 - m^2)^{1/2}$ from the center of coordinates $r = 0$. This ratio b is called the *impact parameter*. Show also that photon orbits that follow from Eq. 11.12 depend *only* on b.

11.3 Prove Eqs. 11.17 and 11.18.

11.4 Plot $\tilde{V}^2$ against r/M for the three cases $\tilde{J}^2 = 25M^2, \tilde{J}^2 = 12M^2, \tilde{J}^2 = 9M^2$ and verify the qualitative correctness of Figures 11.1 and 11.3.

11.5 What kind of orbits are possible outside a star of radius (a) $2.5M$, (b) $4M$, (c) $10M$?

11.6 The centers of most galaxies contain black holes of mass 10^6–$10^9\,\mathrm{M}_\odot$.

(a) Find the radius $R_{0.01}$ at which $-g_{00}$ differs from the 'Newtonian' value $1 - 2\,M/R$ by only 1%. (One may think of this as a kind of limit on the region in which relativistic effects are important.)

(b) A 'normal' star may have a radius of 10^{10} m. Approximately how many such stars could occupy the volume of space between the horizon $R = 2\,M$ and $R_{0.01}$?

11.7 Compute the wavelength of light that gets to a distant observer from the following sources.

(a) Light emitted with wavelength 6563 Å ($H\alpha$ line) by a source at rest where $\Phi = -10^{-6}$. (A typical star.)

(b) Same as (a) for $\Phi = -6 \times 10^{-5}$ (the value for the white dwarf 40 Eridani B).

(c) Same as (a) for a source at rest at radius $r = 2.2\,M$ outside a black hole of mass $M = 1\,\mathrm{M}_\odot = 1.47 \times 10^5$ cm.

(d) Same as (c) for $r = 2.02\,M$.

11.8 A clock is in a circular orbit at $r = 10\,M$ in a Schwarzschild metric.

(a) How much time elapses on the clock during one orbit? (Integrate the proper time $d\tau = |ds^2|^{1/2}$ over an orbit.)

(b) It sends out a signal to a distant observer once each orbit. What time interval does the distant observer measure between receiving any two signals?

(c) A second clock is located at rest at $r = 10\,M$ next to the orbit of the first clock. (Rockets keep it there.) How much time elapses on it between successive passes of the orbiting clock?

(d) Calculate (b) again in seconds for an orbit at $r = 6\,M$ where $M = 14\,M_\odot$. This is the minimum fluctuation time one expects in the X-ray spectrum of Cyg X-1: why?

(e) If the orbiting 'clock' is the twin Artemis, in the orbit in (d), how much does she age during the time her twin Diana lives 40 years far from the black hole and at rest with respect to it?

11.9 (a) Derive Eqs. 11.20 and 11.24.
(b) Derive Eqs. 11.26 and 11.28.

11.10 This problem requires programming a computer or at least working with a spreadsheet.

(a) Integrate numerically Eq. 11.26 or Eq. 11.28 for the orbit of a particle (i.e. for r/M as a function of ϕ) when $E^2 = 0.91$ and $(\tilde{J}/M)^2 = 13.0$. Compare the perihelion shift from one orbit to the next with Eq. 11.37.
(b) Integrate again when $\tilde{E}^2 = 0.95$ and $(\tilde{J}/M)^2 = 13.0$. How much proper time does this particle require to reach the horizon from $r = 10\,M$ if its initial radial velocity is negative?

11.11 (a) For a given value of $\tilde{J}$, what is the minimum value of $\tilde{E}$ that permits a particle with $m \neq 0$ to reach the Schwarzschild horizon?
(b) Express this result in terms of the *impact parameter b* (see Exercise 11.2).
(c) Conversely, for a given value of b, what is the maximum value of $\tilde{J}$ that permits a particle wth $m \neq 0$ to reach the Schwarzschild horizon? Relate your result to Figure 11.3.

11.12 The right-hand side of Eq. 11.28 is a polynomial in u. Trace the u^3 term back through the derivation and show that it would not be present if we had started with the Newtonian version of Eq. 11.9. Interpret this term as a redshift effect on the orbital kinetic energy. Show that it is responsible for the maximum in the curve in Figure 11.1.

11.13 (a) Prove that Eq. 11.32 solves Eq. 11.31.
(b) Derive Eq. 11.33 from Eq. 11.32 and show that it describes an ellipse by transforming to Cartesian coordinates.

11.14 (a) Derive Eq. 11.34 in the approximation that y is small. Compared with what must it be small?
(b) Derive Eqs. 11.35 and 11.36 from Eq. 11.34.
(c) Verify the remark after Eq. 11.36 that $y = 0$ is not the correct circular orbit for the given $\tilde{E}$ and $\tilde{J}$, by using Eqs. 11.20 and 11.21 to find the correct value of y and comparing it with y_0 in Eq. 11.36.
(d) Show from Eq. 11.13 that a particle which has an inner turning point in the 'Newtonian' regime, i.e. for $r \gg M$, has a value $\tilde{J} \gg M$. Use this to justify the step from Eq. 11.37 to Eq. 11.38.

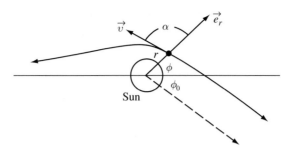

Figure 11.21 The deflection of light by the Sun.

11.15 Compute the perihelion shift per orbit and per year for the following planets, given their distance from the Sun and their orbital period: Venus $(1.1 \times 10^{11}$ m, 1.9×10^7 s); Earth $(1.5 \times 10^{11}$ m, 3.2×10^7 s); Mars $(2.3 \times 10^{11}$ m, 5.9×10^7 s).

11.16 (a) Derive Eq. 11.51 from Eq. 11.49, and show that it describes a straight line passing a distance b from the origin.

(b) Derive Eq. 11.53 from Eq. 11.49.

(c) Integrate Eq. 11.53 to get Eq. 11.54.

11.17 We can calculate the observed deflection of a null geodesic anywhere on its path as follows. See Ward (1970).

(a) Show that Eq. 11.54 may be solved to give

$$bu = \sin(\phi - \phi_0) + \frac{M}{b}[1 - \cos(\phi - \phi_0)]^2 + O\left(\frac{M^2}{b^2}\right). \qquad (11.115)$$

(b) In Schwarzschild coordinates, the vector

$$\vec{v} \rightarrow -(0, 1, 0, d\phi/dr) \qquad (11.116)$$

is tangent to the photon's path as seen by an observer at rest in the metric at the position r. Show that this observer measures the angle α in Figure 11.21 to be

$$\cos\alpha = (\vec{v} \cdot \vec{e}_r)/(\vec{v} \cdot \vec{v})^{1/2}(\vec{e}_r \cdot \vec{e}_r)^{1/2}, \qquad (11.117)$$

where $\vec{e}_r$ has components $(0, 1, 0, 0)$. Argue that $\phi - \pi + \alpha$ is the *apparent* angular position of the star, and show from Eq. 11.115 that if $M = 0$ (no deflection), then $\phi - \pi + \alpha = \phi_0$.

(c) When $M \neq 0$, calculate the deflection

$$\delta\phi := (\phi - \pi + \alpha) - \phi_0 \qquad (11.118)$$

to first order in M/b. Don't forget to use the Schwarzschild metric to compute the dot products in Eq. 11.117. Obtain

$$\delta\phi = \frac{2M}{b}[1 - \cos(\phi - \phi_0)], \qquad (11.119)$$

which is, in terms of the position r of the observer,

$$\delta\phi = \frac{2M}{r}\frac{1 - \cos(\phi - \phi_0)}{\sin(\phi - \phi_0)}. \qquad (11.120)$$

(d) For $M = 1\,M_\odot = 1.47\,\mathrm{km}, r = 1\,\mathrm{AU} = 1.5 \times 10^6\,\mathrm{km}$, how far from the Sun on the sky can this deflection be detected if one can measure angles to an accuracy of 2×10^{-3} arcsec?

11.18 We can use Eq. 11.115 above for a different problem, namely to calculate the expected arrival times at a distant observer of pulses regularly emitted by a satellite in a circular orbit in the Schwarzschild metric. This is a simplified version of the timing problem of the binary pulsar system, which consists of two neutron stars of roughly equal mass in orbit about one another, one of which is a pulsar. See Blandford & Teukolsky (1976) and Epstein (1977).

(a) Show that along the trajectory, Eq. 11.115, coordinate time elapses at the rate

$$dt/d\phi = b\left[(bu)^2\left(1 - \frac{2M}{b}bu\right)\right]^{-1}. \qquad (11.121)$$

(b) Integrate this to find the coordinate travel time for a photon emitted at the position u_E, ϕ_E and received at the position u_R, ϕ_R, where $u_R \ll u_E$.

(c) Since Eq. 11.115 is satisfied at both (u_R, ϕ_R) and (u_E, ϕ_E), show that

$$\phi_R - \phi_0 = (u_R/u_E)\sin(\phi_E - \phi_R)\left\{1 + (u_R/u_E)\cos(\phi_E - \phi_R)\right.$$
$$\left. + Mu_E[1 - \cos(\phi_E - \phi_R)]^2/\sin^2(\phi_E - \phi_R)\right\}, \qquad (11.122)$$

to first order in Mu_E and u_R/u_E and that, similarly,

$$b = (1/u_R)\left\{\phi_R - \phi_0 + Mu_E[1 - \cos(\phi_E - \phi_R)]^2/\sin(\phi_E - \phi_R)\right\}. \qquad (11.123)$$

(d) Use these in your result in (b) to calculate the difference δt in travel time between pulses emitted at (u_E, ϕ_E) and at $(u_E, \phi_E + \delta\phi_E)$, to first order in $\delta\phi_E$. (The receiver is at fixed (u_R, ϕ_R).)

(e) For an emitter in a circular orbit $u_E = $ const., $\phi_E = \Omega t_E$, plot the relativistic corrections to the arrival time interval between successive pulses as a function of observer 'time', Ωt_R. Comment on the use of this graph, in view of the original assumption $M/b \ll 1$.

11.19 Use the expression for distances on a sphere, Eq. 10.2, to show that all the points on the line $\theta = 0$ in Figure 11.10 are the same physical point.

11.20 Derive Eqs. 11.59 and 11.60.

11.21 (a) Using the Schwarzschild metric, compute all the nonvanishing Christoffel symbols:

$$\Gamma^t{}_{rt} = -\Gamma^r{}_{rr} = \frac{M}{r^2}\left(1 - \frac{2M}{r}\right)^{-1}, \qquad \Gamma^r{}_{tt} = \frac{M}{r^2}\left(1 - \frac{2M}{r}\right),$$

$$\Gamma^r{}_{\theta\theta} = \Gamma^r{}_{\phi\phi}/\sin^2\theta = -r\left(1 - \frac{2M}{r}\right),$$

$$\Gamma^\theta{}_{\theta r} = \Gamma^\phi{}_{\phi r} = \frac{1}{r}, \qquad (11.124)$$

$$\Gamma^\phi{}_{\theta\phi} = -\Gamma^\theta{}_{\phi\phi}/\sin^2\theta = \cot\theta.$$

Show that all others vanish or are obtained from these by symmetry. (In your argument that some vanish, you should use the symmetries $t \rightarrow -t, \phi \rightarrow -\phi$, under either of which the metric is invariant.)

(b) Use (a) or the result of Exercise 6.35, § 6.8, to show that the only nonvanishing components of the Riemann tensor are

$$R^t{}_{rtr} = -2\frac{M}{r^3}\left(1 - \frac{2M}{r}\right)^{-1},$$

$$R^t{}_{\theta t\theta} = R^t{}_{\phi t\phi}/\sin^2\theta = M/r,$$

$$R^\theta{}_{\phi\theta\phi} = 2M\sin^2\theta/r, \qquad\qquad (11.125)$$

$$R^r{}_{\theta r\theta} = R^r{}_{\phi r\phi}/\sin^2\theta = -M/r,$$

plus those obtained by symmetries of the Riemann tensor.

(c) Convert these components to an *orthonormal* basis aligned with the Schwarzschild coordinates. Show that all components fall off as r^{-3} for large r.

(d) Compute $R^{\alpha\beta\mu\nu}R_{\alpha\beta\mu\nu}$, which is independent of the basis, and show that it is singular as $r \rightarrow 0$.

11.22 A particle of $m \neq 0$ falls radially toward the horizon of a Schwarzschild black hole of mass M. The geodesic it follows has $\tilde{E} = 0.95$.

(a) Find the proper time required to reach $r = 2\,M$ from $r = 3\,M$.

(b) Find the proper time required to reach $r = 0$ from $r = 2\,M$.

(c) Find, on the Schwarzschild coordinate basis, its four-velocity components at $r = 2.001\,M$.

(d) As it passes $2.001\,M$, it sends a photon out radially to a distant stationary observer. Compute the redshift of the photon when it reaches the observer. Don't forget to allow for the Doppler part of the redshift caused by the particle's velocity.

11.23 The first coordinate system for Schwarzschild that was regular at the horizon was found by P. Painlevé in 1921 and independently by A. Gullstrand in 1922, although at first it was not realised that the solution was Schwarzschild in a different coordinate system. We explore here how it can be derived from Schwarzschild.

(a) Define a new time coordinate T for Schwarzschild by $T = t - f(r)$, where the function $f(r)$ satisfies the differential equation

$$\frac{df}{dr} = -\frac{(2M/r)^{1/2}}{1 - 2M/r}.$$

Show that in the coordinate system (T, r, θ, ϕ) the metric now becomes

$$ds^2 = -\left(1 - \frac{2M}{r}\right)dT^2 + 2\sqrt{\frac{2M}{r}}\,dT\,dr + dr^2 + r^2\,d\Omega^2. \qquad (11.126)$$

(b) This is no longer singular at $r = 2M$, although the time coordinate still goes null there. In the two-dimensional subspace defined by $\theta = \pi/2$ and $\phi = 0$, draw a spacetime diagram with axes T and r, and sketch the light cones ($ds^2 = 0$) at different locations, paying particular attention to the slopes of the ingoing and outgoing null lines.

(c) Argue, on the basis of this diagram, that it is possible for particles outside the horizon $r = 2M$ to fall across the horizon in a finite time ΔT; that the horizon is a null surface; and that any particle or photon starting at $r < 2M$ cannot move outwards and reach the horizon.

11.24 Another coordinate system for Schwarzschild that is regular at the horizon is the Eddington–Finkelstein coordinate system. It was first written in the form we shall use it by Penrose around 1960, but it was based on earlier work by Eddington in 1924 and (independently) Finkelstein in 1958.

(a) Define a new radial coordinate $r*$ for Schwarzschild by the differential equation

$$\frac{dr*}{dr} = \frac{1}{1 - 2M/r}.$$

This is called the *tortoise coordinate*. Next define a new coordinate $v = t + r*$, which begins to remind us of Kruskal's coordinates. Show that in the coordinate system (v, r, θ, ϕ) the metric now becomes

$$ds^2 = -\left(1 - \frac{2M}{r}\right) dv^2 + 2\, dv\, dr + dr^2 + r^2 d\Omega^2. \qquad (11.127)$$

Importantly, we still use r here, not $r*$.

(b) This is no longer singular at $r = 2M$, although the v-coordinate still goes null there. In order to draw a diagram analogous to the one in the previous exercise, we now define a 'time' coordinate $T* = v - r$. In the two-dimensional subspace defined by $\theta = \pi/2$ and $\phi = 0$, draw a spacetime diagram with axes $T*$ and r, and sketch the light cones ($ds^2 = 0$) at different locations, paying particular attention to the slopes of the ingoing and outgoing null lines.

(c) Argue, on the basis of this diagram, that it is possible for particles outside the horizon $r = 2M$ to fall across the horizon in a finite time $\Delta T*$; that the horizon is a null surface; and that any particle or photon starting at $r < 2M$ cannot move outwards and reach the horizon.

11.25 A measure of the tidal force on a body is given by the equation of geodesic deviation, Eq. 6.87. If a human being will be crushed when the acceleration gradient across its body is $400\,\mathrm{m\,s^{-2}}$ per meter, calculate the minimum-mass Schwarzschild black hole that would permit a human being to survive long enough to reach the horizon on the trajectory in Exercise 11.22.

11.26 Prove Eq. 11.67.

11.27 Show that spacetime is locally flat at the center of the Kruskel–Szekeres coordinate system, $u = v = 0$ in Figure 11.12.

11.28 Given a spherical star of radius $R \gg M$ and mean density ρ, estimate the tidal force across it which would be required to break it up. Use this as in Exercise 11.25 to define the tidal radius R_T of a black hole of mass M_H, i.e. the radius at which a star of density ρ near the hole will be torn apart. For what mass M_H is $R_T = 100\, M_H$ if

$\rho = 10^3 \, \mathrm{kg\,m^{-3}}$, typical of our Sun? This illustrates that even some applications of black holes in astrophysical contexts require few 'relativistic' effects.

11.29 Given the area of a Kerr black hole, Eq. 11.85, with r_+ defined in Eq. 11.81, show that any two black holes with masses m_1 and m_2 and angular momenta $m_1 a_1$ and $m_2 a_2$ respectively have a total area less than that of a single black hole of mass $m_1 + m_2$ and angular momentum $m_1 a_1 + m_2 a_2$.

11.30 Show that the *static limit*, Eq. 11.80, is a limit on the region of spacetime in which curves with r, θ, and ϕ constant are timelike.

11.31 (a) Prove Eq. 11.87.
(b) Derive Eq. 11.89.

11.32 In the Kerr metric, show (or argue on symmetry grounds) that a geodesic which passes through a point in the equatorial 'plane' ($\theta = \pi/2$) and whose tangent there is tangent to the plane ($p^\theta = 0$) remains always in the plane.

11.33 Derive Eqs. 11.91 and 11.92.

11.34 Show that a ZAMO has four-velocity components $U^0 = A, U^\phi = \omega A, U^r = U^\theta = 0, A^2 = g_{\phi\phi}/(-D)$, where D is defined in Eq. 11.87.

11.35 Show, as argued in the text, that the Penrose process decreases the angular momentum of a black hole.

11.36 Derive Eq. 11.101 from Eq. 11.100.

11.37 (a) Use the area theorem to calculate the *maximum* energy released when two Schwarzschild black holes of mass M collide to form a Schwarzschild black hole.
(b) Do the same for black holes of mass m_1 and m_2, and express the result as a percentage of m_1 when $m_1 \to 0$ for fixed m_2.

11.38 The Sun rotates with a period of approximately 25 days.

(a) Idealize it as a solid sphere rotating uniformly. Its moment of inertia is $\frac{2}{5}M_\odot R_\odot^2$, where $M_\odot = 2 \times 10^{30}$ kg and $R_\odot = 7 \times 10^8$ m. In SI units compute $J_\odot$.
(b) Convert this to geometrized units.
(c) If the entire Sun suddenly collapsed into a black hole, it would form a Kerr hole of mass $M_\odot$ and angular momentum $J_\odot$. What would be the Kerr parameter, $a_\odot = J_\odot/M_\odot$, in cm? What is the ratio $a_\odot/M_\odot$? Physicists expect that a Kerr hole will *never* be formed with $a > M$, because centrifugal forces will halt the collapse or create a rotational instability. The result of this exercise is that even a quite ordinary star like the Sun needs to get rid of angular momentum before forming a black hole.
(d) Does an electron have too much angular momentum to form a Kerr hole with $a < M$? (Neglect its charge.)

11.39 (a) For a Kerr black hole, prove that for fixed M, the largest area is obtained for $a = 0$ (Schwarzschild).
(b) Conversely, prove that for fixed area, the smallest mass is obtained for $a = 0$.

11.40 (a) An observer sits at constant r, ϕ in the equatorial plane of the Kerr metric ($\theta = \pi/2$) outside the ergoregion. The observer uses mirrors to cause a photon to circle the hole along a circular path of constant r in the equatorial plane. Its world line is thus a null line with $dr = d\theta = 0$, but it is not, of course, a geodesic. How much coordinate time t elapses between the emission of a photon in the direction of increasing ϕ and its receipt after it has circled the hole once? Answer the same question for a photon sent off in the direction of decreasing ϕ, and show that this is a different amount of time. Does the photon return redshifted from its original frequency?

(b) A different observer rotates about the hole on an orbit of $r = $ const. and angular velocity given by Eq. 11.77. Using the same arrangement of mirrors, this observer measures the coordinate time that elapses between emission and receipt of a photon sent in either direction. Show that in this case the two terms are *equal*. (This is a ZAMO, as defined in the text.)

11.41 Consider the equatorial motion of particles with $m \neq 0$ in the Kerr metric. Find the analogs of Eqs. 11.91–11.95 using $\tilde{E}$ and $\tilde{J}$ as defined in Eqs. 11.5 and 11.6. Plot $\tilde{V}_\pm$ for $a = 0.5\,M$ and $\tilde{J}/M = 20$, 12, and 6. Discuss the qualitative features of the trajectories. For arbitrary a determine the relations among $\tilde{E}$, $\tilde{J}$, and r for circular orbits with either sense of rotation. What is the minimum radius of a stable circular orbit? What happens to circular orbits in the ergosphere?

11.42 (a) Given that the Galactic center is about 8.3 kpc away, show that an angular separation of 0.1 arcsec at this distance corresponds to a linear separation of 1.25×10^{10} km.

(b) Compute the Schwarzschild horizon radius for a black hole of mass $4 \times 10^6\,M_\odot$, and show that 0.1 arcsec therefore corresponds to about 10^4 such horizon radii.

11.43 (a) Derive Eq. 11.109 from Eq. 11.105 and the black-body law, luminosity $= \sigma A T^4$, where A is the area and σ is the Stefan–Boltzmann radiation constant, $\sigma = 0.567 \times 10^{-7}\,\mathrm{W m^{-2}\,K^{-4}}$.

(b) How small must a black hole be in order to be able to emit substantial numbers of electron–positron pairs?

Gravitational wave astronomy

12.1 Overview

The long period of development of high-sensitivity interferometric gravitational wave detectors (beginning in the 1970s), and the even longer period of working toward a good theoretical understanding of black holes (starting with Schwarzschild in 1916 and then becoming very active from the 1950s), both finally came together with the first direct detection of gravitational waves on 14 September 2015. The unexpectedly strong burst of waves from the inspiral and merger of two black holes excited both of the LIGO detectors in an almost identical fashion, as shown in Figure 12.1. The excellent agreement of the responses of these two widely separated detectors left little room for doubt that the event was real. Observational gravitational wave astronomy was born with this event.

This is a new way of doing astronomy, a new way of observing the Universe. It enriches our understanding of the Universe especially because gravitational waves are so different from electromagnetic waves – light, radio waves, X-rays, gamma-rays – which were almost our only information carriers before the first LIGO detections. Indeed, the two kinds of observing are strikingly complementary.

Gravitational waves have their source in the mass, momentum, and stress of objects, and they add up constructively over their sources in such a way that they carry information about the global characteristics of the sources, such as the total mass, size, and internal velocities. The typical frequencies of gravitational waves are the orbital frequencies around their astronomical sources, so our detection methods are designed to look for waves at frequencies from nHz to kHz. Electromagnetic waves, by contrast, are emitted by charges, and because matter is typically electrically neutral even on very small scales, the emission regions are tiny: individual particles, atoms, or molecules. They bring us local information about these small scales, such as atomic composition, temperature, density, and local magnetic fields. Their frequencies are high, MHz to EHz (exahertz, 10^{18} Hz) or beyond.

Gravitational waves pass through essentially everything, interacting with matter only in proportion to their amplitude h, which is normally tiny. This allows them to be received from regions which might be opaque to electromagnetic waves. But it also makes them impossible to focus in our detectors, so our detectors need to be, as we have seen, nearly omnidirectional. The weak interaction with our detectors means that we can only detect waves that carry extremely high energy fluxes. The sources, therefore, must be massive and relativistic. Electromagnetic waves, on the other hand, can be screened, scattered, and absorbed on their way to us, but they are also captured by our telescopes, allowing us to

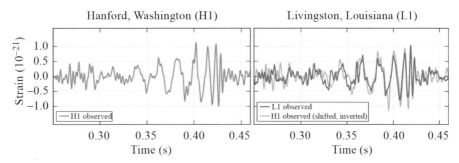

Hanford, Washington (H1)   Livingston, Louisiana (L1)

Figure 12.1 The time-series data from the two LIGO detectors. The left panel is from the Hanford, Washington, detector. The right shows the data from the Livingston, Louisiana, detector, along with (drawn more faintly) the Hanford data multiplied by −1 and shifted in time by a small amount in order to get the the best superposition. The factor −1 arises from the geometry of the two detectors; while aligned, they are effectively rotated by 90° relative to one another. The time shift reflects the location of the event on the sky; the difference in arrival times is the only information we have about where it might have been. The signals are consistent with one another to within the accuracy allowed by the detector noise. (From an image created by Caltech/MIT/LIGO Lab [CC0].)

determine directions to sources with great precision. All the energy of the captured photons is transferred to our detectors, so that we can detect very much less luminous sources. While they can also be highly relativistic, such as neutron stars radiating as pulsars or in X-ray binaries, they need not be.

Of course, there are also similarities between these two kinds of radiation. Both travel at speed c. When emitted from the same source at the same time, they arrive at the same time (provided the electromagnetic waves do not travel through a plasma that slows them down and disperses their frequencies). Because they travel at the speed of light, they both vibrate transversely to the propagation direction. And both can be gravitationally lensed; indeed, essentially the only way that a traveling gravitational wave can be brightened or dimmed is through lensing.

The complementarity between the two kinds of observing means that opportunities where sources emit both electromagnetic and gravitational radiation are highly prized. This is a form of what is called *multimessenger astronomy*. But the differences between the two types of radiation mean that we should not expect electromagnetic counterparts to all of our gravitational wave detections. And we cannot rely on electromagnetic observations to predict accurately the nature of possible gravitational wave sources.

Long before the first detections, physicists and astronomers had made great efforts to understand what kinds of sources of gravitational waves the current detectors might have been able to see. One motivation for this was the need to decide whether the large investment in these detectors was justified: theoretical studies of likely sources and estimates of likely rates of detection were needed. The second motivation for these studies was that the effective sensitivity of detectors depends on the accuracy with which waveforms can be predicted: a signal that can be recognized by pattern-detection methods against detector noise can be detected at lower amplitude and therefore from sources that are further away than a signal that has to stand up well clear of the noise in order to be identified. As with the building of large particle accelerators, which also need much theoretical input to justify their expense and to allow the recognition of new particle phenomena, the path to building LIGO and its partners involved considerable cooperation between experimental and theoretical physicists well versed in the fundamental theory.

While the detailed astrophysics of potential sources is beyond the scope of this textbook, we will review in the next section the basic classes of sources, learning what general relativity says about them, and understanding why they might be interesting to observe. In § 12.3 we will learn something about how the mathematical methods that filter signals out of detector noise use the detailed predictions of signal waveforms, some of which we made in Chapter 9. Finally, in § 12.4 we will review the detections made by LIGO and Virgo so far (as of mid-2021).

12.2 Astrophysical sources of gravitational waves

The principal detected or expected sources are generally grouped into four classes: binary systems, spinning neutron stars, bursts, and the Big Bang. After describing these, we consider whether there might be signals that do not fit these classes.

Binary systems

Binary systems are the simplest radiating systems, at least as long as the components' separation is large enough that they can be treated as 'point' masses, whose internal structure doesn't affect their orbital motion. Since general relativitistic effects scale with total mass, the important parameters governing a binary are the mass ratio $\epsilon = m_2/m_1$ (conventionally defined to lie in the range $0 < \epsilon \leq 0.5$, which means that m_1 is always taken to be the larger of the two masses), plus parameters such as the eccentricity and semi-major axis. Compared with the amount of information needed to model, for example, the interior of a star or the explosion of a supernova, this is simple indeed. That simplicity makes inferences from observations of binary systems more reliable and robust than in most other parts of astronomy.

We saw how to compute the expected wave amplitude from a binary system in Eqs. 9.98 and 9.99, as well as in Exercises 9.38 and 9.46 in § 9.7. These equations suffice to describe the observable radiation from binaries that are far enough apart that the orbital frequency doesn't change during the observation. For systems with solar-mass or larger components, this only applies at low frequencies, below the ground-based observing band. It is expected that the space-based LISA detector will detect many thousands of such systems in our Galaxy, most of them white-dwarf binary systems but including some widely separated binary neutron stars, and possibly even binary black holes. In fact there are expected to be so many white-dwarf binary systems at the lower end of the LISA frequency range, below 1 mHz, that their overlapping signals will create a kind of signal confusion noise larger than LISA's own instrumental noise.

If we consider the even lower band of frequencies, say tens of nHz, that are accessible to pulsar-timing searches for gravitational waves, the most likely signal to be detected first will be the confusion background created by huge numbers of binary systems throughout the Universe, consisting of two supermassive black holes orbiting one another, with masses in the range 10^8–10^9 M$_\odot$. On top of this background it is very possible that there will be one or more individually detectable long-lived binary systems of such supermassive black holes, again with no detectable change in the orbital frequency.

The opposite is true for ground-based detectors: the signals from all binaries in a particular band will be merging in that band, unless there are some objects with unexpectedly low masses. We can estimate the maximum frequency of a binary signal before merger as follows. In the point-particle approximation, the maximum frequency is always formally infinite, but a better estimate of the final frequency f_{final} of the merger is to compute the frequency when the two point particles have just half of one orbit remaining. At this point, there will be at most one cycle of gravitational radiation remaining, terminating at $\omega = \infty$, so any measurement of frequency will not record significant power above the frequency at the start of the last cycle. What is more, real mergers terminate before this. Let us see what this implies for black-hole and neutron-star systems.

We approach this calculation by solving Eq. 9.159 for the gravitational wave frequency $f_{\text{gw}} = \omega/\pi$, and normalizing to a solar-mass chirp mass:

$$f_{\text{gw}} = 8.0 \left(\frac{1 \, \text{M}_\odot}{\mathcal{M}} \right) \psi^{-3/5} \, \text{kHz}. \tag{12.1}$$

Then if we assume that the merger effectively happens for $\psi = \pi$, we get

$$f_{\text{final}} = 4.0 \left(\frac{1 \, \text{M}_\odot}{\mathcal{M}} \right) \text{kHz}. \tag{12.2}$$

The first gravitational wave detection, GW150914, which we will discuss further in § 12.4, had $\mathcal{M} = 30.8 \, \text{M}_\odot$ (Abbott *et al.* 2016b). This gives $f_{\text{final}} = 130$ Hz. The observed maximum was actually at about 250 Hz (Abbott *et al.* 2016a). This is probably as good an agreement as we should expect with this approximation, given that the black-hole system has become very complicated at this point, and it is likely that the horizon vibrations (see later) also play a role.

For two neutron stars of $1.4 \, \text{M}_\odot$ each, with $\mathcal{M} = 1.2 \, \text{M}_\odot$, the frequency at merger is predicted by Eq. 12.2 to be around 4 kHz, which would be challengingly high for LIGO and Virgo to observe. But in fact the signal will start to weaken around 1 kHz as the two fluid stars begin to merge together. This frequency is what the two point particles would emit when they have just 1.6 orbits remaining.

Clearly, LIGO's minimum observing frequency of about 20 Hz sets an upper limit on the masses that can be observed: when $\mathcal{M} = 200 \, \text{M}_\odot$, the merger will happen at a frequency below the LIGO band. Such systems have to be observed from space, as we remarked in § 9.3. Interestingly, as Sesana (2016) pointed out, a system with, for example, $\mathcal{M} = 30 \, \text{M}_\odot$ will in principle be observable by both ground-based detectors and by LISA, at different times. LISA could be observing it when its emission frequency is, say, 30 mHz, provided of course it is near enough to be above LISA's noise. To see how much time will then elapse before a ground-based detector could detect it, we write the time-to-merger expression Eq. 9.158 with an appropriate scaling:

$$\tau_{\text{merger}} = 2.9 \left(\frac{\mathcal{M}}{1\,\text{M}_\odot} \right)^{-5/3} \left(\frac{f_{\text{gw}}}{100\,\text{Hz}} \right)^{-8/3} \text{s}. \tag{12.3}$$

With $\mathcal{M} = 30\,\text{M}_\odot$ and $f_{\text{gw}} = 30\,\text{mHz}$, we get a τ_{merger} value of between 9 and 10 months: the system will leave the LISA band and enter the ground-based band within a year! It turns out that LISA should have enough sensitivity to detect similar systems with regularity, which opens up the exciting possibility that LISA will give advanced notice of mergers – including their sky positions and precise merger times – that ground-based detectors and electromagnetic telescopes can then prepare themselves to observe (Sesana 2016)!

Binary systems observable in gravitational waves by ground-based detectors can contain neutron stars and black holes, in any pairing. Although most stars in the Universe are members of binary or triplet systems, it is not trivial for such a system to survive the evolution of its component stars all the way to such compact objects. Neutron stars and black holes are usually the products of supernova explosions, and such explosions will typically have asymmetries that lead the new object to acquire momentum in its center-of-mass frame: a natal 'kick'. Particularly for neutron stars, this kick can be strong enough to disrupt the binary. In the best case, the kick is directed against the orbital velocity of the exploding star, leaving its remnant in an elliptical orbit with a periastron much closer to the companion.

To form a compact binary whose components are close enough that gravitational radiation will cause the orbit to decay in less than the age of the Universe, the system has to survive the second kick as well, and preferably again in such a way as to slow down the remnant. It should be no surprise, therefore, that merging compact binaries are rare. Yet there has for many years been evidence that at least some systems survive the first supernova. We see such systems as X-ray binaries: binary systems where a normal or giant star is in a close orbit with a black hole or neutron star, whose tidal forces are drawing gas from the normal star onto the compact object. Before reaching the compact object, the gas forms a disk around it, and friction in the disk gradually allows the gas to accrete onto the neutron star or into the black hole. The frictional heating raises the gas temperatures enough to make them strong sources of X-radiation. First discovered in the 1960s using sounding rockets, they have been studied extensively since the early 1970s by a series of satellite X-ray observatories. Riccardo Giacconi shared the 2002 Nobel Prize in Physics for pioneering this field.

Before LIGO, the only evidence for systems that had survived the second supernova was from radio observations of binary pulsars. The known ones were or are all double-neutron-star systems, so there was no direct evidence of binary black holes. In fact, for many years there was serious doubt among astronomers that ordinary stellar evolution would lead to any black-hole binaries. It was thought that X-ray binaries containing black holes and massive companion stars would not last long enough for the second supernova to happen and form the second black hole, because the massive companion would first expand into a giant star so large that it would engulf the first remnant, leading to what astronomers call a 'common envelope' system. In this envelope, gas friction would cause the remnant to spiral into the core of its companion before the giant phase ended. The result would be an exotic star indeed, with a black-hole core, known as a 'Thorne–Zytkow' object. This is not a scenario that leads to a gravitationally radiating binary. But this scenario has a loophole: if the binary stars were formed from low-metallicity gas, not polluted by the nucleosynthesis of previous generations of stars, then the giant would be smaller and consequently less able to provide enough friction to pull the first remnant all the way into the core before the second supernova happens. Low-metallicity gas is rare (but not unknown) in the present Universe, but black holes merging today may well have formed in the early Universe and then just spent the rest of the time spiralling ever closer together. LIGO and Virgo are, of course, detecting many black-hole binaries, so this loophole seems to be a likely way in which they form. See further discussions in the review by Postnov & Yungelson (2014).

Besides this, there are several other ways to form compact-object binaries. They can form by star–star interactions in dense stellar clusters such as globular clusters (Benacquista & Downing 2013) or the so-called nuclear clusters near the supermassive black holes at the centers of galaxies. Or they might form from stellar evolution in triple star systems, which are remarkably common; in these systems, the most distant of the three components can take energy and angular momentum away from the inner pair, encouraging them to form a very close binary. Even more exotic is the possibility that these binaries might be part of the mysterious dark matter that fills the Universe, and which we will discuss in the next chapter. In this model they would have to have formed very early in the Universe, from unusual density fluctuations; it seems a long shot from what we know today but it is not impossible.

Eventually, LIGO and its partners should be able to discriminate among various formation mechanisms for black-hole binaries, in part by using the spins of the black holes, which leave their imprint on the gravitational waveform, as we shall discuss in the next section. Binaries that form from the evolution of bound binary stars probably will end up with black-hole spins that are roughly aligned with each other and with the orbital angular momentum. But black holes that are captured in three-body interactions will show no correlation in spin direction.

Neutron-star binaries are less difficult to explain. In part this is the case because astronomers have been working to understand their formation mechanism ever since the Hulse–Taylor binary pulsar was discovered in 1974, in part because X-ray binaries provide good evidence of systems that are halfway along the road to a two-neutron-star system, in part because the original stars that form neutron stars are less massive than those that form black holes and are not expected to lead so often to the common envelope problem, and finally because for some years there has been a consensus that neutron-star binary mergers are the source of at least some gamma-ray bursts.

Bursts of gamma-rays have been observed since the 1960s, when they were first detected by satellites used by the US military to monitor nuclear weapons tests. Starting in the 1990s, satellite-based gamma-ray observatories have shown that the bursts come from very distant sources, and have recognized two different populations, called short and long bursts. Even before the first satellite observations, an association between bursts and neutron-star mergers had been proposed (Eichler *et al.* 1989). Since then, a consensus has emerged that at least some short bursts originate from binary neutron-star mergers and that they would be accompanied by optical explosions called macronovae or kilonovae. The observation of GW170817, described below, provided convincing verification of this model.

LISA, observing between 0.1 mHz and 100 mHz, will follow the coalescence and merger of black holes of masses in the range 10^4–10^7 M$_\odot$. For two equal-mass 10^6 M$_\odot$ black holes, the time to merger from when they enter the LISA frequency band at a frequency of about 0.1 mHz is of order a month, as shown in Exercise 12.2. Astronomers know that such black holes exist in the centers of most galaxies, including our own Milky Way. LISA will have sufficient sensitivity to see such mergers anywhere in the Universe, even back to the time of the formation of the first stars and galaxies. As we discussed in § 11.4, LISA's observations of black-hole mergers may be tracers of galaxy mergers, informing us about the way in which galaxies themselves formed and merged in the early Universe.

A merger of black holes and/or neutron stars is not finished when the objects first come into contact. Neutron stars may form a very massive star that most likely subsequently collapses to a black hole, but which might be a stable star if the nuclear-physics equation of state allows stars of that mass to become stable. A merger of a neutron star with a black hole might similarly have a short-lived intermediate stage where matter is supported in a disk around the hole but will very quickly become essentially a vacuum black hole. And two black holes will certainly merge to form another, very probably without the accompaniment of matter to send out electromagnetic signals. After the new single black hole has formed, it will oscillate for a very short time until it radiates away all its deformities and settles down as a smooth Kerr black hole (see Chapter 11).

This 'ringdown radiation' carries a distinctive signature that will distinguish the black hole from any neutron star or other material system. Typical gravitational wave frequencies ω are of order $0.5/M$; scaling this as before gives

$$f_{\text{ringdown}} \sim 16 \left(\frac{1\,\text{M}_\odot}{M} \right) \text{kHz}. \tag{12.4}$$

Notice that for the black hole formed in the merger GW150914, which has a mass of about 68 M$_\odot$ in the detector frame, this gives a frequency of about 230 Hz. This is very close to the observed maximum frequency during the merger event. The signal was not strong enough, however, for observers to isolate a ringdown part with confidence against the detector noise. Future observations should, however, allow direct detection of a ringdown.

Another exciting prospect is the detection by LISA of so-called extreme mass-ratio inspiral (EMRI) binaries. These are binary black-hole systems, where one component is a standard LISA-band black hole of mass perhaps 10^6 M$_\odot$, but the other is a stellar-remnant black hole with a mass of 50 or even just 10 M$_\odot$. These are not binaries from birth; rather, the smaller object starts out as a member of the star cluster around the central black hole of a typical galaxy, and random interactions with other stars and black holes near it happen to put it into an orbit that plunges toward the center. It can be captured into a highly eccentric orbit around the central hole, an orbit that subsequent stellar interactions can extract energy from, reducing the eccentricity (Amaro-Seoane 2018). The endpoint of this process, after gravitational radiation emission becomes the dominant way for the orbit to shrink, is that the radiation from the system enters the LISA band when there are still 10^5 or 10^6 orbits remaining before merger. LISA should be able to observer tens of such systems each year, out to 100 Mpc or beyond. By following the details of the orbital evolution, including orbital-plane rotations caused by the gravitational spin–orbit and spin–spin interactions, LISA will be able to make exacting tests of general relativity, well into the strong-field region, which post-Newtonian approximations cannot address.

Spinning neutron stars

We mainly observe neutron stars as pulsars, whose spin sweeps a beam of electromagnetic radiation past the Earth each time they turn. Known pulsars spin rapidly, most of them at frequencies above 20 Hz, and if these are also radiating gravitational waves then they would be in the observing band of ground-based detectors. We call these continuous-wave sources. There could in principle be many other neutron stars that also spin this rapidly but are not known as pulsars, because their beams do not cross the Earth, or because they do not emit such radio beams. And, because radio surveys for pulsars only cover a neighborhood of the Sun in our Galaxy, there could be more distant pulsars in the Galaxy that are not yet known.

Such stars would radiate gravitational waves if they were not symmetric about their rotation axis. Pulsars are clearly not symmetric, since they beam their radiation somehow. But it is not clear how much mass asymmetry is required to produce the beaming. Other sources of asymmetry could include frozen-in irregularities in the semi-solid outer layer of a neutron star (called its 'crust'), or in a possible solid core. It is also known that spinning neutron stars are vulnerable to a gravitational-wave-driven instability called the r-mode instability, which could produce significant radiation (Kokkotas & Schwenzer 2016).

We can compute the radiation due to a given mass asymmetry from Eqs. 9.84–9.86. If the star is nearly axisymmetric then we can approximate the amplitude of either of the polarizations radiated along the spin axis by the formula

$$h \sim 2\varepsilon\Omega^2 I_{NS}/r, \tag{12.5}$$

where I_{NS} is the moment of inertia of the spherical neutron star and ε is a measure of the fractional asymmetry of the star about its spin axis. If we use the typical values $I_{NS} = 10^{38}\,\mathrm{kg\,m^2}$, $r = 1$ kpc, and $\Omega = 2\pi f$ with $f = 60$ Hz, and if we take the asymmetry parameter to have the small value $\varepsilon = 10^{-6}$, then we get $h \sim 10^{-26}$. This is a very small amplitude, but not impossibly small for detection.

One finds such small signals by taking long stretches of data and filtering for them, essentially by performing a Fourier transform, as we discuss in the next section. The Fourier transform concentrates the power of the signal in one frequency band while distributing the noise power of the data stream over the whole observing band. To go from the Advanced LIGO broadband sensitivity of around 10^{-22} to a sensitivity of 10^{-26} for a narrowband signal like the one we are considering here, the number of cycles of the waveform must be equal to at least the square of the ratio of these two numbers, or 10^8. At this frequency, this would take of order a month or longer. Putting this another way, the sensitivity h_{min} of searches for continuous signals improves as the square root of the duration T of the data being searched: $h_{min} \propto T^{1/2}$.

Pulsar searches are the computationally most demanding data analysis tasks for gravitational wave detectors, because the motion of the detector over periods of even a few hours or more induces a time-dependent frequency modulation of the signal that depends on its location on the sky. It is not yet possible to do an all-sky search for as yet unknown neutron stars optimally for such very weak signal amplitudes (see the next section). Searches performed so far (mid-2021) on LIGO and Virgo data, looking for signals across the frequency spectrum from 20 Hz to 2 kHz, have not found any signals. The most recent (Abbott *et al.* 2018b, Dergachev & Papa 2019a, 2019b) set bounds on ε as low as 10^{-7} in some frequency ranges, and on amplitudes as low as 1×10^{-25}.

It is easier to search for gravitational waves from a pulsar whose frequency and sky position are already known from radio observations. The sensitivity is therefore somewhat better. But even so, as yet, there are no detections. The best upper limits can be found in Abbott *et al.* (2017a). For the Crab pulsar, for example, the upper limit on h is below 10^{-25} and the upper limit on ε is a few times 10^{-5}. For some other pulsars the upper limits on amplitude and asymmetry are as low as a few times 10^{-26} and 10^{-7}, respectively. This indicates a remarkable level of axisymmetry of neutron stars.

The gravitational waves emitted by a spinning neutron star will carry away rotational energy and lead to a slowing down of its spin rate. This is called spin-down. Other effects can also spin a pulsar down, such as a loss of energy to radio and other electromagnetic radiation, and the emission of high-energy particles. Most known pulsars do indeed spin down, so a measurement of the spin-down sets an upper limit on the possible amplitude of gravitational waves. Of course, the true amplitudes will be much smaller if other effects are dominant. For the Crab, this seems to be the case: the current upper limit on its amplitude is only about 4% of what it would be if all the spin-down were due to gravitational wave energy loss (Abbott *et al.* 2017a). Since energy is proportional to the square of the amplitude, we can be sure that less than 0.1% of the Crab pulsar's spin energy loss is due to gravitational waves.

As detectors improve, searches will continue to be made for the first continuous-wave source in the Galaxy.

Gravitational wave bursts

The objective that motivated Joseph Weber to develop the first bar detector was to register waves from a supernova. The spectacular optical display of a supernova explosion masks what really happens inside: the compact core of a giant star, having exhausted its supply of energy from nuclear reactions, collapses inward, and the subsequent dynamics can convert some of the energy released into the explosion that blows off the envelope of the star. But what happens to the collapsing core, and how that energy is converted into the explosion, is not well understood because it is impossible to observe the core directly. Gravitational waves, along with neutrinos (which were detected from SN1987a, which happened in the Large Magellanic Cloud, a satellite dwarf galaxy of the Milky Way), provide the only probes that have come to us directly from the core.

However, our ignorance of the details of the presumably chaotic physical conditions in the collapsing core means that it is not possible to predict a detailed waveform, so it is not possible to use the techniques of matched filtering, as described in the next section, that help us find binary and pulsar signals well below the detector noise level. So, searches for signals from gravitational collapse involve looking for bursts of radiation visible above the noise. Such searches could in principle find signals from any unpredicted or un-modeled source, so they are known as burst searches.

Three-dimensional hydrodynamic simulations of gravitational collapse leading to supernova explosions are an active area of research at present, using both the Newtonian approach and full GR for the gravitational field. There seems to be considerable agreement that amplitudes are going to be rather small and some hints that there could be spectral features, some associated with the pulsation modes of the neutron-star core. It is likely that a supernova that occurs in our Galaxy or in one of its companions (like SN1987a) would be detectable by current detectors. Such an explosion would also be very likely to be detected by one or more of several operating neutrino detectors, and this would be crucial to establishing the time of the collapse, since the visible supernova does not appear until the explosion reaches the surface of the star several hours later. Given a coincidence with neutrino detectors, one would be able to have more confidence in a gravitational wave detection even close to the level of the noise.

Other sources of 'bursts' of gravitational waves are rather speculative, since we do not have good astrophysical evidence for their existence. Cosmic strings are frequently suggested. There are plausible theoretical reasons to expect them to have been formed in the early Universe, within certain theories of the fundamental interactions. See the short review by Vachaspati *et al.* (2015). When they collide they reconnect, and the motions of the strings radiate gravitational waves. So far (mid-2021), searches by the LIGO and Virgo teams have not found any (Abbott *et al.* 2018a).

Stochastic backgrounds of gravitational waves

Because essentially every astrophysical body emits gravitational waves constantly or intermittently, there is a random background of gravitational waves in the Universe across all frequencies. Of more fundamental significance is the possibility that there are also random gravitational waves as relics of the Big Bang. The cosmic microwave background (CMB) is an analogous background in the electromagnetic spectrum. But, as we shall see in Chapter 13, the CMB originated hundreds of thousands of years after the Big Bang, when the temperature in the expanding plasma of the Universe fell low enough for the electrons to be captured into hydrogen and helium atoms, allowing photons to move freely for the first time. No such obscuration applies to gravitational waves, to which the Universe has been transparent since the very beginning. So a cosmic gravitational wave background could give us our first glimpse of the Big Bang, and of the poorly known physics of that epoch.

We call these backgrounds *stochastic*: so many different sources contribute to them that the gravitational wave amplitude in any one location will be truly stochastic. But the amplitudes at two different locations can be correlated, provided the spectrum of stochastic waves contains enough power at wavelengths comparable with and longer than the separation between the detectors. The reason is that many of the waves creating the background at such frequencies will be affecting both locations at the same time. The closer these two locations are, the stronger their correlation because higher and higher frequencies become correlated. That means that if we place gravitational wave detectors at those locations, their responses will be correlated. This is the basis for searching for stochastic backgrounds: cross-correlating the outputs of pairs of detectors. We will discuss in more detail how this is done in the next section, but the sensitivity to amplitude h improves with the duration T of the data being correlated, by the fourth root of T: $h_{\min} \propto T^{1/4}$. This is a slow rate of improvement, but nevertheless a steady one.

Because the radiation is random, there is no characteristic value of h that can be assigned to its amplitude. Instead, the amplitude has a characteristic mean square value $\langle h^2 \rangle$, which in the next section we will call $|h|^2$. Since the energy density of the radiation is proportional to this times f^2, it is customary to use its energy density as a measure of its strength. And since we expect this energy density to be the same everywhere, it is also conventional to express the energy density as a fraction Ω_{gw} of the total mass–energy density of the Universe.

In practice, the most sensitive pair of detectors for measuring a stochastic background is the pair of LIGO instruments in the USA. They have a separation of about 10 ms, so that gravitational waves at frequencies below about 100 Hz should be correlated. Higher-frequency waves have to arrive from a direction close to being normal to the separation of the detectors in order to produce a correlated signal. The frequency of 100 Hz is in the middle of the Advanced LIGO frequency band. Other detector pairs are more widely separated and therefore are only correlated at lower frequencies, below the best sensitivity of the detectors; and they are not so well aligned as the two LIGO detectors, so they measure different polarization components.

The frequency of 100 Hz should be good for measuring a background. The binary black holes that LIGO and Virgo have been detecting indicate that the Universe has seen a large number of such coalescences, at frequencies near and below 100 Hz, so many that distant ones blend into one another and create a stochastic background that may well be detected over the next few years. Figure 12.2 shows how strong that background might be and what LIGO's future sensitivity to it might be. The upper curve at the left of the figure shows LIGO's sensitivity in 2017, which sets an upper limit of $\Omega_{\mathrm{gw}} < 1.7 \times 10^{-7}$ (Abbott *et al.* 2017b). The lower curve at the left of the figure shows that ultimately Advanced LIGO should be able either to detect or set an upper limit on a background at a level of about 10^{-9}.

LISA will also be able to make observations of the background. For LISA, the background would have to be stronger than the instrumental noise: correlation gains it nothing, as we explain later in our discussion of detecting stochastic backgrounds. But LISA will have very good sensitivity in its waveband, and it seems likely that it would be able to detect a background below about $\Omega_{\mathrm{gw}} < 10^{-10}$. As mentioned above, LISA will be able to observe a stochastic background stronger than this that is a superposition of radiation from all the close white-dwarf binaries in the Galaxy and is expected to be dominant below 0.1 mHz. Predictions of what LISA might detect at higher frequencies are still uncertain.

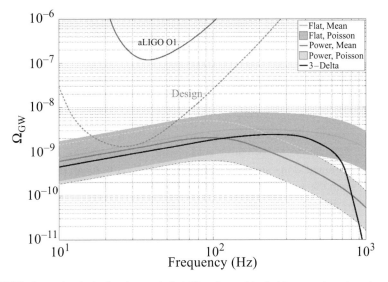

Advanced LIGO's observations in the first data run (called O1) can be combined with assumptions about the frequency-dependence of the spectrum of a stochastic background to make predictions (the various shaded regions) of the strength of a background. The two curves at upper left show LIGO's sensitivity during O1 and its design sensitivity, which should be reached in the early 2020s. (Abbott *et al.* 2017b.) Copyright 2017 by the American Physical Society.

Pulsar timing observations are expected to find, before any single binary, a background generated by the orbits of supermassive black holes in the centers of galaxies throughout the Universe.

The stochastic radiation we have been discussing so far, coming from merging black holes and binary white-dwarf systems, is known as the 'astrophysical background'. It is generated by systems that are part of the normal composition of the Universe and that have emitted their radiation relatively recently, say in the last half of the age of the Universe. Interesting as observing such a background will be for the information it should bring us about populations of radiating relativistic objects, a more fundamental goal is to detect a *cosmological* stochastic gravitational wave background, as we now explain.

The study of the large-scale structure of the Universe, and its history, is called cosmology, and it will be the subject of Chapter 13. Cosmology has undergone a revolution since the 1980s, with a huge increase in data and in our insight into what went on in the early Universe. Part of that revolution impacts on the study of gravitational waves: it seems very probable that the very early Universe was the source of a random sea of gravitational radiation that even today forms a background to our observations of other sources. It is analogous to the cosmic microwave background radiation discovered, by accident, in 1965 by Penzias (1979) and Wilson (1979) in a radio receiver at Bell Labs, a discovery for which they were awarded the Nobel Prize for Physics in 1978.

The cosmological stochastic gravitational radiation originated in a host of individual events too numerous to count. The waves, superimposed now, have very similar character to the random noise that comes from instrumental effects. Although the radiation was intense when it was generated, the expansion of the Universe has cooled it down, and one of the most uncertain aspects of our understanding is how to predict the intensity that it has today.

Theoretical predictions of the energy density of this radiation at LISA and LIGO frequencies vary hugely, from below 10^{-15} up to 10^{-8} and higher. This reflects the uncertainty in theoretical models of the physical conditions and indeed of the laws of physics themselves during the early Big Bang, and demonstrates the importance that detecting a background would play in constraining these models. Indeed, an observation of the random (stochastic) background of gravitational waves would possibly be the most important observation in gravitational wave astronomy.

Perhaps the best way to look for this radiation is through the way in which it affects the polarization of the cosmic microwave background when it was formed. We will return to this in the next chapter. Detecting at least a weak cosmological background with detectors like LIGO, LISA, and pulsar timing seems less likely, because it could easily be masked by a stronger astrophysical background. The best frequencies for searching are those where the Universe's astronomical systems are quietest, and that is probably at high frequencies, say about 0.1 Hz or higher. But the predictions of its strength place the cosmological gravitational radiation background out of reach even of the next-generation 3G ground-based detectors.

Everything else

One cherished hope of gravitational wave detection is to discover a class of sources that had not been anticipated. Most fields of astronomy have made important and unexpected discoveries, from giant radio galaxies to X-ray binary systems to gamma-ray bursts. It would be surprising if the gravitational wave spectrum contained only signals from the classes we have described above: binaries, spinning neutron stars, supernova explosions, a stochastic background.

To be clear, these classes could certainly reveal surprises: deviations from the predictions of general relativity, unexpected properties of nuclear physics that lead to strong bursts from supernovae or to unusual behavior of neutron stars, strong stochastic radiation from an unexpected class of sources distributed through the Universe or from an unexpected process in the very early Universe, and so on. Burst searches do not make strong assumptions about the signal morphology, so they could in principle detect any signal that was of reasonably short duration, even something that had not been anticipated. But it would have to be strong enough to appear against the full detector noise, because burst searching cannot use the advanced filtering methods that we describe in the next section to dig into the noise. This means that the chances of finding unexpected sources gets better as detectors improve in sensitivity.

Where would these totally unexpected signals come from? One obvious possibility is that there are some astrophysical systems that have not revealed themselves in astronomical observations so far but that will turn up in gravitational waves. As we explain in Chapter 13, baryons (protons and neutrons) make up only 4% of the total mass-density of the Universe, yet essentially all the sources we search for are baryonic in origin. The simple reason for this is that some baryons carry charge and radiate the electromagnetic waves that have so far painted our picture of the composition of the Universe. It seems likely that this is an accurate picture, but it is possible that it leaves things out. If the dark matter filling the Universe consists of a variety of constituents, some of which are compact and massive, then there could be some very surprising sources. Cosmic strings, already mentioned above, are a possible exotic constituent, and we will return to them in § 13.1. LIGO and Virgo already conduct regular searches for strings.

12.3 Finding weak signals in noise: what is a detection?

The huge effort and expense of developing and building the current ground-based detectors demand that the analysis of the data should not miss any detectable signal. This raises the question of what is detectable. Any physical measurement is a statement of probability: the outcome is a value X with an uncertainty statement Y, which indicates the probability that the true value of the quantity lies within the uncertainty range about X. Every detection made by LIGO and Virgo is accompanied not only by measurements of parameters (masses, spins, distance, . . .) and their uncertainties, but also by a statement of the probability that it is a 'false alarm' – an artefact of the detection system and not a true gravitational wave.

These probability statements are made with care, and determining the uncertainties takes up most of the computing cycles that go into announcements of detections of binaries. In this section we will look at the fundamentals of how the measurements are made and their reliability assessed. We will first look at the response of interferometers to the incoming wave. Then we discuss the theory of digging into the detector noise in order to make detections and assign uncertainties. Finally we look at how these principles are applied to searches for the different classes of signal that we discussed in the previous section.

How interferometers respond to a gravitational wave

We begin our study of signal analysis by saying more about the data itself. Interferometers have antenna patterns, with different sensitivities in different directions. Incoming waves contain two polarizations, to which the antenna will have different responses. To explain how these two properties are folded together to produce an output response $h(t)$, we follow the treatment in the review by Sathyaprakash & Schutz (2009).

First we establish the coordinates and basis vectors with which to describe the geometry of the detector and the polarization of the wave. Suppose the wave arrives at the detector from a direction (θ, ϕ) on the sky, using spherical coordinates based on the detector, so that the detector lies in the equatorial plane and the detector's zenith direction (straight overhead) is the north pole of the spherical coordinates. The geometry and the definitions of the unit basis vectors are given in Figure 12.3.

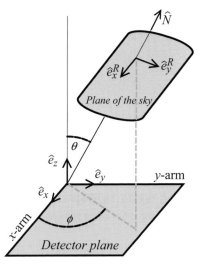

Figure 12.3   Coordinates for the detector plane and the sky. The signal comes from the direction indicated by the unit vector $\hat{N}$, which has coordinates (θ, ϕ) in a spherical coordinate system constructed such that its north pole is the direction perpendicular to the detector's plane (the unit vector $\hat{e}_z$) and its equatorial plane is the plane of the interferometer, whose arms point in the x and y directions. For the incoming wave we set up a local Cartesian coordinate system whose z direction points along $\hat{N}$ and whose other two unit basis vectors $(\hat{e}_x^R, \hat{e}_y^R)$ form a reference system for describing the two polarization components of the wave. The local plane perpendicular to $\hat{N}$ is called the 'plane of the sky'. (Image courtesy of B. Sathyaprakash and *Living Reviews in Relativity*.)

The incoming wave is described in terms of a different basis, with its polarization expressed in terms of vectors in the plane of the sky (perpendicular to the direction of travel of the wave), as shown in the same figure. It is important to keep these wave-related basis vectors (which we call $\hat{e}_x^R$ and $\hat{e}_y^R$) independent of the detector's basis frame, because we will want to consider the detection of a single wave by a network of several detectors, all oriented differently. In its frame, the incoming wave will have two polarization amplitudes $h_+(t)$ and $h_\times(t)$. It is convenient to combine these into a single TT tensor by defining the $+$ and $\times$ basis tensors:

$$\mathbf{e}_+ = (\hat{e}_x^R \otimes \hat{e}_x^R - \hat{e}_y^R \otimes \hat{e}_y^R), \quad \mathbf{e}_\times = (\hat{e}_x^R \otimes \hat{e}_y^R + \hat{e}_y^R \otimes \hat{e}_x^R), \tag{12.6}$$

where $\hat{e}_x^R$ and $\hat{e}_y^R$ are shown in Figure 12.3. (Recall that the outer product $\otimes$ was defined for general tensors in § 6.1.) Then the spatial components of the gravitational wave metric tensor can be written

$$\mathbf{h}(t) = h_+(t)\mathbf{e}_+ + h_\times(t)\mathbf{e}_\times. \tag{12.7}$$

In a similar way, we can define a *detector tensor* that encapsulates the interferometer's geometry:

$$\mathbf{d} = L(\hat{e}_x \otimes \hat{e}_x - \hat{e}_y \otimes \hat{e}_y), \tag{12.8}$$

where L is the arm-length. Then the output displacement $\Delta_{\text{int}}(t)$ of the interferometer (the difference between the stretching of the x-arm and that of the y-arm) is given by the elegant expression

$$\Delta_{\text{int}}(t) = \mathbf{d} : \mathbf{h} := d^{jk} h_{jk}, \tag{12.9}$$

Exercise 12.3 explores this in more detail.

The response $\Delta_{\text{int}}(t)$ is a function of several angular variables as well as of the wave amplitudes. We separate some of these variables explicitly in the following convenient expression for the *dimensionless* strain of the detector, which we shall call δ_{int}:

$$\delta_{\text{int}}(t) := \Delta_{\text{int}}(t)/L = F_+(\theta, \phi)h_+(t) + F_\times(\theta, \phi)h_\times(t), \tag{12.10}$$

where $F_+ := \mathbf{d} : \mathbf{e}_+/L$ and $F_\times := \mathbf{d} : \mathbf{e}_\times/L$ are functions that depend only on the geometry. By separating the h_+ and $h_\times$ contributions we have arrived at coefficients F_+ and $F_\times$ that depend only on θ and ϕ, not on the variables in the plane of the sky. It will be important to remember later, when we discuss networks, that these spherical angles are referred to the detector's zenith, so a given source will be at different values of these angles as seen from different detectors.

The mean dependence of the detector's sensitivity on the incoming direction of the wave can be visualized if one plots $F_+^2 + F_\times^2$ as a function of direction on the sky. This is known as the interferometer's antenna power pattern, because by squaring the response we get something that reflects h^2, proportional to the power of the incoming wave. We plot it in panel (a) of Figure 12.4. It should be noted that this antenna pattern gives the average response to a short burst of gravitational waves. If a signal has a long duration, as from a spinning neutron star, then the motion of the detector as the Earth turns and orbits the Sun can produce a very different sky sensitivity pattern.

A network of detectors has a joint response pattern that, if all the detectors are identical, is just the sum of the antenna power patterns of the individual detectors, provided they are oriented appropriately for their locations on Earth, as we pointed out two paragraphs earlier. In panel (b) of Figure 12.4 we show this for the LIGO–Virgo network. In fact, because Virgo has a lower intrinsic sensitivity than the two LIGO detectors (as we shall discuss next), the true response patten is somewhat different.

Its geometry is only part of the story of the sensitivity of a detector. More fundamental is its intrinsic sensitivity to the strain produced by a gravitational wave. This is usually visualized by plotting a function called $[S_h(f)]^{1/2}$, which shows the r.m.s. amplitude of the detector noise as a function of frequency. It is plotted against frequency because, as we have seen, the frequency window of a detector determines what kinds of sources it will be able to observe. Ideal random noise, such as photon shot noise or thermal vibration noise, is statistically stationary in time but does have frequency dependence, so a frequency plot is informative about where the noise originates.

Most of the noise shown in Figure 12.5 is ideal Gaussian noise: its amplitude at every sampled frequency is a random number drawn from a normal distribution having zero mean and a standard deviation equal to the value of the curve in this figure. This simple noise behavior follows from the central limit theorem, which says that if one has a large enough collection of noise sources, each having its own statistical distribution, their sum will be

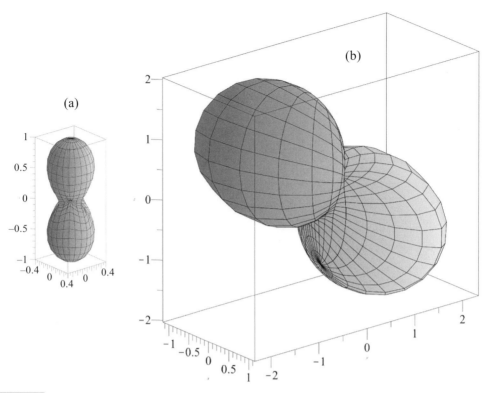

Figure 12.4 Interferometer antenna power patterns. The distance of a plot from its origin in any direction is proportional to the square of the response to a wave coming from that direction, averaged over all polarizations. (a) A single interferometer's response is strongest to waves from the zenith and is zero for waves arriving in its plane along the bisector of its arms. (b) The joint response of a network of two LIGO interferometers and Virgo, assuming all have the same intrinsic sensitivity, and drawn on the same scale. The viewpoint is from midway over the northern Atlantic Ocean. Although the network's sensitivity is greater than that of a single detector, there are still directions where its sensitivity is weak. Adding further detectors in Japan and India will fill in these holes and produce a more nearly spherical pattern, but not one with much larger sensitivity in its best direction. From Schutz (2011) *Class. Quantum Grav.* **28** 125023.

normally distributed. The sources of noise in detectors are so numerous that the central limit theorem is indeed satisfied for most of the noise: the thermal vibrations of the atoms of the mirrors, the random arrivals of the photons in the arms during one data sampling period, and the random seismic waves all jostling the mirror suspensions' foundations at the same time.

For Gaussian noise, what we need to know is how loud it is at different frequencies: its standard deviation. The job of the experimenters is to make this as small as possible. All the curves in Figure 12.5 have a similar shape. They have a minimum flat region around a few hundred Hz, and they rise at lower and higher frequencies. The rise at higher frequencies is due to photon shot noise. High frequencies require higher sampling rates, so that the sampling interval is smaller, and consequently the number of photons N that arrive at the photodetector during one sampling interval scales as $1/f$. The noise then rises as $N^{-1/2}$, or $f^{1/2}$.

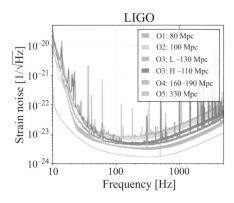
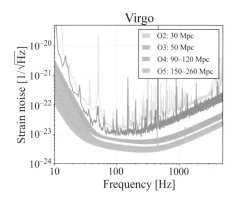

Figure 12.5 Noise amplitude spectrum for LIGO (left) and Virgo (right). Curves are plotted for different observation runs to show how the performances of the detectors have improved. The curves are in the order corresponding to the insert boxes; thus on the left the top curve is O1 and on the right the top curve is O2. As this is being written (mid-2021) we are in the middle of the O3 run. The curves for O4 and O5 (the lowest two curves in each figure) are projections. The insert boxes give the detection range for neutron-star binary systems, which is sometimes taken as a benchmark for sensitivity improvements. The plot is the square root of the mean Fourier transform power spectrum of noise in the measurement of h for a 1-s stretch of data. A plot of the mean power spectrum would give the mean square expected noise per 1-Hz bin on the horizontal axis. Instead, we plot here its square root because we want to compare the noise amplitude with the expected gravitational wave amplitudes. This explains the vertical axis units. Abbott *et al.* (2020h) *Class. Quantum Grav.* **37**, 055002. (CC BY 4.0). DOI: 10.3847/2041-8213/aa920c © AAS. Reproduced with permission.

At low frequencies, the noise rises because it is harder and harder to suppress seismic vibrations. In between, thermal noise from the mirror surfaces and suspensions sets the sensitivity limit at the best observing frequencies. One can also see many 'lines' (narrow peaks at certain frequencies) in the spectra. Some are due to thermal vibrations of the mirror suspensions. These are very narrow and all are plotted with 1-Hz widths, because the data segments used are only one second long. But in fact they are much narrower and so they can be cut out of the data without disturbing significantly any searches for signals with broader spectra. Other lines are broader, some due to electrical interference (60 Hz in LIGO, 50 Hz in Virgo, and harmonics), which is difficult to remove, so the data analysis just has to accept a loss of information in these regions.

Many other kinds of noise lurk below these dominant sources, and they affect the choices of how interferometers are configured. For example, one could reduce shot noise by increasing laser power, and up to a certain point this is the strategy for the LIGO and Virgo detectors. But too much power begins to heat the mirrors, making them vibrate, which adds in a different kind of noise. Furthermore, a high laser power can apply light-pressure to the mirrors, whose motions would add further noise. The KAGRA design, as we mentioned in § 9.3, involves cryogenically cooling the mirrors to reduce noise in the central frequency region; but this sets a limit on the laser power, because mirrors absorb a certain amount of the laser energy incident on them. So then the shot noise rises. Trade-offs like these make the design of interferometers a delicate art, but they also mean that technology improvements and materials-science improvements can suggest design changes.

As detectors on Earth go toward lower frequencies, and especially when third-generation detectors are built, so-called 'Newtonian noise' becomes an issue. This is due to fluctuations in Earth's gravitational field, which couple to detectors through their tidal forces, just as do gravitational waves. Therefore detectors cannot be isolated from them. They can be caused by nearby moving masses, such as people or vehicles, and by changes in the local mass density produced by seismic waves traveling past. The mechanical vibrations due to such waves are suppressed by passive and active (feedback-controlled) systems in current detectors, but at low frequencies the gravitational disturbances are more problematic. The amplitudes of such waves are typically lower deep underground, which is why the KAGRA detector has been built inside a mountain, and why the proposed European third-generation detector is planned to be built deep underground. For a discussion of these issues see the review by Harms (2019). At frequencies below 1 Hz, as we remarked in Chapter 9, Newtonian noise is very large compared with the amplitudes of incoming gravitational waves, and this is the principal reason why LISA is a space mission: the Earth is gravitationally just too noisy.

Adding to the complications, not all the noise is Gaussian. Occasionally there are loud bursts of noise on top of the background Gaussian noise. These are called 'glitches'. They arise from the complexity of these detectors, which have thousands of automated control systems and which contain thousands of electronic circuits that occasionally malfunction. This is a nonstationary noise. Unlike Gaussian noise, whose mean properties are independent of time, glitches come and go and each glitch can have a different morphology. Dealing with glitches and making sure that they are not mistaken for real signals takes up a major part of the data analysis activity of a collaboration. It is likely that, as more detectors join a collaboration, it will be possible to cross-check data streams and reject glitches. At the present time (mid-2021), however, the sensitivity of searches for short bursts of gravitational waves is limited primarily by the glitch background. We will discuss this as part of our next topic: how we actually look for the signals.

Finding signals by filtering for them

The first LIGO detection, in the data segment shown in Figure 12.1, was remarkable not least because it was so strong. By the time of the merger, the signal stood well above the broadband time-series noise that dominates the data at the beginning of the segment. It reached a strain amplitude of 10^{-21}. Most signals that LIGO has subsequently detected are much weaker. The signal of the binary neutron star merger GW170817 never exceeded 10^{-22}. If that had been in the data of Figure 12.1, one would not have noticed it. Yet GW170817 is actually the highest signal-to-noise detection made so far (mid-2021), with the greatest statistical significance. That significance was achieved because it was possible to use prior information about the expected waveform of such a signal in order to filter out most of the noise and expose the signal when it arrived. How that works is our subject here.

As remarked before, the binary waveform depends on very few variables and is very reliably known, so it is possible to predict what it will look like, apart perhaps from the last few cycles, where the merger makes it complicated. So let us suppose, in order to introduce the relevant ideas, that we are looking for a waveform $w(t)$ (called our *template*) and that we know some sources radiate exactly this waveform, but there are two things we don't know: when the signal will arrive, and how far away the source is. So there are two parameters, an arrival time t_0 and an amplitude A, which we cannot control. Let us also assume that the waveform is finite in length with a duration T that is short compared with the total duration of our observation.

Gravitational wave data is sampled at discrete times, with regular spacing Δt. Typical sampling rates might be 8 or 16 kHz. The full sampled data set consists of M values that we call $\{x_j;\ j = 0, \ldots, M - 1\}$. The waveform template of duration T has $N = T/\Delta t$ sampled points, and is represented by the set of values $\{w_j;\ j = 0, \ldots, N-1\}$. For binary-star signals one expects $N \ll M$. Suppose the arrival time of the signal described by this template is t_0, associated with the sampling time t_D for some fixed sample index D. Because $N \ll M$, we extend the template w by defining $w_j = 0$ if $j < 0$ or $j \geq N$. The full sampled detector data containing a signal described by the template can be written as

$$x_j = n_j + A w_{j-D}, \tag{12.11}$$

where $\{n_j\}$ is the Gaussian-distributed noise, with standard deviation σ. If the signal is absent then we simply set $A = 0$. We shall assume that the noise has zero mean (negative and positive fluctuations are equally likely), which is normally the case.

The simplest way to detect the signal is just to watch the data coming from the detector, and to try to recognize the signal if its maximum is above the random noise. We can quantify this as follows. If the largest value of w_j is w_K for some fixed sample index K, then we should expect to detect a signal if its amplitude is somewhere above the threshold

$$A_{\min} = \frac{\sigma}{w_K}. \tag{12.12}$$

If the signals are rare, so that we have to watch a lot of data go by before we catch a signal, then we have to raise the threshold higher to avoid being trapped by a random large noise fluctuation, which we call a false alarm.

Because signals from binaries of massive black holes at any given distance are much stronger than those from neutron-star binaries, the signal detection criterion in Eq. 12.12 disfavors neutron stars. Fortunately, there is a better way to do the data analysis, which is available to us if we have prior knowledge of the waveform, and which improves our sensitivity particularly to the long wave trains we detect from neutron-star binaries. This method is called *matched filtering*.

It is a technique that was devised for radar applications, but it seems to have first been mentioned in the context of gravitational wave data analysis by Kafka (1977) for the analysis of bar detector data. It starts with a convolution between the signal waveform we are looking for (called the template), $\{w_j\}$, and the detector data, $\{x_j\}$, defined by:

$$(x * w)_k = \sum_j x_j w_{j-k}. \tag{12.13}$$

If the data contains a signal matching the one we are convolving with, then the data will look like Eq. 12.11 and the convolution will be

$$(x * w)_k = \sum_j n_j \, w_{j-k} + A \sum_j w_{j-D} \, w_{j-k}. \tag{12.14}$$

This convolution is a Gaussian-distributed zero-mean random variable because it depends linearly on the noise, and its statistical mean (denoted by angle brackets $\langle \rangle$) is simply the second sum, which itself depends only on the difference $k - D$:

$$\langle (x * w)_k \rangle = A \left(\sum_j x_j \, w_{j-k} \right) = A(w * w)_{k-D}. \tag{12.15}$$

We are using here a convolution with the template $\{w_j\}$ as a filter, and the filter's expected output is the convolution of the signal with itself times the unknown amplitude A of the incoming signal (which might of course be zero). Since this signal is of finite duration T, this convolution will vanish if $|k - D| \geq M$. And it clearly will be a maximum when the two times match, $k = D$, which then allows us to measure the arrival index D that corresponds to the arrival time t_0. Then the expected maximum output of the filter will be the integral of the square of the signal:

$$\langle (x * w)_k \rangle_{\max} = A \sum_j [w_j]^2 = A||w||^2, \tag{12.16}$$

where we use the notation $||w||^2$ for the squared norm of the vector w_j.

To know the significance of this, we need to obtain the variance of the random variable in Eq. 12.13. This is the expectation of its square minus the square of its expectation (Eq. 12.15)

$$\text{var}[(x * w)_k] = \left\langle \left[\sum_j x_j \, w_{j-k} \right] \left[\sum_{j'} x_{j'} \, w_{j'-k} \right] \right\rangle - A^2[(w * w)_{k-D}]^2. \tag{12.17}$$

The expectation operates only on terms containing the noise, so that when we replace x_j by $n_j + Aw_j$ we get four terms, the first of which will simply be the noise variance,

$$\text{var}[n] = \sum_j \sum_{j'} \langle n_j n_{j'} \rangle. \tag{12.18}$$

The second and third terms, being linear in the noise, have zero expectation. The fourth term within the expectation contains only the convolution of the template with itself, and this cancels with the final term in Eq. 12.17.

Now we make a simplification, which helps to show the power of this technique more clearly. We assume that the noise is *white*, which means that there is no statistical correlation between the noise at different times. This implies that Eq. 12.18 simplifies to

$$\text{var}[n] = \langle n_j \, n_{j'} \rangle = \sigma^2 \delta_{j,j'},$$

where σ is as above and $\delta_{j,j'}$ is the Kronecker delta symbol. Of course, as Figure 12.5 clearly shows, the noise in our detectors is emphatically *not* white. We will return to how we deal with this below. But for now, the Kronecker delta function allows us to simplify the integral for the noise product, and it is straightforward to show that Eq. 12.17 simply becomes

$$\text{var}[(x * w)_k] = \sigma^2 (w * w)_{k-D} \tag{12.19}$$

The standard deviation of the filter measurement given by Eq. 12.15 is the square root of this.

If we look at the sample $k = D$ where the filter output is a maximum, then we can determine a detection threshold A_{min} by analogy with Eq. 12.12:

$$A_{min} = \frac{\sigma}{\|w\|}, \tag{12.20}$$

where $\|w\|$ is the square root of the squared norm $\|w\|^2$ that we introduced in connection with Eq. 12.16. The change from the simple threshold-crossing criterion in Eq. 12.12 is that we have replaced the maximum value of $\{w_j\}$ by its integrated norm. This is significant, because the norm is an integral over the duration of the signal, and therefore increases with the duration of the signal. When we filter for neutron-star binaries, therefore, we get an advantage because their signal is in-band for much longer, so $\|w\|$ will be larger and the detection threshold A_{min} smaller: we can detect weaker signals in this way.

The gain for a longer signal is easy to estimate. Consider a perfectly sinusoidal signal template, $w_j = B \sin(\omega \Delta t\, j)$. Its squared norm is

$$B^2 \sum_j \sin^2(\omega \Delta t\, j) = \frac{1}{2}B^2 \sum_j [1 - \cos(2\omega \Delta t\, j)].$$

If the signal has many cycles and we sum over its whole duration T then the contribution of $\cos(2\omega \Delta t\, j)$ will be small because of its oscillations, so the result will be dominated by the sum over the constant 1. We will get $B^2 N/2$. Of course, for this simple case the template amplitude B is the maximum value of the template, w_{max}, so when we take the square root of the squared norm we get from Eq. 12.20 a simple comparison with the threshold-crossing method:

$$A_{min} = \frac{\sigma}{w_{max}(N/2)^{1/2}}. \tag{12.21}$$

The detection threshold has been lowered by a factor proportional to $N^{1/2}$, the square root of the duration of the signal. We have improved the sensitivity of our detectors by using theoretical information about the expected signal to construct the template. Since the range of the detector is inversely proportional to the minimum detectable amplitude, we have improved the detection range for this signal by a factor $N^{1/2}$ and the volume of space in which it could be detected (and presumably therefore the number of detections) by $N^{3/2}$.

Although we obtained this by assuming a simple sinusoidal signal, the scaling with $N^{1/2}$ is robust, and it is basically the same random-walk factor that one finds for errors in repeated measurements: they converge as the square root of the number of terms being added. Notice that when the noise is expressed in terms of the strain h in the detector, if $h = \sigma$ a signal has SNR = 1. From Eq. 12.21, it follows that, after matched filtering, a signal with amplitude $h_c = \sigma(N/2)^{-1/2}$ has SNR = 1. Therefore h_c is a kind of threshold for signals that can be found by filtering, an effective noise level for detectors.

It might seem from Eq. 12.21 that one can reduce the detection threshold just by increasing the number of signal data points N, for example by sampling the data faster. This is not the case, because the sampled noise standard deviation σ in this equation will also change. For example, if a data stream is sampled at a certain rate, and then the rate is halved, every point in the new sampled stream is effectively an average over the noise values at two points in the old stream. So, since the noise is Gaussian, the standard deviation in the new stream is smaller by a factor of $\sqrt{2}$. And the number of signal points N has gone down by the same factor, leading to the *same* threshold. Of course, reducing the sampling rate must not be taken too far. If a data stream contains only one data point, then one is not going to detect any signal there! The ideal situation is to have a sampling rate about twice the frequency of the signal. That shows that the improvement factor given in Eq. 12.21 is best interpreted as the square root of the number of cycles in the waveform, rather than the square root of half the number of data points. Long wave trains benefit greatly from matched filtering.

Our assumption that we are looking for a single template waveform w is, of course, not realistic. The components of binaries have a huge range of masses, so we must apply a bank of templates to the data, one for each distinguishable waveform. If a signal from this bank is present, then it will show up not only in the filter that uses the template for this signal but also in other filters that are used for parameter values not so different from the real one. But the output for the filter that matches the real signal will, on average, be bigger than that from any other filter. This is the essence of matched filtering: the biggest response over a template bank tells us which template best matches the signal.

Although at the lowest order of approximation we have only one intrinsic parameter $\mathcal{M}$ for a template bank of compact binary signals, our wavetrains are so long (thousands of cycles for a neutron-star merger picked up at 20 Hz) that we need to incorporate post-Newtonian corrections in order to adequately match the signal, and these bring in the individual masses of the components and also their spins, which affect the waveform through spin–orbit and spin–spin interactions. And since we want to study the merger waveform, we switch to numerically simulated mergers for the last few orbits of our templates, because the post-Newtonian approximation breaks down when the objects are too close. See Blanchet (2014) or Poisson & Will (2014) for details of post-Newtonian approximations for orbits.

The computational cost of applying thousands or even tens of thousands of templates to the data is significant, and makes the quick recognition of signals challenging. The computing cost is made manageable by doing the convolutions in frequency space rather than in the time domain: one starts with the Fourier transforms of the data and template, and it turns out that the convolution can be computed using the Fast Fourier Transform algorithm, which reduces the number of operations dramatically compared with the cost in the time domain. Doing the calculation in the frequency domain also allows one to take into account optimally the frequency-dependent sensitivity of the detectors. This is described in a number of references, such as Andersson (2019) and Sathyaprakash & Schutz (2009); we explore it in Exercise 12.4.

The standard filtering done at present by LIGO and Virgo searches for circular binary systems. Since, as we have seen before, the inspiral phase of a binary shrinks the eccentricity to close to zero in most cases, this seems like a good assumption. However, eccentric systems can form, for example when triple star systems form compact black-hole binaries, so searches for fairly eccentric binaries are planned for the future.

LISA also will be using matched filtering to find binary coalescence signals and signals from long-lived white-dwarf binary systems. There are two complications that do not trouble ground-based detectors. One is that the signals will be of such long duration that LISA will move a significant amount in its orbit around the Sun during the observation. This induces a frequency modulation into the signal waveform that must be taken into account in the filtering. Ground-based searches for gravitational wave pulsars have the same problem, and we will address it in detail when we discuss pulsar detection below.

The other complication LISA has is in its search for EMRI signals, which we described earlier. The problem here is theoretical: building a signal template that is faithful over 10^5 orbits in the relativistic geometry near a spinning black hole, when the orbit is not equatorial, is a considerable challenge. If it cannot be solved by the time LISA launches, then sub-optimal searches will be needed, analogous to those that are presently used in ground-based gravitational wave pulsar searches, where again (as we shall see below) the matched filtering problem cannot be performed optimally.

Extracting information from a detected signal

The use of more parameters brings a big improvement in the science, since it opens the possibility of measuring the individual component masses and spins by finding the template that returns the highest signal-to-noise ratio. This is done in practice by using Bayesian statistics, which allows the data analysts to fold in any existing prior knowledge from astronomy about the masses and spins of stars and black holes, as well as about the performance of the detectors. Because the Bayesian method produces a posterior probability distribution for any measurable, it gives us not only the 'best' value of a parameter but also its uncertainty.

Because the individual masses are parameters at a higher post-Newtonian order than the lowest order, which was adequate for our discussions of binary radiation, they cannot be as accurately measured. The chirp mass is the most accurately measured parameter associated with binary systems, often to better than 1%. The total mass, the individual masses, and the spin components are measured with uncertainties that are typically larger by a factor 5 or more. The mass and spin parameters are called the intrinsic parameters of the system, since they depend only on the intrinsic properties of the radiating system itself. They come from fitting the waveform to the detector's output, so they can be measured by a single detector.

In addition, there are other parameters that we would like to measure, such as the sky location, the degree of elliptical polarization, the orientation of the polarization ellipse, and the distance. These extrinsic parameters cannot be measured by a single detector but, rather, require a network. The sky location can be measured from time delays in the signal arrival time at different detectors, provided we have three or more detectors (two or more independent delays). For neutron-star mergers it will sometimes happen that the associated kilonova will be found by electromagnetic observatories, giving a sky location. In that case, only two detectors are needed in order to resolve the polarization of the signal, provided they are not closely aligned (as the two LIGO detectors in the USA are). In the absence of an electromagnetic localization, three detectors can also of course return the polarization information. The polarization provides a measurement of the inclination of the orbit to the line of sight (equivalently, the degree of elliptical polarization) and of the orientation of the orbital plane on the sky (equivalently, the orientation of the polarization ellipse). Once the inclination of the orbit has been measured the distance to the system can be determined, as we saw in § 9.6.

Let us consider the problem of solving for the gravitational wave polarizations and estimating the inclination angle of a binary's orbit, using just two detectors and a given known sky location, for example from a gamma-ray-burst observation. By doing this in detail we can present some of the subtleties of this kind of analysis.

We will call the detectors 1 and 2. Each of them has a response equation like Eq. 12.10. When we consider a wave from a given source each detector will have different local angular sky position values $\{\theta, \phi\}$. We have mentioned this before. But the detectors will also receive the signal at different times. To combine the data properly one needs to shift the local time to a common time T, referring to the same phase of the wave as it passes through each detector. This time T could be the time that a particular phase of the wave passes through one of the detectors as a standard, or the time at which the phase passes through the center of the Earth or the center of mass of the Solar System. For our purposes this choice is not relevant, but we do assume that some common standard has been chosen, so that T is the common time, shifted from each detector's own time in such a way that it refers to the same wave phase in each detector. Then we can write the two response equations as

$$\left.\begin{aligned}\delta_1(T) &= F_{+,1}h_+(T) + F_{\times,1}h_\times(T), \\ \delta_2(T) &= F_{+,2}h_+(T) + F_{\times,2}h_\times(T),\end{aligned}\right\} \tag{12.22}$$

where the antenna pattern functions F_+ and $F_\times$ must be evaluated for the appropriate direction to the source, which will be at a different angular position in the sky for each of the two detectors.

This can be solved in a simple way for the wave's amplitudes. It is convenient to define the matrix

$$M_{12} = \begin{pmatrix} F_{+,1} & F_{\times,1} \\ F_{+,2} & F_{\times,2} \end{pmatrix}, \tag{12.23}$$

whose inverse is

$$M_{12}^{-1} = \frac{1}{\det(M_{12})}\begin{pmatrix} F_{\times,2} & -F_{\times,1} \\ -F_{+,2} & F_{+,1} \end{pmatrix}. \tag{12.24}$$

Note that the subscript '12' on M and M^{-1} does not denote indices of the components of M; it denotes that M is formed from interferometers 1 and 2. This will be relevant when we discuss three-detector networks below.

Finally, writing Eq. 12.22 together as a matrix equation with M and then applying M^{-1} to it results in

$$\left.\begin{aligned} h_+(T) &= \left[F_{\times,2}\delta_1(T) - F_{\times,1}\delta_2(T)\right] / \det(M_{12}), \\ h_\times(T) &= \left[-F_{+,2}\delta_1(T) + F_{+,1}\delta_2(T)\right] / \det(M_{12}). \end{aligned}\right\} \tag{12.25}$$

This is how two detectors measure the two polarizations of a wave arriving from a known direction.

Now, if we want to determine the inclination of the orbital plane, we have first to understand that the polarization amplitudes we have just derived refer to whatever choice we may have made for the basis vectors in the plane of the sky, as shown in Figure 12.3. This will be the same for both detectors, but it will not generally align with the orientation of the binary's plane projected onto the plane of the sky.

To see how to handle this, consider first the special case where the binary's orbit is seen edge on, which means an inclination $\iota = 90°$. Then the radiation will be linearly polarized, as we can see from Eq. 9.100. But when referred to an arbitrary basis, both polarizations will be present. However, if we rotate the sky-plane basis with which we started then the coefficients F_+ and $F_\times$ for each detector will change appropriately, and for some angle of rotation we will find that Eq. 12.25 will give zero for one of the two polarization amplitudes. This is the most convenient sky-plane basis orientation for inferring the inclination of the orbital plane, which in this case of course was $90°$.

How does this work in the general case, when the incoming radiation is elliptically polarized? We want to get our h_+ to equal h_{xx} in Eq. 9.100, and our $h_\times$ to equal h_{xy}. Note that in Eq. 9.100 the two waveforms are fully out of phase: if we take the real part then the + polarization contains $\cos[2\omega(r-t)]$ while the $\times$ polarization contains $\sin[2\omega(r-t)]$. When they are mixed by the use of a different basis, the phase relationship will be more complicated. So we can take this to be our criterion: rotate the sky basis vectors until $h_+(T)$ and $h_\times(T)$ are exactly out of phase. Then we see from Eq. 9.100 that the larger-amplitude component will be h_+ and the smaller will be $h_\times$. Moreover, the ratio of their amplitudes will be

$$\frac{|h_+(T)|}{|h_\times(T)|} = \frac{1 + \cos^2 \iota}{2\cos \iota}. \tag{12.26}$$

In principle, therefore, measuring the two amplitudes in this special sky-plane basis is enough to determine the inclination ι. However, when one looks into it in detail, one finds that the inclination is poorly determined for nearly face-on systems (Nissanke et al. 2010). This is the case because the radiation amplitude is at its maximum for this orientation, so that small changes in the inclination make only second-order changes in the radiation amplitude. The solution of Eq. 12.26 for ι is explored in Exercise 12.7.

Since face-on systems radiate the loudest signals, they can be heard at the greatest distances, and they will make up a disproportionately large fraction of the detected population. The larger uncertainties in their inclinations feed through into larger uncertainties in, for example, the measurement of their distances. This problem can be alleviated by measuring weaker contributions to the radiation, which originate in post-Newtonian corrections to the equations. These come out at different frequencies and have a different dependence on inclination. A technical discussion of these so-called 'higher harmonics' is beyond our scope in this book, but we shall see below that these harmonics are already affecting (and improving) the parameter estimates in some LIGO and Virgo detections. The LISA detector and the next generation of ground-based detectors should be able to resolve this distance–inclination degeneracy with good precision.

What is a detection? – a question of significance

The large number of templates that one needs to use for a realistic search increases the computational complexity of the search but it also complicates the search in two further ways. First, it gives more opportunities for random noise to masquerade as a detection, so we have to be careful to raise the detection threshold high enough to make this possibility remote. Second, there is a greater chance that some template will match a glitch.

For zero-mean Gaussian noise, the probability that a sample of random noise n_j will exceed a given amplitude threshold $a\sigma$, where σ is the standard deviation of the noise, is

$$p(|n_j| > a\sigma) = 2 \int_a^\infty \phi(z)\, \mathrm{d}z, \tag{12.27}$$

where $\phi(z)$ is the Gaussian normal probability distribution function for unit standard deviation:

$$\phi(z) = \frac{1}{\sqrt{(2\pi)}} e^{-z^2/2}. \tag{12.28}$$

The factor 2 in Eq. 12.27 is there because the noise can fluctuate in a positive or negative sense, and both would give us a false alarm. We are interested in approximating this for large a, because gravitational wave events are rare and so there will always be a large number of random chances for the noise to masquerade as a signal. In Exercise 12.5 it can be shown that this probability is bounded as follows:

$$\frac{a}{1 + a^2}\phi(a) < p(|n_j| > a\sigma) < \frac{\phi(a)}{a}. \tag{12.29}$$

Since the lower bound approaches the upper bound as a gets large, the upper bound becomes a good approximation for the probability of a false alarm. For example, a fluctuation of noise as large as 5σ (i.e. $a = 5$) has probability approximately 6×10^{-7}, which means it will occur by chance once every 1.7 million times. If this seems rare, remember that a data run lasting one year, sampled at 8 kHz, has 2.4×10^{11} data samples, so there will be over 140,000 such noise fluctuations!

In fact, we are more interested in chance fluctuations of the filtered output rather than of the raw data, because the filter might tell us a signal has arrived. Here the relevant 'sample time' is roughly one period of the wave in the template. If we take this to be at about 200 Hz, or a period of 5 ms, then a year-long data run will have 6×10^9 independent filter samples. So a threshold of $a = 5$ is still too low. If we set the threshold at $a = 6$ then the false-alarm rate goes down to about 2×10^{-9}, which would still give an expected 12 false alarms per year, which is a bit too many for comfort. However, the false-alarm rate goes down very rapidly as the threshold increases. If we just set the threshold to $a = 6.5$, then the false-alarm rate is about 8×10^{-11}, which would suggest a more comfortable rate of one false alarm roughly every two years.

Therefore, if we want a single detector's output to make a detection in a given filter that is not likely to be a Gaussian noise fluctuation, we should consider only events where the output of the filter has a signal-to-noise value greater than 6.5. However, if we have 10,000 filters, then each one is essentially an independent 'trial' of the data, so that we should have a false-alarm rate 10,000 times lower than the rate for $a = 6.5$. The probability that the noise will exceed $a = 8$ is 1.3×10^{-15}, more than 10^4 times smaller than for $a = 6.5$. So, if we want to be safe with 10^4 independent filters, we can set the threshold for a single-detector detection at a signal-to-noise ratio greater than 8.

However, single-detector detections are more problematic than this because of glitches. The glitch rate can be quite high (although as the detectors operate for longer periods of time the experimental teams tend to be able to reduce it). So for a single detector one has to understand the glitch population and set thresholds that one is confident are higher than any glitch that might occur in the observing run. This is a difficult judgement to make. For that reason, LIGO was built with two identical detectors, and observing at least up to now (mid-2021) has been mainly in what is called coincidence mode.

In coincidence mode, one looks for signal candidates in each detector that all fall within a coincidence window of time, which is fixed by the light-travel times between detectors: a gravitational wave will have a maximum delay in exciting two different detectors if it arrives from a direction parallel to the line joining the detectors. So glitches, which are not correlated between detectors, only become signal candidates if each detector in the network is excited by a glitch during a coincidence window. This eliminates most of the glitch background.

Most, but not all. The LIGO–Virgo collaboration estimates the rate at which glitches can still occur within this window by doing what are called 'time-slides'. These time-slides shift each detector's data stream by an arbitrary time increment (which can be negative), and then they do signal searches in the same way as with the zero-shifted data. The shifts are arranged so that the true coincidence window is zero, which ensures that all signal candidates in such time-slide searches are false alarms. Nevertheless, since the glitches are uncorrelated with one another, their rate should be the same as in the zero-slide data. By doing this over thousands of different slide combinations, the data analysts can reliably characterize the false-alarm background. This background includes all causes of false alarms, both glitches and Gaussian noise. In practice, the mean glitch coincidence rate is higher than it would be for just Gaussian noise, so the glitches dominate the false background. In this way, analysts can determine a safe threshold signal-to-noise level for recognizing a signal. The threshold must, of course, be high enough to ensure that there is a very low false-alarm rate across all the filters during the observing period.

As more detectors join the network, this threshold naturally goes down, and it may happen that false alarms become dominated by the Gaussian noise. Our discussion here has been in terms of the probabilities of noise, which is called a frequentist approach to statistics. A deeper analysis would use a Bayesian statistical approach, which would allow other information to be folded in, such as astrophysical uncertainties in our signal models, detector and search biases, and so on. This would result in the construction of a likelihood ratio, which would be the ratio of the probability that a signal matching our filter is present to the probability that the data include no signal at all. Then one would set a threshold on this ratio.

Coherent detection by a network

Coincidence detection is safe but generally not optimal. It relies just on the information about time of arrival, and then the amplitude of the response of a given filter in each detector. For two detectors, looking for waves coming from any direction and with any polarization, this is all one can do. But with three or more detectors, it begins to be possible to check for consistency among the amplitude responses of the detectors, a consistency that gravitational wave signals will exhibit but that glitches should not. To exploit this, one combines the outputs of all the detectors before applying any thresholds. This is called coherent analysis. Its advantages were first pointed out by Gürsel & Tinto (1989), and it was systematically explored in a series of papers by Bose *et al.* (1999, 2000) and especially by Finn (2001), who pointed out the similarity of this method to the aperture synthesis technique used in arrays of radio telescopes.

A simple counting argument shows why coherent detection is possible and useful. A gravitational wave is fully described by just two functions of time, the waveforms of its two polarizations. Two detectors that are not aligned can therefore determine the polarization if they know from which direction the wave is coming. Three detectors can determine the direction and then have redundancy in determining the two polarization waveforms. Four or more detectors have even more redundancy. This redundancy allows one, among other things, to detect when there are glitches, and to discover errors of detector calibration. We shall briefly discuss here how that works.

When there is redundancy, that means that, once one determines the sky position, it is possible to find the exact linear combinations of detector outputs that solve for the two polarization data streams (defined by whatever convenient orientation one chooses for the basis vectors in the plane of the sky, as in Figure 12.3), and then construct further linear combinations that contain no signal at all. These are called null streams. A three-detector network has one null stream, a four-detector network two, and so on. The null stream was introduced by Gürsel & Tinto (1989), and its usefulness was emphasized by Wen & Schutz (2005).

Let us examine the three-detector case in detail. The generalization of Eq. 12.22 is

$$\left.\begin{aligned}\delta_1(T) &= F_{+,1}h_+(T) + F_{\times,1}h_\times(T),\\ \delta_2(T) &= F_{+,2}h_+(T) + F_{\times,2}h_\times(T),\\ \delta_3(T) &= F_{+,3}h_+(T) + F_{\times,3}h_\times(T).\end{aligned}\right\} \tag{12.30}$$

Because the three strains depend on just two amplitudes h_+ and $h_\times$, one can find one relationship among the strains that eliminates the amplitudes. In Exercise 12.8 we show that the following particular linear combination of the gravitational-wave-generated detector output strains is exactly zero:

$$D_{12}\delta_3(T) + D_{23}\delta_1(T) + D_{31}\delta_2(T) = 0, \tag{12.31}$$

where D_{ab} is shorthand for the determinant of the matrix M_{ab} we introduced earlier to describe the joint response of any two detectors:

$$D_{ab} := \det(M_{ab}) = \det \begin{pmatrix} F_{+,a} & F_{\times,a} \\ F_{+,b} & F_{\times,b} \end{pmatrix}. \tag{12.32}$$

Now, the output of a detector contains not just the gravitational-wave-induced strain $\delta(T)$ but also the detector noise $n(T)$. So, while it is not possible to construct from observations the exact (noise-free) null combination in Eq. 12.31, what we can construct observationally is the same combination of the outputs $x(T) = n(T) + \delta(T)$. We call this the *null stream* N_{123}:

$$N_{123}(T) := D_{12}x_3(T) + D_{23}x_1(T) + D_{31}x_2(T). \tag{12.33}$$

Because this is a linear equation, the terms involving the gravitational-wave-induced strain δ add up to zero as they do in Eq. 12.31, so that $N_{123}(T)$ is composed of pure noise: the signal content of the various outputs has been filtered out. The null stream noise will normally be dominated by that of the least sensitive detector, the one whose strain noise is larger than that of the others.

One can construct a null stream from the raw detector outputs, as here, or from the outputs of a filter. Filter outputs must be handled with a bit more care, since the projection of a given expected signal on the three detectors may be different, so filtering must be done for these projections. In the case of binaries, the filter is essentially the same: the waveform in all the detectors is the same, only with different amplitudes. But the phase of the binary signal is affected by the projection, as we have already seen: the + and × polarization components are out of phase with each other. So the phase has to be adjusted correctly to suit each detector.

What is the point of having a linear combination of the detectors' noise outputs? A null stream cancels out the incoming gravitational wave signal, but does not cancel anything that originates in individual detectors. It therefore can be used to examine detector problems that might confuse or complicate finding the signal. A key one is glitches: the null stream will contain all the glitches, so it can act as a veto to a candidate signal that is really caused by coincident glitches. This by itself can allow the detection threshold to go below that set by the glitch false-alarm rate, down to that set by Gaussian false alarms.

If a glitch happens to occur during the time a signal is arriving, as in fact did happen in GW170817, the null stream could in principle be used to measure the glitch cleanly, without any superimposed signal, and subtract it from the data. This was not done for GW170817 because the Virgo detector, although it participated in the detection, had by chance a very weak signal due to its orientation with respect to the incoming signal direction. (We will discuss this later in this chapter.)

Another use for the null stream is to optimally find the direction to a signal; this was pointed out by Gürsel & Tinto (1989) and tested in simulations by Wen & Schutz (2005). The coefficients F_+ and $F_\times$ depend on the source direction. If we have the direction wrong and there is a signal present, then the signal will appear in the null stream. By changing the direction one can determine the true direction to the source as the direction that nulls out the signal.

Finally, the null stream can potentially help to find inconsistencies in calibration of the detectors. Calibration is the art of ensuring that the output strain δ truly represents the strain in the detector, across all frequencies. It is one of the most challenging problems of building and operating interferometers. The null stream will be null only if the calibration is correct. So if a signal is detected whose direction is known – as in GW170817, from the associated kilonova explosion – then the null stream is a test of consistency of the calibration. In principle, one would have to calibrate only one detector in a network to high precision, and then propagate that calibration to the others by using the highly predictable and reliably computed waveforms of inspiralling binaries.

A full coherent analysis not only finds the null stream(s), it also constructs the optimum values of the two independent polarizations with the best achievable signal-to-noise ratio. This is the principal output of such an analysis. It is beyond our scope here to discuss this in detail. The interested reader should consult the references in the bibliography at the end of the chapter.

Searching for gravitational wave pulsar signals

Our discussion up to now has assumed that the signal duration is short enough that the accelerated motion of our detectors (due to the rotation of the Earth and its orbital motion induced by the Sun and the Moon) does not introduce significant Doppler modulation of the signal or alterations in the incoming direction relative to the detectors. Signals from spinning neutron stars are continually present, and they are weak enough that they cannot be detected by current detectors in a short enough time to neglect detector accelerations. This makes the detection problem for such sources signficantly different, and more difficult, than it is for short-duration bursts. We shall see that it makes optimal matched filtering searches over large regions of the sky essentially impossible.

It is rather simple to discover how long an observation needs to be before these Doppler effects become significant (Schutz 1991). A rotational motion with angular velocity Ω on a circle of radius R produces a velocity change in a time T of magnitude $\Omega^2 RT$. If this change is along the direction to a source radiating at a frequency f, then the frequency change will be

$$\Delta f_{\text{shift}} = f\Omega^2 RT/c.$$

This will be detectable if the frequency resolution of an observation of duration T, which is simply $\Delta f_{\text{obs}} = 2/T$, is smaller than this. Setting these two frequency increments equal to one another determines the maximum observation time that would not need corrections:

$$T_{\text{max}} = \left(\frac{2c}{\Omega^2 fR} \right)^{1/2}. \tag{12.34}$$

If we use values appropriate to the rotation of the Earth, we get about $70(f/1\,\text{kHz})^{-1/2}$ minutes. If we use values for the orbital motion around the Sun, this goes up to about three and a quarter hours, again for a 1 kHz signal. We noted above that observation times of many months will be needed to detect gravitational wave pulsar signals by matched filtering, so it is clear that the signal analysis method needs to take the Doppler effects into account.

The size and the pattern of Doppler effects depend on the direction to the source. This means that, if the Doppler effects can be measured by an observation, the sky location of the signal can be determined, with great accuracy. Basically, by observing a source over one rotation, the detector synthesizes a telescope with the angular diameter of the orbit, $2R$. Observing gravitational waves with a wavelength λ permits an angular resolution at the diffraction limit of such a telescope, which is (to within factors of order 1) just $\lambda/2R$. From this it is clear that the orbit around the Sun sets the angular resolution, so that if a source is observed for, say, half an orbit, the best angular resolution will be about

$$\Delta\theta = 2 \times 10^{-6} \left(\frac{f}{1\,\text{kHz}}\right)^{-1}. \tag{12.35}$$

This is about 0.4 arcsec, far better than the angular accuracy with which our detectors can determine the positions of short bursts of radiation. It is similar to the accuracy with which radio telescopes can pinpoint the location of a radio pulsar, by making repeated observations over the course of a year. The mathematics for both kinds of positioning is the same, because the radio observations use pulse timing as the coherent signal across the Earth's orbit.

This high accuracy comes at a price, however. To dig these weak signals out of the noise, we need to use matched filtering. The Doppler effects mean that there has to be a separate filtering operation for each square arcsecond box on the sky. The number of such boxes for a six-month all-sky search will be about $4\pi/(\Delta\theta)^2 = 3 \times 10^{12}$. An optimal search for a constant-frequency signal might involve demodulating the data for a particular box on the sky, which effectively creates data as it would be observed by a detector at rest at the Solar System barycenter, and then performing a Fourier transform to look for signals at different frequencies. This would have to be done 10^{12} times, once for each resolution patch on the sky.

But even this is not good enough, because in a six-month observation the frequency resolution is about 10^{-7} Hz, and in this time most pulsars would be expected to change their spin rate by more than this tiny amount: pulsars slow down because of the electromagnetic radiation they emit, the particles they accelerate away, and the gravitational waves they emit. Typical spin-down ages as measured by radio observations are less than a million years (for young pulsars they can be a few thousand years), meaning that a 500-Hz spin rate (corresponding to the 1-kHz gravitational wave frequency with which we have been working) would change by 5×10^{-4} Hz in one year, a figure much larger than the frequency resolution. So a simple Fourier transform of the demodulated data would not find this signal; one would have to search through thousands or even millions of distinguishable spin-down rates.

When one takes into account the numerical computations that an optimal filtered search over half a year would require, it is far more than the computing time available on all the world's computers; see Exercise 12.9. Therefore all-sky searches conducted today are sub-optimal, which means that there is no 'best' algorithm. Different scientific teams compete to obtain the best sensitivity with the computing resources available to them. So far there have been no detections but, as we remarked earlier, this in itself implies an astounding degree of axial symmetry for at least some pulsars.

We have focused the discussion so far on an all-sky ('blind') search, because it is the most challenging. But radio astronomers have identified over 1,000 radio pulsars in known positions, and for these the demodulation is simple to perform. A search still has to be made, however, over spin-down parameters. For the youngest pulsars, one even has to parametrize the second time-derivative of the decreasing frequency. Again, there have been no detections yet.

Some of the most interesting pulsars are X-ray pulsars in binary systems. They are possible strong sources because there is some chance that the gas that they accrete from their companions will trigger instabilities that lead to the emission of gravitational waves (Andersson *et al.* 1999). But the orbital parameters of such systems are not known nearly accurately enough to allow a reliable demodulation of that additional Doppler effect on the signal, so this adds a large parameter space to such a search (Watts *et al.* 2008). The LIGO and Virgo teams have done some sub-optimal searches for emission from the Sco X-1 binary system, but without success so far (Abbott *et al.* 2020g). Reviews of the way in which searches are being performed may be found in Palomba (2017) and Sieniawska & Bejger (2019).

Because a modulated pulsar signal is so complex over a long observation period, it seems very unlikely that any terrestrial artefact will arise that can be confused with it, so the false-alarm rate should depend just on the Gaussian noise. Because of this, most searches take place using the data from just one detector, usually the most sensitive one operating in a given period. Once a signal candidate has been identified, the signal-to-noise ratio can be improved by looking for exactly the same modulation pattern in other detectors. The algorithms that may lead to identifying such candidates, however, are usually sub-optimal, as we have explained, and so if they involve combining searches in many short data segments, then local interference can play a more important role in confusing the analysis. However, the use of the null stream of three detectors could eliminate such intrusions.

Despite the difficulties and uncertainties of searching for gravitational waves from neutron stars in this way, considerable effort is being put into it both by the LIGO–Virgo teams and others. A detection and the consequent measurement of the quadrupolar distortion of a neutron star would contribute enormously to reducing the uncertainties in our understanding of the physics of these exotic and important stars.

Searching for a stochastic background

Theoretical studies of the Big Bang suggest, as we discussed earlier, that there should be an isotropic random background of gravitational waves at all wavelengths filling the Universe. In addition, all the binary systems that have radiated in any frequency band will have left behind them a similar random bath of gravitational waves, which might be isotropic but which might also be dominated by sources in our Galaxy. In both cases the background consists of the superposition of huge numbers of waves from all directions, so we can assume it will be an isotropic zero-mean Gaussian random field.

A single detector could only detect such a signal if the gravitational wave 'noise' was louder than the instrumental noise. But in that case it would be difficult to know what the level of instrumental noise is, because every measurement of the detector output would have the gravitational wave noise in it: the gravitational waves cannot be screened out. And, in any case, theoretical studies suggest that the gravitational wave noise is well below instrumental. So the right way to search for such radiation is by cross-correlation, by multiplying detector data streams and integrating over long periods of time. The idea is that, if the detectors are near to one another then they will respond to the same gravitational wave amplitudes, but their instrumental noise will be independent. So, if their data streams are multiplied and integrated over time, the independent noise cancels out and the correlated gravitational wave noise survives.

Let us see how this works in the simplest case, where we have two co-located detectors with their arms aligned so that they respond to the same wave amplitudes. We record a long data stream of N samples from each detector. Detector 1 has data $x_j = n_j^1 + h_j$, while detector 2 has data $x_j = n_j^2 + h_j$. We assume that there is no correlation between n_j^1 and n_j^2, but that the signal amplitude h_j in both streams is the same. All three variables are zero-mean Gaussian random variables. We call their standard deviations σ_1, σ_2, and $|h|$, respectively.

Then their correlation is defined as

$$\text{Corr}(1, 2) := \sum_j x_j^1 x_j^2. \tag{12.36}$$

We assume, as before, that the noise is white so the samples at different times are uncorrelated. That means that when we add up N terms in this sum, it is the same as taking N random samples from the same distribution and adding them. This is what happens when one calculates the expectation value of a random distribution, except that the sum is then divided by N to get the expectation of a single sample. So our sum will just be N times the expectation value of the summand:

$$\begin{aligned} \text{Corr}(1, 2) &= N\langle x^1 x^2 \rangle \\ &= N\langle n^1 n^2 + n^1 h + h n^2 + hh \rangle \\ &= N|h|^2. \end{aligned} \tag{12.37}$$

This is the expected 'signal' coming out of the correlation, but what is the 'noise' in the measurement? We need to calculate the variance of the measurement in Eq. 12.37. Let us again make our task simpler, this time by assuming that the gravitational wave stochastic amplitude is much smaller than the detector noise, i.e. that $|h| \ll \sigma_1$ and $|h| \ll \sigma_2$. In that case the outputs are dominated by detector noise, and $\text{Corr}(1, 2) \approx \sum_j n_j^1 n_j^2$. Although the expectation of this vanishes, which allows us to get at the smaller terms involving gravitational waves, the variance of it does not and that will dominate any contributions to the variance from the gravitational wave signal. So we have, within our assumption,

$$\text{var}[\text{Corr}(1, 2)] = N\text{var}(n^1 n^2) = N\sigma_1^2 \sigma_2^2, \tag{12.38}$$

the final equality stemming from the fact that n^1 and n^2 are completely uncorrelated.

If this is the variance of the measurement in Eq. 12.37, then the signal-to-noise ratio of that measurement is the result divided by its standard deviation, i.e. by the square root of the variance:

$$\text{SNR}_{\text{stochastic-power}} = \frac{N|h|^2}{N^{1/2}\sigma_1\sigma_2} = N^{1/2}\frac{|h|^2}{\sigma_1\sigma_2}. \tag{12.39}$$

This is a simple result: correlating the two data streams enhances the product of the 'raw' signal-to-noise ratios in each of the streams, $|h|/\sigma_1$ and $|h|/\sigma_2$, by a factor $N^{1/2}$. Our ability to detect a background grows with the square root of the duration of the correlated observation. In principle, even a very small stochastic background is detectable if one observes for long enough. We call this the stochastic power SNR because it uses the square of the signal amplitude, which is proportional to the energy density in the gravitational waves. As we mentioned earlier, the energy density, normalized to the total energy density of the Universe, is a convenient measure of the strength of the stochastic background.

For other detections, however, we talk about our detectors' sensitivity to the gravitational wave amplitude h, not to its power. The square root of Eq. 12.39 gives the amplitude ratio:

$$\text{SNR}_{\text{stochastic-amplitude}} = N^{1/4}\frac{|h|}{\sqrt{(\sigma_1\sigma_2)}}. \tag{12.40}$$

So, if the r.m.s. amplitude of the gravitational wave field $|h|$ is a factor of, say, 10^4 below the noise amplitude of our detectors, $(\sigma_1\sigma_2)^{1/2}$, then we can begin to see it if we correlate a data set with more than 10^{16} samples. At a sampling rate of a few kHz, this would take a mere 100,000 years! Therefore, while looking for a pulsar that is 10^4 times weaker than instrumental noise is routine, looking for a *random* field as weak as that is simply out of the question.

We oversimplified our calculation with our assumption that the detectors are co-located. However, they need to be well separated in order to guarantee that their instrumental noise is truly uncorrelated. In fact the aim of locating burst sources accurately on the sky makes it desirable that their separations have much larger values. As we explained earlier, the two LIGO detectors are the best pair for doing a correlation search, but the search has to be confined to frequencies below about 100 Hz. This also means that, when interpreting the length N of the data set in our expressions, one should assume a maximum effective sampling rate of around 100 Hz. This is the same consideration about interpreting N as we discussed when we introduced matched filtering.

A one-year correlation will therefore involve effectively around 3×10^9 samples, so the search can hope to detect backgrounds with mean amplitudes $|h|$ perhaps as weak as 100 times below the detectors' broadband sensitivity. In energy terms, this is a factor of 10^4 lower energy density than the energy density that could be measured by a single detector. While standard predictions of the energy density of radiation from the Big Bang are much lower than this, some models of the radiation arising from phase transitions can give predictions this large, and it is very possible that astrophysical backgrounds from binary systems will be detectable by LIGO in the next few years. For the most recent upper limits (mid-2021) see Abbott *et al.* (2019b). For an in-depth review of search methods for stochastic backgrounds, see Romano & Cornish (2017).

Third-generation ground-based detectors (as the next generation is being called) are much under discussion as this is being written, and their sensitivity might extend down to 1 or at least a few Hz. That might make correlations practical even if they are separated by intercontinental baselines: their considerable cost makes it likely that they can be built only by large coalitions of countries.

On the other hand, the LISA space-based detector, which is expected to be launched in the first part of the 2030s, can in fact do a search for a stochastic background, because its triangular formation gives it three detectors, allowing a null stream to be composed, which is called the Sagnac mode. The null stream allows an independent measurement of the detector noise, so if the apparent noise in the three standard interferometry signals is larger than this then it can be ascribed to a stochastic background. But there will not be a gain proportional to $N^{1/4}$, because the three LISA interferometers will share arms and so have common noise, which ensures that the correlated noise grows at the same rate as the correlated signal.

Pulsar timing searches for gravitational waves at nHz frequencies may also detect a stochastic background, an astrophysical background from supermassive binary systems. The method is not so different from the correlation method we have described: each monitored pulsar is like a detector that is affected by both gravitational noise and other kinds of noise, and looking for the correlations between the data from different pulsars can in principle find the wave background.

Searching for burst signals with unpredicted waveforms

A signal about which we have no information can only be recognized if it stands well up above the noise. Confidence in a detection would come from setting the coincidence threshold high enough, or by using a null stream to search for artefacts arising in single detectors. It must be borne in mind that a general signal might not look the same in both its polarization components, so the responses in different detectors could look quite different from one another.

A supernova explosion in our Galaxy or local group might be detectable, and it should be accompanied by neutrino emission that would be observed by neutrino detectors on Earth. This coincidence would allow a gravitational wave signal to be identified even if it was only a few times larger than the broadband noise. And if the signal is actually some kind of chaotic, almost stochastic, waveform, then a cross-correlation between two detectors could enhance it.

Similarly, signals from the exotic cosmic string sources have a fairly well-defined spectral shape, which permits a kind of matched filtering. This search is done routinely on LIGO and Virgo data.

However, it could be that there are other signals arriving with unexpected waveforms. To search for those, the LIGO–Virgo collaboration has developed methods that make minimal assumptions about the waveform. It seems likely that all strong gravitational waves will consist of at least a limited number of wave cycles, and so one can do a form of matched filtering with very simple filters, such as so-called sine-Gaussian waveforms. These are the product of a sine-wave with a Gaussian that is only a few wavelengths wide, so that the waveform has significant power only for a few periods of the wave. Using a template bank that contains signals with a range of wavelengths and Gaussian widths, it is possible to capture many kinds of waveforms, even if the match is not exact. For example, the first ever detection, GW150914, which we describe in the next section, was first noticed by an online version of this kind of burst search: the final part of the waveform, which stands well above the noise (shown in Figure 12.1), looks similar to a sine-Gaussian shape. For the most recent (mid-2021) searches, see Abbott *et al.* (2019a).

As detector sensitivities improve, there is a greater likelihood that weak unexpected signals will eventually be identified. This 'discovery space' is one of the most interesting aspects of gravitational wave detection.

12.4 The first LIGO and Virgo detections

The field of observational gravitational wave astronomy began with the first detection in 2015, a full 100 years after Einstein finished the formulation of his field equations. Although detection had been pursued since the 1960s and was confidently expected sometime in the near future, the detection of GW150914[1] came as a considerable surprise. Theoretical predictions of likely sources had suggested that the Advanced LIGO detectors would, when they reached their design sensitivity, detect at least one event per year, and that this most likely would be the coalescence of two neutron stars at a distance of perhaps 100 Mpc. By September 2015, the first stage of the Advanced LIGO upgrade was ready for its first observing run, but it would operate initially with a sensitivity that was a factor of 2 or 3 worse than the design goal. The plan was to alternate observing periods with small increments in sensitivity. Therefore, that first detection still seemed perhaps three years away. What happened next illustrates how full of surprises astronomy can be.

On 15 September, LIGO was still preparing for that first observing run, which was called O1. The Virgo detector's upgrade had started later than LIGO's, and it was not yet ready for observations. LIGO was in an 'engineering run' at the time, which meant that it was operating but being tested and tuned by the hardware specialists. By the early hours of the morning, at each site the scientists had finished their scheduled work, restored (fortunately) the instruments to undisturbed observation mode, locked up the buildings, and gone home.

At 03:50 in Livingston and 01:50 in Hanford, an unexpectedly strong gravitational wave excited both detectors. The collaboration members who had responsibility at that particular moment for monitoring the two output data streams were at the Albert Einstein Institute in Hanover, Germany, where the local time was 10:50 in the morning. This was the laboratory which had built the lasers that were powering LIGO's interferometers and which was operating the smaller detector GEO600, where many of the new technologies that had recently been installed in Advanced LIGO had been tested and proved. What these scientists saw, in both data streams, astonished them.

---

[1] The detected events discussed in this book have simple names constructed as GWyymmdd. The date and all detection times are referred to Universal Time (UT), essentially GMT. As this passage is being written (mid-2021) the collaboration is discussing how to extend the name convention to accommodate more than one event per day.

After performing a number of checks to make sure that the detectors were operating correctly, they realized that they had become the first people to see the waveform of a gravitational wave. And it did not look like the waveform of a binary-neutron-star coalescence at all: the chirp rate was far too fast, and it was unexpectedly strong. It had to be the merger of two black holes. Gravitational wave astronomy had become an observational science, and had already seen its first surprise.

That was just the first of what became a regular series of detections, their rate increasing as the sensitivity of the detector network improved. In this section we discuss a few of the detections that occurred in the period 2015–2019, principally to understand what we have learned from them and to illustrate how diverse and surprising they have been.

GW150914: the first direct detection of gravitational waves

Because GW150914 had arrived even before O1 had officially begun, because it was the first detection, because it was a binary-black-hole signal and because its amplitude was unexpectedly strong, the collaboration held back its announcement until it could perform the tests that it needed in order to be sure it was a real detection, and then wrote its discovery paper. Finally, on 11 February 2016, the announcement was made simultaneously with the publication of the discovery paper (Abbott 2016a).

The plot of the signal in Figure 12.1 shows why this first event was so unexpected. The signal stands well up above the noise, so there is no need for matched filtering in order to detect it, and it is even possible to read off from the trace of the signal approximately what kind of system we are dealing with (see Exercise 12.10). Matching to model waveforms is needed, of course, for the most accurate determination of parameter values. The values of selected parameters from the best-fitting waveform are given in the table in Table 12.1.

The LIGO–Virgo collaboration also tested the fit of the signal to a wider family of model waveforms, which included various violations of general relativity. None of these produced a better fit, which allowed upper limits to be set on any possible violations. A particularly interesting question is whether gravitational waves really travel at the speed of light. One way to violate this would be if the graviton, which is the gravitational analog of the photon, has a very small but nonzero mass. This would produce frequency dispersion as the wave travels, so that shorter wavelengths would travel faster, distorting the signal shape. Because this effect would accumulate over the whole travel, GW150914 was able to set an upper limit on the graviton mass of 1.2×10^{-22} eV. This seems very small, but in fact stricter bounds exist from the effects that such a mass would have on long-range gravitational fields. This is discussed in Abbott *et al.* (2016c).

The reason that this and subsequent detections are called black-hole mergers is that, given the measured component masses and their extreme compactness (testified to by the value of the highest frequency that the orbit reaches), we have no other reasonable physical model. There are speculative models for objects nearly as compact as black holes but without horizons, but they require exotic physics that has not yet been observed. A good proof that the final remnant is a black hole would be to observe directly its quasi-normal mode pulsations (see § 11.4), which should be there but which die away quickly. Different physical objects have different oscillation spectra, so if these oscillations could

Table 12.1 Data from selected detections

Event name	$\mathcal{M}$ ($M_\odot$)	m_1 ($M_\odot$)	m_2 ($M_\odot$)	M_f ($M_\odot$)	D_L (Mpc)	SNR	FAR (yr^{-1})	Ω_{sky} sq. deg.
GW150914	28	36	29	62	410	24	5×10^{-6}	600
GW170814	24	31	25	53	600	18	4×10^{-5}	87
GW170817	1.186	1.44	1.28	?	40	32	1×10^{-5}	16
GW190425	1.44	1.6–2.5	1.1–1.7	?	159	12.9	1.5×10^{-5}	8284
GW190521	66	85	65	142	5.3	14.7	2×10^{-4}	658
GW190814	6.1	23.2	2.6	25.6	241	25	8×10^{-4}	18

Measured parameter values are given here (mostly without uncertainty ranges) from data for selected detections. All masses are the values measured in the system's own local frame, corrected for the cosmological redshift by Eq. 9.163. Similarly, the distance is given as the cosmological luminosity distance D_L. The parameter M_f is the mass of the final merged object, as inferred from numerical simulations; this is not estimated for the two binary-neutron-star events in the table because we do not know how much mass was ejected from the system during and after the merger. The column labeled SNR contains the signal-to-noise ratio against the Gaussian noise, after using the best-matched filter, using all detectors combined. The parameter FAR denotes the false-alarm rate, which is the rate of false coincidences of this strength that would be expected in the detectors' noise stream, including background glitches. For GW190425, detected by a single detector, this is the maximum rate at which such events might be seen in that single data stream. The parameter Ω_{sky} is the size of the sky region within which the source is located, with 90% confidence. Note that the chirp mass in the table may not exactly match the result of computing Eq. 9.156 using the given component masses but is, rather, the most likely value within the range of values that are consistent with the (typically rather large) uncertainty ranges of the component masses. `https://tinyurl.com/79wh5uf4`.

be measured then there would be a direct test of the existence of a remnant black hole. The LIGO–Virgo collaboration did not claim such a test, but others have suggested that it might be possible (Isi *et al.* 2019).

For astronomers, the big surprise of GW150914 was that the component masses were so large. As we mentioned in § 12.2, there was considerable doubt before this that black-hole binaries existed at all. On top of that, the most massive black holes that had been observed before this were around $20\,M_\odot$, in X-ray binaries. The more massive component in GW150914 was almost twice this mass. Both of these facts support the idea that this binary formed from low-metallicity gas, because this leads to the formation of more massive stars in the first place. But of course the other formation methods discussed earlier are not excluded.

Observing run O1 started with GW150914 and ended on 19 January the following year. In this run two further binary black-hole detections were made, GW151012 and GW151226. These resolved any lingering doubts that GW150914 might have been an instrumental fluke or a hack. With chirp masses of 15.2 and 8.9 solar masses, these had component masses more consistent with previously known X-ray binary-black-hole masses.

GW170814: the first three-detector observation

After a long pause for hardware upgrades, LIGO's next observation run, O2, started on 30 November 2016, with a significantly improved sensitivity. There followed four further binary-black-hole detections made by LIGO alone: GW170104, GW170608, GW170729, and GW170809. Each event has its own individual characteristics, but GW170729 is particularly notable because it was the most distant merger thus far detected; at 2.8 Gpc, it was more than six times further away than GW150914.

Virgo joined the run on 1 August 2016, but it did not participate in the event on 09 August. The first three-detector detection was GW170814, another binary-black-hole system (Abbott *et al.* 2017c). Its properties can be found in the table in Table 12.1. Because there were three detectors, the sky region for the source location was 87 square degrees, only 15% of the size of the region for GW150914.

Since there were three independent data streams, the collaboration used the event to test general relativity's polarization property: could the responses of three detectors be consistently fitted using just two polarization waveforms, h_+ and $h_\times$? The analysis was able to exclude a purely scalar polarization (with circular rather than elliptical tidal forces) and a purely vector polarization (analogous to the polarizations of electromagnetic waves). The probability for scalar instead of tensor polarization was 0.1%, and for vector instead of tensor 0.5%. (Technically, these follow from Bayes' factors favoring tensor polarizations of 1,000 and 200, respectively.) Reassuring as these results are, it is important to note that there exist no viable alternative theories to general relativity that have purely scalar or vector polarizations. Instead, most theories modify general relativity by *adding in* extra kinds of polarizations. The data were not of sufficient signal-to-noise ratio to permit a meaningful search for scalar or vector polarizations in addition to the tensor component predicted by general relativity.

GW170817: the first detection of a binary-neutron-star merger

The most important event after GW150914 was the first detection of gravitational waves from a neutron-star merger and the associated observation of the electromagnetic emissions from its subsequent kilonova explosion. As noted in the introduction to this chapter, coordinated observing like this is called multimessenger astronomy, and its inception for gravitational waves was spectacular. It has been estimated that something like 10% of all professional astronomers participated in observations of the kilonova, and observations continue as this is being written (mid-2021).

The gravitational waves were in-band for about 100 s, and all three detectors were in observation mode. The merger itself occurred at 12:41:04.4 UTC on 17 August (Abbott *et al.* 2017d). Just 1.7 s later, the Fermi Gamma-Ray Space Telescope registered a burst of gamma-rays called GRB170817A. (Another identifier for this event is AT 2017gfo, which is what it was called by astronomers searching for the counterpart before they were confident it could be identified with the source of the gamma-ray and gravitational wave bursts.)

Although automatic notifications were operating, at first there was a delay in registering this coincidence. This was caused by an unfortunate hardware-related glitch in the LIGO-Livingston data stream, which caused the event to be automatically vetoed. But the signal in Hanford was so strong that LIGO scientists went back to the Livingston stream and realized that they could see the signal clearly despite the glitch. Initially, by just cutting the glitch out of the data (replacing it simply by zeros), they could use the Livingston and Hanford data together well enough for the first announcements.

Besides the glitch, the collaboration had another difficulty: the Virgo data stream contained an unexpectedly weak excitation from the signal compared with the other two detectors. The only explanation for this was that the event was unluckily located near a zero of Virgo's antenna pattern, coming by chance from a direction to which Virgo was insensitive. This itself, however, provided useful directional information, and when combined with the data from the other two detectors, it led to a sky localization region of about 31 square degrees (at 90% confidence), which was much smaller than the gamma-ray observation had given. Knowing the location then allowed the standard-siren distance to be estimated to 40 ± 8 Mpc, as described in § 9.6. This was an extraordinarily small distance compared with expectations for this kind of event.

The LIGO–Virgo team issued these data about five hours after the event. This gave astronomers a three-dimensional volume to search that contained fewer than 100 galaxies. Astronomers around the world dropped what they were doing and initiated searches for the kilonova. Six teams independently identified the kilonova and its host galaxy, NGC4993, about 11 hours after the event. About 70 observatories worldwide then began extended observations of many different types (Abbott *et al.* 2017f).

The best waveform fit to the gravitational wave signal produced a matched-filter network signal-to-noise ratio of 32, the largest thus far for the collaboration. The measured masses were in the normal range for the components of binary pulsar systems we see in our Galaxy. Best values are given in Table 12.1, and have been taken from the collaboration's final analysis (Abbott *et al.* 2019d). That table does not give a mass for the remnant, because that is difficult to estimate. For binary-black-hole mergers, this mass is estimated from numerical relativity simulations of collapses using the measured component masses. The remnant is typically about 5% smaller than the initial total mass because of the mass loss to gravitational waves. But in a binary-neutron-star system, the merger happens when the orbit is not as relativistic as for black holes, so the mass loss to gravitational waves is much smaller. On the other hand, mass is carried away by the kilonova explosion. For this system, computer modeling is not yet able to predict the total mass loss accurately.

The estimated masses of the two component stars depend to some extent on assumptions about the spins of the neutron stars. The values given in the Table 12.1 assume that the stars had modest spins, similar to those seen in binary pulsar systems. But if the stars were spinning at close to the maximum spin sustainable by a neutron star, then the error ranges on the individual masses get larger. The chirp mass is not affected by this uncertainty. It is possible to place limits on the spins by looking for signatures in the waveform of orbital precession caused by spin–orbit coupling, but in this event no positive detection of this effect was possible.

Another important physical effect that astronomers would like to measure in such mergers is the tidal distortion of the stars. This affects the rate of change of the orbital period as the stars get close to one another. The amount of tidal distortion depends on the deformability of the star, which is captured in a parameter called the Love number (Flanagan & Hinderer 2008, Andersson 2019, Chirenti *et al.* 2020). This in turn depends on the equation of state, so that measurements of tidal effects can give information about the fundamental nuclear physics and strong interactions in a regime that is not accessible to experiment on Earth. Because tidal interactions fall off as $1/r^3$ as one moves away from the star, they show up only in the last few orbits, so again it has only been possible to place limits on the Love number. But in this case, these limits are already good enough to exclude some exceptionally stiff equations of state (Abbott *et al.* 2019d).

GRB170817A was the nearest gamma-ray burst known up to that time, and yet in apparent brightness it was only an average burst. This raised important questions. Gamma-ray bursts generated by neutron-star mergers are believed to arise from highly relativistic jets emerging from the central explosion, presumably along the angular momentum axis of the system. The jets are probably narrow, with opening angles between five and 20 degrees. So the weakness could be explained in one of two ways: either our viewpoint was a bit off-axis (although still fairly close to face-on), or this was an instance of a different class of intrinsically weak bursts. Since the jet direction is linked to the orbital axis, measuring the orbital inclination with gravitational waves could resolve this or alternatively getting good electromagnetic data to pin down the jet's angle could improve the inclination measurement.

This was a key question about the astrophysics of mergers and gamma bursts, but it had additional importance to the gravitational wave collaborators because they wanted to use their standard-siren distance measurement to GW170817 in order to provide the first gravitational wave measurement of the important cosmological parameter called the Hubble–Lemaître constant, H_0. They could do this because the host galaxy NGC4993 had been identified, so its redshift could be measured. As we shall see in Chapter 13, in our expanding Universe distant galaxies recede from us with a velocity that is proportional to their distance from us, and the proportionality constant is H_0. Therefore, converting the galaxy's redshift into a velocity and dividing by the standard-siren luminosity distance measures H_0. However, the accuracy of the distance measurement and therefore of the value of H_0 depends on how well the inclination can be measured. As we have seen in Exercise 12.7, errors in inclination for nearly face-on detections, as one expects from gamma-ray bursts, lead to large errors in distance estimation (also see Ajith & Bose 2009). So a more precise estimate of the jet direction would lead to a better value for H_0.

The collaboration decided to publish its initial measurement of H_0 on the same day as the discovery paper referred to above, without the help of any constraints on the inclination that might come from electromagnetic observations, because these were not expected to be available until after a longer period of observation. The principal result was a probability distribution for the value of H_0, given the various uncertainties (Figure 12.6). Chief among these uncertainties was the inclination angle.

Subsequent observations by gamma-ray, X-ray, and radio astronomers attempted to determine the inclination angle. The most direct determination has come from radio

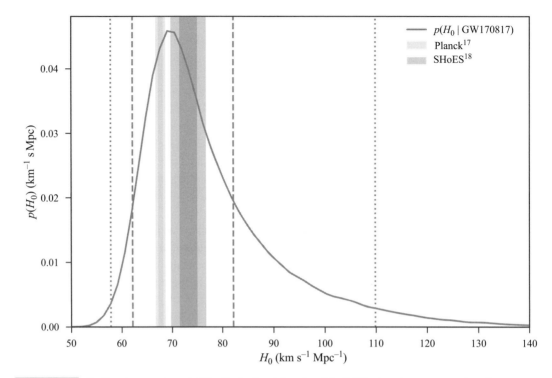

Figure 12.6 The Bayesian posterior probability distribution of the value of the Hubble–Lemaître parameter, given the
uncertainties in various input data and assumptions. The two vertical shaded bars represent the uncertainty
regions of the values obtained from the Planck satellite measurements (Ade *et al.* 2016) of the cosmic microwave
background (the vertical bar to the left of the posterior peak) and from the SHoES Collaboration's measurements
(Riess *et al.* 2016) using Type-Ia supernovae as standard candles (the bar to the right). The gravitational wave peak
falls nicely between the two, but has a much larger uncertainty, at least half of which is due to the poor accuracy of
inclination measurements for nearly face-on systems. Another source of error is estimating how much of the
redshift of NGC4993 is due to cosmological expansion and how much to its random peculiar velocity relative to
other galaxies. Plot taken from Abbott *et al.* (2017e). © AAS. Reproduced with permission.

observations. After the jet (which consists of a relativistic plasma) breaks through the
ejected stellar material, it propagates its radiating cloud at nearly the speed of light.
Radio observations by Hotokezaka *et al.* (2019) have restricted the jet's angle to the
observer's line of sight to between 14 and 28 degrees. With this value, as well as improved
measurements of the recession velocity of the host galaxy, these authors calculated
$H_0 = 70.3 \pm 5\,\mathrm{km\,s^{-1}\,Mpc^{-1}}$, remarkably consistent with the peak value in Figure 12.6,
but with greatly improved and symmetrical errors.

The event GW170817 gave astronomers their first opportunity to follow a kilonova from
its earliest hours until it faded completely from view. The richness of the data confirmed the
basic predictions of the kilonova/macronova models and also confirmed that the heavier
elements of the periodic table are mostly made in such events, not in supernovae as
had once been supposed. Instead of nuclear fusion, which is responsible for producing
the nuclei lighter than iron, the heavier nuclei are formed in kilonova events by fission.

A neutron star is basically one huge nucleus, and when the collision throws off a fraction of that material, it goes through repeated stages of fission, first forming large chunks of neutron matter and then breaking up into smaller chunks, during which the neutrons are decaying into protons and neutrinos. Eventually the chunks fission down to the size of heavy neutron-rich nuclei, and those that are stable or have decay lifetimes comparable with the age of the Universe eventually get incorporated into gas clouds that collapse to form new stars.

This process was first proposed in the context of black-hole–neutron-star mergers by Lattimer & Schramm (1974), even before the discovery of the first binary pulsar system, and then extended to binary-neutron-star mergers by Eichler *et al.* (1989). The conclusion was that these kinds of mergers in binaries could have created all of the Earth's abundance of what are called *r*-process elements, including the precious metals, thorium, uranium and most other heavy nuclei. Spectroscopic observation of the ejecta of GW170817 seems to confirm this (Piran *et al.* 2017). Of these elements, it is worth noting that ^{238}U and ^{232}Th have played a key role in shaping the Earth and assisting the evolution of life on our planet. They are distributed throughout the Earth, and their continuing decays provide a significant part of the heat flow that drives plate tectonics and keeps the Earth's core molten. The molten iron core maintains Earth's magnetic field, which protects the planet from the ionizing radiation of the solar wind.

GW170817 illustrates the potential that gravitational wave multimessenger astronomy has for astronomy, but it also illustrates the challenges. The statistics of binary pulsars suggest that the expected distance to the nearest such event in a one-year observing period should be of order 100 Mpc. GW170817 was in the nearest 7% of this volume, so it was almost certainly a very lucky chance for astronomers. Had it been at a distance of, say, 150 Mpc, its modest gamma-ray burst would not have been detected and the much weaker gravitational wave signal would have had a much larger localization uncertainty region. Identifying the counterpart for such an event, when the kilonova fades from view in a few days, would have been very challenging.

Further observations

A number of other interesting detections have been made in the O2 and O3 observing runs. Because the LIGO data are being made public, a number of detections have also been claimed by non-LSC groups (Venumadhav *et al.* 2019, 2020), which were generally events that the LSC analysis found but to which it did not give sufficiently high statistical significance. Since the statistical significance does depend on the analysis method, it is possible for independent analyses to 'promote' events in order to secure detections.

We summarize briefly a few of the latest detections that illustrate the variety of systems that are being detected and the different kinds of information that gravitational wave observations are able to contribute to astronomy. As the detectors continue to explore the dark Universe, this variety and body of information will only get more interesting.

- *GW190425.* This event is particularly interesting because of its unusual masses and because it was observed only by LIGO-Livingston and Virgo; LIGO-Hanford was off-line at the time (Abbott *et al.* 2020b). Moreover, the signal in Virgo was very weak compared with the signal at Livingston, so only the Livingston data was used to determine the significance and justify the detection. It follows also that the localization of the event on the sky was poor. That was unfortunate, because it was very likely to have been a double-neutron-star merger, but at its distance of about 150 Mpc, astronomers were not able to locate an electromagnetic counterpart.

In fact, the total mass of about $3.4\,M_\odot$ was much larger than that of any binary-neutron-star system known up to this time, so the possibility exists that the larger of the two components was a small black hole. The individual masses were not well enough determined to exclude this. Even in this case, one would have expected some kind of electromagnetic counterpart event. The masses of this system challenge conventional theories of compact binary formation and evolution, which have been tuned to produce the kinds of systems seen as binary pulsars. Whether it consisted of two neutron stars or of a neutron star and an exceptionally light black hole, it is a strange outlier.

- *GW190521.* This exceptionally massive system has also challenged conventional formation theories, in this case of heavy black holes (Abbott *et al.* 2020d, 2020e). With a total mass of about $150\,M_\odot$, its merger frequency was so low that it was in the detection band for only four cycles (two orbits), lasting just 0.1 s. Its network SNR of 14.7 leaves little doubt of its reality. And as this is being written, it has become even more interesting: there is a possibility that it may have been associated with a bright and short-lived optical flare seen by astronomers (Graham *et al.* 2020)!

Although black holes are black, they can live in gas-filled environments, so in principle their interactions with the gas can produce an electromagnetic counterpart signal. One way in which this could happen is through the 'kick' that the merger remnant gets from the asymmetric emission of gravitational radiation at the end of the merger. This can make the remnant travel supersonically through the gas, where its gravitational field could create a luminous shock. Graham *et al.* (2020) suggested that the black holes that radiated GW190521 could have been in a gas disk that orbited the central supermassive black hole of an active galactic nucleus. The flare was first noticed 34 days after the gravitational wave event, and it reached its peak luminosity about 50 days post-merger. The parameters of their model (the masses, etc.) are consistent with those of the detection, some of which had not been made public before the Graham *et al.* (2020) paper was written. Interestingly, Graham *et al.* (2020) suggested that the kick velocity of the remnant is not high enough for it to escape the central black hole, and so the flare might repeat on a timescale of order two years. This could be an important test of the validity of this association.

Regardless of whether this fascinating association survives further discussion and observation, GW190521 will remain an important detection, not least because the merger created an intermediate-mass black hole (IMBH), a class of objects of which we have as yet little evidence, as mentioned in § 11.4. The other puzzle is the large mass of its primary component, $85\,M_\odot$. Theoretical expectations have been that black holes of this mass do not form directly from stellar collapse. If that expectation is not revised, then presumably this component was itself formed in a merger of two smaller black holes. That in turn raises the question of whether this initial merger occurred in a triple system, in which the third component was the secondary in the GW190521 merger, or whether the primary formed from a simple binary and later managed to form a binary with its secondary.

Finally, it is worth mentioning that the signal was strong enough to show evidence in those final four cycles of radiation of other multipoles than just the basic quadrupole. Our calculations of binary evolution and the radiation pattern in this textbook have been based on purely quadrupole radiation, because we have worked to lowest order in the orbital velocity and used the Newtonian approximation to Einstein's gravity to compute the orbits. But when two black holes are so close that they are only two orbits away from merger, the nonlinearity of general relativity becomes important, and other multipole radiation makes an appearance. When waveforms that include such higher harmonics in the radiation were used as filters, they produced a better match than those with just quadrupole radiation. This radiation has a different radiation pattern, and it therefore helps to make the polarization measurement more accurate. In turn, this breaks the degeneracy between the distance measurement and the inclination of the system, making the distance measurement more accurate. Without the higher multipoles, the distance had been estimated at 4.1 Gpc. Including the higher multipoles pushed the system out to 5.3 Gpc.

- *GW190814.* This interesting event (Abbott *et al.* 2020c) was a merger between a black hole of mass $23\,M_\odot$ and a compact object of mass $2.6\,M_\odot$. As far as we understand the physics of compact objects at present, the smaller object could have been either a small black hole or a massive neutron star. Both are unexpected: although the theoretical maximum mass of neutron stars, judging from proposed equations of state, may be around $2.5\,M_\odot$, pulsar observations have not shown us any with masses larger than $2.2\,M_\odot$; at the same time, X-ray binary observations have not shown us any black holes smaller than about $6\,M_\odot$. This 'mass gap' for compact objects between 2.2 and $6\,M_\odot$ has been explained in a number of ways, but the discovery by LIGO and Virgo of an object that is clearly in the gap challenges these explanations. Observations with future such events should help us to resolve this.

  This event is remarkable as well because of the large mass ratio between the components, almost 9:1. This is the most extreme ratio so far detected. One of the characteristics of such systems is that they also emit significant radiation in the higher harmonics, as mentioned in § 12.3 above. These arise from the asymmetry of the masses and are detectable over many orbits, not just the last two as in GW190521. So they allow a more accurate measurement of the orbital inclination. In this case, it was also helpful that the inclination was close to $45°$, so that the polarization and distance errors were not coupled so tightly. Abbott *et al.* (2020c) used the resulting more accurate distance to make another estimation of the Hubble–Lemaître constant, even though the host galaxy had not been identified. At the measured luminosity distance of 241 Mpc, the redshift of each candidate galaxy in the three-dimensional position-uncertainty region was used to produce a candidate value of H_0. These were averaged (as in Schutz 1986) and then combined with the GW170817 measurement of H_0, to obtain $H_0 = 70\,\mathrm{km\,s^{-1}\,Mpc^{-1}}$, with error bars not very different from those obtained with GW170817 alone. This was an in-principle demonstration of the utility of such large-mass-ratio systems, and future observations that combine data from many detections will no doubt have a more marked effect on reducing the errors in these measurements.

- *Catalogs.* The LIGO–Virgo collaboration publishes separate papers on individual events only if the events have an unusual characteristic that merits a paper. To find all the detections during a period, one should consult a Gravitational Wave Transient Catalog (GWTC) for the relevant observing period. As of this writing there have been two releases: GWTC-1 (Abbott *et al.* 2019f) and GWTC-2 (Abbott *et al.* 2020f). GWTC-1 covers the first and second observing runs, O1 and O2. GWTC-2 covers the first half of the O3 run. Run O1 detected three events (including the discovery event GW150914) in a span of 16 months, using just the two LIGO detectors. During O2, the GWTC-1 catalog lists eight events, including GW170817, in a span of just nine months. Virgo participated in the final month of observing. GWTC-2 lists 39 events detected in a period of five months, with all three detectors participating.

The sharply increasing rate of detections through this series of runs principally reflects the improving sensitivity of the detectors: if their amplitude noise level decreases by a factor of two, their range increases by the same factor, and the spatial volume they can search increases by a factor of eight. But other factors are at play: with three detectors, it is easier to reject false positives created by glitches and, as confidence in the detection system has grown, the significance threshold for claiming a detection has gone down. GWTC-2 contains all events that passed the false-alarm threshold of two per year. This means that one would expect about one false positive among the 39 events. It is important to reduce this threshold as far as is reasonable, because otherwise one would be ignoring many real events. Statistical conclusions from the ensemble of events will not be drastically affected if one false positive is included, but they would be degraded if a large number of real events were ignored by setting the threshold too high.

One of the uses of catalogs is to improve the statistics of measurements such as the rates of events and tests of general relativity. In particular, Abbott *et al.* (2019g) examine the improvements of tests of the validity of general relativity that come from GWTC-1. These included testing for the consistency of the model waveforms with the real ones, looking for deviations from the post-Newtonian approximation to general relativity at various orders, constraining differences in the speed of gravitational waves from the speed of light, and testing for polarization modes that are not part of general relativity.

The increased rate of detections as the observing program has gone forward also points toward the future. LIGO is, as of this writing (mid-2021), funded to improve its sensitivity by a factor of perhaps 3–4 from the O3 sensitivity to the A+ level, and LIGO-India will also join at that level. Improvements in Virgo and KAGRA are also to be expected. So, by the middle of the decade 2020–2030, the global five-detector network may be recording a dozen or more events each day.

12.5 Bibliography

Gravitational wave astronomy is a rapidly developing field, so the best way to get an update beyond the material in this text, either for more recent detections or for more depth, is to exploit a number of resources on the web. Most events at the moment have their own Wikipedia page, and Wikipedia also has a summary page listing detections and marginal events, with the title 'List of gravitational wave detections'. LIGO maintains its own page, listing publications associated with different events and topics, at `www.ligo.caltech.edu/page/detection-companion-papers`. The main LIGO webpage is a good place to look for news and summaries of events: `www.ligo.caltech.edu`. The sounds of gravitational waves can be heard at the LIGO–Virgo GW Open Science Center `www.gw-openscience.org/audio/` Readers who wish to look at the real data may want to download open data from `www.ligo.caltech.edu/page/ligo-data`. It is also possible to participate in the data analysis by donating spare processor time on home computers, using the Einstein@Home screensaver available at `https://einsteinathome.org`, or to help to identify and classify glitches in real data via Gravity Spy at `www.zooniverse.org/projects/zooniverse/gravity-spy`. Readers who want to keep track of the development of the LISA mission can do so at the website `www.lisamission.org`.

An accessible introduction to gravitational wave astronomy, with more detail on the detector physics than we have given here, is the short online book by Stuver (2019).

The first detection, GW150914, is described from the point of view of a non-physicist in *Gravity's Kiss* (Collins 2017). Collins, a sociologist, had special access to the collaboration for many years in order to study how such groups of scientists work together.

Readers wanting to learn more about neutron-star physics in the context of gravitational wave observations can consult Rezzolla *et al.* (2018) and Andersson (2019). For more on neutron-star mergers, see the reviews by Faber & Rasio (2021) and Shibata & Taniguchi (2011). A comprehensive review of kilonova physics, written after the lessons from GW170817 were learned, can be found in Metzger (2020).

The post-Newtonian approximation is essential for computing accurate enough waveforms; a good reference is Poisson & Will (2014), and there is a review of the literature in Blanchet (2014). An update on the way that gravitational waves are testing general relativity is given in the book by Will & Yunes (2020). A focus on the way in which LISA will improve such tests is in the review by Gair *et al.* (2013). More generally, LISA's ability to explore fundamental physics is reviewed in Barausse *et al.* (2020). Readers who want to learn more about statistical analysis methods in the context of gravitational wave detection may want to consult the books by Jaranowski & Krolak (2009) and Creighton & Anderson (2011).

Exercises

12.1   Use Eq. 12.2 to find the final gravitational wave frequencies at merger of pairs of equal-mass black holes of masses 10, 10^6, and $10^9 \, M_\odot$, respectively.

12.2   (a) Do the arithmetic to arrive at Eqs. 12.1, 12.2, and 12.3.

(b) Show that the duration of a signal that enters the ground-based frequency band at 20 Hz from a system consisting of two equal-mass neutron stars of mass $1.4 \, M_\odot$ is about two and a half minutes.

(c) Show that, for two black holes of $30 \, M_\odot$ each, the corresponding time is less than one second.

(d) If you were to listen to an audio recording of a chirp signal, you might only begin to distinguish it clearly from instrumental noise when it reaches 40 Hz. How much remaining time does the signal from the two neutron stars have then? From the black holes?

(e) Show that, for a system consisting of two equal-mass black holes each of mass $10^6 \, M_\odot$ that enters the LISA band at 10^{-4} Hz, the remaining time to merger is about 26 days.

12.3 The double inner product given in Eq. 12.9 is a geometrical way of writing the response of the interferometer to waves from any direction and with any polarization. To see this we test it for various special cases.

(a) Suppose the wave arrives from the zenith direction $\theta = 0$. Align the basis vectors in the plane of the sky with those in the (parallel) plane of the detector, so that $\hat{e}_x^R = \hat{e}_x$ and $\hat{e}_y^R = \hat{e}_y$. Show by using Eq. 12.9 that the output of the detector depends only on the incoming + wave amplitude $h_+(t)$.

(b) As in the previous part but turn the basis vectors for the plane of the sky by $45°$. Show that the output now depends only on $h_\times(t)$.

(c) Let the wave arrive along the x-arm of the interferometer, that is from the direction $\theta = \pi/2$, $\phi = 0$. Orient the basis vectors of the plane of the sky (this is now parallel to the detector's y–z plane) so that $\hat{e}_x^R = -\hat{e}_z$ and $\hat{e}_y^R = \hat{e}_y$. Show that the output of the detector depends only on $h_+(t)$ and is only half as large as in part (a).

(d) Let the wave arrive along the direction bisecting the angle between the x- and y-arms of the interferometer, that is from the direction $\theta = \pi/2$, $\phi = \pi/4$. Orient the basis vectors in the plane of the sky so that $\hat{e}_x^R$ is in the detector's x–y plane, and $\hat{e}_y^R = -\hat{e}_z$. Show that the interferometer's response is zero.

12.4 In doing data analysis calculations, it is much more efficient to use Fourier transforms, which means doing the analysis in the frequency domain rather than the time domain. This exercise examines how that works. We assume here that the student is familiar with these powerful tools.

(a) The convolution between two discretely sampled data vectors $\{w_j\}$ and $\{x_j\}$ is defined in the text in Eq. 12.13 as $(x * w)_k = \sum_j x_j w_{j-k}$. The index j runs over all M values of the data set $\{x_j\}$. The template $\{w_j\}$ contains N nonzero values, where $N < M$, and needs to be extended with $M - N$ zeros to make up a set of length M for the summation. As the index j increases, the zeros at the end of the w series get pushed past the last x value and need to be brought around to the beginning of the w series: in this way the convolution is interpreted as a circular operation. Show that the convolution tests the data set for a signal only at its first $M - N$ points, not at all M points.

(b) Show that the number of arithmetic operations needed to compute the convolution in the first part of this question is of order MN.

(c) The (discrete) Fourier transform of the data vector $\{x_j\}$ of length M is computed by

$$\tilde{x}_a = \sum_j x_j e^{-2\pi \iota a j/M} w_j. \tag{12.41}$$

(Don't confuse ι, the imaginary number, with the Latin indices. Our convention here will be that indices $\{j, k, \ldots\}$ refer to time-domain data samples and $\{a, b, \ldots\}$ to frequency-domain data points.) Show that the number of arithmetic operations needed to compute the full set of transform values $\{\tilde{w}_a\}$ term by term is of order M^2 for a data set of length M.

(d) The Fourier transform can be calculated most efficiently using the Fast Fourier Transform (FFT) algorithm, which takes of order only $M \ln M$ arithmetic operations. Show that, for a five-minute data set sampled at 8 kHz, the FFT is more than 10^5 times faster than a term-by-term computation of a Fourier transform.

(e) Show that the Fourier transform of the convolution of two data sets, $\widetilde{x * w}$, is just the product of the transforms of the two data sets, one of them as the complex conjugate:

$$\widetilde{x * w}_a = \tilde{x}_a \tilde{w}_a^*. \tag{12.42}$$

(f) Suppose the data set in the previous part is to be filtered using a bank of templates, each of which is two minutes long. Estimate the speed-up gained by doing the calculation in the frequency domain over that in the time domain, as follows. Ignore the computational cost of computing the Fourier transforms of the filters themselves: once computed they can be stored and used for every five-minute segment data in succession, so their cost averaged over a long data run is negligible. Ignore also the computational cost of transforming the data into the frequency domain, since the transformed data will be reused for every one of the many thousands of filters in the bank, so again the cost per filter is negligible. Include in your estimate of the effort in the frequency domain the operation count for doing the convolution with one filter there, plus the count for transforming the result back to the time domain (again using the FFT). Compare that with the operation count of doing the filter in the time domain, which was calculated in the second part of this problem. By what factor does the frequency-domain method speed up this signal search?

12.5 This exercise proves the double inequality in Eq. 12.29. Note that since ϕ appears in all three terms of the inequality, the factor $2/\sqrt{2\pi}$ in front of the exponential integral is irrelevant. Therefore we will prove the inequality for the integral itself.

(a) First we prove the upper bound, which can be written as

$$\int_a^\infty e^{-z^2/2} \, dz < \frac{1}{a} e^{-a^2/2}.$$

Show this by using integration by parts on this integral, written in the form

$$\int_a^\infty \frac{1}{z} \left[z e^{-z^2/2} \right] dz.$$

(b) For the lower bound, integrate again by parts, starting with the integral

$$\int_a^\infty \frac{z^2}{1+z^2} e^{-z^2/2} \, dz = \int_a^\infty \frac{z}{1+z^2} \left[z e^{-z^2/2} \right] dz.$$

Show that the result of this integration can be rearranged to give the desired lower bound on $\int_a^\infty \exp(-z^2/2) \, dz$.

12.6 In the three paragraphs following Eq. 12.29, a number of false-alarm rates are quoted from the upper bound. Compute the value of the upper bound for the cases $\{a = 5, 6, 6.5, 8\}$ and use them with the sampling rates given in the text to compute the false-alarm rates.

12.7 Here we explore solving Eq. 12.26 for the inclination angle ι.

(a) First we solve for $\cos \iota$, which we denote by z. Let the ratio $|h_+(T)|/|h_\times(T)|$ be denoted by r. Note that we have two constraints on our variables, $r \geq 1$ and $-1 \leq z \leq 1$. Show that Eq. 12.26 leads to the quadratic equation

$$z^2 - 2rz + 1 = 0.$$

(b) Solve this for z. Show, from the constraints mentioned in the previous part, that only one of the two roots is meaningful:

$$z = r - (r^2 - 1)^{1/2}.$$

(c) Explore this in the limit of large r. Show that this is the linear polarization limit. Show also that in this limit

$$r \to \infty : \quad \cos \iota \sim \frac{1}{2r} \quad \Rightarrow \quad \iota \sim \frac{\pi}{2} - \frac{1}{2r}.$$

(d) Explore the solution for z in the limit as r approaches 1, by defining $\epsilon = r - 1$. Show that

$$\epsilon = r - 1 \to 0 : \quad \cos \iota \sim 1 - (2\epsilon)^{1/2}.$$

(e) Expand the cosine function for small ι and show that in this same limit

$$\epsilon = r - 1 \to 0 : \quad \epsilon \sim \frac{\iota^4}{8}.$$

(f) This result shows that ι needs to get fairly large before ϵ becomes measurable, or equivalently r become measurably different from 1. Suppose that r is measured to be 1, with an error of 1%. Show that ι might be as large as 0.53 radians, or $30°$.

(g) The distance to a standard siren can be inferred from a measurement of one of the amplitudes along with knowledge of the inclination. In Eq. 9.100, use the first equation to estimate the variation in the amplitude that follows from the uncertainty in inclination calculated in the previous part. If ι can range from

1 down to 0.53 radians, then show that the percentage change of $1 + \cos^2 \iota$ in this range is about 7%. This will be the uncertainty in the inferred distance for a face-on signal, even though the amplitudes themselves are measured to 1% accuracy.

12.8 We derive the null combination of detector strains given in Eq. 12.31. Begin with the three-detector response in Eq. 12.30. We have already solved the first two equations for the wave amplitudes h_+ and $h_\times$ in our discussion of the two-detector response in Eq. 12.25. Insert these solutions into the third equation of Eq. 12.30 to obtain an equation containing only the strains $\{\delta_1, \delta_2, \delta_3\}$. Then put this equation into the form Eq. 12.31.

12.9 Here we examine how an optimum all-sky search might correct for signal modulation and pulsar spin-down, and we use that to do a rough estimation of the number of arithmetic operations that would be required to search for a gravitational wave pulsar in six months of data. Each arithmetic operation is called, in the language of computing, a floating-point operation. For the purposes of this calculation we assume that the search goes up to 2 kHz in gravitational wave frequency, because theoretically it would be possible for a neutron star to spin at 1 kHz. We assume that the data are sampled at 4 kHz, which is the minimum sampling rate to allow a 2 kHz bandwidth search. The method will be to demodulate the recorded data for a particular patch on the sky, as described in the text, then to compensate for a particular spin-down value by the method described below, and then to do a Fourier transform for each combination of sky patch and spin-down value, in order to find a signal. Since this is a rough estimate, we regularly round our results to one significant figure. It should be emphasized that there can be other approaches to this problem that are more efficient, but not by huge factors. The approach here is straighforward to describe and shows rather readily that an optimum search is far out of reach.

(a) Show that if the duration of the observation is $T = 1.5 \times 10^7$ s, the number of data samples is $N_s = 6 \times 10^{10}$, and the frequency resolution of the observation is

$$\Delta f = 2/T = 1.3 \times 10^{-7} \text{ Hz.} \qquad (12.43)$$

(b) To demodulate the Doppler shifting imposed on the signal by the motion of the detector, one resamples the data set at sampling times that are uniform at the center of mass of the Solar System, called the solar barycenter. One can do this because a clock at the barycenter will not accelerate with respect to the distant stars, and so receives an unmodulated signal from any direction. If the displacement vector from the barycenter to our detector is called d, and if the source direction from the barycenter is denoted by the unit vector n, then show that a wavefront from the source arrives at the detector a time $n \cdot d/c$ earlier than it arrives at the barycenter. To correct for demodulation, we sample the data of the detector at the times that correspond to regular 4 kHz sampling times at the barycenter. This involves interpolating the detector's sampled data set. Assuming that the interpolation requires of order ten operations per data point, show that each demodulation requires $N_{\text{demod}} = 6 \times 10^{11}$ operations. There is, of course, one demodulation for each resolvable patch on the sky.

(c) To correct for the pulsar's changing signal frequency, we will assume that the only parameter we need is the first time-derivative of the frequency, i.e. that the frequency changes linearly in time over the observation period. We shall handle this in the same way as the demodulation, by shifting to a different time measure. We assume that the source's wavefronts are emitted at a uniform rate with respect to a clock that is slowing down. Once we map that clock's time onto barycenter time, we restore the pulsar to a single-frequency signal and we can use a Fourier transform to search for it. Letting $t = 0$ represent the start of the observation time in the barycenter, we can give concrete expression to this by relating the pulsar's 'clock' T to barycentric time t as follows:

$$t_p = t(1 - t/\tau), \tag{12.44}$$

where τ is a parameter known as the slowdown time, a measure of how long it takes the pulsar to change its frequency by a significant amount. Since realistic slowdown times are larger than 10^3 years, we always have $t \leq T \ll \tau$, so that the fractional changes in clock rates are small.

(i) Solve Eq. 12.44 for t as a function of t_p, assuming that $t/\tau \ll 1$, to show that

$$t = t_p(1 + t_p/\tau) + O([t_p/\tau]^2). \tag{12.45}$$

This is the equation showing how to resample the data in the barycentric frame in order to make the signal constant frequency.

(ii) Argue, from our definitions, that the frequency of the pulsar's signal in the barycentric frame, if it starts at $t = 0$ with frequency f_0, is $f(t) = f_0(1 - 2t/\tau)$, so that $df/dt = -2/\tau$.

(iii) Assume that the largest slowdown that we want to search for (the largest value of $1/\tau$) occurs when a pulsar that starts out emitting at 2 kHz loses half its spin frequency in 1,000 years. Show that

$$(1/\tau)_{\max} = 8 \times 10^{-8}\,\text{s}^{-2}.$$

(iv) What is the measurable resolution $\Delta(1/\tau)$ in the parameter $1/\tau$? We can resolve a frequency change of 1.3×10^{-7} Hz, as shown in Eq. 12.43. Set this equal to the frequency difference over the observation period T that is caused by a change in $1/\tau$ by the minimum measurable amount, $2T f_0 \Delta(1/\tau)$ for the fastest pulsar (2 kHz), to show that

$$\Delta(1/\tau) = 2 \times 10^{-18}\,\text{s}^{-1}.$$

(v) Use the two previous results to show that we need to use 4×10^{10} different values of the slowdown parameter $1/\tau$ in order to search for all possible signals of this kind up to frequency 2 kHz.

(vi) We need to do this for each patch on the sky. Show that this means that there need to be effectively 1.2×10^{23} independent searches on the data, each performed with a Fourier transform of the data set after demodulation and removal of the slowdown.

(vii) An N_s-point Fourier transform can be computed using the FFT algorithm in about $N_s \ln N_s$ operations. Show in our case that $\ln N_s \sim 25$. Given that each data set going into the transform was produced by two interpolations, each of which, we have assumed, took $10N_s$ operations, show that this means that the computing effort per independent search is about 3×10^{12} operations. Multiply this by the number of searches (see part (vi)) to show that a full all-sky optimal search up to 2 kHz and six months would require 5×10^{35} operations.

(viii) We would want to do this calculation within six months, in order to keep up with the data stream. Show that this requires a computing speed of 3×10^{28} floating-point operations per second (commonly called flops). The fastest supercomputer today (mid-2021) can compute at a rate of about 400 petaflops, or 4×10^{17} flops. How many such supercomputers would we have to employ to do our search?

12.10 For GW150914, use the waveform in Figure 12.1 to estimate the following:

(a) the wave's amplitude, frequency, and rate of change of frequency just before merger;
(b) the amplitude of the noise;
(c) the number of cycles for which the signal was in-band;
(d) the expected SNR from matched filtering in one detector;
(e) the chirp mass $\mathcal{M}$ of the black holes from

$$\frac{df}{dt} = 12 \left(\frac{\mathcal{M}}{M_\odot} \right)^{5/3} \left(\frac{f}{100\,\text{Hz}} \right)^{11/3} \text{s}^{-2}. \tag{12.46}$$

12.11 Observations of binary mergers often quote the chirp mass quite accurately, the total mass relatively accurately, and the component masses with larger error bars. But if one knows the chirp and total masses, then one can deduce the component masses. This exercise explores why the component masses nevertheless often have larger errors.

(a) Suppose we take $\mathcal{M}$ and M (the total mass) as given. If the component masses are m_1 and m_2, with $m_1 \geq m_2$ by convention, then show that

$$\frac{m_1}{M} = \frac{1}{2} + \left(1 - \frac{4\mu}{M} \right)^{1/2}, \tag{12.47}$$

where μ is the reduced mass. (You may find Exercise 9.42 helpful.) What is the expression for m_2?

(b) Now suppose that M changes by a small amount δM, within the errors. Show that, to first order in δM, the change δm_1 is

$$\frac{\delta m_1}{\delta M} = \frac{1}{2} + \frac{\frac{1}{2} - \frac{\mu}{3M}}{\left[1 - \frac{4\mu}{M}\right]^{1/2}}. \qquad (12.48)$$

What is the corresponding expression for δm_2?

(c) Explain why, in the case where the two components' masses are close to one another, the ratios $\delta m_1/\delta M$ and $\delta m_2/\delta M$ might be large.

(d) In Table 12.1, look at the masses given for GW170817. The collaboration published uncertainty ranges for these values as well (Abbott *et al.* 2019d). The uncertainty in the chirp mass was one part in 10^3, so for the purposes of this exercise, we take it as exactly known. The 90% uncertainty range of the total mass M was asymmetrical, given as $(+0.04, -0.01)M_\odot$. Calculate the corresponding 90% bounds on the component masses m_1 and m_2. (These are only approximate bounds; a full analysis would take into account the uncertainty in the chirp mass and its correlation with the uncertainty in the total mass.)

Cosmology

13.1 What is cosmology?

The Universe in the large

Cosmology is the study of the Universe as a whole: its history, evolution, composition, dynamics. The primary aim of research in cosmology is to understand the large-scale structure of the Universe, but cosmology also provides the arena, and the starting point, for the development of all the detailed small-scale structure that arose as the Universe expanded away from the Big Bang: galaxies, stars, planets, people.

The interface between cosmology and other branches of astronomy, physics, and biology is therefore a rich area of scientific research. Moreover, as astronomers have begun to be able to study the evidence for the Big Bang in detail, cosmology has begun to address deeply fundamental questions of physics: what are the laws of physics at the very highest possible energies, how did the Big Bang happen, what (if anything) came before the Big Bang, how did the building blocks of matter (electrons, protons, neutrons) get made? Ultimately, the origin of every system and structure in the natural world, and possibly even the origin of the physical laws that govern the natural world, can be traced back to some aspect of cosmology.

Our ability to understand the Universe in the large depends in an essential way on general relativity. It is not hard to see why. Newtonian theory is an adequate description of gravity as long as, roughly speaking, the mass M of a system is small compared to its size, R: $M/R \ll 1$. We must replace Newtonian theory with GR if the system changes in such a way that M/R gets close to 1. This can happen if the system's radius R becomes small faster than M does, which is the domain of compact or collapsed objects: neutron stars and black holes have very small radii for the mass they contain. But we can also get to the relativistic regime if the system's mass increases faster than its radius. This is the case for cosmology: if space is filled with matter of roughly the same density everywhere, then, as we consider volumes of larger and larger radius R, the mass increases as R^3, and M/R eventually must get so large that GR becomes important.

What length scale is this? Suppose we begin increasing R from the center of our Sun. The Sun is nowhere relativistic and once R is larger than $R_\odot$, M hardly increases at all until the next star is reached. The system of stars of which the Sun is a minor member is a galaxy, and contains some 10^{11} stars in a radius of about 15 kpc. (Recall from Chapter 9 that one parsec, abbreviated pc, is about 3×10^{16} m.) For this system $M/R \sim 10^{-6}$, similar to that for the Sun itself. So, galactic dynamics has no need for relativity. (This applies to the galaxy as a whole: small regions, including the very center, may be dominated by black holes or other relativistic objects.) Galaxies are observed to form clusters, which often have thousands of members in a volume of the order of a Mpc. Such a cluster could have $M/R \sim 10^{-4}$, but it would still not need GR to describe it adequately.

When we go to scales larger than the size of a typical galaxy cluster, however, we enter the domain of *cosmology*.

In the cosmological picture, galaxies and even clusters are very small-scale structures, mere atoms in the larger Universe. Our telescopes are capable of seeing to distances greater than 10 Gpc. On this large scale, the Universe is observed to be *homogeneous*, to have roughly the same density of galaxies, and roughly the same types of galaxies, everywhere. As we shall see later, the mean density of mass–energy is roughly $\rho = 10^{-26}\,\mathrm{kg\,m^{-3}}$. Taking this density, the mass $M = 4\pi\rho R^3/3$ is equal to R for $R \sim 6$ Gpc, which is well within the observable Universe. So to understand the Universe that our telescopes reveal to us, we need GR.

Indeed, GR has provided scientists with their first consistent framework for studying cosmology. We shall see that metrics exist that describe cosmological models that embody the observed homogeneity: they have no boundaries, no edges, and are homogeneous everywhere. Newtonian gravity could not consistently describe such models, because the solution of Newton's fundamental equation $\nabla^2\Phi = 4\pi G\rho$ is ambiguous if there is no outer edge on which to set a boundary condition for the differential equation. So only after Einstein did cosmology become a branch of physics and astronomy.

We should ask the converse question: if we live in a Universe whose overall structure is highly relativistic, how is it that we can study our local region of the Universe without reference to cosmology? How can we, as in earlier chapters, apply general relativity to the study of neutron stars and black holes as if they were embedded in an empty asymptotically flat spacetime, when actually they exist in a highly relativistic cosmology? How can astronomers study individual stars, geologists study individual planets, biologists study individual cells – all without reference to GR? The answer, of course, is that in GR spacetime is locally flat: as long as your experiment is confined to the local region you don't need to know about the large-scale geometry. This separation of local and global is not possible in Newtonian gravity, where even the local gravitational field within a large uniform-density system depends on the boundary conditions far away, on the shape of the distant 'edge' of the Universe (see Exercise 13.3). So GR not only allows us to study cosmology, it explains why we can study the rest of science without needing GR!

The cosmological arena

In recent years, with the increasing power of ground- and space-based astronomical observatories, cosmology has become a precision science, to which physicists look for answers to some of their most fundamental questions. The basic picture of the Universe that observations reveal is remarkably simple, when averaged over distance scales much larger than, say, 10 Mpc. We see a homogeneous Universe, expanding at the same rate everywhere. The Universe we see is also *isotropic:* it looks the same, on average, in every direction. The Universe is also filled with radiation with a black-body thermal spectrum, with a temperature of 2.725K. We call this the *cosmic microwave background*, or CMB, because at this temperature the peak of the thermal spectrum is in the microwave band.

The expansion means that the Universe has a finite age, or at least that it has expanded in a finite time from a state of very high density. The thermal radiation suggests that the Universe was initially much hotter than today, and has cooled as it expanded. The expansion resolves the oldest of all cosmological conundrums, Olbers' Paradox. The sky is dark at night because we do not receive light from all stars in our infinite homogeneous Universe, but only from stars that are close enough for light to have traveled to us during the age of the Universe.

But the expansion raises other deep questions, about how the Universe evolved to its present state and what it was like much earlier. We would like to know how the first stars formed, why they group into galaxies, why galaxies form clusters: where did the density irregularities come from that have led to the enormously varied structure of the Universe on scales smaller than 10 Mpc? We would like to know how the elements formed, what the Universe was like when it was too dense and too hot to have normal nuclei, and what the very hot early Universe can tell us about the laws of physics at energies higher than we can explore with particle accelerators. We would like to know whether the observed homogeneity and isotropy of the Universe has a physical explanation.

Answering these questions has led physicists to explore some very deep issues at the frontiers of our understanding of fundamental physics, and they have come up with some astonishing answers that raise even deeper questions. The homogeneity problem can be solved if the extremely early Universe expanded exponentially rapidly, in a phase that physicists call *inflation*. We shall see that this could happen if the laws of physics at higher energies than can be explored in the laboratory have a suitable form, and if so this would as a bonus help to explain the density fluctuations that led to the observed galaxies and clusters. As we shall also see, it appears that most of the matter in the Universe is in an unknown form, which physicists call *dark matter* because it radiates no light. Even more strangely, the Universe seems to be pervaded by a relativistic energy density that carries negative pressure and which is driving the expansion faster and faster; physicists call this the *dark energy*. The mysteries of dark energy and of inflation may only ultimately be solved when we have a better understanding of the laws of physics at the highest energies, so theoretical physicists are looking more and more to astronomy for clues to better theories.

Modern cosmology is already providing answers to some of these questions, and the answers are becoming more precise and more definite at a rapid pace. This chapter gives a snapshot of the fundamentals of our understanding at the present time (mid-2021). More than any other area covered in this textbook, the study of the large-scale dynamics of the Universe is a pursuit that promises new insights, surprises, perhaps even a revolution in physics.

13.2 Cosmological kinematics: observing our expanding Universe

Before we can begin to understand the deep questions of cosmology, let alone their answers, we need to be able to describe and work with the notion of an expanding cosmological model. In this section we develop the metric that describes a homogeneous expanding model, we show how astronomical observations measure the expansion history, and we develop the framework for discussing physical processes in our expanding Universe. In § 13.3 we will apply Einstein's equations to our models to see what GR has to tell us about how the Universe expands.

Homogeneity and isotropy of the Universe

The simplest approach to applying GR to cosmology is to use the remarkable observed large-scale uniformity. We see, on scales much larger than 10 Mpc, not only a uniform average density but uniformity in other properties: types of galaxies, their clustering densities, their chemical composition and stellar composition. Of course, when we look very far away we are also looking back in time, the time it took the light we observe to reach us; over sufficiently long look-back times we also see evolution, we see a younger Universe. But the evolution we see is again the same in all directions, even when we look at parts of the early Universe that are very far from one another. We therefore conclude that, on the large scale, the Universe is *homogeneous*.

What is more, on scales much larger than 10 Mpc the Universe seems to be *isotropic* about every point: we see no consistently defined spatial direction. Isotropy is not implied by homogeneity. The Universe *could* be homogeneous but anisotropic, if, for instance, it had a large-scale magnetic field which pointed in one direction everywhere and whose magnitude was the same everywhere. On the other hand, an inhomogeneous Universe could not be isotropic about every point, since most – if not all – places in the Universe would see a sky that is 'lumpy'.

A third feature of the observable Universe is the *uniformity of its expansion*. Galaxies, on average, seem to be receding from us at a speed which is proportional to their distance from us. This recessional velocity is called the *Hubble–Lemaître flow* after the astronomer who discovered it, Edwin Hubble, and the mathematician who first predicted it, Georges Lemaître.[1]

The way in which the Universe expands is easily visualized in the 'balloon' model (see Figure 13.1). Paint uniformly spaced dots on the surface of a spherical balloon and then inflate this two-dimensional 'cosmology'. As it grows, the distance on the surface of the balloon between any two points grows at a rate proportional to that distance. The proportionality of speed to distance preserves the homogeneity of the distribution of dots with time, and that means that *any* point will see all other points receding at a rate

---

[1] Many books and other references simply refer to the ratio of the recessional speed and the distance of the galaxy as the Hubble constant, or the Hubble parameter. It was renamed the Hubble–Lemaître constant by the International Astronomical Union in 2018 to recognize Lemaître's fundamental theoretical contribution.

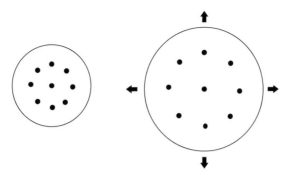

Figure 13.1 As a balloon expands, the distance between *any* two dots increases at a rate proportional to its magnitude. Dots with greater separation separate faster than dots that are closer together.

proportional to their distance. Therefore, our own location in the Universe is not special, even though it appears that everything else is receding away from us. We are no more at the 'center' of the cosmological expansion than any other point is. The Hubble–Lemaître flow is compatible with the *Copernican Principle*, which in modern usage is the idea that we do not live in a particularly special place in the Universe. In particular, the Universe does not revolve around (or expand away from) our location in a way that is different from other locations.

It is important to understand that the Hubble–Lemaître flow velocity at some point in the Universe is not necessarily the velocity of any particular galaxy, since galaxies have random local velocities relative to one another. Instead, it is the velocity of the *local cosmological rest frame* at the point in question. One might think that this would be the frame in which the average of the relative velocities of the ensemble of galaxies near the point is zero. But we know that there are large-scale flows of galaxies, so even averaging the velocities of a specific group of galaxy neighbors might not determine the correct frame. The cosmological microwave background radiation, mentioned earlier, provides the right definition: the local cosmological rest frame is the local inertial frame that is at rest with respect to the background radiation. This local rest frame can in principle be determined by measuring the temperature of the radiation in all directions: any velocity with respect to the rest frame of the radiation would make the temperature in the forward direction a bit hotter and in the backward direction a bit cooler, by the same amount. In the local cosmological rest frame, there are no systematic large-scale temperature variations of that kind across the CMB sky.

This raises the question, what is our own local cosmological rest frame? Earth is a moving platform, and the Sun has its own local random velocity relative to other stars, so it is unlikely that we are at rest with respect to the CMB at any time. To answer this and many other deep questions about the history of the Universe (to which we will return below), astronomers have studied the temperature distribution of the CMB in different directions, making measurements from the ground, from rockets and balloons, and from a succession of three satellites, NASA's COBE and WMAP satellites, and more recently the European Space Agency's Planck mission.

In simple terms, their aim has been to map the microwave background's temperature as a function of angular position on the sky, and then express it as a sum over spherical harmonics $Y_{\ell m}$, which are functions of the direction angles. Normally astronomers use the Milky-Way-centered coordinate system of Galactic longitude l (running from $0°$ to $360°$) and Galactic latitude b (running from $-90°$ to $+90°$), where the direction $l = 0$, $b = 0$ is the direction from the Sun to the centre of our Galaxy (called Sag A*). Then the measurements of the temperature T as a function of ℓ and m lead to the representation of the angular distribution in multipoles:[2]

$$T(l,\, b) = \sum_{\ell=0}^{\infty} \sum_{m=-\ell}^{\ell} T_{\ell m} Y_{\ell m}(l,\, b). \tag{13.1}$$

The multipole moments $T_{\ell m}$ contain the information about the temperature variations on different angular scales. The practical upper limit on the sum over ℓ is determined by the angular resolution of the measurements.

Our velocity relative to the local cosmological rest frame is contained in the dipole moments T_{1m} for $m = (-1, 0, 1)$. The Planck mission has measured these with extraordinary accuracy (Aghanim *et al.* 2019), and they give the following value for the velocity of the Solar System barycenter (center of mass) with respect to the local cosmological rest frame: $v = 369.82 \pm 0.11 \text{ km s}^{-1}$, in the direction $l = 264.021 \pm 0.001$, $b = 48.253 \pm 0.005$ (in degrees).

The main purpose of CMB angular-distribution measurements is not, however, to measure this velocity. The higher multipole moments for $\ell \geq 2$ are the target, because the Universe is of course not perfectly homogeneous. Galaxies and clusters of galaxies illustrate the small-scale inhomogeneity of the mass distribution, and these must have originated in the early Universe somehow. As noted above, inhomogeneities inevitably produce anisotropies, and by measuring the anisotropies of the CMB, which largely reflect the physical state of the Universe at the time the CMB radiation was emitted, one can get an immense amount of information about the early phase, going back to inflation. We will come back to this in detail below, but it is worth noting that the Planck mission was able to measure to beyond $\ell = 2000$. The typical anisotropies are of order 10^{-5} to 10^{-6} kelvins, which reassures us that, if we are interested in the dynamics of the Universe as a whole, rather than in the evolution of any small domain, then the assumption of homogeneity and isotropy is a very good one indeed.

Because of this extraordinary observed simplicity of the Universe on large scales, we can safely describe the relation between recessional velocity and distance with a single constant of proportionality H, independent of direction:

$$v = Hd \tag{13.2}$$

Astronomers call H the *Hubble–Lemaître parameter.* Its present value is known as the *Hubble–Lemaître constant, H_0.* There are several ways of measuring H_0, as we mentioned

---

[2] Be careful not to confuse the angle l with the integer index ℓ.

in Chapter 12, and different methods today give different values. The differences are not large, but they are significant compared with the claimed accuracies of the various measurements. We will discuss this important issue below, but for now let us examine what we can learn from assuming a rough average value of $H_0 = 71 \, \text{km s}^{-1} \, \text{Mpc}^{-1}$, in astronomers' peculiar but useful units. To get its value in normal units, convert 1 Mpc to 3.1×10^{22} m. This gives $H_0 = 2.3 \times 10^{-18} \, \text{s}^{-1}$. In geometrized units, divide this by c to get $H_0 = 7.7 \times 10^{-27} \, \text{m}^{-1}$. Associated with the Hubble–Lemaître constant is the *Hubble–Lemaître time* $t_H = H_0^{-1} = 4.3 \times 10^{17}$ s. This is about 14 billion years, and it sets the timescale for the cosmological expansion. The age of the Universe will not exactly be this, since in the past the expansion speed has varied, but this gives the order of magnitude of the time that has been available for the Universe as we see it to have evolved. By comparison, Earth's age is about 4.54 billion years, inferred from radioactive dating of its rocks.

One may (or *should*) object that the above discussion of expansion ignores the relativity of simultaneity. If the Universe is changing in time – expanding – then it may be possible to find *some* definition of time such that hypersurfaces of constant time are homogeneous and isotropic, but this would not be true for other choices of a time coordinate. Moreover, Eq. 13.2 cannot be exact since, for $d > 1.3 \times 10^{26} \, \text{m} = 4200 \, \text{Mpc}$, the velocity exceeds the velocity of light!

These objections are both correct. Our discussion was a *local* one (applicable for recessional velocities $\ll 1$) and took the point of view of a particular observer, ourselves. Fortunately, the cosmological expansion is slow, so that over regions of dimension 1000 Mpc, large enough to study the average properties of the homogeneous Universe, the relative velocities of galaxies within the region are essentially nonrelativistic. Moreover, the average random velocities of galaxies relative to their near neighbors is typically less than $100 \, \text{km s}^{-1}$, which is certainly nonrelativistic, and is much smaller than the systematic expansion speed over cosmological distances.

Therefore, the correct relativistic description of the expanding Universe is that there exists a *preferred choice of time,* whose hypersurfaces are homogeneous and isotropic, and with respect to which Eq. 13.2 is valid in the neighborhood of the origin of the local inertial frame of *any* observer who is at rest with respect to these hypersurfaces at *any* location. The CMB fixes the choice of constant-time hypersurfaces: they must have constant CMB temperature. We shall derive below the corrections to Eq. 13.2 as one gets further away from the origin of the local frame.

The existence of a preferred cosmological reference frame may at first seem startling: did we not introduce special relativity as a way to get away from preferred reference frames? There is no contradiction, however: the laws of physics themselves are still invariant under a change of observer. But there is only one Universe, and its physical make-up defines a convenient reference frame. Just as when studying the Solar System it would be silly for us to place the origin of our coordinate system at, say, the position of Jupiter on 1 January 1900, so too would it be silly for us to develop the theory of cosmology in a frame that does not take advantage of the simplicity afforded by the large-scale homogeneity. From now on we will, therefore, work in the cosmological reference frame, with its preferred definition of time.

Models of the Universe: the cosmological principle

If we are to construct a large-scale model of the Universe, we must make some assumptions about regions that we have no way of seeing now because they are too distant for our telescope. We should in fact distinguish two different inaccessible regions of the Universe.

The first inaccessible region is the region which is so distant that no information (traveling on a null geodesic) could have reached us from it, no matter how early this information began traveling. This region is everything that is outside our past light cone. Such a region usually exists if the Universe has a finite age, as ours does. This is illustrated in the left-hand part of Figure 13.2. This 'unknown' region is unimportant in one respect: what happens there has no effect on the interior of our past light cone, so how we incorporate it into our model of the Universe has no effect on the way the model describes our observable history. On the other hand, our past light cone is a kind of horizon, which is called the *particle horizon:* as time passes, more and more of the previously unknown region enters the interior of our past light cone and becomes observable. So the unknown regions across the particle horizon can have a real influence on our future. In this sense cosmology is a retrospective science: it reliably helps us understand only our past.

It must be acknowledged, however, that if information began coming in tomorrow that yesterday's 'unknown' region was in fact very different from yesterday's observed Universe, say highly inhomogeneous, then we would be posed difficult physical and philosophical questions regarding the apparently special nature of our history until this moment. It is to avoid these difficulties that we usually assume that the unknown regions are very like what we can observe, and in particular tomorrow's observable Universe will still be homogeneous and isotropic.

There are very good reasons for adopting this idea. Consider the right-hand part of Figure 13.2. Two hypothetical observers are shown within our own past light cone, but at such an early time in the evolution of the Universe that their own past light cones are disjoint. Then they are outside each other's particle horizon. But we, from our location in spacetime, are able to observe the physical conditions near each of them, and what we

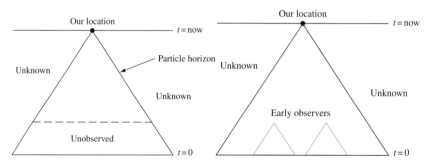

Figure 13.2 On the left, a schematic spacetime diagram showing the past history of the Universe, back to the Big Bang at $t = 0$. The 'unknown' regions have not had time to send us information; the 'unobserved' regions are obscured by intervening matter. On the right, the same history but showing the completely disjoint past histories of two earlier observers.

see is that they are statistically very similar. We can therefore confirm that if astronomers at those events apply the principle that regions outside their particle horizons are similar to regions inside, then they would be right! It seems unreasonable to expect that, if this principle holds for such observers, then it will not also hold for us.

This modern version of the Copernican Principle is called the *Cosmological Principle*, or more informally the *Assumption of Mediocrity*, the ordinary-ness of our own location in the Universe. It is, mathematically, an extremely powerful (i.e. restrictive) assumption. We shall adopt it, but one should bear in mind that predictions about the *future* depend strongly on the assumption of mediocrity.

The second inaccessible region is that part of the interior of our past light cone about which our instruments cannot get information. This includes galaxies so distant that they are too dim to be seen and events that are masked from view, such as those which emitted electromagnetic radiation before the epoch of decoupling (see below) when the Universe ceased to be an ionized plasma and became transparent to electromagnetic waves. The limit of decoupling is sometimes called our *optical horizon* since no light reaches us from beyond it (from earlier times). But gravitational waves did propagate freely before this, and soon we hope to begin to make observations across this 'horizon': the optical horizon is not a fundamental limit in the way that the particle horizon is.

Cosmological metrics

The metric tensor that represents a cosmological model must incorporate the observed homogeneity and isotropy. We shall therefore adopt the following idealizations about the Universe: (i) spacetime can be sliced into spacelike hypersurfaces of constant time which are perfectly homogeneous and isotropic; and (ii) the local rest frame associated with the Hubble–Lemaître flow agrees with this definition of simultaneity. Let us next try to simplify the problem as much as possible by adopting *comoving coordinates*: at each point, the coordinates are at rest in the local cosmological rest frame. We give each point a fixed set of coordinates $\{x^i; i = 1, 2, 3\}$. We choose our time coordinate t to be proper time in the local cosmological rest frame. The expansion of the Universe – the change of proper distance between points that are locally at rest with respect to the CMB – will be represented by time-dependent metric coefficients. Thus, if at one moment, t_0, the hypersurface of constant time has the line element

$$\mathrm{d}l^2(t_0) = h_{ij}(t_0)\, \mathrm{d}x^i\, \mathrm{d}x^j \tag{13.3}$$

(these h's have nothing to do with linearized theory), then the expansion of the hypersurface can be represented by

$$\mathrm{d}l^2(t_1) = f(t_1, t_0)h_{ij}(t_0)\, \mathrm{d}x^i\, \mathrm{d}x^j$$
$$= h_{ij}(t_1)\, \mathrm{d}x^i\, \mathrm{d}x^j. \tag{13.4}$$

This form guarantees that all the h_{ij} increase at the same rate; otherwise the expansion would be anisotropic (see Exercise 13.4). In general, then, Eq. 13.3 can be written

$$\mathrm{d}l^2(t) = R^2(t)h_{ij}\, \mathrm{d}x^i\, \mathrm{d}x^j, \tag{13.5}$$

where R is an overall scale factor which equals 1 at t_0, and where h_{ij} is a constant metric equal to that of the hypersurface at t_0. We shall shortly explore what form h_{ij} can take in detail.

Before that, we shall extend the constant-time hypersurface line element to a line element for the full spacetime. In general, it would be

$$ds^2 = -dt^2 + g_{0i}\, dt\, dx^i + R^2(t)h_{ij}\, dx^i\, dx^j, \qquad (13.6)$$

where $g_{00} = -1$, because we have assumed that t is the proper time along a line $dx^i = 0$. However, if the definition of simultaneity given by $t = $ const. is to agree with that given by the local Lorentz frame of the Hubble–Lemaître flow (idealization (ii) above), then $\vec{e}_0$ must be orthogonal to $\vec{e}_i$ in our comoving coordinates. This means that $g_{0i} = \vec{e}_0 \cdot \vec{e}_i$ must vanish, and we get

$$ds^2 = -dt^2 + R^2(t)h_{ij}\, dx^i\, dx^j. \qquad (13.7)$$

Now we investigate what form h_{ij} can take. Since it is isotropic, it must be spherically symmetric about the origin of the coordinates, which can of course be chosen to be located at any point we like. When we discussed spherical stars we showed that a spherically symmetric metric always has the line element (see the last part of Eq. 10.5)

$$dl^2 = e^{2\Lambda(r)}\, dr^2 + r^2 d\Omega^2. \qquad (13.8)$$

This form of the metric implies only isotropy about one point. We want a stronger condition, namely that the metric is homogeneous. A necessary condition for this is certainly that the Ricci scalar curvature of the three-dimensional metric, $R^i{}_i$, must have the same value at every point: every scalar must be independent of position at a fixed time. We will show below that, remarkably, this is sufficient as well, but for now we just treat it as the next constraint we place on the metric in Eq. 13.8. We can calculate $R^i{}_i$ using Exercise 6.35. Alternatively we can use Eqs. 10.15–10.17 from our discussion of spherically symmetric spacetimes in Chapter 10, realizing that G_{ij} for the line element in Eq. 13.8 is obtainable from G_{ij} for the line element in Eq. 10.7 for a spherical star, by setting Φ to zero. One obtains

$$G_{rr} = -\frac{1}{r^2}e^{2\Lambda}(1 - e^{-2\Lambda}),$$
$$G_{\theta\theta} = -re^{-2\Lambda}\Lambda', \qquad (13.9)$$
$$G_{\phi\phi} = \sin^2\theta\, G_{\theta\theta}.$$

The trace of this tensor is also a scalar, and must also therefore be constant. So instead of computing the Ricci scalar curvature, we simply require that the trace G of the three-dimensional Einstein tensor be a constant. (In fact this trace is just $-1/2$ the Ricci scalar.) The trace is

$$G = G_{ij}g^{ij}$$
$$= -\frac{1}{r^2}\, e^{2\Lambda}(1 - e^{-2\Lambda})e^{-2\Lambda} - 2re^{-2\Lambda}\Lambda' r^{-2}$$

$$= -\frac{1}{r^2} + \frac{1}{r^2}e^{-2\Lambda}(1 - 2r\Lambda')$$

$$= -\frac{1}{r^2}[1 - (re^{-2\Lambda})']. \tag{13.10}$$

Demanding homogeneity means setting G to some constant κ:

$$\kappa = -\frac{1}{r^2}[r(1 - e^{-2\Lambda})]'. \tag{13.11}$$

This is easily integrated to give

$$g_{rr} = e^{2\Lambda} = \frac{1}{1 + \frac{1}{3}\kappa r^2 - A/r}, \tag{13.12}$$

where A is a constant of integration. As in the case of spherical stars, we must demand local flatness at $r = 0$ (compare with § 10.5): $g_{rr}(r = 0) = 1$. This implies $A = 0$. Defining the more conventional constant $k = -\kappa/3$ gives

$$g_{rr} = \frac{1}{1 - kr^2}$$

$$dl^2 = \frac{dr^2}{1 - kr^2} + r^2\,d\Omega^2. \tag{13.13}$$

We have not yet proved that this space is isotropic about every point; all we have shown is that Eq. 13.13 is the unique space which satisfies the necessary condition that this curvature scalar be homogeneous. Thus, if a space that is isotropic and homogeneous exists at all, it must have the metric Eq. 13.13 for at least some k.

In fact, the converse *is* true: the metric of Eq. 13.13 is homogeneous and isotropic for *any* value of k. We will demonstrate this explicitly for positive, negative, and zero k separately in the next subsection. General proofs not depending on the sign of k can be found in, for example, Weinberg (2008) or Schutz (1980b). Assuming this result for the time being, therefore, we conclude that the full cosmological spacetime has the metric

$$ds^2 = -dt^2 + R^2(t)\left[\frac{dr^2}{1 - kr^2} + r^2\,d\Omega^2\right]. \tag{13.14}$$

This is called the *Friedmann–Robertson–Walker metric*. Notice that, without loss of generality, we can scale the coordinate r in such a way as to allow k to take one of the three values $+1, 0, -1$. To see this, consider for definiteness $k = -3$. Then redefine $\tilde{r} = \sqrt{3}r$ and $\tilde{R} = 1/\sqrt{3}R$, and the line element becomes

$$ds^2 = -dt^2 + \tilde{R}^2(t)\left[\frac{d\tilde{r}^2}{1 - \tilde{r}^2} + \tilde{r}^2\,d\Omega^2\right]. \tag{13.15}$$

What one cannot do with this rescaling is to change the sign of k. Therefore there are only three spatial hypersurfaces that we need to consider: $k = (-1, 0, +1)$. We call k the *curvature parameter*.

Three types of cosmological model

Here we prove that all three kinds of spatial hypersurfaces represent homogeneous and isotropic metrics that have different large-scale geometries.

Consider first $k = 0$. At any moment t_0, the line element of the hypersurface (obtained from the full spacetime metric by setting $dt = 0$) is

$$dl^2 = R^2(t_0)\left[dr^2 + r^2\,d\Omega^2\right] = d(r')^2 + (r')^2\,d\Omega, \qquad (13.16)$$

with $r' = R(t_0)r$. (Remember that $R(t_0)$ is constant on the hypersurface.) This is clearly the metric of flat Euclidean space, corresponding to the *flat* Friedmann–Robertson–Walker cosmological model. That it is homogeneous and isotropic is obvious.

Consider, next, $k = +1$. Let us define a new coordinate $\chi(r)$ such that

$$d\chi^2 = \frac{dr^2}{1 - r^2} \qquad (13.17)$$

and $\chi = 0$ where $r = 0$. This integrates to

$$r = \sin\chi, \qquad (13.18)$$

so that the line element for the space $t = t_0$ is

$$dl^2 = R^2(t_0)[d\chi^2 + \sin^2\chi\,(d\theta^2 + \sin^2\theta\,d\phi^2)]. \qquad (13.19)$$

We showed in Exercise 6.33 that this is the metric of a three-sphere of radius $R(t_0)$, i.e. of the set of points in four-dimensional Euclidean space that are all at a distance $R(t_0)$ from the origin. This model is called the *spherical* Friedmann–Robertson–Walker metric. It is a *closed* hypersurface, in the sense that it has only a finite three-volume. Being closed does not mean that it has a boundary, though: when moving along any curve in the three-sphere one never meets an edge to the space. The balloon analogy of cosmological expansion (Figure 13.1) is particularly appropriate for $k = 1$. The three-sphere is clearly homogeneous and isotropic: no matter where one stands in the three-sphere, it looks the same in all directions. Remember that the fourth spatial dimension – the radial direction to the center of the three-sphere – has *no* physical meaning for us: all our measurements are confined to our three-space so we can have no physical knowledge about the properties or even the existence of that dimension. At this point one should perhaps think of it as simply a tool for making it easy to visualize the three-sphere, not as an extra real dimension, although see the final section of this chapter for a potentially different point of view on this.

The third kind of cosmological model has $k = -1$. An analogous coordinate transformation (Exercise 13.8) gives the line element

$$dl^2 = R^2(t_0)(d\chi^2 + \sinh^2\chi\,d\Omega^2) \qquad (13.20)$$

This is called the *hyperbolic*, or *open*, Friedmann–Robertson–Walker model. Notice one peculiar property. As the proper radial coordinate χ increases away from the origin, the circumferences of spheres increase as $\sinh\chi$. Since $\sinh\chi > \chi$ for all $\chi > 0$, it follows that these circumferences increase *more* rapidly with proper radius than in flat space. For this reason this hypersurface is *not* realizable as a three-dimensional hypersurface in a four- or higher-dimensional Euclidean space. That is, there is no picture which we can easily draw like that for the three-sphere. The space is called 'open' because circumferences of spheres increase monotonically with χ: they do not close off as they do for $k = 1$.

In fact, as we show in Exercise 13.8, this geometry is the geometry of a particular three-dimensional hypersurface in four-dimensional *Minkowski* spacetime. Specifically, it is a hypersurface of events that all have the same timelike interval from the origin. As with the representation of the three-sphere as a subspace of four-dimensional Euclidean space, here also the fourth dimension of the Minkowski spacetime (in this case the time dimension) is not accessible to us physically and so is not real in any measurable sense. It is simply a device to allow us to visualize the hypersurface. Notice that this hypersurface is invariant under Lorentz transformations in the larger Minkowski spacetime, which effectively move the origin of the hypersurface to different points without changing the interval that all the events have from the origin of the spacetime. This demonstrates that the hypersurface is indeed homogeneous and isotropic, exactly analogously to the way in which the hypersurfaces for $k = 1$ are invariant under rotations in the four-dimensional Euclidean space that preserve the four-dimensional distances of all points from the four-dimensional origin.

Cosmological redshift as a distance measure

When studying small regions of the Universe around the Sun, astronomers measure proper distances to stars and other objects and express them in parsecs, as we have seen, or in multiples such as kpc and Mpc. But if an object is at a cosmological distance in our expanding Universe, then what we mean by distance is a little ambiguous, due to the long time it takes light to travel from the object to us. Its separation from our location when it emitted the light that we receive today may have been much less than its separation at present, i.e. on the present hypersurface of constant time. Indeed, the object may not even exist any more: all we know about it is that it existed at the event on our past light cone when it emitted the light we receive today. But between then and now it might have exploded, collapsed, or otherwise changed dramatically. So the notion of the separation between us and the object *now* is not as important as it might be for local measurements.

Instead, astronomers commonly use a different measurement of separation: the redshift z of the spectrum of the light emitted by the object, let us say a galaxy. In our expanding Universe, the Hubble–Lemaître law (Eq. 13.2) states that the further away the galaxy is, the faster it is receding from us, so the redshift is a nice monotonic measure of separation: larger redshifts imply larger distances. Of course, as we noted in the discussion following Eq. 13.2, the galaxy's redshift contains a contribution from its random local velocity; over cosmological distances this is a small uncertainty. For parts of the Universe that are so nearby that random velocities dominate the Hubble–Lemaître flow, astronomers use conventional distance measures, mainly Mpc, instead of redshift.

To compute the redshift in our cosmological models, let us assume for simplicity that the galaxy moves exactly with the Hubble–Lemaître flow, which means that it has a fixed coordinate position on some hypersurface at the cosmological time t at which it emits the light that we eventually receive at time t_0. Recall our discussion of conserved quantities in § 7.4: if the metric is independent of a coordinate, then the associated covariant component of momentum is constant along a geodesic. In the cosmological case, the homogeneity of the hypersurfaces ensures that the covariant components of the spatial momentum of a photon that we receive from this galaxy are constant along its trajectory. Suppose that we place ourselves at the origin of the cosmological coordinate system (since the cosmology is homogeneous, we can put the origin anywhere we like), so that light from this galaxy travels along a radial line θ = const., ϕ = const. towards us. In each of the cosmologies the line element restricted to the trajectory has the form

$$0 = -\mathrm{d}t^2 + R^2(t)\,\mathrm{d}\chi^2. \tag{13.21}$$

(To obtain this for flat hypersurfaces, simply rename the coordinate r in the first part of Eq. 13.16 to χ.) It follows that the relevant conserved quantity for the photon is p_χ. Now, the cosmological time coordinate t is proper time, so the energy as measured by a local observer at rest in the cosmology anywhere along the trajectory is $-p^0$. We argue in Exercise 13.9 that conservation of p_χ implies that p^0 is inversely proportional to $R(t)$. It follows that the wavelength as measured locally (in proper distance units) is proportional to $R(t)$, and hence that the redshift z of a photon emitted at time t and observed by us at time t_0 is given by

$$1 + z = R(t_0)/R(t). \tag{13.22}$$

It is important to keep in mind that this is just the cosmological part of any overall redshift: if the source or observer is moving radially relative to the cosmological rest frame then there will be a further factor of $1 + z_{\mathrm{motion}}$ multiplied into the right-hand side of Eq. 13.22 for each of those motions.

We now show that the *Hubble–Lemaître parameter* $H(t)$ is the instantaneous relative rate of expansion of the Universe at time t:

$$H(t) = \frac{\dot{R}(t)}{R(t)}. \tag{13.23}$$

The galaxy that we are observing at a fixed coordinate location χ is being carried away from us by the cosmological expansion. At the present time t_0 its proper distance d from us (in the constant-time hypersurface) is the same for each of the cosmologies when expressed in terms of χ:

$$d_0 = R(t_0)\chi. \tag{13.24}$$

It follows by differentiating this that the current rate of change of proper distance between the observer at the origin and the galaxy at fixed χ is

$$v = \dot{R}(t_0)\chi = (\dot{R}/R)_0 d_0. \tag{13.25}$$

By comparison with Eq. 13.2, we see that $(\dot{R}/R)_0$ is just the present value of the Hubble–Lemaître parameter. Since our present time is not special, it follows that Eq. 13.23 will hold at any time t.

We show in Exercise 13.10 that the velocity in Eq. 13.25 is just $v = z$, which is what is required to give the redshift z, provided that the galaxy is not far away. In our cosmological neighborhood, therefore, the cosmological redshift is a true Doppler shift. Moreover, the redshift is proportional to proper distance in our neighborhood, with the Hubble–Lemaître constant as the constant of proportionality.

We shall now investigate how various measures of distance depend on redshift when we leave our cosmological neighborhood. The scale factor of the Universe $R(t)$ is related to the Hubble–Lemaître parameter by Eq. 13.23. Integrating this for R gives

$$R(t) = R_0 \exp\left[\int_{t_0}^{t} H(t')\,dt'\right]. \tag{13.26}$$

The Taylor expansion of this about the present time t_0 is

$$R(t) = R_0[1 + H_0(t - t_0) + \tfrac{1}{2}(H_0^2 + \dot{H}_0)(t - t_0)^2 + \cdots], \tag{13.27}$$

where the subscript zeros denote quantities evaluated at t_0. The time-derivative of the Hubble–Lemaître parameter contains information about the acceleration or deceleration of the expansion. Cosmologists in the past normally replaced $\dot{H}_0$ with the dimensionless *deceleration parameter* q_0, defined as

$$q_0 = -R_0\ddot{R}_0/\dot{R}_0^2 = -\left(1 + \dot{H}_0/H_0^2\right). \tag{13.28}$$

The minus sign in the definition and the name 'deceleration parameter' reflect the assumption, when this parameter was first introduced, that gravity would be slowing down the cosmological expansion, so that q_0 would be positive. However, astronomers now believe that the expansion of the Universe is accelerating, so the use of q_0 has gone out of fashion. Of course, any formula containing $\dot{H}_0$ can be converted to one in terms of q_0 and vice versa.

What does the Hubble–Lemaître law, Eq. 13.2, look like to this accuracy? The recessional velocity v is deduced from the redshift of spectral lines, so it is more convenient to work directly with the redshift. Combining Eq. 13.26 with Eq. 13.22 we get that the redshift measured by us today (t_0) of a galaxy that emitted its light at the earlier time t is

$$1 + z(t) = \exp\left[-\int_{t_0}^{t} H(t')dt'\right]. \tag{13.29}$$

The Taylor expansion of this is

$$z(t) = H_0(t_0 - t) + \tfrac{1}{2}(H_0^2 - \dot{H}_0)(t_0 - t)^2 + \cdots. \tag{13.30}$$

This is not directly useful yet, since we have no independent information about the time t at which the galaxy emitted its light. Perhaps Eq. 13.30 is more useful when inverted to give an expansion for the look-back proper time to an event with redshift z:

$$t_0 - t(z) = H_0^{-1} \left[z - \tfrac{1}{2}(1 - \dot{H}_0/H_0^2)z^2 + \cdots \right].$$ (13.31)

From the simple expansion

$$H(t) = H_0 + \dot{H}_0(t - t_0) + \cdots,$$

one can use the previous equation to replace $(t - t_0)$ and get an expansion for H as a function of z. Keeping only the term linear in z gives

$$H(z) = H_0 \left(1 - \frac{\dot{H}_0}{H_0^2} z + \cdots \right).$$ (13.32)

Note that Eq. 13.29 can also be inverted to give the exact and very simple relation

$$H(t) = -\frac{\dot{z}}{1 + z}.$$ (13.33)

Although cosmology is self-consistent only within a relativistic framework, it is nevertheless useful to ask how the expansion of the Universe looks from a Newtonian point of view. One imagines a spherical region uniformly filled with galaxies, starting at some time with radially outward velocities that are proportional to the distance from the center of the sphere. If we are not near the edge – and of course the edge may be much too far away for us to see today – then one can show that the expansion is homogeneous and isotropic about every point. The galaxies just fly away from one another, and the Hubble–Lemaître constant is the scale for the initial velocity: it is the radial velocity per unit distance away from the origin.

The problem with this Newtonian model is not that it cannot describe the local state of the Universe; rather, it is that, with gravitational forces that propagate instantaneously, the dynamics of any bit of the Universe depends on the structure of this cloud of galaxies arbitrarily far away, i.e. the dynamics depends on the edge, if it exists. In a relativistic theory of gravity, on the other hand, the dynamical evolution of the Universe depends only on conditions inside our past light cone, and these conditions are in principle observable. Only in a relativistic theory of gravity, therefore, can we make sense of the dynamical evolution of the Universe. This is a subject we will study below.

When light is redshifted, it loses energy. Where does this energy go? The fully relativistic answer is that it just goes away: since the metric depends on time, there is no conservation law for energy along a geodesic. Interestingly, in the Newtonian picture of the Universe just described, the redshift is just caused by the different velocities of the diverging galaxies relative to one another. As a photon moves outward in the expanding cloud, it finds itself passing galaxies that are moving faster and faster relative to the center. It is not surprising that observers in these galaxies measure the energy of the photon to be smaller and smaller as it passes them and continues to move outwards. This reinforces our earlier remark that, at least in our cosmological neighborhood, the cosmological redshift is a Doppler shift.

Cosmography: measures of distance in the Universe

Cosmography refers to the description of the expansion of the Universe and its history. In cosmography we do not yet apply the Einstein equations to explain the dynamics of the Universe; instead we simply measure its expansion history. The language of cosmography is the language of distance measures and the evolution of the Hubble–Lemaître parameter.

By analogy with Eq. 13.2, we would like to replace t in Eq. 13.30 with distance. But what measure of distance is suitable over vast cosmological separations? Not coordinate distance, which would be unmeasurable. What about proper distance? The proper distance between the events of emission and reception of light is zero, since light travels on null lines. The proper distance between an emitting galaxy and us at the present time is also unmeasurable: in principle, the galaxy may not even exist now, as we remarked earlier. To get out of this difficulty, let us ask how distance crept into Eq. 13.2 in the first place.

Distances to nearby galaxies are almost always inferred from luminosity measurements. Consider an object whose distance d is known, which is at rest, and which is near enough to us that we can assume that space is Euclidean. Then a measurement of its flux F leads to an inference of its absolute luminosity:

$$L = 4\pi d^2 F, \qquad (13.34)$$

provided we assume that its flux is isotropic, i.e. that it radiates the same flux F in all directions. Alternatively, if the object's absolute luminosity L is known, then a measurement of F leads to the distance d, again assuming isotropy. The role of d in Eq. 13.2 is, then, as a replacement for the observable $(L/F)^{1/2}$.

Astronomers have used brightness measurements to build up a carefully calibrated *cosmological distance ladder* to measure the scale of the Universe. For each step on this ladder they identify a *standard candle*, which we defined in § 9.6 to be a class of objects whose absolute luminosity L is known (say from a theory of their nature or from reliably calibrated distances to nearby examples of this object). As their ability to see to greater and greater distances has developed, astronomers have found new standard candles that they could calibrate from previous standard candles but that were bright enough to be seen to greater distances than the previous ones. Indeed, there are a number of different distance ladders using different selections of standard candles. These distance ladders start at the nearest stars, whose distances can be measured by parallax (independently of luminosity), and some ladders continue all the way to very distant high-redshift galaxies.

In the spirit of such measurements, cosmologists define the *luminosity distance* d_L to any object, no matter how distant, by inverting Eq. 13.34:

$$d_L = \left(\frac{L}{4\pi F} \right)^{1/2}. \qquad (13.35)$$

The luminosity distance is often the observable that can be directly measured by astronomers: if the intrinsic luminosity L of the object is known or can be inferred, then a measurement of its brightness F determines the luminosity distance. The luminosity distance is the proper distance the object would have in a Euclidean cosmology if it were at rest with respect to us, if it had an intrinsic luminosity L, and if we received an energy flux F from it. However, in an expanding cosmology this will not generally be the proper distance to the object at the present time.

If an object does not radiate isotropically, then to infer its luminosity distance we must know the radiated flux in our direction. Standard sirens, which we introduced in § 9.6, are an example of this. We mentioned there that the distance we obtain from a gravitational wave measurement of a standard siren is the cosmological luminosity distance. But the gravitational wave amplitude radiated perpendicular to the orbital plane is twice that emitted in directions in the plane, so the radiated flux up the orbital axis is four times larger than in the plane. In order to use standard sirens as a distance measure, we need to know the inclination angle of the orbital plane with respect to the direction to the source. Fortunately, this information can be deduced from the degree of elliptical polarization of the signal, which can be measured by a network of detectors. (See Exercise 12.7.)

We shall now find the relation between luminosity distance and the cosmological scales we have just introduced. We shall assume isotropic emission just to keep the discussion simple, but the results can also be applied to anisotropic emitters whose flux in our direction is known. Consider an object emitting with luminosity L at a time t_e. What flux do we receive from it at the later time t_0? Suppose for simplicity that the object gives off only photons of frequency ν_e at time t_e. (This frequency will drop out in the end, so our result will be perfectly general.) In a small interval of time δt_e the object emits

$$N = L\delta t_e/h\nu_e \tag{13.36}$$

photons in a spherically symmetric manner. To find the flux we receive, we must calculate the area of the sphere that these photons occupy at the time we observe them, the energy of the photons relative to us, and the time it takes this group of photons to pass us.

For this calculation we place the object at the origin of the coordinate system, and suppose that we sit at coordinate position r in this system, as given in Eq. 13.14. Then when the photons reach our coordinate distance from the emitting object, the proper area of the sphere they occupy is given by integrating over the sphere the solid-angle part of the line element in Eq. 13.14, which is $R_0^2 r^2 d\Omega^2$. The integration is just over the spherical angles and produces the area:

$$A = 4\pi R_0^2 r^2. \tag{13.37}$$

Now, the photons have been redshifted by the amount $(1 + z) = R_0/R(t_e)$ to frequency ν_0:

$$h\nu_0 = h\nu_e/(1 + z). \tag{13.38}$$

Moreover, they arrive spread out over a time δt_0 which is also stretched by the redshift:

$$\delta t_0 = \delta t_e(1 + z). \tag{13.39}$$

The energy flux at the observation time t_0 is thus $Nh\nu_0/(A\delta t_0)$, from which it follows that

$$F = L/[A(1 + z)^2]. \tag{13.40}$$

From Eq. 13.35 we then find that

$$d_\mathrm{L} = R_0 r (1 + z).$$
(13.41)

To use this we need to know the comoving source coordinate location r as a function of the redshift z of the photon emitted by the source. This comes from solving the equation of motion of the photon. In this case, all we have to do is use Eq. 13.14 with $ds^2 = 0$ (a photon world line) and $d\Omega^2 = 0$ (a photon traveling on a radial line from its emitter to the observer at the center of the coordinates). This leads to the differential equation

$$\frac{dr}{(1 - kr^2)^{1/2}} = -\frac{dt}{R(t)} = \frac{dz}{R_0 H(z)},$$
(13.42)

where the last step follows from differentiating Eq. 13.22. This equation involves the curvature parameter k, but for small r and z the curvature will come into the solution only at second order. If we ignore this at present and work only to first order beyond the Euclidean relations, it is not hard to show that

$$d_L = R_0 r (1 + z) = \left(\frac{z}{H_0}\right)\left[1 + \left(1 + \frac{1}{2}\frac{\dot{H}_0}{H_0^2}\right)z\right] + \cdots .$$
(13.43)

If one can measure the luminosity distances and redshifts of a number of objects, then one can in principle measure both H_0 and $\dot{H}_0$. Measurements of this kind led to the discovery of the accelerating expansion of the Universe (below).

Another convenient measure of distance is the *angular-diameter distance*. This is based on another way of measuring distances in a Euclidean space: the angular size θ of an object at a distance d can be inferred if we know the proper diameter D of the object transverse to the line of sight, $\theta = D/d$. This leads to the definition of the angular diameter distance d_A to an object anywhere in the Universe:

$$d_A = D/\theta.$$
(13.44)

The dependence of d_A on redshift z is explored in Exercise 13.17. The result is

$$d_\mathrm{A} = R_\mathrm{e} r = (1 + z)^{-2} d_\mathrm{L},$$
(13.45)

where R_e is the scale factor of the Universe when the photon was emitted. The analogous expression to Eq. 13.43 is

$$d_A = R_0 r / (1 + z) = \left(\frac{z}{H_0}\right)\left[1 + \left(-1 + \frac{1}{2}\frac{\dot{H}_0}{H_0^2}\right)z\right] + \cdots .$$
(13.46)

There are situations where one has in fact an estimate of the comoving diameter D of an emitter. In particular, the temperature irregularities in maps of the cosmic microwave background radiation (see below) have an intrinsic length scale that is determined by the physics of the early Universe.

Although we have provided small-z expansions for many interesting measures, it is important to bear in mind that astronomers today can observe objects out to very high redshifts. Some galaxies and quasars are known at redshifts greater than $z = 7$. The cosmological microwave background, which we will discuss below, originated at redshift $z \sim 1090$, and is our best tool for understanding the Big Bang. To make sense of these measurements, one has to understand the evolution of the size of the Universe over long times, which comes from solving Einstein's equations. We will address that in § 13.3 below.

Although a redshift of 1090 is large, the Universe was already some 380,000 years old at that time. Today, astronomers are studying the microwave background radiation to look for evidence imprinted on it of gravitational waves that originated in the Big Bang itself. If these can be measured then we will have direct evidence about the nature of our Universe when it was just a fraction of a second old. We will return to this below.

The derivations of Eqs. 13.43 and 13.46 illustrate a point which we have encountered before: in the attempt to translate the nonrelativistic formula $v = Hd$ into relativistic language, we were forced to re-think the meaning of all the terms in the equations and to go back to the quantities which one can directly measure. If the study of GR teaches us only one thing, it should be that physics rests ultimately on *measurements*: concepts like distance, time, velocity, energy and mass are derived from measurements, but they are often not the quantities directly measured and, one's assumptions about their global properties must be guided by a careful understanding of how they are related to measurements.

Measuring the Hubble–Lemaître constant – not a simple matter

The importance of the Hubble–Lemaître constant for understanding the age and particularly the evolution of our Universe means that measuring it accurately has been a major preoccupation of many astronomers from Hubble's time onwards. Their principal difficulty has been to arrive at reliable distances to objects at cosmological distances. Hubble himself derived a value of about $500 \, \text{km} \, \text{s}^{-1} \, \text{Mpc}^{-1}$, underestimating true distances by a factor of about 7. The distance ladder was improved over the years, but even as late as 1990 there were two different groups claiming values around 50 and 100. But by the end of that decade it had become clear that the value was near $70 \, \text{km} \, \text{s}^{-1} \, \text{Mpc}^{-1}$.

The most reliable type of standard candle has usually been taken to be the Cepheid variable stars. These stars pulsate radially, causing corresponding changes in their brightness. Henrietta Swan Leavitt, studying stars in the Large Magellanic Cloud in 1908, observed that the average apparent brightness of a star was strongly correlated with the pulsation period. Since all the stars were at about the same distance from us, that meant that there is a remarkable fundamental relationship between the mean luminosity of the star and its pulsation period. Since there are a number of Cepheid variables in our Galaxy as well, these stars could be used as standard candles anywhere if the absolute distances to a few of them could be determined.

Calibrating the Cepheid distance scale therefore became a major activity in cosmological astronomy, and changes in the scale largely account for the changes in the estimated values of H_0 that we noted above. Today the best calibration comes from direct parallax measurements of Cepheids by the Gaia satellite (Brown *et al.* 2018). However, there is some evidence that the Cepheid law changes systematically with time by a small amount, as the Universe evolves and the nucleosynthesis of heavier elements results in stars formed at different epochs having different chemical compositions (Gieren *et al.* 2018).

More recently it was realized that supernovae of Type Ia are good standard candles. These are systems in which a white-dwarf star explodes. The explosion is usually triggered by accretion of matter from a binary companion star. Normally one would expect that when the accretion pushes the mass of the white dwarf above the Chandrasekhar mass (recall the discussion in § 10.7), it will simply collapse to form a neutron star. But for certain compositions of the white dwarf (which depend on its own formation history), just before the mass reaches the Chandrasekhar limit, nuclear reactions are triggered that incinerate the entire star, creating an explosion that is even brighter than a typical Type-II supernovae, which is the kind that does form neutron stars and black holes. Because the mass of the progenitor is determined by the fundamental physics of electron degeneracy for the narrow range of compositions thought to lead to this kind of explosion, there is good reason to expect that the luminosity of such supernovae will all be very similar, and even more importantly will not change systematically with changes over cosmic time of the composition of the stars from which the white dwarfs form. Such composition changes may affect how many white dwarfs of the right composition form through stellar evolution, but once such a dwarf exists in a binary system, the resulting explosion depends mainly on the fundamental physics.

But Type-Ia supernovae still need to be calibrated: the luminosity is difficult to compute reliably from numerical simulations of the explosions. This is normally done by using Cepheid variables in our cosmological neighborhood, thus avoiding the problem of their systematic luminosity changes with redshift. If a Type-Ia supernova occurs in a nearby galaxy, astronomers search that galaxy for Cepheids, to compute its distance and thereby to infer the absolute luminosity of the supernova. Since the supernovae are intrinsically much brighter than Cepheids, this calibration allows the supernovae to be used to estimate cosmological distances (Riess *et al.* 2019).

Other standard candles have also been proposed, both as calibrators of Type Ia supernovae or as standard candles for H_0 measurements themselves. For example it is thought that red giant stars have a maximum luminosity that is again independent of the composition of the star and of cosmic time; this method is known as the 'tip of the red giant branch' (TRGB) candle. Gravitational lensing has also been used: if a system with multiple images of a distant object like a quasar is accurately enough modeled, then it is possible to estimate the Hubble–Lemaître constant. But at present there is considerable divergence among the values that follow from different ways of doing the modeling.

Gravitational wave binary mergers are, as we saw in § 9.6, standard sirens, whose luminosity distances can be read from the signal. These do not need calibration, and so they are expected to offer an independent test of the systematics of these various distance ladders. The first measurement of H_0 using this method was done with GW170817, as described in § 12.4. But the uncertainty was large because the inclination of the orbit could not be determined accurately. It seems likely that a value of H_0 with an accuracy of a few percent will be possible only once the number of binaries that can be used is large enough to average down these kinds of uncertainties. While GW170817 was a binary-neutron-star merger with an identified counterpart, such mergers are rare. Far more common are binary-black-hole events, which have no observable counterpart. These can also be used to find H_0, by a statistical method first suggested by Schutz (1986). In the upcoming LVK observing run O4, it is possible that 200–300 such binaries will be detected in a one-year observing run, so that this could lead to the first standard-siren determination of H_0 with a usefully small uncertainty. This might of course be too optimistic, but in the long run, and especially after the third-generation detectors are built, the standard-siren method is likely to provide the most accurate value of H_0. LISA may be able to take this to much higher redshifts, as we shall see below.

Most importantly, there are also ways to measure H_0 that use the physics of the early Universe. These are measurements of the anisotropies of the CMB, and measurements of the early density irregularities that led to galaxy formation. These density irregularities are called baryon acoustic oscillations (BAOs). Because they use the angular sizes of anisotropies, they measure H_0 using the angular-diameter distance rather than the luminosity distance. We will discuss how they work in § 13.4 below, after we have introduced the physics of the early Universe.

We saw in Figure 12.6 in the previous chapter that the two principal methods for measuring H_0 are the Type Ia supernovae and the anisotropies of the CMB. These give values for H_0 that, while not very different from each other, are separated by several standard deviations in their measurement accuracies. At the current time (mid-2021) the supernovae give $74.03 \pm 1.42\,\mathrm{km\,s^{-1}\,Mpc^{-1}}$ (Riess *et al.* 2019) and the CMB gives $67.4 \pm 0.5\,\mathrm{km\,s^{-1}\,Mpc^{-1}}$ (Aghanim *et al.* 2019). This discrepancy might be explained by unknown systematic errors in either observing method, including the possible evolution of supernova luminosities with time, or it might be that there is some important physics being left out of our cosmological models. Measurements made with gravitational waves may, over the next few years, become accurate enough to shed light on how to resolve this so-called 'tension' between the two kinds of measurements.

The Universe is accelerating!

The most remarkable cosmographic result since Hubble's original work has been the discovery that the expansion of the Universe is not slowing down, but rather speeding up. This result was found by essentially performing the Type Ia measurement of the Hubble–Lemaître constant at much larger redshifts. Although the current value of H_0 is somewhat uncertain, the acceleration can be detected reliably using the same methods because it has to do with how $H(z)$ changes with redshift, and that will be visible in the data even if the overall scale $H_0 = H(z = 0)$ is changed.

Two teams of astronomers, called the High-z Supernova Search Team (Riess *et al.* 1998) and the Supernova Cosmology Project (Perlmutter *et al.* 1999), respectively, used Type Ia supernovae as standard candles out to redshifts of order 1. Although there was considerable scatter among the data points, both teams found that the best fit to the data was given by the assumption that the Universe is speeding up and not slowing down.

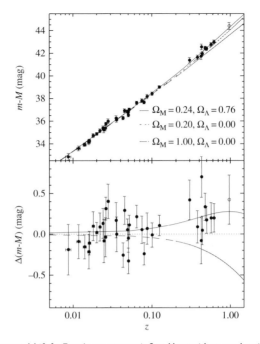

Figure 13.3 The trend of luminosity versus redshift for Type Ia supernovae is fitted best with an accelerating cosmology. The slope of the low-*z* part of this curve determines H_0; the bending of the high-*z* part demonstrates acceleration. See the text for a full explanation of what is plotted here. High-*z* Supernova Search Team: Riess *et al.* (1998) 'Observational evidence from supernovae for an accelerating Universe and a cosmological constant'. *Astron. J.* **116**,1009. © AAS. Reproduced with permission. DOI: 10.1088/0004-637X/746/1/85.

Figure 13.3 shows the data and error bars from the High-*z* Team, plotted logarithmically as luminosity distance against redshift, but where the luminosities are given in magnitudes. (The difference between the apparent magnitude *m* and the intrinsic magnitude *M* in this diagram is proportional to the logarithm of the ratio of the observed flux to the intrinsic luminosity. This in turn is proportional to the logarithm of the luminosity distance, as in Eq. 13.35. Because larger magnitudes represent *dimmer* objects, the vertical scale represents decreasing apparent brightness, or increasing distance.) This kind of diagram is called a *magnitude–redshift* diagram, and we derive its low-*z* expansion in Exercise 13.13.

Three possible fits are shown, which are described by values of two parameters Ω_M and Ω_Λ. These are measures of the energy densities of matter and so-called dark energy, respectively, scaled to the so-called critical energy density of the Universe now. The critical density is the total energy density our Universe would have if it were well described by the flat Friedmann–Robertson–Walker spatial hypersurfaces, $k = 0$ in Eq. 13.14. These terms will make more sense after our discussion of the dynamical evolution of the Universe in the next section. For now, we just need to know that the matter density Ω_M is a parameter that causes the expansion to slow down and the dark energy density Ω_Λ is a parameter that causes it to accelerate.

The upper diagram in Figure 13.3 shows the measured flux, and the trend seems to curve upwards, meaning that at high redshifts the supernovae are dimmer than expected. This would happen if the Universe were speeding up, because the supernovae would simply be further away than expected. The lower diagram shows the same data but plots only the residuals from the fit to the middle set of parameters, a model of our Universe with zero acceleration parameter and matter with a density of 20% of the critical density. This shows more clearly how the data favor the curve for the accelerating Universe with nonzero dark energy ($\Omega_\Lambda = 0.76$).

Astronomers and physicists initially resisted this conclusion, because it undermines a basic assumption we have always made about gravity, that it is universally attractive. If the energy density of the Universe exerts attractive gravity, the expansion should be slowing down. Instead it is speeding up. What can be the cause of this repulsion? This is still an unsolved mystery, to which we shall return repeatedly through the rest of this chapter. But we shall see in the next section that Einstein already gave us a tool that might help to explain it: the cosmological constant.

The data shown in Figure 13.3 were only the first strong evidence for acceleration. Since then, several other lines of investigation have led to the same conclusion, so that now there is little doubt that the acceleration is real. The figure also reinforces the point made earlier, that a rescaling of the present value of H_0 cannot explain the acceleration seen at redshifts of order 1.

13.3 Cosmological dynamics: understanding the expanding Universe

In the last section we saw how to describe a homogeneous and isotropic cosmological model and how to measure its expansion. In order to study the evolution of the Universe, to understand the creation of the huge variety of structures that we see, and indeed to make sense of the accelerating expansion measured today, we need to apply Einstein's equations to the problem, and to marry them with enough physics to explain what we see. In this section we will study Einstein's equations, with relatively simple perfect-fluid physics and with the cosmological constant that seems to be implied by the expansion. In the following sections we will study more and more of cosmological physics.

Dynamics of Friedmann–Robertson–Walker cosmologies: Big Bang and dark energy

We have seen that a homogeneous and isotropic cosmology must be described by one of the three Friedmann–Robertson–Walker metrics given by Eq. 13.14. For each choice of the curvature parameter $k = (-1, 0, +1)$, the evolution of the cosmological model depends on just one function of time, the scale factor $R(t)$. Einstein's equations will determine its behavior.

As in earlier chapters, we idealize the Universe as being filled with a homogeneous perfect fluid. The fluid must be at rest in the preferred cosmological frame, for otherwise its velocity would allow us to distinguish one spatial direction from another: the Universe would not be isotropic. Therefore the stress–energy tensor will take the form of Eq. 4.36 in the cosmological rest frame. Because of homogeneity, all fluid properties depend only on time: $\rho = \rho(t), p = p(t)$, etc.

First we consider the equation of motion for matter, $T^{\mu\nu}{}_{;\nu} = 0$, which follows from the Bianchi identities of Einstein's field equations. Because of isotropy, the spatial components of this equation must vanish identically. Only the time component $\mu = 0$ is nontrivial. It is easy to show (see Exercise 13.24) that it gives

$$\frac{\mathrm{d}}{\mathrm{d}t}(\rho R^3) = -p\frac{\mathrm{d}}{\mathrm{d}t}(R^3),\qquad(13.47)$$

where $R(t)$ is the cosmological expansion factor. This is easily interpreted: R^3 is proportional to the volume of any fluid element, so the left-hand side is the rate of change of the element's total energy, while the right-hand side is the work it does as it expands ($-p\,\mathrm{d}V$).

There are two simple cases of interest, a *matter-dominated cosmology* and a *radiation-dominated cosmology*. In a matter-dominated era, which includes the present epoch, the main energy density of the cosmological fluid is in cold nonrelativistic matter particles, whose random velocities are small and which therefore behave like dust: $p = 0$. So we have

$$\text{Matter-dom:}\quad \frac{\mathrm{d}}{\mathrm{d}t}(\rho R^3) = 0.\qquad(13.48)$$

In a radiation-dominated era (as we shall see, in the early Universe) the principal energy density of the cosmological fluid is in radiation or hot, highly relativistic particles, which have an equation of state $p = \frac{1}{3}\rho$ (Exercise 4.22). Then we get

$$\text{Radiation-dom:}\quad \frac{\mathrm{d}}{\mathrm{d}t}(\rho R^3) = -\frac{1}{3}\rho\frac{\mathrm{d}}{\mathrm{d}t}(R^3),\qquad(13.49)$$

or

$$\text{Radiation-dom:}\quad \frac{\mathrm{d}}{\mathrm{d}t}(\rho R^4) = 0.\qquad(13.50)$$

The Einstein equations are also not hard to write down for this case. Isotropy will guarantee that $G_{tj} = 0$ for all j, and also that $G_{jk} \propto g_{jk}$. This means that only two components are independent, G_{tt} and (say) G_{rr}. But the Bianchi identity will provide a relationship between them, which we have already used in deriving the matter equation in the previous paragraph. (The same happened for the spherical star.) Therefore we need to compute only one component of the Einstein tensor (see Exercise 13.16):

$$G_{tt} = 3(\dot{R}/R)^2 + 3k/R^2.\qquad(13.51)$$

Therefore, besides Eqs. 13.48 or 13.50, we have only one further equation, the Einstein equation with cosmological constant Λ:

$$G_{tt} + \Lambda g_{tt} = 8\pi T_{tt}. \tag{13.52}$$

Physicists today hope that they will eventually be able to compute the value of the cosmological constant from first principles in a consistent theory where gravity is quantized along with all the other fundamental interactions. From this point of view, the cosmological constant will represent just another contribution to the whole stress–energy tensor, which can be given the notation

$$T_\Lambda^{\alpha\beta} = -(\Lambda/8\pi)g^{\alpha\beta}. \tag{13.53}$$

From this point of view, the energy density and pressure of the cosmological constant 'fluid' are

$$\rho_\Lambda = \Lambda/8\pi, \qquad p_\Lambda = -\rho_\Lambda. \tag{13.54}$$

Cosmologists call ρ_Λ the *dark energy:* an energy that is not associated with any known matter field. Its associated dark pressure p_Λ has the opposite sign. Physicists generally expect that the dark energy will be positive, so that most discussions of cosmology today are in the framework of $\Lambda \geq 0$. We will return below to the implications of the associated negative value for the dark pressure. Notice that, as the Universe expands, the dark energy density and dark pressure remain *constant*. In these terms the tt-component of the Einstein equations can be written

$$\tfrac{1}{2}\dot{R}^2 = -\tfrac{1}{2}k + \tfrac{4}{3}\pi R^2(\rho_m + \rho_\Lambda), \tag{13.55}$$

where now we write ρ_m for the energy density of the matter (including radiation), to distinguish it from the dark energy density. This equation makes it easy to understand the observed acceleration of our Universe. It appears (see below) that $k = 0$, or at least that the k-term is negligibly small. Then, since we are at present in a matter-dominated epoch, the term $R^2\rho_m$ decreases as R increases, while the term $R^2\rho_\Lambda$ increases rather strongly. Since today, as we shall see below, $\rho_\Lambda > \rho_m$, the result is that $\dot{R}$ increases as R increases. This trend must continue now forever, provided the acceleration is truly propelled by a cosmological constant, and not by some physical field that will go away later.

How, physically, can a positive dark energy density drive the Universe into accelerated expansion? Is not positive energy gravitationally attractive, which would simply make the deceleration even greater? To answer this it is helpful to look at the spatial part of Einstein's equations, where the acceleration $\ddot{R}$ explicitly appears. Rather than derive this from the Christoffel symbols, we can use the fact that (as remarked above) it follows from the two basic equations we have already written down: Eq. 13.47 and the time-derivative of Eq. 13.55. In Exercise 13.17 we show that the combination of these two equations implies the following simple 'equation of motion' for the scale factor:

$$\frac{\ddot{R}}{R} = -\frac{4\pi}{3}(\rho + 3p), \tag{13.56}$$

where ρ and p are the *total* energy density and pressure, including both the normal matter and the dark energy.

This shows us an important fact: in general relativity, the acceleration is produced, not by the energy density alone, but by $\rho + 3p$. We have met this combination before, in Exercise 8.20, where we called it the *active gravitational mass*. We showed there that, in general relativity, when pressure cannot be ignored, the source of the far-away Newtonian field is $\rho + 3p$, not just ρ. In the cosmological context, the same combination generates the cosmic acceleration or deceleration. It is clear that the negative pressure associated with the cosmological constant can, if it is large enough, make this sum negative, and that is what drives the Universe faster and faster. Einstein's gravity with a cosmological constant has a kind of in-built anti-gravity!

Notice that a negative pressure is not by any means unphysical. Negative stress is called *tension*, and in a stretched rubber band, for example, the component of the stress tensor along the band is negative. Interestingly, our analogy using a balloon to represent the expanding Universe also introduces a negative pressure, the tension in the stretched rubber. What is remarkable about the dark energy is that its tension is so large, and that it is isotropic. See Exercise 13.25 for a further discussion of the tension in this 'fluid'.

Equation 13.55 is written in a form that suggests studying the expansion of the Universe in a way analogous to the energy methods physicists use for particle motion (as we did for orbits in Schwarzschild in Chapter 11). The left-hand side looks like a 'kinetic energy' and the right-hand side contains a constant $(-k/2)$ that plays the role of the 'total energy' and a 'potential' term proportional to $R^2(\rho_m + \rho_\Lambda)$. The potential's dependence on R requires us to know how ρ_m depends on R. The dynamics of R will be constrained by this energy equation.

We can use this constraint to explore what might happen to our Universe in the far distant future, assuming of course the Cosmological Principle, that nothing significantly different comes over our particle horizon. If $\rho_\Lambda \geq 0$ (see above) and if the matter content of the Universe also has positive energy density, then one conclusion from Eq. 13.55 is immediate: *an expanding hyperbolic Universe (k = −1) will never stop expanding.* For the flat Universe ($k = 0$), an expanding Universe will also never stop if $\rho_\Lambda > 0$; however, if $\rho_\Lambda = 0$ then it could asymptotically slow down to a zero expansion rate as R approaches infinity, since the matter density will decrease at least as fast as R^{-3}. An expanding closed cosmology ($k = 1$) will, if $\rho_\Lambda = 0$, always reach a maximum expansion radius and then turn around and re-collapse, again because $R^2\rho_m$ decreases with R. A re-collapsing Universe eventually reaches another singularity, called the *Big Crunch!* But if $\rho_\Lambda > 0$ then the ultimate fate of an expanding closed Universe depends on the balance of ρ_m and ρ_Λ.

We can ask similar questions about the *history* of our Universe: was there a Big Bang, where the scale factor R had the value 0 at a finite time in the past? First we consider for simplicity $\rho_\Lambda = 0$. Then Eq. 13.55 shows that, as R gets smaller, the matter term gets more and more important compared to the curvature term $-k/2$. Again this is because $R^2 \rho_m$ is proportional either to R^{-1} for matter-dominated dynamics or, even more extremely, to R^{-2} for the radiation-dominated dynamics of the very early Universe. Since we are assuming $\rho_\Lambda = 0$ at the moment, then, since our Universe is expanding now, it could not have been at rest with $\dot{R} = 0$ at any time in the past. The question of the existence of a Big Bang, i.e. of whether we reach $R = 0$ at a finite time in the past, depends only on the behavior of the matter; the curvature term is not important, and all three kinds of cosmology have qualitatively similar histories.

Let us do the computation for a cosmological model that is radiation-dominated, as ours would have been at an early enough time, and keeping our assumption that $\Lambda = 0$. We write $\rho = BR^{-4}$ for some constant B, and we neglect k in Eq. 13.55. This gives

$$\dot{R}^2 = \tfrac{8}{3}\pi B R^{-2},$$

or

$$\frac{dR}{dt} = (\tfrac{8}{3}\pi B)^{1/2} R^{-1}. \tag{13.57}$$

This has the solution

$$R^2 = (\tfrac{32}{3}\pi B)^{1/2}(t - T), \tag{13.58}$$

where T is a constant of integration. So, indeed, $R = 0$ was achieved at a finite time in the past, and we conventionally adjust our zero of time so that $R = 0$ at $t = 0$, which means we redefine t so that $T = 0$.

Note that we have found that a radiation-dominated cosmology with no cosmological constant has an expansion rate where $R(t) \propto t^{1/2}$. If we had done this computation for a matter-dominated cosmology with $\rho_\Lambda = 0$ we would have found $R(t) \propto t^{2/3}$. (See Exercise 13.19.)

What happens if there is a cosmological constant? If the dark energy is positive, there is no qualitative change in the conclusion, since the term involving ρ_Λ simply increases the value of $\dot{R}$ at any value of R, and this brings the time where $R = 0$ closer to the present epoch. *If the matter density has always been positive, and if the cosmological constant is nonnegative, then Einstein's equations make the Big Bang inevitable: the Universe began with $R = 0$ at a finite time in the past.* This is called the *cosmological singularity*: the curvature tensor is singular, tidal forces become infinitely large, and Einstein's equations do not allow us to continue the solution to earlier times. Within the Einstein framework we cannot ask questions about what came before the Big Bang: time simply began there.

How certain, then, is our conclusion that the Universe began with a Big Bang? First, one must ask whether isotropy and homogeneity were crucial; the answer is no. The 'singularity theorems' of Penrose and Hawking (see Hawking & Ellis 1973) have shown that our Universe certainly had a singularity in its past, regardless of how asymmetric it may have been. But the theorems predict only the *existence* of the singularity: the nature of the singularity is unknown, except that it has the property that at least one particle in the present Universe must have originated in it. Nevertheless, the evidence is strong indeed that we *all* originated in it. Another consideration however is that we don't know the laws of physics at the incredibly high densities ($\rho \rightarrow \infty$) which existed in the early Universe. The singularity theorems of necessity assume (1) something about the nature of $T^{\mu\nu}$, and (2) that Einstein's equations (without cosmological constant) are valid at all R.

The assumption about the positivity of the energy density of matter can be challenged if we allow quantum effects. As we saw in our discussion of the Hawking radiation in Chapter 11, fluctuations can create negative energy for short times. In principle, therefore, our conclusions are not reliable if we are within one *Planck time* $t_{Pl} = GM_{Pl}/c^2 \sim 10^{-43}$ s of the Big Bang! (Recall the definition of the Planck mass in Eq. 11.111.) This is the domain of quantum gravity, and it may well turn out that, when we have a quantum theory of the gravitational interaction, we will find that the Universe has a history before what we call the Big Bang.

Philosophically satisfying as this might be, it has little practical relevance to the Universe we see today. We might not be able to start our Universe model evolving from $t = 0$, but we can presumably start it from, say, $t = 100t_{Pl}$ and rely on the Einstein framework from then on. The primary uncertainties about understanding the physical cosmology that we see around us are, as we will discuss below, to be found in the physics of what we call the early Universe, which is at a time much later than a few Planck times after the Big Bang.

So far we have restricted our attention to the case of a positive cosmological constant. While this seems to be the most relevant to the evolution of our Universe, cosmologies with negative cosmological constants are also interesting. We leave their exploration to Exercise 13.21.

Einstein introduced the cosmological constant in order to allow his equations to have a static solution, $\dot{R} = 0$. Hubble had not measured the expansion at the time, and Einstein followed the standard assumption of astronomers of his day that the Universe was static. Even in the framework of Newtonian gravity a static Universe would have presented problems, but no-one seems to have tried to find a solution until Einstein addressed the issue within general relativity. One has to do more than just set $\dot{R} = 0$ in Eq. 13.55: one has to guarantee that the solution is an equilibrium one, that the dynamics won't change $\dot{R}$, i.e. that the Universe is at a minimum or maximum of the 'potential' we discussed earlier. The reader can show in Exercise 13.20 that the static solution requires

$$\rho_\Lambda = \frac{1}{2}\rho_0.$$

Thus, for Einstein's static solution, the dark energy density has to be exactly half of the matter energy density. But this is an unstable situation: a small decrease in the matter density would make the dark energy take over, expanding the Universe faster and faster; a small increase in the matter density would overwhelm the dark energy, causing the Universe to collapse inexorably. So Einstein's way of avoiding a dynamical model for the Universe was unsatisfactory. Famously, when Hubble discovered the expansion, Einstein disavowed the cosmological constant, calling it his biggest-ever mistake.

It is indeed ironic, then, that the cosmological constant is back on the table! We shall see below that in our Universe the measured value of the dark energy density is about twice that of the matter energy density, so we are near to but not exactly at Einstein's static solution.

Critical density and the parameters of our Universe

If we divide Eq. 13.55 by $4\pi R^2/3$, we obtain a version that is instructive for discussions of the physics of the Universe:

$$\frac{3H^2}{8\pi} = -\frac{3k}{8\pi R^2} + \rho_m + \rho_\Lambda, \tag{13.59}$$

where we have substituted the Hubble–Lemaître parameter H for $\dot{R}/R$. Since the last two terms on the right are energy densities, it is useful to interpret the other terms in that way. Thus, the Hubble–Lemaître expansion has associated with it an energy density $\rho_H = 3H^2/8\pi$, and the spatial curvature parameter contributes an effective energy density $\rho_k = -3k/8\pi R^2$. This equation becomes

$$\rho_H = \rho_k + \rho_m + \rho_\Lambda.$$

Now, if in the Universe today the 'physical' energy density $\rho_m + \rho_\Lambda$ is less than the Hubble–Lemaître energy density ρ_H, then (as we have seen before), the curvature energy density must be positive, the curvature parameter k must be negative, and the Universe has hyperbolic hypersurfaces. Conversely, if the physical energy density is larger than the Hubble–Lemaître energy density, the Universe will be described by the closed model. The Hubble–Lemaître energy density is therefore a threshold, and we call it the *critical energy density* ρ_c:

$$\rho_c = \frac{3}{8\pi}H_0^2. \tag{13.60}$$

The ratio of any energy density to the critical value is called Ω, with an appropriate subscript. Thus, we can divide the earlier energy-density equation, evaluated at the present time, by ρ_c to get

$$1 = \Omega_k + \Omega_m + \Omega_\Lambda. \tag{13.61}$$

These are the quantities used to label the curves in Figure 13.3. The data from supernovae, the cosmic microwave background, and studies of the evolution of galaxy clusters (below) all suggest that our Universe at present has values approximately (Aghanim *et al.* 2019)

$$\Omega_\Lambda = 0.69, \quad \Omega_m = 0.31, \quad \Omega_k = 0. \tag{13.62}$$

These mean that we live in a flat Universe, dominated by a positive cosmological constant.

What sizes do these numbers have? Because the currently measured values of H_0 cluster near $70\,\mathrm{km\,s^{-1}\,Mpc^{-1}}$, we shall simplify the equations by scaling H_0 to that value, with the definition (nothing to do with gravitational wave amplitudes!):

$$h_{70} := H_0/(70 \, \text{km s}^{-1} \, \text{Mpc}^{-1}). \qquad (13.63)$$

Since h_{70} will be very close to 1, in most equations it can just be ignored if one wants a rough physical idea of the parameters of the cosmology. Using this scaling, the critical energy density is

$$\rho_c = 9.2 \times 10^{-27} h_{70}^2 \, \text{kg m}^{-3}.$$

As we have noted, the matter energy density is about 0.3 times this value, which is much more than astronomers can account for by counting stars and galaxies. In fact, studies of the formation of elements in the early Universe (below) tell us that the density of *baryonic* matter (normal matter made of protons, neutrons, and electrons) has $\Omega_b = 0.04$. So *most* of the matter in the Universe is nonbaryonic, does not emit light, and can be studied astronomically only indirectly, through its gravitational effects. This is called *dark matter*. So we can split Ω_m into its components:

$$\Omega_m = \Omega_b + \Omega_d, \quad \Omega_b = 0.04, \quad \Omega_d = 0.27. \qquad (13.64)$$

We will return in § 13.4 to a discussion of the nature and distribution of the dark matter.

The variety of possible cosmological evolutions and the data that constrain them are captured in the diagram in Figure 13.4. The evidence is strong that the dark energy is present, and dominant. That raises new, important questions. The deepest is, where in physics does this energy come from? We will mention below some of the speculations, but at present there is simply no good theory of it. In such a situation, better data might help. For example, astronomers could try to determine whether the dark energy density really is constant in time (as it would be if it comes from a cosmological constant) or variable, which would indicate that it comes from some physical field masquerading as a cosmological constant.

From the point of view of general relativity, one of the most intriguing ways of studying the dark energy is with the LISA gravitational wave detector. As mentioned in Section 12.2, LISA will be able to observe coalescences of black holes at high redshifts and measure their distances. What will be measured from the signal is the luminosity distance d_L to the binary, since it is based on an inference of the luminosity of the system in gravitational waves from the information contained in the signal. LISA should be able to make this measurement with great accuracy, perhaps with errors at the few percent level. To do cosmography, we have to combine these luminosity distance measures with redshifts, and that might be tricky: black-hole coalescences do not give off any electromagnetic radiation directly, so it may not be easy to identify the galaxy in which the event has occurred. But the galaxies hosting the mergers will not be normal galaxies, probably themselves having recently undergone a galaxy–galaxy merger that brought the two black holes together in the first place. The merger event might also be accompanied by other signs, such as an alteration of X-ray luminosity, the existence of unusual jets, or a disturbed morphology.

The European Space Agency is planning to launch its flagship Athena X-ray Observatory to overlap with LISA, which may sometimes allow rapid post-merger identification

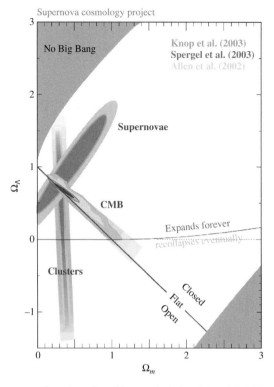

Supernova cosmology project

Figure 13.4   In the Ω_m–Ω_Λ plane one can see the variety of possible cosmological models and their histories and futures. The constraints from studies of supernovae (Knop *et al.* 2003), the cosmic microwave background radiation (Spergel 2003), and galaxy clustering (Allen *et al.* 2002) are consistent with one another and all overlap in a small region of parameter space centered on $\Omega_\Lambda = 0.7$ and $\Omega_m = 0.3$. This means that $\Omega_k = 0$ to within the errors. This conclusion has only been strengthened by the results from the Planck mission (Aghanim *et al.* 2019). Figure courtesy of the Supernova Cosmology Project. N. Suzuki *et al.* (2012) *Astrophys. J.* **746**, 85. The American Astronomical Society. All rights reserved. © AAS. Reproduced with permission.

of the galaxy. Therefore, it may well be possible in many cases to identify the host galaxy of a LISA merger, leading to a precise redshift measurement. This may lead not just to an accurate value of H_0 at high redshifts, but also to a better measurement of the rate of change of H_0 and therefore the possibility of testing whether the dark energy comes simply from a cosmological constant or from a physical field whose density and pressure change with time in some way.

It should be noted that, at these large cosmological distances, the Hubble flow velocity far exceeds the random velocities of galaxies, so that it will not be necessary to estimate the consequent corrections. However, over these distances gravitational lensing does begin to be important: the gravitational waves are likely to experience weak lensing (§ 11.1), which by distorting the amplitude of the wave will create errors in the luminosity distance inferred from the waveform.

As for measurements of H_0, gravitational wave measurements would be a desirable complement to other studies of the dark energy, because they need no calibration other than that of the detectors themselves: they would be independent of the assumptions of the cosmic distance ladder. In the case of LISA, such measurements might be able to reach to higher redshifts than other methods can.

13.4 Physical cosmology: the evolution of the Universe we observe

The observations described in the last section confirm the reliability of using a general relativistic cosmological model with dark energy to describe the evolution of the Universe, starting as far back as our observations can take us. During the last few decades, astrophysicists have developed a deep and rich understanding of how the observable Universe, with all its structure and variety, evolved out of a homogeneous hot expanding plasma. It is fair to say that there is now a consistent and reliable story that goes from the moment that protons and neutrons became identifiable particles right up to the formation of stars like our Sun and planets like our Earth. Many of the details are still poorly understood, especially where observations are difficult to perform, but the physical framework for understanding them is not in doubt. In this section we will try to sketch the main themes of this fascinating story.

The expansion of the Universe was accompanied by a general cooling off of its matter: the energy of the photons was redshifted, the random velocities of gas particles lessened, and structures such as galaxies and stars condensed out. The early history of the Universe is a thermal history. Over much of that history, the contents of the Universe were a plasma of radiation and matter in local thermodynamic equilibrium. Therefore, instead of using cosmological time t or the scale factor R to mark different stages of evolution of this plasma, we will use its temperature, or the equivalent energy, converting between them by $E = k_B T$, where k_B is Boltzmann's constant, not to be confused with the curvature parameter k that we introduced above. Using this monotonically decreasing energy as our time coordinate brings us closer to the physics.

Our understanding of the history of the Universe rests on our understanding of its physical laws, and these are tested up to energies of order 1 TeV in modern particle colliders. So, our physical picture of the evolution of the Universe can go back reliably as far as the time when the expanding plasma had that sort of energy.

Decoupling: forming the cosmic microwave background radiation

At the present time, of course, the Universe is far from being an equilibrium plasma. Therefore our first question is, when did thermal equilibrium begin to fail? If we start at the present moment and go backwards in time then the matter energy density increases as the scale factor R decreases but the dark energy density remains constant, so (unless the dark energy comes from some exotic physics) we can safely ignore the dynamical effect of the dark energy at redshifts larger than, say, 3. The density of ordinary matter (dark and baryonic) increases as R^{-3}, while the energy density of the photons of the cosmic microwave background increases as R^{-4}. Since the energy density of the cosmic microwave background today is $\Omega_\gamma \sim 10^{-5}$ and the matter density is $\Omega_m = 0.3$, they will have been equal when the scale factor was a factor 3×10^4 smaller than today. Since redshift and scale factor go together, this is the redshift when the expanding Universe changed from radiation-dominated to matter-dominated. The temperature was about 10^5 K and the energy scale was about 10 eV. This happened about 3,000 years after the Big Bang.

Now, this energy is near the ionization energy of hydrogen, which is 13.6 eV. This is an important number because hydrogen is the principal constituent of the baryonic matter. If the temperature is high enough to ionize hydrogen, the Universe will be filled with a plasma that is opaque to electromagnetic radiation. Once the hydrogen cools off enough to become neutral, the remaining photons in the Universe will be able to move through it with a low probability of scattering. This epoch of *decoupling* (also sometimes called recombination) defines the time at which the cosmic microwave background radiation was created. This actually occurs at a rather smaller energy than 13.6 eV, since there are enough free electrons to stop the photons even when only a small portion of the hydrogen is ionized. The epoch of decoupling occurred at a temperature a little below 1 eV, at a time when the Universe had already become matter-dominated. The redshift was about 2000, and the time was about 4×10^5 years after the Big Bang.

Notice that we have ignored the dark matter, even though it dominated the matter energy density. The usual assumption is that it consists of inert particles that hardly interact with normal matter or with photons. If this assumption were to fail before we got back to the decoupling era, then we would see the effects of, say, ionized dark matter in the Sun and even in high-temperature laboratory experiments. Even though the dark matter is not interacting with other matter, its energy density is increasing as R^{-3} just because of the changing proper volume of each small fluid element, in step with the energy density of baryonic matter up to decoupling.

Observations of the cosmic microwave background reveal that it has an almost perfect black-body spectrum with a temperature of $T = 2.725$ K. But they also show that it has small but significant temperature irregularities, departures from strict homogeneity that are the harbingers of the formation of galaxy clusters and galaxies. A map of these is shown in Figure 13.5. The temperature irregularities are of order 10^{-5} of the background temperature, and they are caused by irregularities in the matter distribution having the same relative size.

Because the dominant form of matter is the dark matter, the density fluctuations that are seen in Figure 13.5 and that led to galaxy formation were in its distribution. As we shall discuss in the next subsection, numerical simulations show that only if the random velocities (and therefore the pressure) of the dark matter particles were very small could the self-gravity of these only slightly over-dense regions halt their expansion, allowing the small irregularities to grow in size fast enough to trap baryonic gas and make it form galaxies. The dark matter had, therefore, to be *cold* at decoupling, and we call this model of galaxy formation the *cold dark matter model*. The standard cosmological model is called ΛCDM: cold dark matter with a cosmological constant.

The parameters of our cosmology – the cosmological constant, matter fraction, even the element abundances in the primordial gas – leave their imprint on these fluctuations.

Figure 13.5 A map of the small-scale temperature inhomogeneities of the cosmic microwave background, made by the Planck satellite (Aghanim *et al.* 2019). The range of fluctuations is $\pm 300\,\mu\text{K}$. Figure courtesy of the European Space Agency and the Planck Collaboration.

The fluctuations occur on all length scales, but they do not have the same size on different scales. The positions of the fluctuations, as shown in Figure 13.5, are completely random, but their sizes are affected by the physical conditions in the early Universe, such as the distance that an acoustic (compressional) wave could travel in the time since the Big Bang. This is reflected in the sizes of the different spherical harmonic terms in Eq. 13.1. The locations on the sky of the fluctuations affect the phases of the (complex) coefficients $T_{\ell m}$, but the physical information is in their amplitudes, whose squares are plotted in Figure 13.6. This angular spectrum of fluctuations contains a rich amount of information about the cosmological parameters, and it is here that one finds the constraints shown in Figure 13.4.

Dark matter and galaxy formation: the Universe after decoupling

Going forward from the time of decoupling, physicists have simulated the evolution of galaxies and clusters of galaxies from initial perturbations that have the same statistical distribution as those observed in the CMB. The density perturbations in the dark matter grow slowly, as they have been doing since before decoupling. However, before decoupling the baryonic matter could not respond very much to them, because it remained in equilibrium with the photons, which cannot get trapped. Once the baryonic matter took the form of neutral atoms, it could begin to fall into gravitational wells created by the dark matter.

Unlike the dark matter, the baryonic matter had the ability to concentrate itself at the bottoms of these wells. The reason for this is that the baryonic matter was charged: as the atoms fell into the potential wells, they collided with one another and excited their electrons into higher energy levels. The density was low enough that the typical time to the

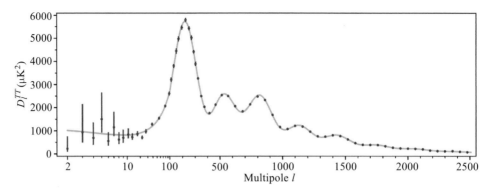

Figure 13.6 Plot of the power in the temperature fluctuations in Figure 13.5, as a function of their angular scale, as represented by the index ℓ in the spherical harmonic representation of the fluctuations, Eq. 13.1. At low values of l (which represent large angular scales) the errors are large because of the need to subtract the microwave radiation coming from our Galaxy. But on small scales (large l) the error bars are smaller than the dots used to plot them here. Plot made from data from the Planck satellite (Aghanim *et al.* 2019). Figure courtesy of the European Space Agency and the Planck Collaboration.

next collision was longer than the time it took for the electron to decay back to the ground state by radiating away the excess energy. The radiated photons simply left the gravitational well, because after decoupling the mean free path of a photon was enormous. This meant that there was a constant drain of energy away from the baryonic matter, imprisoning it deeper and deeper into the dark-matter well. Astronomers call this process *cooling*, even though the net effect of radiating energy away is to make the baryonic matter hotter as it falls deeper into the potential wells and as more baryonic matter falls in from the outer regions of the original density perturbation. We met this phenomenon of negative specific heat when we discussed stellar evolution in § 10.7.

The dark matter itself is not charged, so it cannot form such a strong density contrast. It forms extended 'halos' around galaxies today, as we shall see below. Extensive numerical simulations using supercomputers show that the clumps of baryonic gas eventually began to condense into basic building-block clumps of a million solar masses or so, and then these began to merge together to form galaxies. So, although images of the Universe seem to show a lot of well-separated galaxies, the fact is that most of the objects we see were formed hierarchically from many hundreds or more of mergers. Mergers are still going on: astronomers believe they have discovered a fragment of several million solar masses that is currently being integrated into our own Milky Way galaxy, on the other side of the center from our location. The unusual star cluster Omega Centauri seems to be the core of such a mini-galaxy that was absorbed by the Milky Way long ago (Noyola *et al.* 2008). As mentioned in § 11.4, astronomers have found a massive black hole in its center. And the Magellanic Clouds, easily visible in the sky in the southern hemisphere, are on their way to merging into the Milky Way.

In December 2013 the European Space Agency launched the astrometry satellite GAIA, which has been measuring the positions and proper motions of about a billion stars, most of them within our Milky Way galaxy, with unprecedented accuracy. For about 150 million

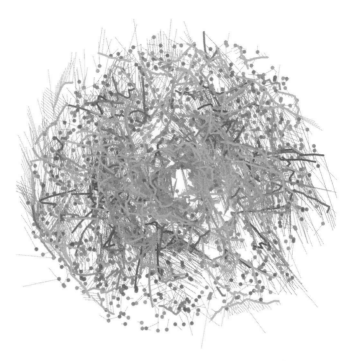

Figure 13.7 The Galaxy viewed from above, looking down on its plane. Each string is a collection of stars that originated in the same formation event, but which since then have followed different trajectories in the Galaxy. This shows that the Galaxy is not made of a random Maxwellian distribution of stars, but rather of a complex intertwined network of stellar systems. Unravelling this ball of strings can help us understand the history of the Galaxy. Figure courtesy of ESA, credit M. Kounkel and K. Covey, based on data in Kounkel & Covey (2019).

of these stars, GAIA has also been measuring radial velocities. When combined with the measured proper motion, the radial velocity reconstructs the full three-dimensional velocity of the object. In their second data release (Brown *et al.* 2018), the team was able to map the velocity streams of the Galaxy, which reflect the history of the diverse ways in which stars have been formed and introduced into the Galaxy. This history is preserved because stars in the Galaxy are basically collisionless objects which conserve their phase-space density (according to the Liouville theorem). The map shown in Figure 13.7 illustrates the richness of this information.

From this data it seems that our Galaxy had its last collision with another massive galaxy about 10 billion years ago (Helmi *et al.* 2018). Dubbed the Gaia-Enceladus Sausage, this galaxy disturbed the stars of our Galaxy so much that the disk remains thicker than is normal for spiral galaxies. It is also worth remarking that the nearest spiral galaxy to our own, Andromeda (M31), will merge with our Galaxy in about 3 billion years from now!

During the hierarchy of major and minor mergers that built the Galaxy, there were times when the pristine gas of hydrogen and helium that was formed in the Big Bang got dense enough to collapse under its own gravitational pull, leading to the formation of the first generation of stars. These are called Population III stars, and they were unlike anything we see today. Since the gas from which they formed was composed only of hydrogen and helium, with none of the heavier elements that were made by this generation of stars and incorporated into the next, these stars needed to be much more massive in order to stop the photons generated by nuclear reactions leaking out. Hydrogen and helium have smaller opacity than the heavier elements, whose outer electrons are more weakly bound and which therefore more readily absorb passing photons.

Population III stars, with masses between 100 and 1000 $M_\odot$, accordingly became denser inside compared with stars like our Sun, became very hot, evolved quickly, generated heavier elements, blew much of their outer layers and much of the new elements away with strong stellar winds, and then are very likely to have left behind a large population of black holes. The ultraviolet light emitted by these hot stars seems to have reionized much of the hydrogen in the Universe that had not yet been pulled into star formation, and which of course had been neutral since decoupling. All this happened sometime between redshifts of 8 and 20, and is called the epoch of reionization. This was the epoch of first light for galaxies, the first time that the expanding Universe would have looked optically a bit like it does today, if there had been anyone there to observe it!

The time between decoupling and reionization is sometimes called the Universe's dark ages, because there was little new generation of electromagnetic radiation. Because the reionized hydrogen was opaque to visible light, observations with electromagnetic telescopes and satellites cannot see anything earlier than this: light originating in the dark ages was absorbed during the reionization epoch. The CMB, however, by this time was at such a low frequency that it was not absorbed. Gravitational waves, too, were not affected by this gas and, as we have seen, one of the principal goals of the LISA gravitational wave observatory is to detect black-hole mergers at redshifts out to 20 or more, if they were occurring then.

There is good reason to expect that mergers were indeed happening, at least at redshifts between 10 and 20. One of the puzzles in the scenario of hierarchical galaxy formation is the fact that, as we saw in Chapter 11, apparently all galaxies contain massive black holes in their centers. Astrophysicists do not yet know whether these formed directly by the collapse of huge gas clouds as the baryonic matter was accumulating in the potential wells, or if they arose later by the growth and merger of intermediate-mass black holes left behind by Population III stars. The fact that galaxy evolution is dominated by mergers suggests that LISA will have an abundance of black-hole mergers to study.

After reionization and the generation of heavier elements by the first stars, the continued expansion of the Universe meant that galaxies were more and more isolated from each other, and they became nurseries for one generation of new stars after another, each with a little more of the heavier elements. Our Sun, whose age is about 5 billion years, was formed at a redshift between 0.3 and 0.4, and we are made of the elements synthesized by earlier generations of stars.

What is the dark matter?

Given the central role played by dark matter in this story of how the fluctuations seen in Figure 13.5 led to our own existence, and given that physicists still don't know the composition of the dark matter, it is important to ask what evidence we have for dark matter, apart from the galaxy formation process. The first suggestion that there could be considerable dark matter was, in fact, made from observations of galaxy clusters. In 1933, the pioneering astronomer Fritz Zwicky used the virial theorem to show that the random velocities of galaxies in the Coma Cluster were far too large for them to be held together by the mass that could be inferred from their starlight. He named the missing mass 'dark cold matter' (Zwicky 1933; see also Zwicky 1937).[3]

With more modern equipment and a better cosmic distance scale, Vera Rubin and collaborators in the 1970s and later measured the rotation speeds of stars in galaxies, and thereby provided a great deal of convincing evidence that there is about six times as much dark matter around galaxies as the mass in visible stars (see Rubin & Ford 1970). The most powerful survey telescope of the early twenty-first century has been named the Vera C. Rubin Observatory in her honor. The evidence for dark matter today is even richer, coming from studies of gravitational lensing by the dark matter haloes around clusters of galaxies (Figure 11.8) and by individual interacting galaxy systems, among other systems.

Despite this strong but circumstantial evidence, physicists do not have a good idea of what the dark matter is composed of, and direct searches for these particles passing through the laboratory have so far proved fruitless. Although it seems reasonable to assume that they are particles that are somehow related to known elementary particles but that have not yet been discovered because they don't participate in normal interactions, they may conceivably instead be black holes formed primordially, shortly after the Big Bang, or some other compact objects of a more exotic nature. Direct searches are continuing.

Using the CMB to measure H_0

Having learned how the temperature fluctuations seen in the CMB arise from small perturbations in the density of the dark matter at the time of decoupling, we can now return to the question of how the CMB measures the Hubble–Lemaître constant. Such a measurement requires a known length scale, which for Type Ia supernovae or binary gravitational wave mergers is their luminosity distance. The temperature fluctuations in the CMB appear to us as characteristic angular sizes on the sky, so it would be natural to use the angular-diameter distance here. But to do that, one needs the physical size of the angular fluctuations, represented by D in Eq. 13.44.

These distances are the typical sizes that acoustic waves in the dark matter could have traveled in the time available since the Big Bang. While one might try to compute this, the values reported by Planck and other CMB observations use a different method. To understand this, consider what happens to these density fluctuations after decoupling. We have described earlier that the over-dense regions tend to get even denser, through their self-gravity. As they get denser they naturally also get hotter. Immediately after decoupling, any heating of the neutral hydrogen in these regions will reionize it, coupling it back to the

---

[3] Remarkably, Zwicky's suggestion came in the same year as he and Walter Baade proposed the existence of neutron stars at a December meeting of the American Physical Society (Baade & Zwicky 1934). We discussed this in Chapter 10.

hot photon gas. This will stop the hydrogen from following the dark matter. After more time has passed, the cosmic expansion will have cooled off the hydrogen enough that it can be heated by compression without being ionized, and then the baryons will begin to collapse into the gravitational potential wells created by the dark matter, as we described earlier. This initial resistance to their collapse is called 'photon drag'. Quantitatively, the Planck satellite measured decoupling to have taken place at $z = 1090$, while photon drag stops being effective at $z = 1060$. This is the epoch at which structure formation really begins.

Astronomers have extensively investigated the resulting structure by looking for traces of baryon acoustic oscillations (BAOs). These are the early phase of the collapse, while the perturbations in the baryon component of matter are still small and can be treated as linear waves. It is possible to calculate what signature they leave behind in the distribution of galaxy clusters in the present epoch. The BAO observations measure the linear sizes of these traces today, and from them it is possible to infer the linear sizes they had at $z = 1060$. It is then a small step to extrapolate them back to $z = 1090$, giving us a value for D in Eq. 13.44. From this, one can infer the angular diameter distance back to the decoupling epoch, and from that the present value of the expansion rate, H_0. As we have noted before, the value from the Planck satellite, $H_0 = 67.4 \pm 0.5 \, \mathrm{km \, s^{-1} \, Mpc^{-1}}$ (Aghanim *et al.* 2019), is nearly five standard deviations different from the value obtained by using Type Ia supernovae as standard candles.

If the Achilles heel of the Type-Ia measurement is that there could be systematic errors in the distance ladder calibrating these supernovae, then the corresponding concern for the CMB measurement is that the extrapolation from BAO measurements to the standard fluctuation size on the CMB sky depends on understanding the way in which perturbations evolve in the matter sector after decoupling. This involves assumptions about the composition of the Universe which affect the sound speed that is used. As of this writing (mid-2021), it is not clear how to reconcile these two measurements. Perhaps something is wrong with the way in which the astronomical systems in the distance ladder have been modeled, or perhaps there is something crucial that we do not understand about the physics of the early Universe. Future gravitational wave standard-siren measurements may shed some light on this, because they are independent of both the distance ladder and of assumptions about the early Universe.

How a stochastic gravitational wave background affects the CMB

Measuring the gravitational wave content of the very early Universe is one of the most fundamental goals of gravitational wave science. Unlike the density perturbations that lead to images like Figure 13.5, which are the result of local physics responding to the effects of initial density perturbations at a much earlier time, any gravitational waves generated from the Big Bang will travel to us today essentially unaffected by anything in between, apart from the universal redshift of their frequencies. It is possible that gravitational wave detectors like LISA or ET might detect these waves as a stochastic signal in their data, but the most likely first way in which we can observe these waves is by using the CMB as a gravitational wave detector. We examine here how this can be done.

At the epoch of decoupling, the pre-existing density fluctuations are not the only effect that modifies the CMB. Random gravitational waves of sufficient strength can also leave a detectable fingerprint in the observations. Gravitational waves act, as we saw, transversally to their propagation direction, and they act in an area-conserving way: if one direction stretches, the perpendicular direction compresses, and the resulting ellipse has the same area as the original circle. So gravitational waves do not produce density or temperature variations in the CMB. Instead, they induce velocity perturbations in the plasma that affect the way in which electromagnetic radiation scatters, and in particular affect the polarization of the scattered radiation. Hidden in the polarization pattern of the CMB is the signature of the gravitational waves that were present at the time of decoupling.

The polarization of the CMB is actually affected by both density perturbations and gravitational waves. We shall first look at how a density inhomogeneity creates polarization. We shall always assume that there was no significant magnetic field at the time of decoupling. Every photon carries an electric field that is perpendicular to its velocity: just like gravitational waves, electromagnetic waves are transverse. So when a photon is scattered (by a free electron, for example), its final polarization must be perpendicular to its direction of motion, and it would have come from an electric field that was perpendicular to its original direction of motion. These two velocity directions form a plane, to which the final polarization is perpendicular.

Let us consider a photon that originally was traveling through the Universe in a direction perpendicular to our line of sight, and which was scattered through a 90° angle toward us. Then it will have a polarization in the direction perpendicular to its original direction of motion. If the Universe at the time was completely homogeneous and isotropic, then there will be as many photons traveling one way as the opposite way, and the net polarization of the radiation coming toward us would be zero. But if there were density perturbations, then these would have been accompanied by temperature perturbations: regions of higher density also had higher temperature. These regions radiated more photons than lower-temperature regions. Imagine a simple pattern of alternating hot and cold spots on the sky. Then between a hot spot and a cold spot, the net flux of photons will be away from the hot spot, and so we will observe a net polarization in a direction perpendicular to the line joining the spots. There will be no net polarization at the center of either spot.

The temperature fluctuations in the CMB are caused by acoustic waves traveling through the plasma. A plane wave traveling across the sky will create a pattern of polarization that is linear, aligned with its wavefront. Importantly, this polarization field on the sky has no curl in the plane of the sky. Now, the observed temperature fluctuation pattern is caused by the superposition of lots of random waves, but because the perturbations are (as we have seen) very weak, then they superpose linearly, and the complicated pattern will still have no curl. This will turn out to be the way in which we can separate the polarization caused by the temperature fluctuations we are already measuring and those caused by the gravitational waves that we would like to measure.

Consider a gravitational wave traveling directly towards us through the primordial plasma. It is pushing the particles of the plasma apart along the plane of the sky in one direction (let's call that the x direction on the sky at this location) and pushing them together in the perpendicular y direction. There is no net density perturbation, just a velocity pattern on the sky. From a reference point at the center of the gravitational wave ellipse, the material being pushed toward it along the y axis acquires a higher temperature through the Doppler effect, and this results in a higher intensity of scattering, producing polarization along the x direction. Conversely, gas being pushed out along the x axis produces less scattering that has polarization along the y axis. That means that the net polarization of the scattered radiation lies along the x axis, the major axis of the ellipse.

Of course, this would average out over time, after the gravitational wave has changed its phase by π. But we are interested in the last scattering of our photons, so the phase reversal of the gravitational wave occurs at a later time, after decoupling, and there is no longer any scattering to change the polarization imprinted earlier. The result is that the polarization we observe reflects the polarization of the gravitational wave. Since the gravitational wave background is random, the net polarization pattern on the sky consists essentially of a collection of vectors with random orientations.

Such a pattern will not be curl-free. About half the polarization created by gravitational waves coming toward us will contribute to a divergence-free pattern on the sky. This is called the B-mode of the polarization. Taking the curl of the random polarization vector field on the sky will eliminate the polarization caused by the acoustic effects (the effects of temperature fluctuations) and the curl-free part of the polarization caused by gravitational waves, which together make up the so-called E-mode of the polarization pattern. What is left after taking the curl is the B-mode polarization, caused only by the gravitational waves. This is the observational signature of the random gravitational waves present in the primordial plasma at the time of decoupling.

A little thought should reveal that this discussion is a bit oversimplified. We have not discussed possible velocities in the primordial plasma that just come from hydrodynamic effects. Vorticity in the plasma also will induce a polarization pattern, but it turns out to be curl-free. Similarly, gravitational waves that are present in the plasma but are traveling across the sky rather than toward us also induce a curl-free pattern. The inference of the amount of gravitational radiation present at the time of decoupling needs to take into account that the waves are traveling in all directions, not just toward us. A good accessible discussion can be found on the website by Hu & White (1997).

The result of all of this is that the divergence-free part of the pattern is small, and hunting for it requires high sensitivity to the CMB. It also requires great care, because the CMB radiation can acquire a polarization as it travels through the Universe to us. Dust grains scatter light, and the resulting polarization depends on the orientation of the grains. If a local magnetic field lines them up, then this will polarize, or change the polarization of, the CMB as it passes through. The primary location of dust along the line of sight of the CMB is in our Galaxy, so mapping the dust in the Galaxy is an important part of searching for the B-mode. This dust also affects the microwaves of the CMB, so the CMB observations themselves can help with the mapping.

At the time of writing (mid-2021) there are a number of CMB experiments searching for the B-mode signal, and more are planned and proposed. It is a little strange to reflect that, when the microwaves that carry a detectable B-mode signal are received by a CMB antenna, the gravitational waves that created their polarization signal are also passing through the antenna at the same time!

The early Universe: fundamental physics meets cosmology

Thus far, our discussion of physical cosmology has treated the Universe after decoupling. Our maps of the CMB contain the signature of the density perturbations that led to the Universe we observe all around us, and indeed that led to our own existence. But where did these density perturbations come from? And where, for that matter, did the electrons and protons as well as the dark matter at the time of decoupling come from? In this subsection, we will discuss the origin of matter as we know it. In what follows we will treat our best model for understanding where the CMB perturbations arose: inflation.

If we go back in time from the epoch when the radiation and matter densities were equal, then we are in a radiation-dominated Universe. Eventually, when we are only about about 200 s away from the Big Bang, the temperature has risen to about 50 keV, the mass difference between a neutron and a proton. This is the temperature at which nuclear reactions among protons and neutrons come into equilibrium with each other. At earlier times, at temperatures above this critical one, all the baryons were free; there were no compound nuclei. And of course the photons and electrons were in equilibrium with the protons because they would be scattering all the time. Any neutrinos would also have been in equilibrium: they would have scattered freely from baryons at even higher temperatues and densities, where the weak and strong interactions were of comparable strength. Then, as the Universe cooled, their temperature would have kept pace with that of photons because the neutrino rest mass is so small that at these temperatures it behaves essentially like a massless particle. The baryons, too, were in equilibrium with the photons.

Let's go forward in time from this epoch. As the Universe cooled through 50 keV, some heavier elements were formed: mainly ^{4}He, but also small amounts of ^{3}He, Li, B, and traces of other light elements. Essentially all the lithium and helium that we see distributed around the Universe today was formed at this time: processes inside stars tend to destroy these very light elements, not make them. Their final abundances are very sensitive to the rate at which the Universe was expanding at this time. Heavier nuclei hardly formed at all, because the Universe expanded too quickly, cooling off before they could build up from the lighter ones.

The expansion rate at this epoch, in turn, depends on the number of species of neutrinos. The more neutrino species there were, the higher would have been the density of (effectively) massless particles, since each species would have the same density as the photons. This means that the more neutrino species there were, the stronger gravity would have been, and the greater would have been the deceleration of the Universe's expansion at this time. This would have affected the abundances of helium and lithium formed after the temperature fell below 50 keV, since these abundances depended on the density and on the amount of time available before the density dropped so low that the reactions were quenched.

From extensive computations of the reaction networks, astrophysicists have been able to show that the background neutrino 'gas' of the Universe should contain only the three that are known from particle-physics experiments today. In addition, if there were other kinds of massless particles then they cannot have changed the self-gravity of the Universe by more than a small amount. That means that, today, their energy density must be significantly smaller than that of the photons, which have $\Omega_\gamma \sim 10^{-5}$. Stochastic gravitational waves from the Big Bang, in particular, must satisfy this nucleosynthesis upper bound.

What about the dark matter? This bound also applies to it, which means that the dark matter cannot have been relativistic at this epoch: the dark matter particles must have rest mass significantly greater than 50 keV. Considering that this is a factor of 10 smaller than the rest mass of the electron, it is perhaps not a very strong constraint on the kind of particle we are looking for!

Notice that we are already within 200 s of the Big Bang in this discussion, and still we are in the domain of well-understood physics. Let us reverse our clock again for a moment, and go back in time from the epoch of nucleosynthesis. At about 1 s, the temperature was around 500 keV, the mass of the electron. In this plasma, therefore, there was an abundance of electrons and positrons, constantly annihilating one another and being created again by photons. Much earlier than this the rest mass of the electrons can be ignored, so the numbers and energy densities of photons and of electrons and positrons were similar. Going forward in time from the 500 keV epoch, as the Universe expanded and cooled, the electrons and positrons continued to annihilate, but no more were produced. After a few seconds, there were apparently essentially no positrons, and yet observations show that there remained about one electron for every 10^9 photons. This ratio of 10^9 is called the specific entropy of the Universe, a measure of its disorder.

Why were there any electrons left at all after this annihilation phase? Why, in other words, was there any matter left over to build into planets and people? Extensive observational programs, coupled to numerical simulations, have convincingly established that there is no 'missing' antimatter hidden somewhere, no anti-stars or anti-galaxies: significant amounts of antimatter just do not exist any more. Clearly, during the equilibrium plasma phase, electrons and positrons were not produced in equal numbers. The same must also have happened at a much earlier time, when protons and antiprotons were in equilibrium with the photon gas, at temperatures above a few hundred MeV (only 10 μs after the Big Bang): something must have favored protons over antiprotons in the same ratio as for electrons over positrons, so that the overall plasma remained charge-neutral. This is one of the central mysteries of particle physics. Something in the fundamental laws of physics gave a slight preference to electrons. *Nature has a matter-antimatter asymmetry.*

At times earlier than 10^{-5} s there are no protons or neutrons, just a plasma of quarks and gluons, the fundamental building blocks of baryons. According to particle theory, quarks are 'confined', so that we never see a free one detached from a baryon. But at high enough temperatures and densities, the protons overlap so much that the quarks can stay confined and still behave like free particles.

We can even push our perspective another step higher in temperature, to around 10 TeV, which is the frontier for current accelerators. The Large Hadron Collider (LHC) at CERN in Geneva put the standard model of the nuclear interactions on a firm foundation when it discovered the Higgs boson in 2012, leading to the award of the 2013 Nobel Prize in physics to Peter Higgs and Françoise Englert. However, as of this writing (mid-2021), the LHC has not found any evidence for supersymmetry, an idea that was proposed in order (among other things) to make it easier within particle physics theories to arrive at a value of the cosmological constant in the range of what we observe.

In this scheme there is one important thing that is missing: we have so far mentioned no mechanism in any of this physics for generating the density irregularities in the dark matter that led to galaxy formation. We observe them in the microwave background, and we know that they are needed in order to trigger all the processes that eventually led to our own evolution. But even at 10^{-14} s, they are simply an initial condition: they have to be there, at a much smaller amplitude than we see them in the microwave background because the irregularities grew as the Universe expanded. But they must be there. And standard physics has no explanation for how they got there. The exciting answer to this problem lies in the scenario of *inflation,* which we are about to describe.

However, the density perturbations are not the only feature of the early Universe that is not explained. Right from the start we have assumed homogeneity and isotropy, based on observations. The perturbations introduced a small amount of inhomogeneity in the density distribution at the time of decoupling, but that is not a problem: our assumption of homogeneity does not hold on small scales, where galaxies and planets form. But on the large scale, why is the Universe so smooth?

This is particularly challenging to the standard Big Bang model, which offers no physical process that could smooth things out. Consider the primordial abundance of helium. It was fixed when the Universe was only five minutes old. When we look with our telescopes in opposite directions on the sky, we can see distant quasars and galaxies that appear to have the same element abundances as we do, and yet they are so far away from each other and from us that they could not have been in communication when the helium was being made: they are outside each other's particle horizon in the standard cosmological model. (This is illustrated in the right-hand panel in Figure 13.2.) One way to 'explain' this is simply to postulate that the initial conditions for the Big Bang were the same everywhere, even in causally disconnected regions. But it would be more satisfying physically if some process could be found that enabled these regions to communicate with each other at a very early time, even though they became disconnected later. Again, inflation offers such a mechanism, so we turn to that subject now.

Inflation

The basic idea of inflation (Starobinsky 1980, Guth 1981, Linde 1982) is that, at a very early time such as 10^{-35} s, the Universe was dominated by a large positive cosmological constant, much larger than we have today, but one that was only temporary: it turned on at some point and then turned off again, for reasons we will discuss below. But during the time when the Universe was dominated by this constant, the matter and curvature were unimportant, and the Universe expanded according to the simple law

$$H^2 = \frac{8}{3}\pi\rho_\Lambda \qquad \Longrightarrow \qquad \frac{\dot{R}}{R} = \frac{1}{\tau_\Lambda}, \tag{13.65}$$

which is an exponential law with a growth time

$$\tau_\Lambda = \left(\frac{3}{8\pi\rho_\Lambda}\right)^{1/2}. \tag{13.66}$$

If this exponential expansion lasted twenty or thirty e-foldings, then a region of very small size could have been inflated into the size of a patch that would be big enough to become the entire observable Universe today. The idea is that, before inflation, this small region had been smoothed out by some physical process, which was possible because it was small enough to do this even in the time available. Then inflation set in and expanded it into the initial data for our Universe.

This would explain the homogeneity of what we see: everything did indeed come from the same patch. What is more, inflation also explains the fluctuation in the cosmic microwave background. To understand this we have to go into more detail about the mechanism for inflation. Attempts to compute the cosmological constant today focus on the vacuum energy of quantum fields, which we used in order to explain the Hawking radiation in Chapter 11. The vacuum energy is useful for this purpose because the vacuum must be invariant under Lorentz transformations: there should not be any preferred observer for empty space in quantum theory. This means that any stress–energy tensor associated with the vacuum must be Lorentz invariant. Now, the only Lorentz-invariant symmetric tensor field of type $\binom{0}{2}$ is the metric tensor itself, so any vacuum-energy explanation of dark energy will automatically produce something like a cosmological constant, proportional to the metric tensor.

In some models of the behavior of the physical interactions at very high energies, beyond the TeV scale, it is postulated that there is a phase transition in which the nature of the vacuum changes, and a large amount of vacuum energy is released in the form of a cosmological constant, powering inflation. But this is a dynamical process, which sets in when the phase change occurs and then stops when the energy is converted into the real energy that eventually becomes the particles and photons in our Universe. So for a limited time the Universe inflates rapidly. Now, at the beginning there are the usual vacuum fluctuations, and the remarkable thing is that the exponential inflation amplifies these fluctuations in much the same way as a nonlinear oscillator can pump up its oscillation amplitude. When inflation finishes, what were small density perturbations on the quantum scale have become much larger, classical perturbations.

These perturbations include the density perturbations revealed by the CMB, and they also include the gravitational wave fluctuations that could turn up in the polarization of the CMB. As we mentioned before, any such waves would have traveled unhindered from the time of their creation (about 10^{-35} s) to the time they left their imprint on the CMB. We would be using the CMB as a detector observing as close to the Big Bang as we could hope to get.

When physicists perform computations within the inflation model, it does remarkably well. The amplitude of the fluctuations is reasonable, their spectrum matches that which is inferred from the cosmic microwave background, and the physical assumptions are consistent with modern views of unification among the various interactions of fundamental physics. Inflation will, however, remain a 'model' and not a 'theory' until either a theory extending the standard model of the nuclear forces to much higher energies is found, or until some key observation reveals the fields and potential that are postulated within the model. However, it is a powerful and convincing paradigm, and it is currently the principal framework within which physicists address the deepest questions about the early Universe.

Beyond general relativity

Inflation goes beyond standard physics, making assumptions about the way that the laws governing the nuclear interactions among particles behave at the very high energies that obtained in the early Universe. But it does not modify gravity: it works within classical general relativity. Nevertheless, as we have remarked before, the classical theory must eventually be replaced by a quantum description of gravitation, and the search for this theory is a major activity in theoretical physics today.

Although no consistent theory has yet emerged, the search has produced a number of exciting ideas that offer the possibility of new kinds of observations, new kinds of explanations. One approach, called loop quantum gravity, directly attacks the problem of how to quantize spacetime, ignoring at first the other forces in spacetime, such as electromagnetism. On a fine scale, presumably the Planck scale, it postulates that the manifold nature of spacetime breaks down, and the smaller-scale structure is one of nested, tangled loops. There are a number of variants on this approach, with different structures, but the common idea is that spacetime is a coarse-grained average over something that has a much richer topology. These ideas come from the mathematics in a natural way. A remarkable consequence of loop quantum gravity is that the Big Bang was not singular after all, that going backwards in time the Universe is able to pass through the Big Bang and become a classical collapsing cosmology on the other side (Bojowald 2008).

Even more active, in terms of the number of physicists working in it, is the string-theory approach to quantum gravity. Here the aim is to unify all the interactions, including gravity, so that the theory includes the nuclear and electromagnetic interactions from the start. String theory seems to be consistent, in the sense of not having to do artificial things to get rid of infinite energies, only this requires eleven spacetime dimensions. String theory also seems to depend strongly on supersymmetry. Regarding the large number of dimensions, we of course exist in just four of these, so it is interesting to ask questions about the remaining ones.

The first assumption was that these remaining dimensions never got big: that attached to each point is a Planck-sized seven-sphere offering the possibility of exiting from our four-dimensional Universe only to things that are of order the Planck length. This would not be easy to observe. But it is also possible that some of these extra dimensions are big, and that our four-dimensional Universe is simply a four-surface in these five-or-more-dimensional surroundings. This 4-surface has come to be called a *brane*, from the word 'membrane'. String theory on branes has a special property: electromagnetism and the nuclear forces are confined to our brane, but gravitation can act in the extra dimensions too. This would lead to a modification of the inverse square law of Newtonian gravity on short distances, on scales comparable to a relevant length scale in the surrounding space. All we can say is that this scale must be smaller than about 0.04 millimeter, from experiments on the inverse square law (Lee *et al.* 2020). But there are many decades between $40\mu m$ and the Planck scale, and new physics might be waiting to be discovered anywhere in between (Maartens 2004).

The new physics could take many different forms. Some kind of collision with another brane might have triggered the Big Bang. A nearby extra brane might have a parallel world of stars and galaxies, interacting with us only through gravity: shadow matter. There might be extra amounts of gravitational radiation, due either to shadow matter or to unusual brane-related initial conditions at the Big Bang.

Although these ideas sound like science fiction, they are firmly grounded in model theories, which are deliberate oversimplifications of the full equations of string theory and which involve deliberate choices of the values of certain constants in order to get these strange effects. They should be treated as neither predictions nor idle speculation, but rather as harbingers of the kind of revolutionary physics that a full quantum theory of gravity might bring us. Experimental hints, from high-precision physics, or observational results, perhaps from gravitational waves or from cosmology, might at any time provide key clues that could point the way to the right theory.

Since the time of Hubble, astronomy has made steady progress in understanding our neighborhood in the Universe: galaxies, stars, planets, even exotica like black holes and neutron stars. Cosmology is a more recent frontier, having become a precision science only with the advent of satellite observations of the CMB in the 1990s. It may not be surprising, therefore, that cosmology has turned up mysteries that our local physics did not even hint at: dark matter, dark energy, inflation. These problems on the scale of the Universe as a whole are young problems, and will no doubt eventually be understood. What is more unexpected, however, is that understanding these problems may involve understanding the fundamentals of physics at the shortest distance scales, right down to the Planck length. It is too soon to know if this will turn out to be correct. There are, after all, many decades of energy between the 10 TeV scale that we can presently experiment with and the Planck scale of 10^{16} TeV!

13.5 Bibliography

The literature on cosmology is vast. In the body of the chapter I have given the principal references to original results, so I list here some recommended books on the subject.

Standard cosmology is treated in great detail in the classic text by Weinberg (2008). Cosmological models in general relativity become somewhat more complex when the assumption of isotropy is dropped, but they retain the same overall features: the Big Bang, open versus closed. See Ryan & Shepley (1975). Beginners in cosmology are often confused by the different kinds of horizons, and worried by the apparently faster-than-light expansion that is outside our horizon; see Davis & Lineweaver (2004) for a good and clear discussion.

The first precision observations of the black-body nature and temperature anisotropies of the CMB were made by the COBE satellite, launched in 1989. See Mather *et al.* (1994), Smoot *et al.* (1992), and Bennett *et al.* (1996). These observations led to the award of the Nobel Prize for Physics in 1996 to John Mather and George Smoot.

The study of physical cosmology was pioneered by James Peebles, who also predicted the CMB. For his work he received the Nobel Prize in Physics in 2019. He has written classic introductions to the subject: Peebles (1980) and Peebles (1993).

Other early discussions on physical cosmology that remain classics include Liang & Sachs (1980), Heidmann (1980), and Balian *et al.* (1980).

For greater depth on modern physical cosmology, see the excellent texts by Mukhanov (2005) and Padmanabhan (2002). More up to date is the recently revised Dodelson & Schmidt (2020). For a different point of view on 'why' the Universe has the properties it does, see the book by Barrow & Tipler (1986) on the anthropic principle.

As we have noted, there is at present (mid-2021) significant disagreement among different measurements of the value of the Hubble–Lemaître constant. For reviews, see Jackson (2015), Riess (2020), and references that these reviews cite.

For information on the planned Athena X-ray mission, due for launch in the early 2030s, see `www.the-athena-x-ray-observatory.eu`. For information about the planned LISA mission, see Chapter 9 and Chapter 12 and particularly the bibliographic information in § 9.7.

String theory and loop quantum gravity are not the only approaches to quantum gravity. Another interesting one is called causal set theory. For a review, see Surya (2019). Cosmology is also an arena in which general relativity can be tested. See Ishak (2019) for a review.

For popular-level cosmology articles written by research scientists, see the Einstein Online website: `www.einstein-online.info/en/`.

Exercises

13.1   Use the metric of a two-sphere to prove the statement associated with Figure 13.1 that the rate of increase of the distance between any two points as the sphere expands (as measured *on* the sphere!) is proportional to that distance.

13.2   The astronomer's distance unit, the parsec, is defined to be the distance from the Sun to a star whose parallax is exactly one second of arc. (The parallax of a star is half the maximum change in its angular position as measured from Earth, as Earth orbits the Sun.) Given that the radius of Earth's orbit is $1 \, \text{AU} = 10^{11} \, \text{m}$, calculate the length of one parsec.

13.3   Newtonian cosmology.

   (a) Apply Newton's law of gravity to the study of cosmology by showing that the general solution of $\nabla^2 \Phi = 4\pi\rho$ for $\rho = \text{const.}$ is a quadratic polynomial in Cartesian coordinates, but is not necessarily isotropic.

   (b) Show that if the Universe consists of a region where $\rho = \text{const.}$, outside of which there is vacuum, then if the boundary is not spherical the field will not be isotropic: the field will show significant deviations from sphericity throughout the interior, even at the center.

   (c) Show that, in such a Newtonian cosmology, an experiment done locally could determine the shape of the boundary, even if the boundary is far outside our particle horizon.

13.4 Show that if $h_{ij}(t_1) \neq f(t_1,t_0)h_{ij}(t_0)$ for all i and j in Eq. 13.4, then distances between galaxies would increase anisotropically: the Hubble–Lemaître law would have to be written as

$$v^i = H^i{}_j \tag{13.67}$$

for a matrix $H^i{}_j$ not proportional to the identity.

13.5 Show that if galaxies are assumed to move along the lines $x^i = $ const., and if the local Universe is to be seen as homogeneous, then g_{0i} in Eq. 13.6 must vanish.

13.6 (a) Prove the statement leading to Eq. 13.9, that we can deduce G_{ij} for our three-spaces by setting Φ to zero in Eqs. 10.15–10.17.
(b) Derive Eq. 13.10.

13.7 Show that the metric given by Eq. 13.8 is not locally flat at $r = 0$ unless $A = 0$ in Eq. 13.12.

13.8 (a) Find the coordinate transformation leading to Eq. 13.20.
(b) Show that the intrinsic geometry of a hyperbola $t^2 - x^2 - y^2 - z^2 = $ const. > 0 in Minkowski spacetime is identical with that of Eq. 13.20 in appropriate coordinates.
(c) Use the Lorentz transformations of Minkowski space to prove that the $k = -1$ cosmological model is homogeneous and isotropic.

13.9 (a) Show that a photon which propagates on a radial null geodesic of the metric given by Eq. 13.14 has energy $-p_0$ inversely proportional to $R(t)$.
(b) Show from this that a photon emitted at time t_e and received at time t_r by observers at rest in the cosmological reference frame is redshifted by

$$1 + z = R(t_r)/R(t_e). \tag{13.68}$$

13.10 Show from Eq. 13.25 that the relationship between velocity and cosmological redshift for a nearby object (small light travel time to us) is $z = v$, as one would expect for an object with a recessional velocity v.

13.11 (a) Prove Eq. 13.30 and deduce Eq. 13.31 from it.
(b) Fill in the indicated steps leading to Eq. 13.32.

13.12 Derive Eq. 13.43 from Eq. 13.32.

13.13 Astronomers usually do not speak in terms of intrinsic luminosity and flux. Rather, they use absolute and apparent magnitude. The (bolometric) *apparent magnitude* of a star is defined by its flux F relative to a standard flux F_s:

$$m = -2.5 \log_{10}(F/F_s) \tag{13.69}$$

where $F_s = 3 \times 10^{-8} \, \mathrm{J\,m^{-2}\,s^{-1}}$ is roughly the flux of visible light at Earth from the brightest stars in the night sky. The *absolute magnitude* is defined as the apparent magnitude that the object would have at a distance of 10 pc:

$$M = -2.5 \log_{10}[L/(4\pi(10\,\mathrm{pc})^2 F_s)]. \tag{13.70}$$

Using Eq. 13.43, with Eq. 13.28, rewrite Eq. 13.35 in astronomer's language as:

$$m - M = 5 \log_{10}(z/10\,\mathrm{pc}H_0) + 1.09(1 - q_0)z. \tag{13.71}$$

Astronomers call this the *redshift–magnitude relation*.

13.14 (a) For the 'Friedmann–Robertson–Walker' metric, Eq. 13.14, compute all the Christoffel symbols $\Gamma^\mu{}_{\alpha\beta}$. In particular show that the nonvanishing ones are:

$$\Gamma^0{}_{jk} = \frac{\dot{R}}{R} g_{jk}, \quad \Gamma^j{}_{0k} = \frac{\dot{R}}{R} \delta^j{}_k, \quad \Gamma^r{}_{rr} = \frac{kr}{1 - kr^2},$$

$$\Gamma^r{}_{\theta\theta} = -r(1 - kr^2), \quad \Gamma^r{}_{\phi\phi} = -r(1 - kr^2)\sin^2\theta, \qquad (13.72)$$

$$\Gamma^\theta{}_{r\theta} = \Gamma^\phi{}_{r\phi} = \frac{1}{r}, \quad \Gamma^\theta{}_{\phi\phi} = \sin\theta\cos\theta, \quad \Gamma^\phi{}_{\theta\phi} = \cot\theta.$$

(b) Using these Christoffel symbols, show that the time component of the divergence of the stress–energy tensor of the cosmological fluid is

$$T^{0\alpha}{}_{;\alpha} = \dot{\rho} + 3(\rho + p)\frac{\dot{R}}{R}. \qquad (13.73)$$

(c) By multiplying this equation by R^3, derive Eq. 13.47.

13.15 Show from Eq. 13.50 that if the radiation has a black-body spectrum of temperature T, then T is inversely proportional to R.

13.16 Use the Christoffel symbols computed in Exercise 13.14 to derive Eq. 13.51.

13.17 Use Eq. 13.47 and the time-derivative of Eq. 13.55 to derive Eq. 13.56 for $\ddot{R}$. Make sure you use the fact that $p_\Lambda = -\rho_\Lambda$.

13.18 In this chapter we saw that the negative pressure (tension) of the cosmological constant is responsible for accelerating the Universe. But is this a contradiction to ordinary physics? Does not a tension pull inward, not push outward? Resolve this apparent contradiction by showing that the net pressure *force* on any local part of the Universe is zero. Refer to the discussion at the end of Section 4.7.

13.19 Assuming the Universe to be matter-dominated and to have zero cosmological constant, show that at times early enough for one to be able to neglect k in Eq. 13.55, the scale factor evolves with time as $R(t) \propto t^{2/3}$. Show also that if in addition the Universe is flat ($k = 0$), as ours seems to be, and if we assume it was matter-dominated back to the Big Bang, then the present age of the Universe is $2/3H_0$. Show that this is about 10^{10} years. Of course, this is only an approximation because we have neglected the fact that the Universe was radiation-dominated at the beginning.

13.20 Assume that the Universe is matter-dominated and find the value of ρ_Λ that permits the Universe to be static.

(a) Because the Universe is matter-dominated at the present time, we can take $\rho_m(t) = \rho_0[R_0/R(t)]^3$ where the subscript '0' refers to the static solution for which we are looking. Differentiate the 'energy' equation Eq. 13.55 with respect to time to find the dynamical equation governing a matter-dominated cosmological model:

$$\ddot{R} = \frac{8}{3}\pi\rho_\Lambda R - \frac{4}{3}\pi\rho_0 R_0^3 R^{-2}. \qquad (13.74)$$

Set this to zero to find the solution

$$\rho_\Lambda = \frac{1}{2}\rho_0.$$

For Einstein's static solution, the cosmological constant energy density has to be half the matter energy density.

(b) Put our expression for ρ_m into the right-hand side of Eq. 13.55 to get an energy-like expression whose derivative has to vanish for a static solution. Verify that the above condition on ρ_Λ does indeed make the first derivative vanish.

(c) Compute the second derivative of the right-hand side of Eq. 13.55 with respect to R and show that, at the static solution, it is positive. This means that the 'potential' is a maximum and *Einstein's static solution is unstable*.

13.21 Explore the possible futures and histories of an expanding cosmology with *negative* cosmological constant. You may wish to do this graphically, by drawing figures analogous to Figure 11.1. See also Figure 13.4.

13.22 (Parts of this exercise are suitable only for students who can program a computer.) Construct a more realistic equation of state for the Universe as follows.

(a) Assume that, today, the matter density is $\rho_m = m \times 10^{-27}\ \mathrm{kg\,m^{-3}}$ (where m is of order 1) and that the cosmic radiation has black-body temperature 2.7 K. Find the ratio $\varepsilon = \rho_r/\rho_m$, where ρ_r is the energy density of the radiation. Find the number of photons per baryon, $\sim \varepsilon m_p c^2/k_B T$.

(b) Find the general form of the energy-conservation equation, $T^{0\mu}{}_{,\mu} = 0$, in terms of $\varepsilon(t)$ and $m(t)$.

(c) Numerically integrate this equation and Eq. 12.2 back in time from the present, assuming $\dot{R}/R = 75\ \mathrm{km\,s^{-1}\,Mpc^{-1}}$ today, and assuming there is no exchange of energy between matter and radiation. Do the integration for $m = 0.3$, 1.0, and 3.0. Stop the integration when the radiation temperature reaches $E_i/26.7 k_B$, where E_i is the ionization energy of hydrogen (13.6 eV). This is roughly the temperature at which there are enough photons to ionize all the hydrogen: there is roughly a fraction 2×10^{-9} photons above energy E_i when $k_B T = E_i/26.7$, and this is roughly the fraction needed to give one such photon per H atom. For each m, what is the value of $R(t)/R_0$ at that time, where R_0 is the present scale factor? Explain this result. What is the value of t at this epoch?

(d) Determine whether the pressure of the matter is still negligible compared with that of the radiation. (You will need the temperature of the matter, which equals the radiation temperature now because the matter is ionized and therefore strongly coupled to the radiation.)

(e) Integrate the equations backwards in time from the decoupling time, now with the assumption that radiation and matter exchange energy in such a way as to keep their temperatures equal. In each case, how long ago was the time at which $R = 0$, the Big Bang?

13.23 Calculate the redshift of decoupling by assuming that the cosmic microwave radiation has temperature 2.7 K today and had the temperature $E_i/20 k_B$ at decoupling, where $E_i = 13.6$ eV is the energy needed to ionize hydrogen (see Exercise 13.22(c)).

13.24 If the Hubble–Lemaître constant is $75\ \mathrm{km\,s^{-1}\,Mpc^{-1}}$, what is the minimum present density for a $k = +1$ Universe?

13.25 Estimate the times earlier than which our uncertainty about the laws of physics prevents us drawing firm conclusions about cosmology as follows.

(a) Deduce that, in the radiation-dominated early Universe, where the curvature parameter $k = (+1, 0, -1)$ does not have much effect, the temperature T behaves as follows:

$$T = \beta t^{-1/2}, \quad \beta = (45\, \hbar^3/32\pi^3)^{1/4} k_B^{-1}.$$

(b) Assuming that our knowledge of particle physics is uncertain for $k_B T > 10^3$ GeV, find the earliest time t at which we can have confidence in the physics.

(c) Quantum gravity is probably important when a photon has enough energy $k_B T$ to form a black hole within one wavelength ($\lambda = h/k_B T$). Show that this gives $k_B T \sim h^{1/2}$. This is the *Planck temperature*. At what time t is this an important worry?

Appendix A **Summary of linear algebra**

For the convenience of the student we collect those aspects of linear algebra that are important in our study. We hope that none of this is new to the reader.

Vector space. A collection of elements $V = \{A, B, \ldots\}$ forms a *vector space* over the real numbers if and only if they obey the following axioms (with a, b real numbers).

(1) V is an abelian group with operation $+$ ($A + B = B + A \in V$) and identity 0 ($A + 0 = A$).
(2) Multiplication of vectors by real numbers is an operation which gives vectors and which is

    (i) distributive over vector addition, $a(A + B) = a(A) + a(B)$;
    (ii) distributive over real number addition, $(a + b)(A) = a(A) + b(A)$;
    (iii) associative with real number multiplication, $(ab)(A) = a(b(A))$;
    (iv) consistent with the real number identity, $1(A) = A$.

This definition could be generalized to vector spaces over complex numbers or over any field, but we shall not need to do so.

A set of vectors $\{A, B, \ldots\}$ is said to be *linearly independent* if and only if there do not exist real numbers $\{a, b, \ldots, f\}$, not all of which are zero, such that

$$aA + bB + \cdots + fF = 0.$$

The dimension of the vector space is the largest number of *linearly independent* vectors one can choose. A basis for the space is any linearly independent set of vectors $\{A_1, \ldots, A_n\}$, where n is the dimension of the space. Since for any B the set $\{B, A_1, \ldots, A_n\}$ is linearly dependent, it follows that B can be written as a linear combination of the basis vectors:

$$B = b_1 A_1 + b_2 A_2 + \cdots + b_n A_n.$$

The numbers $\{b_1, \ldots, b_n\}$ are called the components of B on $\{A_1, \ldots, A_n\}$.

An *inner product* may be defined on a vector space. It is a rule associating with any pair of vectors, A and B, a real number $A \cdot B$, which has the properties:

(1) $A \cdot B = B \cdot A$,
(2) $(aA + bB) \cdot C = a(A \cdot C) + b(B \cdot C)$.

By (1), the map $(A, B) \rightarrow (A \cdot B)$ is symmetric; by (2), it is bilinear. The inner product is called positive-definite if $A \cdot A > 0$ for all $A \neq 0$. In that case the *norm* of the vector A is

$|A| \equiv (A \cdot A)^{1/2}$. In relativity we deal with inner products that are indefinite: $A \cdot A$ has one sign for some vectors and another for others. In this case the norm, or magnitude, is often defined as $|A| \equiv |A \cdot A|^{1/2}$. Two vectors A and B are said to be orthogonal if and only if $A \cdot B = 0$.

It is often convenient to adopt a set of basis vectors $\{A_1, \ldots, A_n\}$ that are *orthonormal*: $A_i \cdot A_j = 0$ if $i \neq j$ and $|A_k| = 1$ for all k. This is not necessary, of course. The reader unfamiliar with nonorthogonal bases should try the following. In the two-dimensional Euclidean plane with Cartesian (orthogonal) coordinates x and y and associated Cartesian (orthonormal) basis vectors e_x and e_y, define A and B to be the vectors $A = 5e_x + e_y$, $B = 3e_y$. Express A and B as linear combinations of the nonorthogonal basis $\{e_1 = e_x, e_2 = e_y - e_x\}$. Notice that, although e_1 and e_x are the same, the 1 and x components of A and B are *not* the same.

Matrices. A matrix is an array of numbers. We shall only deal with square matrices, e.g.

$$\begin{pmatrix} 1 & 2 \\ 3 & 1 \end{pmatrix} \quad \text{or} \quad \begin{pmatrix} 1 & 2 & 5 \\ -6 & 3 & 18 \\ 10^5 & 0 & 0 \end{pmatrix}.$$

The *dimension* of a matrix is the number of its rows (or columns). We denote the elements of a matrix by A_{ij}, where the value of i denotes the row and that of j denotes the column; for a 2×2 matrix we have

$$\mathbf{A} = \begin{pmatrix} A_{11} & A_{12} \\ A_{21} & A_{22} \end{pmatrix}.$$

A column vector W is a set of numbers W_i, for example $\begin{pmatrix} W_1 \\ W_2 \end{pmatrix}$ in two dimensions. (Column vectors form a vector space in the usual way.) The following rule governs multiplication of a column vector by a matrix to give a column vector $V = \mathbf{A} \cdot W$:

$$\begin{pmatrix} V_1 \\ V_2 \end{pmatrix} = \begin{pmatrix} A_{11} & A_{12} \\ A_{21} & A_{22} \end{pmatrix} \begin{pmatrix} W_1 \\ W_2 \end{pmatrix} = \begin{pmatrix} A_{11}W_1 + A_{12}W_2 \\ A_{21}W_1 + A_{22}W_2 \end{pmatrix}.$$

In index notation this is clearly

$$V_i = \sum_{j=1}^{2} A_{ij} W_j.$$

For n-dimensional matrices and vectors, this generalizes to

$$V_i = \sum_{j=1}^{n} A_{ij} W_j.$$

Notice that the sum is on the *second* index of $\mathbf{A}$.

Matrices form a vector space themselves, with addition and multiplication by a number defined by:

$$\mathbf{A} + \mathbf{B} = \mathbf{C} \Rightarrow C_{ij} = A_{ij} + B_{ij}.$$

$$a\mathbf{A} = \mathbf{B} \Rightarrow B_{ij} = aA_{ij}.$$

For $n \times n$ matrices, the dimension of this vector space is n^2. A natural inner product may be defined on this space:

$$\mathbf{A} \cdot \mathbf{B} = \sum_{i,j} A_{ij} B_{ij}.$$

One can easily show that this is positive-definite. More important than the inner product, however, for our purposes, is *matrix multiplication*. (A vector space with multiplication is called an algebra, so we are now studying the matrix algebra.) For 2×2 matrices, the product is

$$\mathbf{AB} = \mathbf{C} \Rightarrow \begin{pmatrix} C_{11} & C_{12} \\ C_{21} & C_{22} \end{pmatrix}$$

$$= \begin{pmatrix} A_{11} & A_{12} \\ A_{21} & A_{22} \end{pmatrix} \begin{pmatrix} B_{11} & B_{12} \\ B_{21} & B_{22} \end{pmatrix}$$

$$= \begin{pmatrix} A_{11}B_{11} + A_{12}B_{22} & A_{11}B_{12} + A_{12}B_{22} \\ A_{21}B_{11} + A_{22}B_{21} & A_{21}B_{12} + A_{22}B_{22} \end{pmatrix}.$$

In index notation this is

$$C_{ij} = \sum_{k=1}^{2} A_{ik} B_{kj}.$$

Generalizing to $n \times n$ matrices gives

$$C_{ij} = \sum_{k=1}^{n} A_{ik} B_{kj}.$$

Notice that the index summed is the second index of A and the first index of B. Multiplication is associative but not commutative; the identity is the matrix whose elements are δ_{ij}, the Kronecker delta symbol ($\delta_{ij} = 1$ if $i = j$, 0 otherwise).

The *determinant* of a 2×2 matrix is

$$\det \mathbf{A} = \det \begin{pmatrix} A_{11} & A_{12} \\ A_{21} & A_{22} \end{pmatrix}$$

$$= A_{11}A_{22} - A_{12}A_{21}.$$

Given any $n \times n$ matrix $\mathbf{B}$ and an element B_{lm} (for fixed l and m), we call $\mathbf{S}_{lm}$ the $(n-1) \times (n-1)$ submatrix defined by excluding row l and column m from $\mathbf{B}$, and we call D_{lm} the determinant of $\mathbf{S}_{lm}$. For example, if $\mathbf{B}$ is the 3×3 matrix

$$\mathbf{B} = \begin{pmatrix} B_{11} & B_{12} & B_{13} \\ B_{21} & B_{22} & B_{23} \\ B_{31} & B_{32} & B_{33} \end{pmatrix},$$

then the submatrix S_{12} is the 2×2 matrix

$$S_{12} = \begin{pmatrix} B_{21} & B_{23} \\ B_{31} & B_{33} \end{pmatrix}$$

and its determinant is

$$D_{12} = B_{21}B_{33} - B_{23}B_{31}.$$

Then the determinant of $\mathbf{B}$ is defined as

$$\det(\mathbf{B}) = \sum_{j=1}^{n} (-1)^{i+j} B_{ij} D_{ij} \quad \text{for any } i.$$

In this expression one sums only over j for fixed i. The result is independent of which i was chosen. This enables one to define the determinant of a 3×3 matrix in terms of that of a 2×2 matrix, and that of a 4×4 in terms of a 3×3, and so on.

Because matrix multiplication is defined, it is possible to define the multiplicative inverse of a matrix, which is usually just called its inverse:

$$(\mathbf{B}^{-1})_{ij} = (-1)^{i+j} D_{ji} / \det(\mathbf{B})$$

The inverse is defined if and only if $\det(\mathbf{B}) \neq 0$.

References

Aasi, J., Abadie, J., Abbott, B. P., Abbott, R., Abbott, T. D., *et al.* (LIGO Scientific Collaboration) (2013) 'Enhanced sensitivity of the LIGO gravitational wave detector by using squeezed states of light'. *Nature Photonics* **7**, 613.

Abadie, J., Abbott, B. P., Abbott, R., Abbott, T. D., Abernathy, M., *et al.* (LIGO Scientific Collaboration) (2011) 'A gravitational wave observatory operating beyond the quantum shot-noise limit'. *Nature Phys.* **7**, 962.

Abbott, B. P., Abbott, R., Abbott, T. D., Abernathy, M. R., Acernese, F., *et al.* (LIGO Scientific Collaboration & Virgo Collaboration) (2016a) 'Observation of gravitational waves from a binary black hole merger'. *Phys. Rev. Lett.* **116**, 061102. DOI: https://doi.org/10.1103/PhysRevLett.116.061102

Abbott, B. P., Abbott, R., Abbott, T. D., Abernathy, M. R., Acernese, F., *et al.* (LIGO Scientific Collaboration & Virgo Collaboration) (2016b) 'Improved analysis of GW150914 using a fully spin-precessing waveform model'. *Phys. Rev. X* **6**, 041014. https://doi.org/10.1103/PhysRevX.6.041014

Abbott, B. P., Abbott, R., Abbott, T. D., Abernathy, M. R., Acernese, F., *et al.* (LIGO Scientific Collaboration & Virgo Collaboration) (2016c) 'Tests of general relativity with GW150914', *Phys. Rev. Lett.* **116**, 221101. https://doi.org/10.1103/PhysRevLett.116.221101

Abbott, B. P., Abbott, R., Abbott, T.D., Abernathy, M.R., Acernese, F., *et al.* (LIGO Scientific Collaboration, Virgo Collaboration, & radio astronomy partners) (2017a) 'First search for gravitational waves from known pulsars with Advanced LIGO'. *Astrophys. J.* **839**, 12.

Abbott, B. P., Abbott, R., Abbott, T. D., Abernathy, M. R., Acernese, F., *et al.* (LIGO Scientific Collaboration & Virgo Collaboration) (2017b) 'Upper limits on the stochastic gravitational-wave background from advanced LIGO's first observing run'. *Phys. Rev. Lett.* **118** 121101.

Abbott, B. P., Abbott, R., Abbott, T. D., Acernese, F., Ackley, K., *et al.* (LIGO Scientific Collaboration & Virgo Collaboration) (2017c) 'GW170814: A three-detector observation of gravitational waves from a binary black hole coalescence'. *Phys. Rev. Lett.* **119**, 141101. https://doi.org/10.1103/PhysRevLett.119.141101.

Abbott, B. P., Abbott, R., Abbott, T. D., Acernese, F., Ackley, K., *et al.* (LIGO Scientific Collaboration & Virgo Collaboration) (2017d) 'GW170817: Observation of gravitational waves from a binary neutron star inspiral'. *Phys. Rev. Lett.* **119**, 161101. https://doi.org/10.1103/PhysRevLett.119.161101.

Abbott, B. P., Abbott, R., Abbott, T. D., Acernese, F., Ackley, K., *et al.* (LIGO Scientific Collaboration, Virgo Collaboration, several multimessenger partner collaborations) (2017e) 'A gravitational-wave standard siren measurement of the Hubble constant'. *Nature* **551**, 85–88. https://doi.org/10.1038/nature24471.

Abbott, B. P., Abbott, R., Abbott, T. D., Acernese, F., Ackley, K., *et al.* (LIGO Scientific Collaboration, Virgo Collaboration, many multimessenger partner collaborations) (2017f) 'Multi-messenger observations of a binary neutron star merger'. *Astrophys. J.* **848**, L12. https://doi.org/10.3847/2041-8213/aa91c9.

Abbott, B. P., Abbott, R., Abbott, T. D., Acernese, F., Ackley, K., *et al.* (LIGO Scientific Collaboration & Virgo Collaboration) (2018a) 'Constraints on cosmic strings using data from the first advanced LIGO observing run'. *Phys. Rev. D* , **97**, 102002.

Abbott, B. P., Abbott, R., Abbott, T. D., Acernese, F., Ackley, K., *et al.* (LIGO Scientific Collaboration & Virgo Collaboration) (2018b) 'Full band all-sky search for periodic gravitational waves in the O1 LIGO data'. *Phys. Rev. D* **97**, 102003.

Abbott, B. P., Abbott, R., Abbott, T. D., Acernese, F., Ackley, K., *et al.* (LIGO Scientific Collaboration & Virgo Collaboration) (2018c) 'GW170817: Measurements of neutron star radii and equation of state'. *Phys. Rev. Lett.* **121**, 161101. https://doi.org/10.1103/PhysRevLett.121.161101.

Abbott, B. P., Abbott, R., Abbott, T. D., Abraham, S., Acernese, F. *et al.* (LIGO Scientific Collaboration & Virgo Collaboration) (2019a) 'All-sky search for short gravitational-wave bursts in the second advanced LIGO and advanced Virgo run'. *Phys. Rev. D* **100**, 024017.

Abbott, B. P., Abbott, R., Abbott, T. D., Abraham, S., Acernese, F. *et al.* (LIGO Scientific Collaboration & Virgo Collaboration) (2019b) 'Search for the isotropic stochastic background using data from advanced LIGO's second observing run'. *Phys. Rev. D* **100**, 061101.

Abbott, B. P., Abbott, R., Abbott, T. D., Acernese, F., Ackley, K., *et al.* (LIGO Scientific Collaboration & Virgo Collaboration) (2019c) 'Tests of general relativity with GW170817'. *Phys. Rev. Lett.* **123**, 011102. https://doi.org/10/gf43hf.

Abbott, B. P., Abbott, R., Abbott, T. D., Acernese, F., Ackley, K., *et al.* (LIGO Scientific Collaboration & Virgo Collaboration) (2019d) 'Properties of the binary neutron star merger GW170817'. *Phys. Rev. X* **9**, 011001. https://doi.org/10/gfvs73.

Abbott, B. P., Abbott, R., Abbott, T. D., Acernese, F., Ackley, K., *et al.* (LIGO Scientific Collaboration & Virgo Collaboration) (2019e) 'Tests of general relativity with GW170817'. *Phys. Rev. Lett.* **123**, 011102. https://doi.org/10/gf43hf.

Abbott, B. P., Abbott, R., Abbott, T. D., Abraham, S., Acernese, F., *et al.* (LIGO Scientific Collaboration & Virgo Collaboration) (2019f) 'GWTC-1: A gravitational-wave transient catalog of compact binary mergers observed by LIGO and Virgo during the first and second observing runs'. *Phys. Rev. X* **9**, 031040. https://doi.org/10/gf8dgf.

Abbott, B. P., Abbott, R., Abbott, T. D., Abraham, S., Acernese, F., *et al.* (LIGO Scientific Collaboration & Virgo Collaboration) (2019g) 'Tests of general relativity with the binary black hole signals from the LIGO–Virgo Catalog GWTC-1'. *Phys. Rev. D* **100**, 104036. https://doi.org/10/gg4hpx.

Abbott, B. P., Abbott, R., Abbott, T. D., Abernathy, M. R., Acernese, F., *et al.* (LIGO Scientific Collaboration & Virgo Collaboration) (2020a) 'Prospects for observing and localizing gravitational-wave transients with Advanced LIGO, Advanced Virgo and KAGRA'. *Liv. Rev. Relativity*, to be published. (Update of 2018 review.)

Abbott, B. P., Abbott, R., Abbott, T. D., Abraham, S., Acernese, F., *et al.* (LIGO Scientific Collaboration & Virgo Collaboration) (2020b) 'GW190425: Observation of a compact binary coalescence with total mass ∼ 3.4 M$_\odot$'. *Astrophys. J.*, **892**, L3. https://doi.org/10/ggs45m.

Abbott, R., Abbott, T. D., Abraham, S., Acernese, F., Ackley, K., *et al.* (LIGO Scientific Collaboration & Virgo Collaboration) (2020c) 'GW190814: Gravitational waves from the coalescence of a 23 solar mass black hole with a 2.6 solar mass compact object'. *Astrophys. J.* **896**, L44. https://doi.org/10/gg2283.

Abbott, R., Abbott, T. D., Abraham, S., Acernese, F., Ackley, K., *et al.* (LIGO Scientific Collaboration & Virgo Collaboration) (2020d) 'GW190521: A binary black hole merger with a total mass of 150 M$_\odot$'. *Phys. Rev. Lett.* **125**, 101012.

Abbott, R., Abbott, T. D., Abraham, S., Acernese, F., Ackley, K., *et al.* (LIGO Scientific Collaboration & Virgo Collaboration) (2020e) 'Properties and astrophysical implications of the 150 M$_\odot$ binary black hole merger GW190521'. *Astrophys. J.* **900**, L13.

Abbott, R., Abbott, T. D., Abraham, S., Acernese, F., Ackley, K., *et al.* (LIGO Scientific Collaboration & Virgo Collaboration) (2020f) 'GWTC-2: Compact binary coalescences observed by LIGO and Virgo during the first half of the third observing run'. *Phys. Rev. X* **11**, 021053.

Abbott, B. P., Abbott, R., Abbott, T. D., Abraham, S., Acernese, F., *et al.* (LIGO Scientific Collaboration & Virgo Collaboration) (2020g) 'Search for gravitational waves from Scorpius X-1 in the second advanced LIGO observing run with an improved hidden Markov model'. *Phys. Rev. D* **100**, 122002. https://doi.org/10/gg4wnw.

Abbott, B. P., Abbott, R., Abbott, T. D., Abraham, S., Acernese, F., *et al.* (2020h) *Class. Quantum Grav.* **37**, 055002.

Abraham, R., Marsden, J. E. & Ratiu, T. (1988) *Manifolds, Tensor Analysis, and Applications* (Applied Mathematical Sciences, vol. 75) (2nd edn) (Springer).

Abuter, R. Amorim, A., Anugu, N., Bauböck, M., Benisty, M., *et al.* (GRAVITY Collaboration) (2018a) 'Detection of the gravitational redshift in the orbit of the star S2 near the Galactic centre massive black hole'. *Astron. Astrophys.* **615**, L15.

Abuter, R. Amorim, A., Bauböck, M., Berger, J. P., Bonnet, H., *et al.* (GRAVITY Collaboration) (2018b) 'Detection of orbital motions near the last stable circular orbit of the massive black hole SgrA*'. *Astron. Astrophys.* **618**, L10.

Abuter, R., Amorim, A., Bauböck, M., Berger, J. P., Bonnet, H., *et al.* (GRAVITY Collaboration) (2020) 'Detection of the Schwarzschild precession in the orbit of the star S2 near the Galactic centre massive black hole'. *Astron. & Astrophys.* **636**, L5.

Ade, P. A. R., Aghanim, N., Arnaud, M., Ashdown, M., Aumont, J., *et al.* (Planck Collaboration) (2016) 'Planck 2015 results: XIII. Cosmological parameters'. *Astron. & Astrophys.* **594**, A13. https://doi.org/10/f9scmm.

Adler, R., Bazin, M. & Schiffer, M. (1975) *Introduction to General Relativity*, 2nd edn (McGraw-Hill, New York).

Aghanim, N., Akrami, Y., Arroja, F., Ashdown, M., Aumont, J., *et al.* (Planck Collaboration) (2019) 'Planck 2018 results. I. Overview and the cosmological legacy of Planck'. *Astron. & Astrophys.* https://doi.org/10/ggxrm5.

Ajith, P. & Bose, S. (2009) 'Estimating the parameters of nonspinning binary black holes using ground-based gravitational-wave detectors: Statistical errors', *Phys. Rev. D* **79**, 084032.

Akiyama, K., Alberdi, A., Alef, W., Asada, K., Azulay, R., *et al.* (2019) 'First M87 event horizon telescope results. I. The shadow of the supermassive black hole'. *Astrophys. J. Lett.* **875**, L1. https://doi.org/10/gfx8zm.

Akutsu, T., Ando, M., Arai, K., Arai, Y., Araki, S., *et al.* (KAGRA Collaboration) (2019) 'KAGRA: 2.5 generation interferometric gravitational wave detector'. *Nature Astron.* **3**, 35.

Alcubierre, M. (2008) *Introduction to 3+1 Numerical Relativity* (Oxford University Press).

Allen, S. W., Schmidt, R. W. & Fabian, A. C. (2002) 'Cosmological constraints from the X-ray gas mass fraction in relaxed lensing clusters observed with *Chandra*'. *Mon. Not. Roy. Astron. Soc.* **334**, L11.

Amaro-Seoane, P. (2018) 'Relativistic dynamics and extreme mass ratio inspirals'. *Liv. Rev. Relativity* **21**, 1. https://doi.org/10/gdkrz8.

Andersson, N. (2017) *A Gentle Wizard* (Speed of Think Publishing).

Andersson, N. (2019) *Gravitational-Wave Astronomy: Exploring the Dark Side of the Universe* (Oxford University Press).

Andersson, N. & Comer, G. L. (2007) 'Relativistic fluid dynamics: Physics for many different scales' *Liv. Rev. Relativity*, **10**, 1.

Andersson, N., Kokkotas, K. D. & Schutz, B. F. (1999) 'Gravitational radiation limit on the spin of young neutron stars'. *Astrophys. J.* **510**, 846–853.

Andréasson, H. (2011) 'The Einstein–Vlasov system/kinetic theory', *Liv. Rev. Relativity* **14**, 4. www.livingreviews.org/lrr-2011-4.

Ansorg, M. (2005) 'Black holes surrounded by uniformly rotating rings'. *Phys. Rev.* **D 72**, 024019.

Armano, M., Audley, H., Baird, J., Binetruy, P., Born, M., *et al.* (2018) 'Beyond the required LISA free-fall performance: New LISA pathfinder results down to 20 μHz'. *Phys. Rev. Lett.* **120**, 061101.

Armstrong, J. W. (2006) 'Low-frequency gravitational wave searches using spacecraft Doppler tracking'. *Liv. Rev. Relativity* **9**, 1. www.livingreviews.org/lrr-2006-1.

Arzeliès, H. (1966) *Relativistic Kinematics* (Pergamon Press). Translation of a 1955 French edition.

Ashby, N. (2003) 'Relativity in the Global Positioning System', *Liv. Rev. Relativity* **6**, 1. www.livingreviews.org/lrr-2003-1

Ashtekar, A. & Bojowald, M. (2006) 'Quantum geometry and the Schwarzschild singularity'. *Class. Q. Grav.* **23**, 391.

Ashtekar, A., Berger, B. K., Isenberg, J. & MacCallum, M. (2015) *General Relativity and Gravitation: A Centennial Perspective* (Cambridge University Press).

Auger, G. & Plagnol, E., eds. (2017) *An Overview of Gravitational Waves: Theory, Sources and Detection* (World Scientific).

Baade, W. & Zwicky, F. (1934) 'Remarks on super-novae and cosmic rays'. *Phys. Rev.* **46**, 76–77.

Bahcall, J. N. (1978) 'Masses of neutron stars and black holes in X-ray binaries'. *Ann. Rev. Astron. Astrophys.* **16**, 241.

Bailes, M., Berger, B. K., Brady, P. R., Branchesi, M., Danzmann, K., *et al.* (2021) 'Gravitational-wave physics and astronomy in the 2020s and 2030s.' *Nature Rev. Phys.* **3**, 344–366. https://doi.org/10.1038/s42254-021-00303-8.

Baker, J. G., Centrella, J., Choi, D.-I., Koppitz, M. & van Meter, J. (2006) 'Gravitational-wave extraction from an inspiraling configuration of merging black holes'. *Phys. Rev. Lett.* **96**, 111102.

Balian, R., Audouze, J. & Schramm, D. N. (1980) *Physical Cosmology. Proceedings of 1979 Les Houches Summer School* (North-Holland).

Barausse, E., Berti, E., Hertog, T., Hughes, S. A., Jetzer, P., *et al.* (2020) 'Prospects for fundamental physics with LISA'. *Gen. Rel. Gravitation* **52**, 81. https://doi.org/10/gg9nf8.

Barrow, J. & Tipler, F. (1986) *The Anthropic Cosmological Principle* (Oxford University Press).

Bekenstein, J. D. (1973) 'Black holes and entropy'. 'Extraction of energy and charge from a black hole'. *Phys. Rev. D* **7**, 2333–2346.

Bekenstein, J. D. (1974) 'Generalized second law of thermodynamics in black-hole physics'. *Phys. Rev. D* **9**, 3292–3300.

Benacquista, M. J. & Downing, J. M. B. (2013) 'Relativistic binaries in globular clusters'. *Liv. Rev. Relativity* **16**, 1. https://doi.org/10/gbcn9q.

Bennett, C. L., Banday, A. J., Gorski, K. M., Hinshaw, G., Jackson, P., *et al.* (1996) 'Four-year COBE DMR cosmic microwave background observations: Maps and basic results'. *Astrophys. J.* **464** L1–L4.

Berger, B. K. (2002) 'Numerical approaches to spacetime singularities', *Liv. Rev. Relativity* **5**, 1. www.livingreviews.org/lrr-2002-1.

Bertotti, B. (1974) *Experimental Gravitation* (Academic Press). Lectures at Course LVI of the Enrico Fermi Summer School, Varenna, 1972.

Bertotti, B. (1977) *Experimental Gravitation* (Academica Nazionale dei Lincei, Rome). Proceedings of a symposium in Pavia, 1976.

Birrell, N. D. & Davies, P. C. W. (1984) *Quantum Fields in Curved Space* (Cambridge University Press).

Bishop, R. L. & Goldberg, S. I. (1981) *Tensor Analysis on Manifolds* (Dover).

Blair, D. G., ed. (1991) *The Detection of Gravitational Waves* (Cambridge University Press).

Blair, D. G., Howell, E. J., Ju, L. & Zhao C., eds. (2012) *Advanced Gravitational Wave Detectors* (Cambridge University Press).

Blanchet, L. (2014) 'Gravitational radiation from post-Newtonian sources and inspiralling compact binaries'. *Liv. Rev. Relativity* **17**, 2. https://doi.org/10.12942/lrr-2014-2.

Blandford, R. & Teukolsky, S. A. (1976) 'Arrival-time analysis for a pulsar in a binary system'. *Astrophys. J.* **205**, 580.

Blandford, R. D. & Znajek, R. L. (1977) 'Electromagnetic extraction of energy from Kerr black holes'. *Mon. Not. Roy. Astron. Soc.* **179** 433–456.

Blum, A., Lalli, R. & Renn, J. (2018) 'Gravitational waves and the long relativity revolution'. *Nature Astron.*, **2**, 534–543.

Bohm, D. (2008) *The Special Theory of Relativity*, new edition with preface by J. Barrow (Routledge).

Bojowald, M. (2008) 'Loop quantum cosmology'. *Liv. Rev. Relativity* **11**, 4. www .livingreviews.org/lrr-2008-4.

Bona, C. & Palenzuela-Luque, C. (2005) *Elements of Numerical Relativity* (Springer).

Bond, C., Brown, D., Freise, A. *et al.* (2016) 'Interferometer techniques for gravitational-wave detection'. *Liv. Rev. Relativity* **19**, 3. https://doi.org/10.1007/s41114-016-0002-8.

Boriakoff, V., Ferguson, D. C., Haugan, M. P., Terzian, Y. & Teukolsky, S. A. (1982) 'Timing observations of the binary pulsar PSR 1913+16'. *Astrophys. J.* **261**, L97.

Bose, S., Dhurandhar, S. V. & Pai, A. (1999) 'Detection of gravitational waves using a network of detectors'. *Pramana* **53**, 1125–1136.

Bose, S., Pai, A. & Dhurandhar, S. V. (2000) 'Detection of gravitational waves from inspiraling compact binaries using a network of interferometric detectors'. *Int. J. Mod. Phys. D* **9**, 325.

Bourne, D. E. & Kendall, P. C. (1992) *Vector Analysis and Cartesian Tensors* 3rd edn (CRC Press).

Braginsky, V. (1980) 'Quantum nondemolition measurements'. *Science* **209**, 547.

Bramberger, S. F., Hertog, T., Lehners, J.-L. & Vreys, Y. (2017) 'Quantum transitions through cosmological singularities'. *J. Cosmol. Astroparticle Phys.* **1707**, 007.

Brenneman, L. & Reynolds, C. (2006) 'Constraining black hole spin via X-ray spectroscopy'. *Astrophys. J.* **652**, 1028.

Brown, A. G. A., Vallenari, A., Prusti, T., de Bruijne, J. H. J., Babusiaux, C., *et al.* (Gaia Collaboration) (2018) 'Gaia data release 2: Summary of the contents and survey properties'. *Astron. Astrophys.* **616**, A1.

Buchdahl, H. A. (1959) 'General relativistic fluid spheres'. *Phys. Rev.* **116**, 1027.

Buchdahl, H. A. (1981) *Seventeen Simple Lectures on General Relativity Theory* (Wiley).

Butterworth, E. M. & Ipser, J. R. (1976) 'On the structure and stability of rapidly rotating fluid bodies in general relativity. I. The numerical method for computing structure and its application to uniformly rotating homogeneous bodies'. *Astrophys. J.*, **204**, 200.

Campanelli, M., Lousto, C. O., Marronetti, P. & Zlochower, Y. (2006) 'Accurate evolutions of orbiting black-hole binaries without excision'. *Phys. Rev. Lett.* **96**, 111101.

Campanelli, M., Lousto, C. O., Zlochower, Y. & Merritt, D. (2007) 'Maximum gravitational recoil' *Phys. Rev. Lett.* **98**, 231102.

Cardoso, V., Gualtieri, L., Herdeiro, C. & Sperhake, U. (2015) 'Exploring new physics frontiers through numerical relativity'. *Liv. Rev. Relativity* **18**, 1.

Carlotto, A. (2021) 'The general relativistic constraint equations'. *Liv. Rev. Relativity* **24**, 2. www.livingreviews.org/lrr-2021-2.

Carroll, S. (2019) *Spacetime and Geometry* (Cambridge University Press).

Cartan, E. (1923) 'Sur les variétés è connexion affine et la théorie de la relativité généralisée'. *Ann. Ecole Norm. Sup.* **40** 325.

Carter, B. (1969) 'Killing horizons and orthogonally transitive groups in space-time'. *J. Math. Phys.* **10** 70.

Carter, B. (1973) In DeWitt, B. & De Witt, C. eds., *Black Holes* (Gordon and Breach).

Carter, B. (1979) 'The general theory of the mechanical, electromagnetic, and thermodynamic properties of black holes'. In Hawking & Israel (1979), p. 294.

Caves, C. M., Thorne, K. S., Drever, R. W. P., Sandberg, V. D. & Zimmerman, M. (1980) 'On the measurement of a weak classical force coupled to a quantum-mechanical oscillator. I. Issues of principle'. *Rev. Mod. Phys.*, **52** 341.

Centrella, J., Baker, J. G., Kelly, B. J. & van Meter, J. R. (2010) 'Black-hole binaries, gravitational waves, and numerical relativity' *Rev. Mod. Phys.* **82**, 3069.

Chamel, N. & Haensel, P. (2008) 'Physics of neutron star crusts'. *Liv. Rev. Relativity* **11**, 10.

Chandrasekhar, S. (1939) *An Introduction to the Study of the Stellar Structure* (University of Chicago Press).

Chandrasekhar, S. (1957) *Stellar Structure* (Dover).

Chandrasekhar, S. (1980) In Wayman, P., ed., *Highlights of Astronomy* (Reidel). Proceedings of the 1979 IAU meeting in Montreal, p. 45.

Chandrasekhar, S. (1983) *The Mathematical Theory of Black Holes* (Oxford University Press).

Chandrasekhar, S. (1987) *Truth and Beauty* (University of Chicago Press).

Chirenti, C., Posada, C. & Guedes, V. (2020) 'Where is love? Tidal deformability in the black hole compactness limit'. *Classical and Quantum Gravity* **37**, 195017.

Choquet-Bruhat, Y. & York, J. W. (1980) 'The Cauchy problem'. In Held (1980a), p. 99.

Choquet-Bruhat, Y., DeWitt-Morette, C. & Dillard-Bleick, M. (1977) *Analysis, Manifolds, and Physics* (North-Holland).

Chow, T. L. (2007) *Gravity, Black Holes, and the Very Early Universe* (Springer).

Chruściel, P. T. (1996) 'Uniqueness of stationary, electro-vacuum black holes revisited'. *Helv. Phys. Acta* **69**, 529–552.

Ciufolini, I., Paolozzi, A., Pavlis, E. C., Koenig, Riess, J., *et al.* (2016) 'A test of general relativity using the LARES and LAGEOS satellites and a GRACE Earth gravity model'. *Eur. Phys. J. C* **76**, 120.

Collins, H. M. (2004) *Gravity's Shadow: the Search for Gravitational Waves* (University of Chicago Press).

Collins, H. M. (2017) *Gravity's Kiss: The Detection of Gravitational Waves* (University of Chicago Press).

Comins, N. & Schutz, B. F. (1978) 'On the ergoregion instability'. *Proc. Roy. Soc. A* **364**, 211.

Cook, G. (2000) 'Initial data for numerical relativity', *Liv. Rev. Relativity* **3**, 5. www.livingreviews.org/lrr-2000-5.

Cordes, J. M., Kramer, M., Lazio, T. J. W., *et al.* (2004) 'Pulsars as tools for fundamental physics & astrophysics'. *New Astr Rev* **48**, 1413–1438.

Creighton, J. D. E. & Anderson, W. G. (2011) *Gravitational-Wave Physics and Astronomy: An Introduction to Theory, Experiment and Data Analysis* (Wiley).

Cromartie, H. T., Fonseca, E., Ransom, S. M., Demorest, P. B., Arzoumanian, Z., *et al.* (2020) 'Relativistic Shapiro delay measurements of an extremely massive millisecond pulsar'. *Nature Astron.* **4**, 72–76.

Cumming, A. V., Sorazu, B., Daw, E., Hammond, G. D., Hough, J., *et al.* (2020) 'Lowest observed surface and weld losses in fused silica fibres for gravitational wave detectors'. *Classical and Quantum Gravity* **37**, 195019.

Dafermos, M. & Luk, J. (2017) 'The interior of dynamical vacuum black holes I: The C^0-stability of the Kerr Cauchy horizon', *arXiv e-print server*, 1710.01722.

Dafermos, M. & Rodnianski, I. (2013) 'Lectures on black holes'. In Ellwood, D., Rodnianski, I, Staffilani, G. & Wunsch, J., eds. *Evolution Equations: Clay Mathematics Proceedings*, vol. 17 (American Mathematical Society).

Dafermos, M., Holzegel, G. & Rodnianski, I. (2016) 'The linear stability of the Schwarzschild solution to gravitational perturbations', *arXiv e-print server*, 1601.06467.

Dafermos, M., Holzegel, G. & Rodnianski, I. (2017) 'Boundedness and decay for the Teukolsky equation on Kerr spacetimes I: the case $|a| \ll M$', *arXiv e-print server*, 1711.07944.

Damour, T. (1987) 'The problem of motion in Newtonian and Einsteinian gravity'. In S. Hawking & W. Israel, eds., *300 Years of Gravitation* (Cambridge University Press), pp. 128–198.

Davis, T. & Lineweaver, C. (2004) 'Expanding confusion: Common misconceptions of cosmological horizons and the superluminal expansion of the universe'. *Publ. Astronom. Soc. Australia* **21**, 97–109. doi:10.1071/AS03040.

Delva, P., Puchades, N., Schönemann, E., Dilssner, F., Courde, C., *et al.* (2018) 'Gravitational redshift test using eccentric Galileo satellites'. *Phys. Rev. Lett.* **121**, 231101.

Dergachev, V. & Papa, M. A. (2019a) 'Sensitivity improvements in the search for periodic gravitational waves using O1 LIGO data' *Phys. Rev. Lett.* **123**, 101101. https://doi.org/10/ggf35p.

Dergachev, V. & Papa, M. A. (2019b) 'Results from an extended Falcon all-sky survey for continuous gravitational waves', arXiv:1909.09619 [astro-ph, physics:gr-qc].

DeWitt, B. S. (1979) 'Quantum gravity: the new synthesis'. In Hawking & Israel (1979), p. 680.

Dicke, R. H. (1964) In Dewitt, B. & Dewitt, C. eds., *Relativity, Groups and Topology* (Gordon and Breach, N.Y.). Lectures at the 1963 Les Houches summer school.

Dixon, W. G. (1978) *Special Relativity, The Foundation of Macroscopic Physics* (Cambridge University Press).

Dodelson, S. & Schmidt, F. (2020) *Modern Cosmology*, 2nd edn (Academic Press).

Eardley, D. M. & Press, W. H. (1975) 'Astrophysical processes near black holes'. *Ann. Rev. Astron. Astrophys.* **13**, 381.

Eardley, D. M. & Smarr, L. (1979) 'Time functions in numerical relativity: Marginally bound dust collapse', *Phys. Rev. D* **19**, 2239.

Eckart, A. & Genzel, R. (1997) 'Stellar proper motions in the central 0.1 pc of the Galaxy' *Mon. Not. Roy. Astron. Soc.* **284**, 576.

Ehlers, J. & Kundt, W. (1962) In Witten, L., ed., *Gravitation: An Introduction to Current Research* (Wiley), pp. 49–101.

Eichler, D., Livio, M., Piran, T., and Schramm, D. N. (1989) 'Nucleosynthesis, neutrino bursts and γ-rays from coalescing neutron stars'. *Nature* **340**, 126.

Eisenhauer, F., Genzel, R., Alexander, T., Abuter, R., Paumard, T., *et al.* (2005) 'SINFONI in the Galactic center: Young stars and infrared flares in the central light-month', *Astrophys. J.*, **628**, 246.

Epstein, R. (1977) 'The binary pulsar – Post-Newtonian timing effects'. *Astrophys. J.*, **216**, 92.

Everitt, C. W. F., DeBra, D. B., Parkinson, B. W., Turneaure, J. P. *et al.* (2011) 'Gravity probe B: Final results of a space experiment to test general relativity'. *Phys. Rev. Lett.* **106**, 221101.

Faber, J. A. & Rasio, F. A. (2012) 'Binary neutron star mergers', *Liv. Rev. Relativity* **15**, 8.

Fermi, E. (1956) *Thermodynamics* (Dover).

Filippenko, A. (2008) *Dark Energy and Supernovae* (Princeton University Press).

Finn, L. S. (2001) 'Aperture synthesis for gravitational-wave data analysis: deterministic sources'. *Phys. Rev. D* **63**, 102001.

Flanagan, E. E. & Hinderer, T. (2008) 'Constraining neutron-star tidal Love numbers with gravitational-wave detectors'. *Phys. Rev. D* **77**, 021502. https://doi.org/10/d4txxd.

Font, J. A. (2008) 'Numerical hydrodynamics and magnetohydrodynamics in general relativity'. *Liv. Rev. Relativity* **11**, 7.

Forward, R. L. & Berman, D. (1967) 'Gravitational-radiation detection range for binary stellar systems'. *Phys. Rev. Lett.* **18**, 1071.

French, A. P. (1968) *Special Relativity* (W. W. Norton).

Friedman, J. L. (1978) 'Ergosphere instability'. *Commun. Math. Phys.*, **63**, 243.

Friedman, J. L. & Schutz, B. F. (1978) 'Secular instability of rotating Newtonian stars'. *Astrophys. J.* **222**, 281.

Friedman, J. L. & Stergioulas, N. (2013) *Rotating Relativistic Stars* (Cambridge University Press).

Fulling, S. A. (1989) *Aspects of Quantum Field Theory in Curved Spacetime* (Cambridge University Press).

Futamase, T. (1983) 'Gravitational radiation reaction in the Newtonian limit'. *Phys. Rev. D* **28**, 2373.

Futamase, T. & Itoh, Y. (2007) 'The post-Newtonian approximation for relativistic compact binaries'. *Liv. Rev. Relativity* **10**, 2. www.livingreviews.org/lrr-2007-2.

Futamase, T. & Schutz, B. F. (1983) 'Newtonian and post-Newtonian approximations are asymptotic to general relativity'. *Phys. Rev. D* **28**, 2363.

Gair, J. R., Vallisneri, M., Larson, S. L. & Baker, J. G. (2013) 'Testing general relativity with low-frequency, space-based gravitational-wave detectors'. *Liv. Rev. Relativity* **16**, 7.

Geroch, R. (1978) *General Relativity from A to B* (University of Chicago Press).

Geroch, R. & Horowitz, G. T. (1979) 'Global structure of spacetimes'. In Hawking & Israel (1979), p. 212.

Ghez, A. M., Klein, B. L., Morris, M. & Becklin, E. E. (1998) 'High proper-motion stars in the vicinity of Sagittarius A*: Evidence for a supermassive black hole at the center of our galaxy' *Astrophys, J.* **509**, 678.

Ghez, A. M., Salim, S., Weinberg, N. N., Lu, J. R., Do, T., *et al.* (2008) 'Measuring distance and properties of the Milky Way's central supermassive black hole with stellar orbits'. *Astrophys. J.* **689**, 1044–1062.

Gieren, W., Storm, J., Konorski, P., Górski, M., Pilecki, B., *et al.* (2018) 'The effect of metallicity on Cepheid period–luminosity relations from a Baade-Wesselink analysis of Cepheids in the Milky Way and Magellanic Clouds'. *Astron. & Astrophys.* **620**, A99. https://doi.org/10/gjtxch.

Glendenning, N. K. (2007) *Special and General Relativity: With Applications to White Dwarfs, Neutron Stars and Black Holes* (Springer).

Graham, M. J., Ford, K. E. S., McKernan, B., Ross, N. P., Stern, D., *et al.* (2020) 'Candidate electromagnetic counterpart to the binary black hole merger gravitational-wave event S190521g'. *Phys. Rev. Lett.* **124**, 251102. https://doi.org/10/gg3dp4.

Gron, O. & Hervik, S. (2007) *Einstein's General Theory of Relativity: With Modern Applications in Cosmology* (Springer).

Gürsel, Y. & Tinto, M. (1989) 'Near optimal solution to the inverse problem for gravitational-wave bursts'. *Phys. Rev. D* **40**, 3884–3938.

Guth, A. (1981) 'Inflationary universe: A possible solution to the horizon and flatness problems'. *Phys. Rev. D* **23**, 347.

Häfner, D., Hintz, P. & Vasy, A. (2020) 'Linear stability of slowly rotating Kerr black holes'. *Inventiones Mathematicae* https://doi.org/10/ghf4w2.

Hagedorn, R. (1963) *Relativistic Kinematics* (Benjamin).

Hansen, C. J., Kawaler, S. D. & Trimble, V. (2004) *Stellar Interiors – Physical Principles, Structure, and Evolution* (Springer).

Harms, J. (2019) 'Terrestrial gravity fluctuations'. *Liv. Rev. Relativity* **22**, 6. https://doi.org/10.1007/s41114-019-0022-2.

Harrison, B. K., Thorne, K. S., Wakano, M. & Wheeler, J. A. (1965) *Gravitation Theory and Gravitational Collapse* (University of Chicago Press).

Hartle, J. B. (2003) *Gravity: An Introduction to Einstein's General Relativity* (Benjamin Cummings).

Hawking, S. W. (1972) 'Black holes in general relativity'. *Commun. Math. Phys.*, **25**, 152.

Hawking, S. W. (1975) 'Particle creation by black holes'. *Commun. Math. Phys.*, **43**, 199.

Hawking, S. W. & Ellis, G. F. R. (1973) *The Large-Scale Structure of Space–Time* (Cambridge University Press).

Hawking, S. W. & Israel, W. (1979) *General Relativity: An Einstein Centenary Survey* (Cambridge University Press).

Heidmann, J. (1980) *Relativistic Cosmology* (Springer).

Held, A. (1980a) *General Relativity and Gravitation*, vol. 1 (Plenum).

Held, A. (1980b) *General Relativity and Gravitation*, vol. 2 (Plenum).

Helmi, A., Babusiaux, C., Koppelman, H. H., Massari, D., Veljanoski, J., *et al.* (2018) 'The merger that led to the formation of the Milky Way's inner stellar halo and thick disk'. *Nature* **563**, 85–88.

Herrmann, S., Finke, F., Lülf, M., Kichakova, O., Puetzfeld, D., *et al.* (2018) 'Test of the gravitational redshift with Galileo satellites in an eccentric orbit'. *Phys. Rev. Lett.* **121**, 231102.

Heusler, M. (1998) 'Stationary black holes: Uniqueness and beyond'. *Liv. Rev. Relativity* **1**, 6. www.livingreviews.org/lrr-1998-6.

Hewish, A., Bell, S. J., Pilkington, J. D. H., Scott, P. F. & Collins, R. A. (1968) 'Observation of a rapidly pulsating radio source'. *Nature* **217**, 709–713.

Hoffmann, B. (1983) *Relativity and its Roots* (Dover).

Höflich, P., Kumar, P. & Wheeler, J. C. (2004) *Cosmic Explosions in Three Dimensions: Asymmetries in Supernovae and Gamma-Ray Bursts* (Cambridge University Press).

Holz, D. E., Hughes, S. A. & Schutz, B. F. (2018) 'Measuring cosmic distances with standard sirens'. *Phys. Today* **71**, 34–40. https://doi.org/10.1063/PT.3.4090

Hotokezaka, K., Nakar, E., Gottlieb, O., Nissanke, S., Masuda, K., *et al.* (2019) 'A Hubble constant measurement from superluminal motion of the jet in GW170817'. *Nature Astron.* **3** 940–944. https://doi.org/10.1038/s41550-019-0820-1.

Hough, J. L. & Rowan, S. (2000) 'Gravitational wave detection by interferometry (ground and space)'. *Liv. Rev. Relativity* **3**, 3. www.livingreviews.org/lrr-2000-3.

Hu, W. & White, M. (1997) 'A CMB polarization primer'. http://background.uchicago.edu/~whu/polar/webversion/polar.html.

Hulse, R. A. & Taylor, J. H. (1975) 'Discovery of a pulsar in a binary system'. *Astrophys. J.* **195**, L51–L53.

Ishak, M. (2019) 'Testing general relativity in cosmology'. *Liv. Rev. Relativity* **22**, 1.

Isham, C. J. (1999) *Modern Differential Geometry for Physicists* (World Scientific).

Isi, M., Giesler, M., Farr, W. M., Scheel, M. A. & Teukolsky, S. A. (2019) 'Testing the no-hair theorem with GW150914'. *Phys. Rev. Lett.* **123**, 111102.

Israel, W. & Stewart, J. (1980) 'Progress in relativistic thermodynamics and electrodynamics of continuous media'. In Held (1980a), p. 491.

Jackson, J. D. (1998) *Classical Electrodynamics*, 3rd edn (Wiley).

Jackson, N. (2015) 'The Hubble constant'. *Liv. Rev. Relativity* **18** (2). https://doi.org/10.1007/lrr-2015-2.

Jaranowski, P. & Krolak, A. (2009) *Analysis of Gravitational-Wave Data* (Cambridge University Press).

Kafka, P. (1977) 'Optimal detection of signals through linear devices with thermal noise sources and application to the Munich–Frascati Weber-type gravitational wave detectors'. In V. De Sabbata & J. Weber, eds., *Topics in Theoretical and Experimental Gravitation Physics*. NATO Advanced Study Institutes Series vol. B27 (Reidel), p. 161.

Kaspi, V. & Kramer, M. (2016) 'Radio pulsars: The neutron star population & fundamental physics'. In Blandford, R., Gross, D. & Sevrin, A. eds. *Proc. 26th Solvay Conf. on Physics in Astrophysics and Cosmology*, (World Scientific, Singapore), pp. 22–61.

Kennefick, D. (2007) *Traveling at the Speed of Thought: Einstein and the Quest for Gravitational Waves* (Princeton University Press).

Kerr, R. P. (1963) 'Gravitational field of a spinning mass as an example of algebraically special metrics'. *Phys. Rev. Lett.* **11**, 237–238.

Kilmister, C. W. (1970) *Special Theory of Relativity* (Pergamon).

Knop, R.A., Aldering, G., Amanullah, R., Astier, P., Blanc, G., *et al.* (2003) 'New constraints on Ω_M, Ω_Λ, and w from an independent set of eleven high-redshift supernovae observed with HST'. *Astrophys. J.* **598**, 102.

Kobayashi, S. & Nomizu, K. (2009) *Foundations of Differential Geometry*, vols. 1 and 2 (Wiley).

Kokkotas, K. D. & Schmidt, B. (1999) 'Quasi-normal modes of stars and black holes'. *Liv. Rev. Relativity* **2**, 2. www.livingreviews.org/lrr-1999-2.

Kokkotas, K. & Schwenzer, K. (2016) 'r-Mode astronomy'. *Eur. Phys. J. A* **52**, 38.

Komossa, S., Zhou, H. & Lu, H. (2008) 'A recoiling supermassive black hole in the quasar SDSS J092712.65+294344.0?'. *Astrophys. J. Lett* **678**, L81–L84.

Kopeikin, S., ed. (2014a) *Frontiers in Relativistic Celestial Mechanics*, vol. 1, *Theory* (De Gruyter).

Kopeikin, S., ed. (2014b) *Frontiers in Relativistic Celestial Mechanics*, vol. 2, *Applications and Experiments* (De Gruyter).

Kounkel, M. & Covey, K. (2019) 'Untangling the Galaxy. I. Local structure and star formation history of the Milky Way'. *Astron. J.* **158**, 122.

Kramer, M. (2016) 'Pulsars as probes of gravity and fundamental physics'. *Int. J. Mod. Phys. D* **25**, 1630029.

Kramer, M. & Wex, N. (2009) 'Topical review: The double pulsar system: A unique laboratory for gravity'. *Class. Quant. Grav.* **26**, 073001.

Kramer, M., Stairs, I. H., Manchester, R. N., McLaughlin, M. A., Lyne, A. G., *et al.* (2006) 'Tests of general relativity from timing the double pulsar'. *Science* **314**, 97–102.

Krolak, A. & Schutz, B. F. (1987) 'Coalescing binaries - probe of the universe'. *Gen. Rel. and Grav.* **19**, 1163–1171.

Landau, L. D. & Lifshitz, E. M. (1980) *The Classical Theory of Fields* (Butterworth-Heinemann).

Landau, L. D. & Lifshitz, E. M. (1987) *Fluid Mechanics* (Butterworth-Heinemann).

Lattimer, J. M. (2010) 'Neutron star equation of state'. *New Astronomy Reviews* **54**, 101–109.

Lattimer, J. M. (2012) 'The nuclear equation of state and neutron star masses'. *Ann. Rev. Astron. Astrophys.* **62**, 485–515.

Lattimer, J. M. & Prakash, M. (2000) 'Nuclear matter and its role in supernovae, neutron stars and compact object binary mergers'. *Phys. Rep.* **333**, 121–146.

Lattimer, J. M. & Prakash, M. (2016) 'The equation of state of hot, dense matter and neutron stars'. *Phys. Rep.* **621**, 127–164.

Lattimer, J. M. & Schramm, D. N. (1974) 'Black-hole–neutron-star collisions'. *Astrophys. J.* **192**, L145–L147.

Lee, J. G., Adelberger, E. G., Cook, T. S., Fleischer, S. M. & Heckel, B. R. (2020) 'New test of the gravitational $1/r^2$ law at separations down to 52 μm'. *Phys. Rev. Lett.* **124**, 101101.

Liang, E. P. T. & Sachs, R. K. (1980) 'Cosmology'. In Held (1980b), p. 329.

Liddle, A. (2003) *An Introduction to Modern Cosmology*, 2nd edn (Wiley).

Liebling, S. L. & Palenzuela, C. (2017) 'Dynamical boson stars'. *Liv. Rev. Relativity* **20**, 5. https://doi.org/10/gck9k8.

Lightman, A. P., Press, W. H., Price, R. H. & Teukolsky, S. A. (2017) *Problem Book in Relativity and Gravitation* (Princeton University Press).

Linde, A. (1982) 'A new inflationary universe scenario: A possible solution of the horizon, flatness, homogeneity, isotropy and primordial monopole problems'. *Phys. Lett. B.* **108**, 389.

Lorimer, D. R. (2008) 'Binary and millisecond pulsars'. *Liv. Rev. Relativity* **11**, 8. https://doi.org/10.12942/lrr-2008-8.

Lorimer, D. R. & Kramer, M. (2004) *Handbook of Pulsar Astronomy* (Cambridge University Press).

Lovelace, G., Chen, Y., Cohen, M., Kaplan, J.D., Keppel, D., *et al.* (2010) 'Momentum flow in black-hole binaries. II. Numerical simulations of equal-mass, head-on mergers with antiparallel spins'. *Phys. Rev. D* **82**, 064031.

Lovelock, D. & Rund, H. (1990) *Tensors, Differential Forms, and Variational Principles* (Dover).

Lundgren, E., Bondarescu, D., Bondarescu, M., Bondarescu, R. & Bondarescu, M. (2016) *A Child's First Book on Gravitational Waves: You, Me and the Dancing Black Holes* (self-published, available as an e-book).

Lyne, A.G. & Graham-Smith, F. (1998) *Pulsar Astronomy* (Cambridge University Press).

Maartens, R. (2004) 'Brane-world gravity'. *Liv. Rev. Relativity* **7**, 7. www.livingreviews.org/lrr-2004-7.

MacCallum, M. A. H. (1979) 'Anisotropic and inhomogeneous relativistic cosmologies'. In Hawking & Israel (1979), p. 533.

Maggiore, M. (2007) *Gravitational Waves*, vol. 1, *Theory and Experiments* (Oxford University Press).

Maggiore, M. (2018) *Gravitational Waves*, vol. 2, *Astrophysics and Cosmology* (Oxford University Press).

Manchester, R. N. (2015) 'Pulsars and gravity'. *Int. J. Mod. Phys. D* **24**, 1530018.

Marder, L. (1971) *Time and the Space-Traveller* (George Allen and Unwin).

Mather, J. C., Cheng, E. S., Cottingham, D. A., Eplee, R. E., Jr., Fixsen, D. J., *et al.* (1994) 'Measurement of the cosmic microwave background spectrum by the COBE FIRAS instrument'. *Astrophys. J.* **420**, 439.

Mathews, J. & Walker, R. L. (1965) *Mathematical Methods of Physics* (Benjamin).

McCrea, W. H. (1979) *Quart. J. Roy. Astron. Soc.* **20**, 251.

Melia, F. (2009) *Cracking the Einstein Code* (University of Chicago Press).

Mermin, N. D. (1989) *Space and Time in Special Relativity* (Waveland Press).

Merritt, D. & Milosavljevic, M. (2005) 'Massive black hole binary evolution', *Liv. Rev. Relativity* **8**, 8. www.livingreviews.org/lrr-2005-8.

Metzger, B. D. (2020) 'Kilonovae'. *Liv. Rev. Relativity* **23**, 1. https://doi.org/10.1007/s41114-019-0024-0.

Miller, M. C. & Colbert, E. J. M. (2004) 'Intermediate-mass black holes'. *Int. J. Mod. Phys. D* **13**, 1–64.

Misner, C. W., Thorne, K. S. & Wheeler, J. A. (1973) *Gravitation* (Freeman).

Møller, C. (1972) *The Theory of Relativity* (Clarendon Press).

Morin, D. J. (2017) *Special Relativity: for the Enthusiastic Beginner* (self-published through CreateSpace).

Mukhanov, V. (2005) *Physical Foundations of Cosmology* (Cambridge University Press).

Musoke, N., Hotchkiss, S. & Easther, R. (2020) 'Lighting the dark: Evolution of the postinflationary universe'. *Phys. Rev. Lett.* **124**, 061301.

Nahmad-Achar, E. (2018) *Differential Topology and Geometry with Applications to Physics* (Institute of Physics).

Needham, T. (2021) *Visual Differential Geometry and Forms: A Mathematical Drama in Five Acts* (Princeton University Press).

Nissanke, S., Holz, D. E., Hughes, S. A., Dalal, N. & Sievers, J. L. (2010) 'Exploring short gamma-ray bursts as gravitational-wave standard sirens'. *Astrophys. J.* **725**, 496–514.

Noyola, E., Gebhardt, K. & Bergmann, M. (2008) 'Gemini and Hubble Space Telescope evidence for an intermediate-mass black hole in Alpha Centauri'. *Astrophys. J.* **676**, 1008–1015.

O'Connor, J. J. & Robertson, E. F. www.history.mcs.st-andrews.ac.uk/Biographies/Schwarzschild.html.

Özel, F. & Freire, P. (2016) 'Masses, radii, and the equation of state of neutron stars'. *Ann. Rev. Astron. Astrophys.* **54**, 401–440.

Padmanabhan, T. (2002) *Theoretical Astrophysics*, vol. 3, *Galaxies and Cosmology* (Cambridge University Press).

Pais, A. (1982) *Subtle is the Lord ..., The Science and the Life of Albert Einstein* (Oxford University Press).

Palomba, C. (2017) 'The search for continuous gravitational waves in LIGO and Virgo data'. *Nuovo Cimento* **40**, 129.

Parikh, M. K. & Wilczek, F. (2000) 'Hawking radiation as tunneling'. *Phys. Rev. Lett.* **85**, 5042–5045.

Parker, L. & Toms, D. (2009) *Quantum Field Theory in Curved Spacetime* (Cambridge University Press).

Peebles, P. J. E. (1980) *Large Scale Structure of the Universe* (Princeton University Press).

Peebles, P. J. E. (1993) *Principles of Physical Cosmology* (Princeton University Press).

Penrose, R. (1965) 'Gravitational collapse and space-time singularities'. *Phys. Rev. Lett.* **14**, 57–59 https://link.aps.org/doi/10.1103/PhysRevLett.14.57.

Penrose, R. (1969) 'Gravitational collapse: The role of general relativity'. *Rev. Nuovo Cimento* **1**, 252.

Penrose, R. (1979) In Hawking & Israel (1979), 581.

Penzias, A. A. (1979) 'The origin of the elements'. *Rev. Mod. Phys.* **51**, 425.

Perlick, V. (2004) 'Gravitational lensing from a spacetime perspective'. *Liv. Rev. Relativity* **7**, 9. www.livingreviews.org/lrr-2004-9.

Perlmutter, S., Aldering, G., Goldhaber, G., Knop, R.A., Nugent, P., *et al.* (1999) 'Measurements of Ω and Λ from 42 high-redshift supernovae'. *Astrophys. J.* **517**, 565–586.

Peters, P. C. (1964) 'Gravitational radiation and the motion of two point masses'. *Phys. Rev. B* **136**, 1224.

Peters, P. C. & Mathews, J (1963) 'Gravitational radiation from point masses in a Keplerian orbit' *Phys. Rev.* **131**, 435.

Philippov, A., *et al.* (2019) 'Pulsar radio emission mechanism: Radio nanoshots as a low-frequency afterglow of relativistic magnetic econnection'. *Astrophys. J. Lett.* **876**, L6.

Piran, E., D'Avanzo, P., Benetti, S., Branchesi, M., Brocato, E., *et al.* (2017) 'Spectroscopic identification of R-process nucleosynthesis in a double neutron-star merger'. *Nature* **551**, 67–70.

Poisson, E. & Will, C. (2014) *Gravity: Newtonian, Post-Newtonian, Relativistic* (Cambridge University Press).

Postnov, K. A. & Yungelson, L. R. (2014) 'The evolution of compact binary star systems'. *Liv. Rev. Rel.* **17**, 3.

Pound, R. V. & Rebka, G. A. (1960) 'Apparent weight of photons'. *Phys. Rev. Lett.* **4**, 337.

Pound, R. V. & Snider, J. L. (1965) 'Effect of gravity on gamma radiation'. *Phys. Rev. B* **140**, 788.

Pretorius, F. (2005) 'Evolution of binary black hole spacetimes'. *Phys. Rev. Lett.* **95**, 121101.

Price, R. H. (1972a) 'Nonspherical perturbations of relativistic gravitational collapse. I. Scalar and gravitational perturbations'. *Phys. Rev. D* **5**, 2419.

Price, R. H. (1972b) 'Nonspherical perturbations of relativistic gravitational collapse. II. Integer-spin, zero-rest-mass fields'. *Phys. Rev. D* **5**, 2439.

Reitze, D., Saulson, P. & Grote, H., eds. (2019) *Advanced Interferometric Gravitational-Wave Detectors*, vols. 1 and 2 (World Scientific).

Rendall, A. D. (2005) 'Theorems on existence and global dynamics for the Einstein equations'. *Liv. Rev. Relativity* **8**, 6. www.livingreviews.org/lrr-2005-6.

Renn, J., ed. (2007) *The Genesis of General Relativity: Sources and Interpretations* (Springer, four vols.).

Rezzolla, L., Pizzochero, P., Jones, D. I., Rea, N. & Vidaña, I., eds. (2018) *The Physics and Astrophysics of Neutron Stars* (Springer Nature).

Rickles, D. & DeWitt, C. M., eds. (2011) *The Role of Gravitation in Physics: Report from the 1957 Chapel Hill Conference* (Edition Open Sources, Berlin). ISBN 978-3-945561-29-4. www.edition-open-sources.org/sources/5/index.html.

Riess, A. G. (2020) 'The expansion of the Universe is faster than expected'. *Nature Rev. Phys.* **2**, 10–12.

Riess, A. G., Filippenko, A.V., Challis, P., Clocchiatti, A., Diercks, A., *et al.* (1998) 'Observational evidence from supernovae for an accelerating universe and a cosmological constant'. *Astron. J.,* **116**, 1009–1038.

Riess, A. G., Macri, L. M., Hoffmann, S. L., Scolnic, D., Casertano, S., *et al.* (2016) 'A 2.4% determination of the local value of the Hubble constant'. *Astrophys J.* **826**, 56. https://doi.org/10/gf33jd.

Riess, A. G., Casertano, S., Yuan, W., Macri, L. M., Scolnic, D., *et al.* (2019) 'Large Magellanic Cloud Cepheid standards provide a 1% foundation for the determination of the Hubble constant and stronger evidence for physics beyond ΛCDM'. *Astrophys. J.* **876**, 85.

Rindler, W. (1991) *Introduction to Special Relativity*, 2nd edn (Oxford University Press).

Rindler, W. (2006) *Relativity: Special, General, Cosmological*, 2nd edn (Oxford University Press).

Robinson, D. C. (1975) 'Uniqueness of the Kerr black hole'. *Phys. Rev. Lett.* **34**, 905.

Romano, J. D. & Cornish, N. J. (2017) 'Detection methods for stochastic gravitational-wave backgrounds: A unified treatment'. *Liv. Rev. Relativity* **20** (1). https://doi.org/10/f92dn5.

Rovelli, C. & Vidotto, F. (2020) *Covariant Loop Quantum Gravity: An Elementary Introduction to Quantum Gravity and Spinfoam Theory* (Cambridge University Press).

Rubin, V. C., Ford, W. K, Jr. (1970) 'Rotation of the Andromeda Nebula from a spectroscopic survey of emission regions'. *Astrophys. J.* **159**, 379–403.

Ryan, M. P. & Shepley, L. C. (1975) *Homogeneous Relativistic Cosmologies* (Princeton University Press).

Sathyaprakash, B. S. & Schutz, B. F. (2009) 'Physics, astrophysics and cosmology with gravitational waves'. *Liv. Rev. Relativity* **12**, 2.

Sato, S., Kawamura, S., Ando, M., Nakamura, T., Tsubono, K., *et al.* (2017) 'The status of DECIGO'. *J. Phys. Conference Series*, **840**, 012010.

Saulson, P. (1997) 'If light waves are stretched by gravitational waves, how can we use light as a ruler to detect gravitational waves?' *Am. J. Phys.* **65**, 501.

Saulson, P. (2017) *Fundamentals of Interferometric Gravitational Wave Detectors*, 2nd edn (World Scientific).

Schäfer, G. & Jaranowski, P. (2018) 'Hamiltonian formulation of general relativity and post-Newtonian dynamics of compact binaries'. *Liv. Rev. Relativity* **21** (1). https://doi.org/10/gg4kpn.

Schild, A. (1967) In Ehlers, J., ed., *Relativity Theory and Astrophysics: I. Relativity and Cosmology* (American Mathematical Society).

Schneider, P. (2006) *Extragalactic Astronomy and Cosmology* (Springer).

Schneider, P., Ehlers, J., Falco, E. (1992) *Gravitational Lenses* (Springer).

Schoedel, R., Ott, T., Genzel, R., Eckart, A., Mouawad, N., *et al.* (2003) 'Stellar dynamics in the central arcsecond of our Galaxy'. *Astrophys. J.* **596**, 1015–1034.

Schouten, J. A. (2011) *Tensor Analysis for Physicists*, 2nd edn (Dover).

Schrödinger, E. (1950) *Space-Time Structure* (Cambridge University Press).

Schutz, B. F. (1980a) 'Statistical formulation of gravitational radiation reaction'. *Phys. Rev. D* **22**, 249.

Schutz, B. F. (1980b) *Geometrical Methods of Mathematical Physics* (Cambridge University Press).

Schutz, B. F. (1984) 'Gravitational waves on the back of an envelope'. *Am. J. Phys.* **52**, 412.

Schutz, B. F. (1986) 'Determining the Hubble constant from gravitational wave observations'. *Nature* **323**, 310–311.

Schutz, B. F. (1991) 'Data processing analysis and storage for interferometric antennas'. In Blair, D. G., ed., *The Detection of Gravitational Waves* (Cambridge University Press), p. 406.

Schutz, B. F. (2003) *Gravity from the Ground Up* (Cambridge University Press).

Schwarz, P. M. & Schwarz, J. H. (2004) *Special Relativity: From Einstein to Strings* (Cambridge University Press).

Sedda, M. A., Berry, C., Jani, K., Amaro-Seoane, P., Auclair, P., *et al.* (2019) 'The missing link in gravitational-wave astronomy: Discoveries waiting in the decihertz range'. arXiv:1908.11375.

Sesana, A. (2016) 'Prospects for multiband gravitational-wave astronomy after GW150914'. *Phys. Rev. Lett.* **116**, 231102. https://doi.org/10/gf2kcx.

Shapiro, I. I. (1980) 'Experimental tests of the general theory of relativity'. In Held (1980b), p. 469.

Shapiro, S. L. & Teukolsky, S. A. (1983) *Black Holes, White Dwarfs, and Neutron Stars* (Wiley).

Shibata, M. (2015) *Numerical Relativity: 100 Years of General Relativity*, vol. 1 (World Scientific).

Shibata, M. & Taniguchi, K. (2011) 'Coalescence of black hole-neutron star binaries'. *Liv. Rev. Relativity* **14**, 6.

Sieniawska, M. & Bejger, M. (2019) 'Continuous gravitational waves from neutron stars: Current status and prospects'. *Universe* **5**, 217.

Smarr, L. L., ed. (1979) *Sources of Gravitational Radiation* (Cambridge University Press).

Smoot, G. F., Bennett, C. L., Kogut, A., Wright, E. L., Aymon, J., *et al.* (1992) 'Structure in the COBE differential microwave radiometer first-year maps'. *Astrophys. J. Lett.* **396**, L1–L5.

Spergel, D. N., *et al.* (2003) *Astrophys. J. Suppl.* **148**, 175.

Spivak, M. (1999) *A Comprehensive Introduction to Differential Geometry*, 3rd edn, vols. 1–5 (Publish or Perish B).

Stairs, I. H. (2003) 'Testing general relativity with pulsar timing'. *Liv. Rev. Relativity* **6**, 5. www.livingreviews.org/lrr-2003-5.

Starobinsky, A. (1980) 'A new type of isotropic cosmological models without singularity'. *Phys. Lett. B* **91**, 99.

Stephani, H. (2004) *Relativity: An Introduction to Special and General Relativity*, 3rd edn. (Cambridge University Press).

Stephani, H., Kramer, D., MacCallum, M., Hoenselaers, C. & Herlt, E. (2003) *Exact Solutions of Einstein's Field Equations*, 2nd edn (Cambridge University Press).

Stergioulas, N. (2003) 'Rotating stars in relativity'. *Liv. Rev. Relativity* **6**, 3. www.livingreviews.org/lrr-2003-3.

Stuver, A. L. (2019) *Gravitational Waves* (IOP Publishing). https://doi.org/10.1088/978-0-7503-1393-3.

Surya, S. (2019) 'The causal set approach to quantum gravity'. *Liv. Rev. Relativity* **22** (1).

Synge, J. L. (1960) *Relativity: The General Theory* (North-Holland).

Synge, J. L. (1965) *Relativity: The Special Theory* (North-Holland).

Tayler, R. J. (1994) *The Stars: Their Structure and Evolution*, 2nd edn (Cambridge University Press).

Taylor, E. F. & Wheeler, J. A. (1992) *Spacetime Physics*, 2nd edn (Freeman). Available open-access online at www.eftaylor.com/spacetimephysics/.

Taylor, J. H. & Weisberg, J. M. (1982) 'A new test of general relativity – gravitational radiation and the binary pulsar PSR 1913+16'. *Astrophys. J.* **253**, 908.

Terletskii, Y. P. (1968) *Paradoxes in the Theory of Relativity* (Plenum).

Teukolsky, S. A. (1972) 'Rotating black holes: Separable wave equations for gravitational and electromagnetic perturbations'. *Phys. Rev. Lett.* **29**, 114.

Thiemann, T. (2007) *Modern Canonical Quantum General Relativity* (Cambridge University Press).

Thorne, K. S. (1987) 'Gravitational radiation'. In Hawking, S. W. & Israel, W., eds., *300 Years of Gravitation* (Cambridge University Press), p. 330.

Thorne, K. S., Caves, C. M., Sandberg, V. D. & Zimmerman, M. (1979) 'The quantum limit for gravitational-wave detectors and methods of circumventing it'. In Smarr (1979a), p. 49.

Tinto, M. & Dhurandhar, S. V. (2014) 'Time-delay interferometry'. *Liv. Rev. Relativity* **17** (1). https://doi.org/10/f6cvzc.

Tipler, F. J., Clarke, C. J. S. & Ellis, G. F. R. (1980) 'Singularities and horizons – A review article'. In Held (1980b), p. 97.

Vachaspati, T., Pogosian, L. & Steer, D. A. (2015) 'Cosmic strings'. *Scholarpedia* **10**, 31682. www.scholarpedia.org/article/Cosmic_strings.

Venumadhav, T., Zackay, B., Roulet, J., Dai, L. & Zaldarriaga, M. (2019) 'New search pipeline for compact binary mergers: Results for binary black holes in the first observing run of advanced LIGO'. *Phys. Rev. D* **100**, 023011. https://doi.org/10/gg3dqd.

Venumadhav, T., Zackay, B., Roulet, J., Dai, L. & Zaldarriaga, M. (2020) 'New binary black hole mergers in the second observing run of advanced LIGO and advanced Virgo'. *Phys. Rev. D* **101**, 083030. https://doi.org/10/gg3dp7.

Verbiest, J. P. W., Lentati, L., Hobbs, G., van Haasteren, R., Demorest, P. B., *et al.* (2016) 'The international pulsar timing array: First data release'. *Mon. Not. Roy. Astron. Soc.* **458**, 1267.

Wald, R. M. (1984) *General Relativity* (Chicago University Press). *Hole Thermodynamics* (Chicago University Press).

Wambsganss, J. (1998) 'Gravitational lensing in astronomy'. *Liv. Rev. Relativity* **1**, 12. www.livingreviews.org/lrr-1998-12.

Ward, W. R. (1970) 'General-relativistic light deflection for the complete celestial sphere'. *Astrophys. J.* **162**, 345.

Watts, A. L., Krishnan, B., Bildsten, L. & Schutz, B. F. (2008) 'Detecting gravitational wave emission from the known accreting neutron stars'. *Mon. Not. Roy. Astron. Soc.* **389**, 839–868.

Weber, J. (1961) *General Relativity and Gravitational Waves* (Interscience).

Weinberg, S. (2008) *Gravitation and Cosmology* (Oxford University Press).

Weisberg, J. M. & Huang, Y. (2016) 'Relativistic measurements from timing the binary pulsar PSR B1913+16.' *Astrophys. J.* **829** (1).

Wen, L. & Schutz, B. F. (2005) 'Coherent network detection of gravitational waves: The redundancy veto'. *Class. Quantum Grav.*, **22**, S1321-S1335.

Wex, N. (2014) 'Testing relativistic gravity with radio pulsars'. In Kopeikin (2014b), 39–102.

Wiedemann, H. (2007) *Particle Accelerator Physics*, 3rd edn (Springer).

Will, C. M. (1974) 'Perturbation of a slowly rotating black hole by a stationary axisymmetric ring of matter. I. Equilibrium configurations'. *Astrophys. J.* **191**, 521.

Will, C. M. (1975) 'Perturbation of a slowly rotating black hole by a stationary axisymmetric ring of matter. II. Penrose processes, circular orbits, and differential mass formulae'. *Astrophys. J.* **196**, 41.

Will, C. M. (2006) 'The confrontation between general relativity and experiment'. *Liv. Rev. Relativity* **9**, 3.

Will, C. M. (2014) 'The confrontation between general relativity and experiment'. *Liv. Rev. Relativity* **17**, 4. www.livingreviews.org/lrr-2014-4.

Will, C. M. (2018) *Theory and Experiment in Gravitational Physics*, 2nd edn (Cambridge University Press).

Will, C. M. & Yunes, N. (2020) *Is Einstein Still Right?: Black Holes, Gravitational Waves, and the Quest to Verify Einstein's Greatest Creation* (Oxford University Press).

Williams, L. P. (1968) *Relativity Theory: Its Origins and Impact on Modern Thought* (Wiley).

Wilson, R. W. (1979) 'The cosmic microwave background radiation'. *Rev. Mod. Phys.* **51**, 433.

Woodhouse, N. M. J. (2003) *Special Relativity* (Springer).

Woodhouse, N. M. J. (2007) *General Relativity* (Springer).

Woosley, S. E. & Janka, H.-T. (2005) 'The physics of core-collapse supernovae'. *Nature Phys.*, **1**, 147–154.

Yano, K. (1955) *The Theory of Lie Derivatives and its Applications* (North-Holland).

Yao, W.-M., *et al.* (2006) 'Review of particle physics'. *J. Phys. G* **33**, 1. This is kept up to date on the Particle Data website, http://pdg.lbl.gov/2006/reviews/contents_sports.html.

Zel'dovich, Ya. B. & Novikov, I. D. (1971) In Thorne, K. S & Arnett, W. D., eds., *Stars and Relativity* (*Relativistic Astrophysics*, vol. 1) (University of Chicago Press).

Zwicky, F. (1933) 'Die Rotverschiebung von extragalaktischen Nebeln'. *Helvetica Physica Acta* **6**, 110–127.

Zwicky, F. (1937) 'On the masses of nebulae and of clusters of nebulae'. *Astrophys. J.* **86**, 217–246.

Index